DYNAMICS OF
TROPICAL COMMUNITIES

Dynamics of Tropical Communities

THE 37TH SYMPOSIUM
OF THE BRITISH ECOLOGICAL SOCIETY
CAMBRIDGE UNIVERSITY

1996

EDITED BY D. M. NEWBERY
Geobotanisches Institut,
Universität Bern, Switzerland

H. H. T. PRINS
Department of Terrestrial Ecology
and Nature Conservation,
Wageningen Agricultural University,
The Netherlands

N. D. BROWN
Oxford Forestry Institute,
South Parks Road, Oxford, UK

b

Blackwell
Science

© 1998 by
Blackwell Science Ltd
Editorial Offices:
Osney Mead, Oxford OX2 0EL
25 John Street, London WC1N 2BL
23 Ainslie Place, Edinburgh EH3 6AJ
350 Main Street, Malden
   MA 02148 5018, USA
54 University Street, Carlton
   Victoria 3053, Australia
10, rue Casimir Delavigne
   75006 Paris, France

Other Editorial Offices:
Blackwell Wissenschafts-Verlag GmbH
Kurfürstendamm 57
10707 Berlin, Germany

Blackwell Science KK
MG Kodenmacho Building
7–10 Kodenmacho Nihombashi
Chuo-ku, Tokyo 104, Japan

The right of the Authors to be
identified as the Authors of this Work
has been asserted in accordance
with the Copyright, Designs and
Patents Act 1988.

All rights reserved. No part of
this publication may be reproduced,
stored in a retrieval system, or
transmitted, in any form or by any
means, electronic, mechanical,
photocopying, recording or otherwise,
except as permitted by the UK
Copyright, Designs and Patents Act
1988, without the prior permission
of the copyright owner.

First published 1998

Set by Excel Typesetters Co., Hong Kong
Printed and bound in Great Britain by
MPG Books Ltd, Bodmin, Cornwall

The Blackwell Science logo is a
trade mark of Blackwell Science Ltd,
registered at the United Kingdom
Trade Marks Registry

A catalogue record for this title
is available from the British Library
ISBN 0-632-04944-8

DISTRIBUTORS

Marston Book Services Ltd
PO Box 269
Abingdon, Oxon OX14 4YN
(*Orders*: Tel: 01235 465500
         Fax: 01235 465555)
USA
Blackwell Science, Inc.
Commerce Place
350 Main Street
Malden, MA 02148 5018
(*Orders*: Tel: 800 759 6102
              781 388 8250
         Fax: 781 388 8255)
Canada
Login Brothers Book Company
324 Saulteaux Crescent
Winnipeg, Manitoba R3J 3T2
(*Orders*: Tel: 204 224-4608)
Australia
Blackwell Science Pty Ltd
54 University Street
Carlton, Victoria 3053
(*Orders*: Tel: 3 9347 0300
         Fax: 3 9347 5001)

Library of Congress
Cataloging-in-publication Data

British Ecological Society.
   Symposium (37th:1996: Cambridge University,
Cambridge)
      Dynamics of tropical communities/
   the 37th Symposium of the
   British Ecological Society,
   Cambridge University,
   Cambridge, 1996;
   edited by D.M. Newbery, H.H.T. Prins,
   N.D. Brown.
         p.   cm.
      Includes bibliographical references
   and index.
      ISBN 0-632-04944-8
      1. Rain forest ecology—Congresses.
   I. Newbery, D.M.
   II. Brown, N.
   III. Prins, H.H.T. (Herbert H.T.)   IV. Title.
   QH541.5.R27B765   1997
   577.34—dc21   97-28569
                                                  CIP

CONTENTS

Preface ix

1 Seeds and fruits of tropical rainforest plants: interpretation of the range in seed size, degree of defence and flesh/seed quotients 1
P. J. GRUBB *Department of Plant Sciences, University of Cambridge, Downing Street, Cambridge CB2 3EA, UK and Cooperative Research Centre for Tropical Rainforest Ecology and Management, CSIRO Tropical Forest Research Centre, Atherton, Queensland 4883, Australia*

2 Patterns in post-dispersal seed removal by neotropical rodents and seed fate in relation to seed size 25
P.-M. FORGET*, T. MILLERON[†] and F. FEER* *Muséum National d'Histoire Naturelle, Laboratoire d'Ecologie Générale, CNRS URA 1183, 4 Avenue du Petit Château, F-91800 Brunoy, France; [†]Department of Rangeland Resources and The Ecology Center, Utah State University, Logan, Utah 84322, USA*

3 Disturbance, phenology and life-history characteristics: factors influencing distance/density-dependent attack on tropical seeds and seedlings 51
D. S. HAMMOND*[†] and V. K. BROWN* *International Institute of Entomology, 56 Queen's Gate, London SW7 5JR, UK; [†]Tropenbos Guyana Programme, 12E Garnett Street, Campbellville, Georgetown, Guyana*

4 Gap-size niche differentiation by tropical rainforest trees: a testable hypothesis or a broken-down bandwagon? 79
N. D. BROWN and S. JENNINGS *Oxford Forestry Institute, South Parks Road, Oxford, OX1 3RB, UK*

5 Differential effects of small-scale fishing on predatory and prey fishes on Fijian reefs 95
N. V. C. POLUNIN* and S. JENNINGS[†] *Department of Marine Sciences & Coastal Management, University of Newcastle,*

Newcastle upon Tyne NE1 7RU, UK; †School of Biological Sciences, University of East Anglia, Norwich NR4 7TJ, UK

6 Architecture and development of rainforest trees: responses to light variation 125
F. BONGERS and F. J. STERCK *Department of Forestry, Agricultural University, P.O. Box 342, 6700 AH Wageningen, The Netherlands*

7 Limits to tree species distributions in lowland tropical rainforest 163
E. M. VEENENDAAL and M. D. SWAINE *Department of Plant and Soil Science, University of Aberdeen, Cruickshank Building, St Machar Drive, Aberdeen AB24 3UU, UK*

8 Community structure and the demography of primary species in tropical rainforest 193
R. J. ZAGT*† and M. J. A. WERGER* *Department of Plant Ecology and Evolutionary Biology, University of Utrecht, P.O. Box 80.084, 3508 TB Utrecht, The Netherlands; †Tropenbos Guyana Programme, 12 E Garnett Street, Campbellville, Georgetown, Guyana*

9 Risk-spreading and risk-reducing tactics of West African anurans in an unpredictably changing and stressful environment 221
K. E. LINSENMAIR *Theodor-Boveri-Institut für Biowissenschaften (Biozentrum), Lehrstuhl für Tierökologie und Tropenbiologie, Am Hubland, Universität Würzburg, D-97074 Würzburg, Germany*

10 Limits to exploitation of Serengeti wildebeest and implications for its management 243
S. MDUMA*†, R. HILBORN‡ and A. R. E. SINCLAIR* *Centre for Biodiversity Research, Department of Zoology, 6270 University Boulevard, University of British Columbia, Vancouver, B.C. Canada; †Serengeti Wildlife Research Centre, P.O. Box 3134, Arusha, Tanzania; ‡School of Fisheries WH-10, University of Washington, Seattle WA 98195, USA*

11 Phenology and dynamics of an African rainforest at Korup, Cameroon 267

D. M. NEWBERY*, N. C. SONGWE† and G. B. CHUYONG†
*Department of Biological and Molecular Sciences, University of Stirling, Stirling FK9 4LA, Scotland, UK and Geobotanisches Institut, Universität Bern, Alternbergrain 21, CH-2013 Bern, Switzerland; †Institute de la Recherche Agronomique du Cameroun, Kumba Forestry Research Station, PMB 29 Kumba, SW Province, Cameroon

12 Primates, phenology and frugivory: present, past and future patterns in the Lopé Reserve, Gabon   309
C. E. G. TUTIN* and L. J. T. WHITE† *Centre International de Recherches Médicales de Franceville, Gabon, and Department of Biological and Molecular Sciences, University of Stirling, Stirling FK9 4LA, Scotland, UK; †The Wildlife Conservation Society, New York and Institute of Cell, Animal and Population Biology, University of Edinburgh, King's Buildings, Edinburgh EH9 3JT, UK

13 Effects of habitat fragmentation on plant-pollinator interactions in the tropics   339
S. S. RENNER Department of Biology, University of Missouri—St Louis, St Louis, MO 63121–4499, USA

14 A spatial model of savanna function and dynamics: model description and preliminary results   361
J. GIGNOUX*‡, J.-C. MENAUT†*, I. R. NOBLE‡ and I. D. DAVIES‡ *Laboratoire d'Écologie, Ecole Normale Supérieure, 46 Rue d'Ulm, 75230 Paris Cedex 05, France; †ORSTROM, 213 Rue la Fayette, 75480 Paris Cedex 10, France; ‡Ecosystem Dynamics, Research School of Biological Sciences, Australian National University, Canberra ACT 0200, Australia

15 Evolution and diversity in Amazonian floodplain communities   385
P. A. HENDERSON*, W. D. HAMILTON* and W. G. R. CRAMPTON* *Animal Behaviour Research Group, Department of Zoology, University of Oxford, South Parks Road, Oxford OX1 3PS, UK

16 Community dynamics of arboreal insectivorous birds in African savannas in relation to seasonal rainfall patterns and habitat change   421

	P. JONES *Institute of Cell, Animal and Population Biology, University of Edinburgh, King's Buildings, Edinburgh EH9 3JT, UK*	
17	Species-richness of African grazer assemblages: towards a functional explanation H. H. T. PRINS* and H. OLFF* *Department of Terrestrial Ecology and Nature Conservation, Wageningen Agricultural University, Bornesteeg 69, 6708 PD Wageningen, The Netherlands*	449
18	Niche specificity among tropical trees: a question of scales P. S. ASHTON *The Arnold Arboretum of Harvard University, 22 Divinity Avenue, Cambridge, Massachusetts 02138, USA*	491
19	Disturbance and succession on the Krakatau Islands, Indonesia S. F. SCHMITT* and R. J. WHITTAKER* *School of Geography, Oxford University, Mansfield Road, Oxford OX1 3TB, UK*	515
20	Major disturbances in tropical rainforests T. C. WHITMORE* and D. F. R. P. BURSLEM† *Geography Department, University of Cambridge, Cambridge CB2 3EN, UK; †Department of Plant and Soil Science, University of Aberdeen, Aberdeen AB24 3UU, UK*	549
21	The impact of traditional and modern cultivation practices, including forestry, on Lepidoptera diversity in Malaysia and Indonesia J. D. HOLLOWAY *Department of Entomology, The Natural History Museum, Cromwell Road, London SW7 5BD, UK*	567
22	Tropical forests—spatial pattern and change with time, as assessed by remote sensing E. V. J. TANNER*, V. KAPOS† and J. ADAMS‡ *Department of Plant Sciences, University of Cambridge, Downing Street, Cambridge CB2 3EA, UK; †World Conservation Monitoring Centre, 219 Huntingdon Road, Cambridge CB3 0DL, UK; ‡Department of Geological Sciences, University of Washington, Seattle WA 98195, USA*	599
	List of reviewers	617
	Index	619
	Colour Plates 12.1 and 17.1 appear facing pages 310 and 470, respectively	

PREFACE

In this volume we consider the problems confronting the advancement of our understanding of the dynamics of tropical communities when there is stochasticity and change at all scales in time and space. This is the current forum for much contemporary ecological research but with the difference that the communities on which we focus are among the most species-rich and dynamic in the world. Unpredictability at the individual and population level, the uniqueness of location and of time period studied and the importance of major chance environmental events affecting communities are all becoming much more apparent to tropical ecologists. These processes must also be assessed against the undeniable long-term directional changes that are happening regionally and globally. How much of the variation within and between communities can be accounted for by deterministic factors? What then can we predict with confidence? What do stochastic processes imply? Are there indeed general principles or rules which can simplify (perhaps as models) our ideas? Herein lies an enormous challenge not met in other scientific disciplines in the same way. The demands on tropical ecologists are to be increasingly prescriptive, not just descriptive, and to achieve this requires a large step forward in our conceptual thinking. A new framework is called for and to this end we hope that the readers of our volume will find the assembled chapters a stimulation.

The 22 chapters were presented as papers at a symposium held at Cambridge University on 1–3 April 1996. We are grateful to the local organizer, E. V. J. Tanner for making the event possible. The chapters have been arranged in an increasing order of scale of the processes principally addressed. Each paper was reviewed by at least two independent persons and we are very grateful to them for their time and comments.

D. M. Newbery
H. H. T. Prins
N. D. Brown

1. SEEDS AND FRUITS OF TROPICAL RAINFOREST PLANTS: INTERPRETATION OF THE RANGE IN SEED SIZE, DEGREE OF DEFENCE AND FLESH/SEED QUOTIENTS

P. J. GRUBB

Department of Plant Sciences, University of Cambridge, Downing Street, Cambridge CB2 3EA, UK and Cooperative Research Centre for Tropical Rainforest Ecology and Management, CSIRO Tropical Forest Research Centre, Atherton, Queensland 4883, Australia

SUMMARY

1 Previous studies of seed mass have concentrated on differences between growth forms and between light-demanding and shade-tolerant species, but emphasis here is focused on the very wide range (up to six orders of magnitude) within a single regeneration class and growth form, for example shade-tolerant tall trees.

2 The smallest seeds (<1 mg) are interpreted in terms of their being advantaged at steep microsites and in not being damaged by the teeth of dispersers, while the largest (>2000 mg) are interpreted in terms of attracting large dispersers (mammals and cassowaries).

3 The variety in the mid-range of seed size among species with the same physiological tolerance of shade, determined by rates of uptake and loss of CO_2, is interpreted in terms of species making either many 'risky' seeds or few 'safe', the former producing seedlings at greater risk of death through drought, physical disturbance by animals, herbivory or damage by falling branches, and from overtopping by larger-seeded neighbours where growth is rapid. By hypothesis, the larger-seeded species fail to take over the forest because they suffer greater predation and there are limits to their seed production.

4 Species may have the 'inappropriate' seed size for their regeneration class as a result of (i) phylogenetic inertia, and (ii) seed size being less critical than the C balance of the seedling.

5 In Venzuelan caatinga, a forest type on extremely nutrient-poor soil, trees have smaller seeds than relatives on less nutrient-poor soils. This may be because the advantage of increased seed number is linear up to some

point, while the advantages of seed size seem all to scale with log seed mass. Paradoxically, dominance by relatively large-seeded species is often greater on extremely poor soils, perhaps because large predators are scarcer there.

6 In Australian rainforest, seeds held in fruits providing strong protection through physical means, poisons, stings or unpalatable material have markedly higher concentrations of N and P but not K, Ca or Mg.

7 In the same forest, fruits containing very small individual seeds, and very small total mass of seeds, have notably high flesh/seed dry mass quotients, so that if total fruit production is constant across species, the trade-off between seed number and seed size will be far from linear.

INTRODUCTION

Most papers about seed size in tropical rainforest have concerned the differences between species of different growth form and/or the differences between species that require high irradiance at the time of establishment and those that are shade-tolerant at that stage (Ng 1978, Foster & Janson 1985, Rockwood 1985, Putz & Appanah 1987, Kelly & Purvis 1993, Hammond & Brown 1995, Kelly 1995, Metcalfe & Grubb 1995, Grubb & Metcalfe 1996). Other ecological papers about seeds in tropical rainforest have concerned the nature of the species with persistent seeds in the soil seed bank, and the signals that lead to germination of these latter seeds (Kennedy & Swaine 1992, many papers reviewed by Vázquez-Yanes & Orozco-Segovia 1993, 1996).

Ecological studies on tropical fruits have concerned chiefly the kinds of fruits attractive to particular types of dispersers, with rather few studies on the effectiveness of different dispersers for particular species of fruit or on the types of seeds preferred by specific predators; many papers were reviewed in the symposia edited by Estrada and Fleming (1986) and Fleming and Estrada (1993), while Jordano (1995) has shown the value of allowing for phylogenetic trends in fruit characters when comparing taxa dispersed by different agents.

Very little indeed has been written about the mineral nutrient contents of seeds in tropical rainforest. I have recently shown that the concentration of N falls linearly and appreciably with the logarithm of embryo dry mass across species in the Lauraceae in NE Queensland, and that species in which the embryo rather than the flesh is the attractant to dispersers have a high concentration of N relative to their seed size, both in the Lauraceae and other families (Grubb 1996).

The great majority of tropical fruits are fleshy, and the part of this chapter that deals with fruits is concerned only with fleshy fruits. The dry mass quotient of flesh/seed is of particular importance in the context of the seed mass — seed number trade-off. If fruits containing larger seeds have, in general, a different flesh/seed quotient from fruits containing smaller seeds, and the total productivity of fruits is constant across a number of species, the trade-off will not be linear in form. Most of the available data on flesh/seed quotients are based on fresh mass values (cf. Wheelwright, Haber, Murray & Guindon 1984), and any trend in quotient with mean seed mass may be confounded with a parallel trend in proportional water content.

In this chapter I tackle four issues:
1 the need to think more about the huge range of seed size within any one functional group, for example shade-tolerant tall trees, rather than the differences in mean size between functional groups;
2 the basis for forests on extremely nutrient-poor soils having smaller seeds than those on less nutrient-poor soils, while often being dominated by one species with relatively large seeds;
3 the reason why some species have fruits which give their seeds great physical or chemical protection while others do not, and
4 the relationship between the flesh/seed dry mass quotient and seed mass, and its bearing on the seed mass–seed number trade-off.

This chapter reviews work that has been published recently, is in press or is about to be published in full elsewhere. It is based mainly on work done in lowland rainforests of NE Queensland, but also covers studies undertaken in Singapore and the Venezuelan Amazonas.

A 'seed' it taken to be a single mature fertilized ovule including the testa plus the fibrous inner part of the fruit wall where present. When seed mass was determined, any fleshy part of the fruit wall was removed, as also any aril or sarcotesta except where that was very thin and difficult to remove (e.g. in *Melicope* spp.). For dry-seeded species any wing or plume was removed.

THE GREAT RANGE IN SEED SIZE

Since the time of Salisbury's classic study of the ecological significance of seed mass (1942) it has been known that there is commonly a very wide range of values among the species in any one functional group; for example shade-tolerant herbs or light-demanding trees in a given vegetation type. Yet the emphasis in the literature has been almost uniformly on the differences between the mean values for different functional groups (cf. Westoby, Jurado & Leishman 1992). Among 'shade-tolerant' tall tree species of NE Australia or Malesia—that is, those able to establish in deep shade (say <2% daylight) and persist for many months to years, irrespective of their requirements for onward growth—there are six orders of magnitude difference between the smallest and largest seeds (0.32–48000 mg in Australia; Grubb 1996). The same very wide range is found among the shade-tolerant short trees and shrubs in both regions (Fig. 1.1). The wide range may be compared with the difference of only one to two orders of magnitude found generally between the mean values for groups of species that are respectively light-demanding and shade-tolerant at the stage of establishment (Foster & Janson 1985, Metcalfe & Grubb 1995, Grubb 1996). Even within families we may find two to three orders of magnitude difference in seed mass between species of one functional group; for example in NE Queensland shade-tolerant tall trees in Cunoniaceae and Grossulariaceae or shade-tolerant tall vines in Araceae (Grubb & Metcalfe 1996). Exceptionally, a range of more than one order of magnitude among species of the same functional group may be found in one genus, notably *Syzygium* (three orders).

How are we to interpret the great range in seed size? I suggest that we must first consider the extremes (<1 mg and >2000 mg) and then the middle range. The seeds below 1 mg dry mass are unable to establish upward or downward through a layer of litter (Molofsky & Augspurger 1992) but are advantaged on steep microslopes — including those on standing and fallen trunks — where litter falls away and larger seeds do not readily lodge in the available 'pico-sites' (Floyd 1990, Kohyama & Grubb 1994). Another advantage of the tiny size is that it makes possible production of a very large number of seeds that can persist in the soil until a litter-gap appears. Kennedy and Swaine (1992) found that there were high densities of seeds of some of the genera involved (*Pternandra*, Melastomataceae; *Urophyllum*, Rubiaceae) in the soil of a forest in Sabah. Metcalfe (1996) has shown for 10 species of this kind in four families in Singapore that they germinate only or almost only in light at the time of dispersal, but unlike strict canopy-gap demanders such as *Melastoma malabathricum* they come to germinate in a high far red/red ratio and/or in darkness over a matter of months. Metcalfe and Grubb (1997) have demonstrated that the growth rates of such species show only a modest response to an increase in irradiance from 1 to 7.5% daylight, much like those of larger-seeded shade-tolerators such as *Dysoxylum schiffneri* (Osunkoya, Ash, Hopkins & Graham 1994) or *Guarea cedrata* (Veenendaal, Swaine, Lecha *et al*. 1996). Examples of the tiny-seeded species minimally responsive to increased irradiance are *Ficus chartacea*, *Pternandra echinata* and *Urophyllum hirsutum*.

In the eastern palaeotropics (from Malaysia to Australia) at least 16 families are represented among the species which have seeds of <1 mg and which routinely establish in the shade: Actinidiaceae, Araceae, Cunoniaceae, Gesneriaceae, Grossulariaceae, Melastomataceae, Moraceae, Myrsinaceae, Myrtaceae, Pandanceae, Pentaphragmataceae, Philydraceae, Piperaceae,

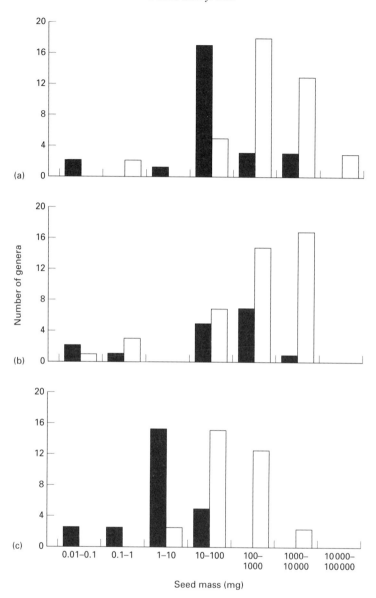

FIG. 1.1. The numbers of genera in logarithmic mean seed dry mass classes for species that are shade-tolerant at the stage of establishment in (a) lowland rainforest on various soils in NE Queensland (from Grubb 1996), (b) lowland rainforest on granite in SE Asia (from Metcalfe & Grubb 1995), and (c) caatinga in the Venezuelan Amazonas (Grubb & Coomes 1997). Empty bars for tall trees (>9 m in Queensland, >15 m in Singapore and Venezuela), and filled bars for short trees and shrubs. The seeds of the tall trees in the caatinga are significantly smaller than those of the tall trees in the Queensland forest (Mann-Whitney U-test; $P = 0.02$) and almost significantly smaller than those in the SE Asian forest ($P = 0.05$).

Rhizophoraceae, Rubiaceae and Urticacae (personal observations). If the limit is raised to 2 mg the Araliaceae join this group. The extremely small seeds of these species, which establish in the shade, contrast with the so-called 'small' seeds (Swaine & Whitmore 1988) of the major genera of trees which become established naturally at forest edges and in large gaps, and which commonly dominate secondary forest in the palaeotropics; for example Araliaceae (*Arthrophyllum, Polyscias*), Euphorbiaceae (*Endospermum, Glochidion, Macaranga, Omalanthus*, some *Mallotus*, etc.), Rhamnaceae (*Alphitonia*), Rubiaceae (*Timonius*), Sterculiaceae (*Commersonia*), Ulmaceae (*Trema*) and Verbenaceae (*Clerodendron*) whose seed dry mass values are almost all in the range 2–30(–50) mg (Metcalfe & Grubb 1995, Grubb & Metcalfe 1996). Among the widespread secondary forest trees and shrubs only the *Ficus* and *Trichospermum* species have seeds of <1 mg.

In the neotropics the clearest examples of families including extremely small-seeded shade-tolerators are the Melastomataceae (Ellison, Denslow, Loiselle *et al.* 1993) and Rubiaceae (*Alseis*, Foster 1983), but almost certainly there are examples among the Araceae, Gentianaceae (*Besleria*), Gesneriaceae and Urticaceae (*Pilea*). In the neotropics there is no contrast in seed mass between these shade-tolerators with extremely small seeds and the conventional 'small-seeded' species of gaps and secondary forest. Mean dry mass values given by Vázquez-Yanes (1976), Foster (1983), Fleming (1985), Fleming, Williams, Bonaccorso & Herbst (1985), Foster and Janson (1985), Ellison *et al.* (1993) and Hammond and Brown (1995) for genera that commonly dominate neotropical secondary forests are 0.1–1.0 mg for *Cecropia* and *Piper*, 0.025 mg for *Muntingia* and from <0.007 to 7 mg for various Melastomataceae. The principal explanation for the difference between the neotropics and the eastern palaeotropics in this respect may relate to the fact in the palaeotropics the seeds of trees of secondary forest in the mass range 2–30 mg are dispersed by birds (Taylor 1982, Mitchell 1994), while in the neotropics those of mass <2 mg are dispersed wholly or at least substantially by bats and, to a lesser extent, other mammals (Estrada, Coates-Estrada & Vázques-Yanes 1984, Fleming 1985, 1986, Charles-Dominique 1986). There may be a great advantage in having seeds small enough to avoid being crushed by the bats' teeth. The idea that very small seed size might be selected as avoiding damage by dispersers appears to have been put forward first by Janzen (1969) in a very general context, and was suggested by Stiles and Roselli (1993) specifically for the Melastomataceae. They adduced evidence that in the neotropics melastomes first diversified in association with tanagers, birds that 'mash' the fruits they eat rather than 'gulp' them, and are thus liable to damage all but the smallest seeds. The work of Stiles and Roselli (1993) incidentally provides an explanation for the Melastomataceae being the one major group of tiny-seeded light-demanding species in the neotropics that are not mainly bat dispersed. I suggested elsewhere (Grubb 1996) that very small-seeded light demanders (<1 mg) are not a viable option on average soils because they would be competed out, but the data from the neotropics negate that argument.

The species that can establish in deep shade from extremely small seeds, like those with large seeds, vary a great deal in the amount of light needed for onward growth into flowering individuals. Most of the herbs and shrubs can complete their life-cycles in deep shade (examples in Gesneriaceae, Melastomataceae, Pentaphramataceae, Piperaceae, Rubiaceae; Kiew 1988, Metcalfe & Grubb 1997) while many of the trees and some shrubs certainly need either small or large canopy-gaps to reach maturity (examples in Cunoniaceae, Grossulariaceae, Moraceae, Myrsinaceae, Myrtaceae and Rhizophoraceae; Grubb & Metcalfe 1996, Metcalfe & Grubb 1997).

I now consider seeds at the other end of the dry mass range—say >2000 mg. In respect of tolerance of the shade of the forest floor as such (as opposed to tolerance of associated risks such as drought and damage considered below) there seems to be no a priori case that their seedlings are greatly advantaged compared with species whose seed mass values fall in the middle range. It is important to make the distinction between the position in low light and that in almost complete darkness, where the properties of the photosynthetic system cease to be relevant, so that the extent of the seed store and/or the rate of respiration do become overwhelmingly important (Boot 1996, Grubb & Metcalfe 1996, Saverimuttu & Westoby 1996). It is possible that if ability to resprout numerous times after destruction of the main shoot by herbivores or fallen branches is a major advantage, and if that ability is confined to the largest seeds, selection might have acted in many cases through that variable.

However, as in the case of the extremely small seeds of many neotropical light-demanders, so in the case of the very large seeds of many species in both neotropics and palaeotropics (both light-demanding and shade-tolerant) I shall emphasize the idea that their size may be determined more by constraints imposed by dispersers than by advantages at the stage of establishment. I have suggested elsewhere that the species with very large seeds are adapted primarily to dispersal by large animals rather than to survive especially deep shade (Grubb 1996, Grubb & Metcalfe 1996, see also Foster & Janson 1985, Kelly 1995). Larger animals need to eat larger food items to gain enough energy per unit time (Martin 1985); strictly the size of the item concerns the fruit and not the seed in the first instance, but in phyletic lines with only one or a very few seeds per fruit the size of the seed size covaries with the fruit size. In addition, it must be the size of the seed rather than that of the fruit which matters where the seed is itself the attractant and source of nourishment—as in the case of nuts, and pyrenes (or seeds from capsulate fruits) with very thick walls able to be breached only by large-jawed rodents (Grubb 1996).

In West Irian, Papua New Guinea and Australia the cassowary (*Casuarius casuarius*), a large flightless bird, disperses many species with large fleshy fruits; in NE Queensland at least 23 families are involved (Stocker & Irvine 1983; unpublished observations by me and by colleagues). In the same area the non-marsupial white-tailed rat (*Uromys caudimaculatus*) is the main species taking large seeds in which the embryo is the source of nourishment, while the marsupial musky rat-kangaroo (*Hypsiprimnodon*) takes some (references in Grubb 1996). In other parts of the tropics the last class of seeds may be taken by squirrels or ruminants as well as ground-based rodents (Gautier-Hoin, Duplantier, Quris et al. 1985). Perhaps the largest seeds (50–200g dry mass) are not dispersed by any present-day animal, for example *Idiospermum australiense* in NE Queensland, or *Parinari excelsa* in Panama (R. Condit personal communication).

There remains the key question of understanding the variety of seed size in the mid-range of 1–1000 mg, in practice 10–1000 mg for most trees and shrubs. Grubb and Metcalfe (1996) have suggested that plants with the same physiological tolerance of shade (e.g. sensitivity of productivity to the irradiance) may make many 'risky' small seeds or fewer larger 'safe' seeds. The smaller, risky seeds give rise to seedlings liable to suffer high mortality in the face of drought, physical disturbance by animals or soil instability or partial consumption by herbivores or partial destruction by falling debris, and to suffer overtopping if they are adjacent to seedlings from larger seeds under favourable conditions. The species with few larger safe seeds fail to take over the forest because they can never occupy all the available microsites, partly as a result of limitations on seed production and dispersal and partly because of post-dispersal predation by larger animals, which is the one hazard from which they may suffer more (on average) than the smaller seeds (Boman & Casper 1995, Hulme 1996). There is still little critical evidence on this last point for tropical rainforest, but Adler (1995) found in cage experiments with Central American spiny rats (*Proechimys semispinosus*) that they did indeed eat larger seeds preferentially. The battle between the species with many risky seeds and those with few safe seeds is like that between colonizers and competitors in the model of Skellam (1951). The species with larger seeds fail in the dimension of space to occupy sites that can be utilized by smaller-seeded species, much as the competitors in Skellam's model fail in the dimension of time to reach the sites used by colonizers, and so fail to prevent them reproducing and colonizing elsewhere. Rees and Westoby (1997) have produced a game-theoretical model that encapsulates a somewhat similar interpretation of the variety of seed size in one functional group. The competition between larger- and smaller-seeded species is taken up again later in this chapter.

There is another level at which we can understand the array of seed sizes among, say, the shade-tolerant trees in a forest. We may imagine that evolution of shade-tolerators from light-demanders, and/or the reverse, continues to this day and did not stop at some time in the past. In so far as the key properties enabling plants to grow in deep shade are the rates of photosynthesis and respiration (Field 1986, Walters & Field 1987, Fredeen & Field 1996), and associated leaf longevity (King 1994), and in so far as there is phylogenetic inertia in seed size, we may expect to find species with 'inappropriate seed size'. Grubb and Metcalfe (1996) have found in NE Queensland contradictory trends when shade-tolerators and light-demanders are compared within genera and within families. In 13 out of 14 families in which they could compare genera that are consistently shade-tolerant with ones that are consistently light-demanding the mean seed dry mass was found to be greater in the shade-tolerators (in nine cases by a wide margin) but in eight out of nine cases where shade-tolerant species and light-demanding species could be compared within genera there was virtually no difference in seed mass or the shade-tolerator had smaller seeds (Fig. 1.2).

This result was interpreted as follows. The species within genera have diverged more recently than the genera within families. Larger seed size provides appreciable additive fitness for shade-tolerators but is not essential. The shade-tolerant small-seeded species have evolved relatively recently from light-demanders, and in most cases have not yet evolved larger seeds— they show inertia rather than adaptation.

Although the few species of *Mallotus* in Australia do not show larger seed mass in shade-tolerators (Grubb & Metcalfe 1996), the position is different in Sarawak where the genus is highly speciose, and Primack and Lee (1991) found a strong trend to larger seed volume in shade-tolerators. It seems that the one tropical genus that has consistently larger seeds in shade-tolerators is *Piper*; the respective logarithmic means of the values given by Vázquez-Yanes (1976, p. 340) for species in one part of Mexico are 1.2 mg (three species) and 0.16 mg (six species). *Piper* has rather few species in Queensland; the only widespread and common species are the shade-tolerant tall climbers *P. caninum* and *P. novae-hollandiae*, which are bird dispersed (mean seed dry mass values of 33–45 mg). Among the much less

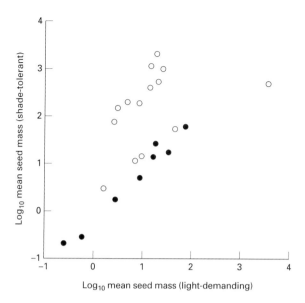

FIG. 1.2. The results of taxonomically controlled comparisons of mean seed dry mass (mg) in shade-tolerant and light-demanding plants at different taxonomic levels in the flora of lowland rainforest in NE Queensland: ○, comparisons of genera within families; ●, comparisons of species within genera. (From Grubb & Metcalfe 1996.)

common species, probably dispersed by bats, the shade-tolerant tall climber *P. rothianum* has a value of 1.2 mg; while the light-demanding small shrub or treelet *P. subpeltatum* has a value of 0.09 mg (D.J. Metcalfe & P.J. Grubb unpublished data; result obtained after submission of Grubb & Metcalfe 1996). It is not clear why *Piper* should show this major difference in seed size, while so many other genera do not. Possibly it has been present for a longer time in both neotropics and paleaeotropics, and thus had more time to diversify in seed mass; that idea would fit with the pantropical distribution of the genus, but fails to explain the lack of a parallel development of larger seed size in shade-tolerators in *Ficus*, which is at least as widespread.

THE SMALL SEEDS OF TREES IN FORESTS ON EXTREMELY NUTRIENT-POOR SOIL OR AT HIGH ALTITUDE

The seeds of trees in an Amazonian caatinga, a kind of forest exceptionally short of available nitrogen, are appreciably smaller than those in normal-looking lowland rainforest on various soils in Singapore and NE

Queensland (Fig. 1.1) and than those on various soils in Peru and Guyana reported by Foster and Janson (1985) and Hammond and Brown (1995). The trees of the caatinga also have smaller seeds than their near relatives in adjacent forest on soil less poor in available nitrogen; six taxonomically controlled contrasts (five within genera, and one within a family) are shown in Fig. 1.3.

The caatinga reaches a height of 18–23 m, whereas the other rainforests reach 35–50 m. It may thus be argued that the primary finding is of smaller seeds on shorter trees. I pursue that interpretation here rather than, for example, the fact that more light reaches the floor of the caatinga than that of most lowland rainforests (1–2% as opposed to 0.5–1.0% of daylight under cloudy conditions). There is clearly an analogy with the successively smaller mean seed sizes of tall trees, short trees, shrubs and herbs, found generally in any one tropical rainforest (Foster & Janson 1985, Rockwood 1985, Kelly 1995, Metcalfe & Grubb 1995) and in various kinds of communities in temperate regions (Mazer 1989, 1990, Thompson & Rabinowitz 1989, Westoby et al. 1992, Leishman & Westoby 1994a). I believe that in order to interpret this general trend we have to work from the smaller amount of material that the less tall species have to invest in seeds, and that for plants which are short of resources (nutrients or light) the pressure to reduce seed size rather than seed number can be understood in the following way.

None of the various advantages of larger seed size is likely to be linearly related to seed mass. That much was assumed in the theoretical treatment of

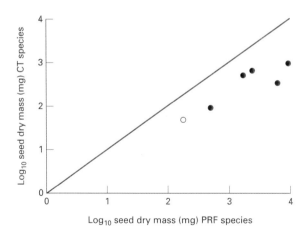

FIG. 1.3. The results of taxonomically controlled comparisons of mean seed dry mass values (mg) of shade-tolerant trees in a forest on extremely nutrient-poor soil (caatinga, CT) and in adjacent forest on less poor soil (palm-rich forest on sandy loam, PRF): ●, comparisons within genera; ○, comparison within a family. (From Grubb & Coomes 1997.)

Smith and Fretwell (1974) for the evolution of offspring size in general. Experiments with non-rainforest plants have found resistance to defoliation, drought and darkness to be related approximately linearly to log seed mass (Armstrong & Westoby 1993, Leishman & Westoby 1994b,c). Also, any advantage of larger seed mass in competition among seedlings is likely to be proportional to a function of seed mass that is much less than linear. As shading of one seedling by another is a major part of competition between species, one might consider as a guide the relationship between height at a given age and seed mass. Seiwa and Kikuzawa (1991) found for a collection of tree species from temperate deciduous forest kept in 9% daylight for a season that the height attained increased curvilinearly with log seed mass; for species with mean seed dry mass values in the ranges 100–1000 mg and 1000–10 000 mg seedling height increased two- to three-fold for a 10-fold increase in seed mass. In contrast, fitness *is* likely to be proportional to the number of seeds produced of a particular type, up to some point that depends on the incidence of intraspecific density-dependent effects. In this way one can understand qualitatively how selection acts more strongly to increase seed number than seed size when the total mass of potential seed material is strongly limited. There will be no single solution regarding 'optimal size' because of the spectrum from 'many risky' to 'few safe' seeds for any one functional type emphasized on p. 8.

It is especially interesting that the effect of soil fertility is seen only on extreme soils. In NE Queensland there is no clear difference in seed size between the forests on the more fertile soils derived from basalt and those on much less fertile soils derived from granite and metamorphics. A clear parallel is seen in the way in which seed size declines with altitude, but only at the highest levels. In NE Queensland no marked reduction in seed size is found between lowland rainforest (LRF, to *c*. 800 m) and lower montane rainforest (LMRF, *c*. 800–1300 m), but there is a marked reduction between lower montane and upper montane rainforest (UMRF, >1300 m). Only one species of tree with seeds >20 mm across is found in UMRF whereas there are many in both LRF and LMRF (records of Hyland & Whiffin 1993). In the four most speciose genera which have species occurring over 1300 m the largest dimension for the largest-seeded species is almost the same in LRF and in LMRF, but it is consistently much greater in these two than in the UMRF. The ranges in maximum seed dimension (mm) for species with upper altitudinal limits of <800 m, >800 m but <1300 m, and >1300 m (numbers of species in parentheses) are recorded by Hyland (1989) and Hyland and Whiffin (1993) as follows: *Cryptocarya* 7.5–35 (16), 8–38 (15) and 10 (1), *Elaeocarpus* 11–60 (5), 12–50 (15) and 17–20 (2), *Pouteria* 10–45 (8), 13–40 (7) and 16 (1), and *Syzygium* 8–55 (14), 6–50 (10) and 3–20 (7).

Rockwood (1985), working with species of Costa Rica where the mountains are much taller and the altitudinal limits of the forest types are higher, did not find a significant difference in seed size between species occurring below 1500 m and those found above that altitude (there including much LMRF as well as UMRF). His result is consistent with lack of appreciable difference between LRF and LMRF in Queensland, and if he had separated the structurally and physiognomically distinct and species-poor UMRF as I have done, he might well have found a lower mean seed size. I suggest that the same explanation of smaller seed size applies to upper montane rainforest as to Amazonian caatinga.

Paradoxically, the rainforests on extremely poor soils are among those that show the most marked dominance by a single species with relatively large seeds, notably *Eperua* spp. (Caesalpinioideae) in northern South America (Richards 1952, Coomes & Grubb 1996). I suggest that this situation comes about because of the combination of two conditions. First, in the absence of human influence, regeneration occurs wholly in rather small gaps — the trees are narrow crowned and make small gaps when they fall; there are no tree species present which need high light for establishment (Coomes & Grubb 1996). Second, competition favours the larger-seeded over the smaller-seeded (for the reasons set out on p. 8 above) as in all forests, but the general paucity of animal life in forests on extremely poor soils (Janzen 1974) includes a lack of large seed predators so that the larger-seeded tree species are enabled to dominate the smaller to a degree not generally seen in forests on less nutrient-poor soils where the pressure from large seed predators is higher. Thus, by hypothesis, the richness in species of shade-tolerant trees of most tropical rainforests arises partly from continuous preferential predation on larger-seeded species, much as the species-richness of a grassland is enhanced by preferential grazing of the taller herbs.

It is important to emphasize that in the tropical rainforests on extremely poor soils with dominance by one or two relatively large-seeded tree species, the absolute size of the seeds in question is smaller than of the largest species found in rainforests on less nutrient-poor soil (i.e. 3–25 g in dry mass, rather than 30–200 g). Examples are *Eperua falcata* 5.7 g and *Catostemma fragrans* 25 g (Hammond & Brown 1995) in wallaba forest on well-drained white sand in Guyana, and *E. obtusata* 3.4 g (Grubb & Coomes 1997) in caatinga on generally waterlogged white sand in the Venezuelan Amazonas. In single-dominant forests on less extremely infertile soils nearby, the dominants have notably larger seeds, for example *Chlorocardium rodiei* (39 g) on brown sand, *Mora excelsa* (62 g) on silt, and *M. gonggrijpii* (*c.* 60 g) on clay in Guyana (Hammond & Brown 1995). Likewise, the forests in Borneo dominated by the huge-seeded

Eusideroxylon zwaggeri are not found on white sand (Richards 1952). In the forests dominated by these very large-seeded species, some factor other than extreme paucity of mineral nutrients in the whole community must lead to a failure of predation on the biggest seeds.

A key role for selective predation on larger seeds was also suggested by Leigh, Wright, Herre & Putz (1993). They hypothesized that in forest fragments isolated on islands within a large man-made lake, the larger animals would be lost, including those responsible for predation on the largest seeds, and the tree species producing the latter would tend to take over.

PHYSICAL AND CHEMICAL PROTECTION OF SEEDS BY FRUIT TISSUES

In all parts of the tropics some species invest heavily in fruit tissues that offer strong physical and/or chemical protection of the seeds inside, while others do not. In the Australian tropics this phenomenon is particularly marked in the form of the thick woody walls of the capsules of many Proteaceae as well as Leguminosae, the thick leathery walls of many *Dysoxylum* species (Meliaceae), the strongly spiny capsules of *Flindersia* (Rutaceae), the irritant hairs on or in the woody capsules of *Brachychiton* (Sterculiaceae), *Mucuna* (Leguminosae) and *Lethedon* (Thymelaeaceae), the stinging hairs of *Dendrocnide* (Urticaceae), the very sticky fruits of *Pisonia* (Nyctaginaceae), the latex-bearing walls of Apocynaceae, the burning taste left by the rhaphides of *Rhaphidophora* (Araceae) of *Freycinetia* and *Pandanus* (Pandanaceae) and the poisonous flesh of *Phaleria* (Thymelaeaceae) and *Solanum* (Solanaceae). Illustrations of these many types are provided by Cooper and Cooper (1994). Other genera have thin-walled capsules, drupes with only modestly developed endocarps or berries with seeds having a thin testa, all of them without irritant, stinging or poisonous structures. For simplicity I refer to these latter types as having 'unprotected' seeds.

In the case of vegetative tissues, there are many examples where protection by physical or chemical means is associated with higher N concentrations in the plants concerned than in their neighbours (Grubb 1992). P.J. Grubb, D.J. Metcalfe, E.A.A. Grubb and G.D. Jones (in preparation) have proposed the hypothesis that the species with protected seeds have seeds that are particularly rich in protein, a resource likely to be limiting to the predatory animals in the system. It is impossible to know which animals might have been the main predators during the evolution of the species in question, but the selection pressure may be continued now by parrots (e.g. the king parrot, *Alisterus scapularis*) and those pigeons with thick-walled gizzards that destroy seeds rather than disperse them (e.g. the white-headed pigeon, *Columba norfolciensis*). Insects may also be important.

The hypothesis of greater richness in N in the protected species has been found correct. For both 'protected' and 'unprotected' species there is a significant linear negative relationship between N concentration and the logarithm of mean seed dry mass. Analysis by the method of Zar (1984, pp. 292–298) shows that the slopes for the two groups are not significantly different, but the intercepts are significantly and markedly different (Fig. 1.4). There is a similar difference between the two groups of species in the concentration of P, but not in that of K, Ca or Mg.

In theory, it might be that the protected seeds are richer in energy as well as nitrogen, and it might even be that possession of a very high concentration of energy-rich compounds is more important than richness in protein, but we have not yet had the chance to test that hypothesis.

Why should certain plants have higher concentrations of N (and P) in their seeds? For certain families that have high seed N concentrations a specific reason can be suggested: the need for an appreciable initial capital for setting up specialized root systems that are expensive in N (and P?) to establish; that is, the nitrogen-fixing nodules of the Leguminosae and coralloid roots of the Zamiaceae. There is abundant evidence of the need for 'starter

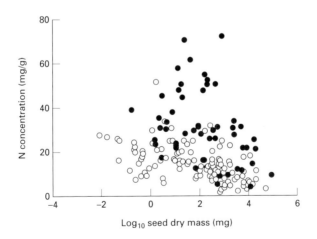

FIG. 1.4. The mean concentration of N in the seed as a function of the logarithm of mean seed dry mass for seeds from two types of fruit in lowland rainforest in NE Queensland: ●, species with notable physical and/or chemical protection of the seeds by the fruits; ○, seeds not so protected. (From P.J. & E.A.A. Grubb, G.D. Jones & D.J. Metcalfe unpublished data.) The regression for the protected species is $y = -9.89 \, (\pm 3.33 \text{ SE}) \log x + 63.8 \, (\pm 9.6)$, and for the unprotected $y = 6.97 \, (\pm 1.80) \log x + 37.5 \, (\pm 4.68)$; the slopes are not significantly different but the intercepts (i.e. the y-values when $x = 1$ mg) are ($P < 0.0001$).

nitrogen' for legumes (Sprent & Thomas 1984). It is tempting to suggest the same type of explanation for the Proteaceac, as there is parallel evidence that proteoid roots form more readily where there is a modest capital of available N and P rather than a totally impoverished substratum (Lamont 1972). However, the root systems of seedlings of rainforest Proteaceae that I have dug up have not developed marked proteoid root systems. The foliar concentrations of N and P in seedlings are low relative to those of most families (P.J. Grubb, unpublished observations); it therefore seems most likely that the seed N and P move mainly into the roots (or stems).

For some other families with high seed N concentrations there is fragmentary evidence from analyses of leaves that I have made in various parts of the tropics which suggests that they spend their whole lives with higher N concentrations than their neighbours in at least their leaves, for example Meliaceae, Nyctaginacae, Solanacae, some Sterculiaceae, Thymelaeaceae and Urticaceae.

For other families (notably Apocynaceae and Rutaceae) there is not evidence that either of the above arguments applies, and it is hard to understand why the seed N concentration is notably high.

FLESH/SEED MASS QUOTIENT IN FRUITS OF DIFFERENT SIZE

Without doubt fleshy fruits dispersed by animals are the dominant type of fruit in all tropical lowland rainforests, at least judged by the number of species, even where a dominant family has dry fruits (e.g. the Dipterocarpacae in Malaysia). If the dry mass quotient of flesh to seed per fruit were found to differ systematically across the range of fruit sizes, it could mean a considerable departure from a linear trade-off between seed number and seed mass. The quotient is certainly low in most plants with arillate or sarcotestal seeds, but they are not discussed here. I confine my attention to species with berries or drupes.

When Herrera (1981) tried to determine whether or not tropical plants with fleshy fruits differ from temperate ones in the extent of the reward they offer to the disperser, he could find published data on flesh/seed mass ratio for only 15 tropical species. Wheelwright *et al.* (1984), working in lower montane rainforest in Costa Rica, published values for the flesh/seed fresh mass quotient for 100 species (21 of them Lauraceae); they found values of 0.5–19(–32) for species with seed-per-fruit fresh mass of <300 mg, and values of 0.25–3.4 for seed mass values >300 mg (Fig. 1.5a). Interpretation in terms of the seed mass—seed number trade-off is difficult because the flesh of the smaller fruits could be much richer in water.

Seeds and fruits

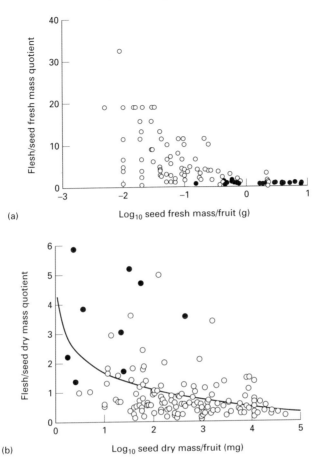

FIG. 1.5. The per-fruit flesh/seed mass quotient for two collections of tropical rainforest species with berries and drupes: (a) fresh mass quotients for species in lower montane forest in Costa Rica (from Wheelwright et al. 1984: ●, Lauraceae; ○, other families), and (b) dry mass quotients for species in the lowland rainforest of NE Queensland. (P.J. & E.A.A. Grubb, A.K. Irvine, G.D. Jones and D.J. Metcalfe unpublished data; ●, species with a mean seed dry mass of ≤1 mg; ○, species with mean seed dry mass >1 mg). In (b) the values for *Davidsonia pruriens* (8.8 at 540 mg) have been omitted; the regression shown is $y = -1.94 \log x + 1.66$ ($r^2 = 0.181; P < 0.001$).

The data for the dry mass quotient for a wide range of fruits in NE Queensland show that although the quotient can vary greatly among species at any one dry mass value for seeds per fruit, it declines markedly across the range of fruits from extremely small-seeded to very large-seeded — from *Abrophyllum ornans* to *Syzygium gustavioides* (Fig. 1.5b). A similar decline is found if the flesh/seed-per-fruit dry mass quotient is considered relative to the mean dry mass of a single seed (Table 1.1). The task of determining the quotient accurately is so tedious that I have data for only 1–4 fruits for the tiny-seeded species. Taking these determinations at face value, there is a steep decline from >5 for two genera with seeds of <0.1 mg (*Abrophyllum*, *Saurauia*) via about 3 for those with seeds of 0.1–1.0 mg (e.g. *Callicarpa*, *Dendrocnide* and *Ficus*) to 1 or less for the range of 1–100000 mg in which the mean value falls much less. If two very exceptional species are discounted, the mean value for the range 100–1000 changes from being more than that of the range 10–100 mg (1.2) to a value that is similar to that for the range 1000–10000 mg (0.74 vs. 0.65; Table 1.1). There are few values in the literature for tropical fruits with which to compare those in Table 1.1, but the data of Fleming *et al.* (1985) for *Muntingia* (Elaeocarpaceae or Flacourtiaceae; cf. Mabberley 1987), with a mean seed mass of 0.025 mg, are in agreement, yielding a quotient of 2.1.

If a parallel decline in flesh/seed quotient with seed mass per fruit (or with mean seed mass) is sought within taxonomic groups (supposed phyletic lines), it is found not to represent the commonest case (7/14 genera with two or more species sampled, and 12/28 families with two or more genera), with the single exception that genera with seeds of mean dry mass <1 mg consistently have higher flesh/seed quotients than their intrafamilial counterparts with seeds of mean dry mass >10 mg. So far we have exact data on which to make the comparison in only four families (Grossulariaceae, Moraceae, Myrtaceae and Verbenaceae) but it is probably true of the Araceae, Myrsinaceae and Rosaceae too. The other families with fleshy fruits containing seeds of dry mass <1 mg (Actinidiaceae, Gesneriaceae, Melastomataceae and Urticaceae) are not represented in Australia by genera with much larger seeds.

It is hard to know how to interpret the modest drift down in the mid-to-upper range of fruit size; at face value it suggests a sorting of phyletic lines

TABLE 1.1. *The mean flesh/seed-per-fruit mass quotient for logarithmic classes of mean individual seed mass, giving genera equal weight except that genera containing species with mean seed dry mass in more than one class are included in each of the relevant classes.*

Seed mass class (mg)	0.001–0.1	0.1–1.0	1–10	10–100	100–1000	1000–10000	10000–100000
Mean Quotient	5.5	3.0	1.3	0.92	0.74*	0.65	0.44
SE	0.35	0.51	0.20	0.15	0.10	0.08	0.06
Number of genera	2	5	15	26	25	19	10

* 1.2 ± 0.35 if two very exceptional species are included (*Davidsonia pruriens* 8.8 and *Tetrasynandra laxiflora* 5.0).

rather than evolution within them — much as families and genera are sorted by habitat in respect of seed size in temperate North America, according to the analysis of Mazer (1990). However, in the case of plants producing extremely small seeds in fleshy fruits, their greatly increased flesh/seed mass quotients do suggest that they have a problem in attracting dispersers. Many of them meet this partly by packing large numbers of seeds (hundreds to thousands) in each fruit (true of Actinidiaceae, Araceae, Gesneriaceae, Pandanaceae but not of Grossulariaceae, Myrtaceae or Urticaceae which have only 15–50). Nevertheless, it seems they still have to pay a penalty by investing a lot of capital in attractant flesh. If the capital available for fruit production is limited, the penalty of having to invest relatively more in flesh must reduce the total number of seeds produced — perhaps by as much as 50–75%. Interestingly, the highly fleshy fruits do not need to be especially rich in protein or oil, judging by the data marshalled by Fleming (1986).

Only a minority of rainforest genera with seeds of <1 mg dry mass develop the seeds in dry capsules with a censer or rain-drop dispersal mechanism. In Australia this type is found in *Geissois* (Cunoniaceae) and *Ophiorhiza* and *Wendlandia* (Rubiaceae) among shade-tolerators, and in *Ludwigia* (Onagraceae) among light-demanders. *Helmholtzia* (Philydraceae, shade-tolerant) has indehiscent capsules, and *Pollia* (Commelinaceae, light-demanding) has minute shining blue nuts.

The problem of attracting dispersers to small-seeded fruits, where 'small' is applied to the range 1–10 mg, is met in several cases by massing of fruits in large prominent infructescences, combined with a moderately high flesh/seed quotient (say 1.2), as in *Polyscias*, a major genus in Australia, and one that even attracts large fruit-eaters such as some pigeons (Floyd 1989).

CONCLUSIONS

This contribution has been partly about new facts and partly about a proposed new perspective. The data on seed size in extremely nutrient-poor forest, on seed nutrient concentrations and on flesh/seed dry mass quotients are among the first available. For the most part the new data can be interpreted with some confidence; for example the greater development of physical and/or chemical defence in fruits containing seeds especially rich in N and therefore potentially of especially high value to predators, the smaller seeds of trees in forests where total productivity is much reduced, and the greater flesh/seed mass quotients in fruits containing extremely small seeds.

The revised perspective emphasizes the following: the need to investigate the range of seed mass values among species of any one functional group rather than the differences in mean values between functional groups;

the value of making taxonomically controlled comparisons at different levels (particularly genus vs. family); constraints on seed size imposed by particular groups of dispersers rather than by what might be advantageous to the seedling; constraints on seed size imposed by the allocation available to the mother plant rather than what might be ideal for the seedling; and due attention to the effect that the association of extremely small seed size with much higher flesh/seed quotients may have on the seed number–seed mass trade-off.

The hypothesis that within any functional group of plants the larger-seeded species will be competitively superior leads to a solution to the long-standing problem of monodominant forests being found on exceptionally nutrient-poor soils via a second hypothesis that there is reduced predation on large seeds in such systems. The maintenance of species-richness in most rainforests is attributed partly to preferential predation on larger-seeded species in a way that is analogous to the maintenance of species-richness in grassland by preferential predation on the vegetative parts of taller species. New long-term experimental studies exclud-ing predators are needed to test these hypotheses, but interpretation may be made difficult by the fact that very often the same animals act as predators and dispersers (Forget, Milleron & Feer, Chapter 2, this volume).

ACKNOWLEDGEMENTS

I am greatly indebted to my wife Anne and to Dr D.J. Metcalfe for their patient support in the collection and processing of innumerable seeds in NE Queensland, to Mr G.D. Jones for very many chemical analyses, and to Dr D.A. Coomes for his co-operation in the study carried out in Venezuela, and assistance with statistical analysis. Seeds were collected in Queensland under permits 1685, 1995 and FO/000782/95/SAB from the Queensland National Parks and Wildlife Service, and under permits 489, 824 and 948 from the Department of Primary Industries Forest Service. The work in Queensland and Venezuela has been financed at different times by CSIRO, the Natural Environment Research Council, the Royal Society of London, Cambridge University and Magdalene College Cambridge. For critical comments on the first version of this paper I thank Drs C.J. Burrows, D.F.R.P. Burslem, E.V.J. Tanner, K. Thompson and I.M. Turner.

REFERENCES

Adler, G.H. (1995). Fruit and seed exploitation by Central American spiny rats, *Proechimys semispinosus*. *Studies on Neotropical Fauna and Environment*, **30**, 237–244.

Armstrong, D.P. & Westoby, M. (1993). Seedlings from large seeds tolerate defoliation better: a test using phylogenetically independent contrasts. *Ecology*, **74**, 1092–1100.
Boman, J.S. & Casper, B.B. (1995). Differential postdispersal seed predation in disturbed and intact temperate forest. *American Midland Naturalist*, **134**, 107–116.
Boot, R.G.A. (1996). The significance of seedling size and growth rate of tropical rain forest tree seedlings for regeneration in canopy openings. In *The Ecology of Tropical Forest Tree Seedlings* (Ed. by M.D. Swaine), pp. 267–283. UNESCO, Paris and Parthenon, Carnforth, UK.
Charles-Dominique, P. (1986). Inter-relations between frugivorous vertebrates and pioneer plants: *Cecropia*, birds and bats in French Guyana. In *Frugivores and Seed Dispersal* (Ed. by A. Estrada & T.H. Fleming), pp. 119–135. Junk, Dordrecht.
Coomes, D.A. & Grubb, P.J. (1996). Amazonian caatinga and related communities at La Esmeralda, Venezuela: forest structure, physiognomy and floristics, and control by soil factors. *Vegetatio*, **122**, 167–191.
Cooper, W. & Cooper, W.T. (1994). *Fruits of the Rain Forest*. Geo, Sydney.
Ellison, A.M., Denslow, J.S., Loiselle, B.A. & Danilo Brenes, M. (1993). Seed and seedling ecology of neotropical Melastomataceae. *Ecology*, **74**, 1733–1749.
Estrada, A. & Fleming, T.H. (Eds) (1986). *Frugivores and Seed Dispersal*. Junk, Dordrecht.
Estrada, A., Coates-Estrada, R. & Vázquez-Yanes, C. (1984). Observations on fruiting and dispersers of *Cecropia obtusifolia* at Las Tuxtlas, Mexico. *Biotropica*, **16**, 315–318.
Field, C.B. (1986). On the role of photosynthetic responses in constraining the habitat distribution of rainforest plants. *Australian Journal of Plant Physiology*, **15**, 343–358.
Fleming, T.H. (1985). Coexistence of five sympatrical *Piper* (Piperaceae) species in a tropical dry forest. *Ecology*, **66**, 688–700.
Fleming, T.H. (1986). Opportunism versus specialization: the evolution of feeding strategies in frugivorous bats. In *Frugivores and Seed Dispersal* (Ed. by A. Estrada & T.H. Fleming), pp. 105–118. Junk, Dordrecht.
Fleming, T.H. & Estrada, A. (Eds) (1993). *Frugivory and Seed Dispersal: Ecological and Evolutionary Aspects*. Kluwer, Dordrecht, *Vegetatio*, **107/108**.
Fleming, T.H., Williams, C.F., Bonaccorso, F.J. & Herbst, L.H. (1985). Phenology, seed dispersal, and colonization in *Muntingia calabura*, a neotropical pioneer tree. *American Journal of Botany*, **72**, 383–391.
Floyd, A.G. (1989). *Rainforest Trees of Mainland South-eastern Australia*. Inkata Press, Melbourne.
Floyd, A.G. (1990). *Australian Rainforests in New South Wales*, Vol. 1. Surrey Beatty, Chipping Norton, New South Wales.
Foster, R.B. (1983). The seasonal rhythm of fruitfall on Barro Colorado Island. In *The Ecology of a Tropical Rainforest: Seasonal Rhythms and Long-Term Changes* (Ed. by E.G. Leigh, A.S. Rand & D.M. Windsor), pp. 151–172. Oxford University Press, Oxford.
Foster, S.A. & Janson, C.H. (1985). The relationship between seed size and establishment conditions in tropical woody plants. *Ecology*, **66**, 773–780.
Fredeen, A.L. & Field, C.B. (1996). Ecophysiological constraints on the distributions of *Piper* species. In *Tropical Forest Plant Ecophysiology* (Ed. by S.S. Mulkey, R.L. Chazdon & A.P. Smith), pp. 597–618. Chapman & Hall, New York.
Gautier-Hion, A., Duplantier, J.-M., Quris, R., Feer, F., Sourd, C., Decoux, J.P. et al. (1985). Fruit characters as a basis of fruit choice and seed dispersal in a tropical forest vertebrate community. *Oecologia*, **65**, 324–337.
Grubb, P.J. (1992). A positive distrust in simplicity—lessons from plant defences and from competition among plants and among animals. *Journal of Ecology*, **80**, 585–610.

Grubb, P.J. (1996). Rainforest dynamics: the need for new paradigms. In *Tropical Rainforest Research: Current Issues* (Ed. by D.S. Edwards, S.C. Choy & W.E. Booth), pp. 215–233. Kluwer, Dordrecht.

Grubb, P.J. & Coomes, D.A. (1997). Seed mass and nutrient content in nutrient-starved tropical rainforest in Venezuela. *Seed Science Research*, **7**, 269–280.

Grubb, P.J. & Metcalfe, D.J. (1996). Adaptation and inertia in the Australian tropical lowland rainforest flora: contradictory trends in intergeneric and intrageneric comparisons of seed size in relation to light demand. *Functional Ecology*, **10**, 512–520.

Hammond, D.S. & Brown, V.K. (1995). Seed size of woody plants in relation to disturbance, dispersal, soil type in wet neotropical forests. *Ecology*, **76**, 2544–2561.

Herrera, C.M. (1981). Are tropical fruits more rewarding to dispersers than temperate ones? *American Naturalist*, **118**, 896–907.

Hulme, P.E. (1996). Herbivory, plant regeneration and species coexistence. *Journal of Ecology*, **84**, 609–615.

Hyland, B.P.M. (1989). A revision of Lauraceae in Australia (excluding *Cassytha*). *Australian Systematic Botany*, **2**, 135–367.

Hyland, B.P.M. & Whiffin, T. (1993). *Australian Tropical Rain Forest Trees: An Interactive Identification System*, vol. 2. CSIRO, Melbourne.

Janzen, D.H. (1969). Seed-eaters versus seed size, number, toxicity and dispersal. *Evolution*, **23**, 1–27.

Janzen, D.H. (1974). Tropical blackwater rivers, animals and mast fruiting by Dipterocarpaceae. *Biotropica*, **6**, 69–103.

Jordano, P. (1995). Angiosperm fleshy fruits and seed dispersers: a comparative analysis of adaptation and constraints in plant-animal interactions. *American Naturalist*, **145**, 163–191.

Kelly, C.K. (1995). Seed size in tropical trees: a comparative study of factors affecting seed size in Peruvian angiosperms. *Oecologia*, **102**, 377–383.

Kelly, C.K. & Purvis, A. (1993). Seed size and establishment conditions in tropical trees. On the use of taxonomic relatedness in determining ecological patterns. *Oecologia*, **94**, 356–360.

Kennedy, D.N. & Swaine, M.D. (1992). Germination and growth of colonizing species in artifical gaps of different sizes in dipterocarp forest. *Philosophical Transactions of the Royal Society*, **B335**, 357–367.

Kiew, R. (1988). Herbaceous flowering plants. In *Malaysia* (Ed. by the Earl of Cranbrook), pp. 56–76. Pergamon, Oxford.

King, D.A. (1994). Influence of light levels on the growth and morphology of saplings in a Panamanian forest. *American Journal of Botany*, **81**, 948–957.

Kohyama, T. & Grubb, P.J. (1994). Above- and below-ground allometries of shade-tolerant seedlings in a warm-temperate rain forest. *Functional Ecology*, **8**, 229–236.

Lamont, B. (1972). The effect of soil nutrients on the production of proteoid roots by *Hakea* species. *Australian Journal of Botany*, **20**, 27–40.

Leigh, E.G., Wright, J.S., Herre, E.A. & Putz, F.E. (1993). The decline in tree diversity on newly isolated tropical islands: a test of a null hypothesis and some implications. *Evolutionary Ecology*, **7**, 76–102.

Leishman, M. & Westoby, M. (1994a). Hypotheses on seed size: tests using the semi-arid flora of western New South Wales, Australia. *American Naturalist*, **143**, 890–906.

Leishman, M. & Westoby, M. (1994b). The role of large seed size in shaded conditions: experimental evidence. *Functional Ecology*, **8**, 205–214.

Leishman, M. & Westoby, M. (1994c). The role of seed size in seedling establishment in dry soil conditions — experimental evidence from semi-arid species. *Journal of Ecology*, **82**, 249–258.

Mabberley, D.J. (1987). *The Plant-Book*. Cambridge University Press, Cambridge.

Martin, T.E. (1985). Resource selection by tropical frugivorous birds: integrating multiple functions. *Oecologia*, **66**, 563–573.

Mazer, S. (1989). Ecological, taxonomic and life history correlates of seed mass among Indiana dune angiosperms. *Ecological Monographs*, **59**, 153–175.

Mazer, S. (1990). Seed mass of Indiana dune genera and families: taxonomic and ecological correlates. *Evolutionary Ecology*, **4**, 326–357.

Metcalfe, D.J. (1996). Germination of small-seeded tropical rain forest plants exposed to different spectral compositions. *Canadian Journal of Botany*, **74**, 516–520.

Metcalfe, D.J. & Grubb, P.J. (1995). Seed mass and light requirement for regeneration in Southeast Asian rain forest. *Canadian Journal of Botany*, **73**, 817–826.

Metcalfe, D.J. & Grubb, P.J. (1997). The responses to shade of seedlings of very small-seeded tree and shrub species from tropical rain forest in Singapore. *Functional Ecology*, **11**, 215–221.

Mitchell, T.C. (1994). The ecology of *Macaranga* (Euphorbiaceae) trees in primary lowland mixed dipterocarp forest, Brunei. PhD dissertation, University of Cambridge.

Molofsky, J. & Augspurger, C.K. (1992). The effect of leaf litter on early seedling establishment in a tropical forest. *Ecology*, **73**, 68–77.

Ng, F.S.P. (1978). Strategies of establishment in Malayan forest trees. In *Tropical Trees as Living Systems* (Ed. by P.B. Tomlinson & M.H. Zimmermann), pp. 129–162. Cambridge University Press, Cambridge.

Osunkoya, O.O., Ash, J.E., Hopkins, M.S. & Graham, A.W. (1994). Influence of seed size and seedling ecological attributes on shade-tolerance of rain-forest tree species in northern Queensland. *Journal of Ecology*, **82**, 149–163.

Primack, R.B. & Lee, H.S. (1991). Population dynamics of pioneer (*Macaranga*) trees and understorey (*Mallotus*) trees (Euphorbiaceae) in primary and selectively logged Bornean rain forests. *Journal of Tropical Ecology*, **7**, 439–458.

Putz, F.E. & Appanah, S. (1987). Buried seeds, newly dispersed seeds, and the dynamics of a lowland forest in Malaysia. *Biotropica*, **19**, 326–333.

Rees, M. & Westoby, M. (1997). Game-theoretical evolution of seed mass in multi-species ecological models. *Oikos*, **78**, 116–126.

Richards, P.W. (1952). *The Tropical Rain Forest*. Cambridge University Press, Cambridge.

Rockwood, L.L. (1985). Seed weight as a function of life form, elevation and life zone in neotropical forests. *Biotropica*, **17**, 32–39.

Salisbury, E.J. (1942). *The Reproductive Capacity of Plants*. Bell, London.

Saverimuttu, T. & Westoby, M. (1996). Seedling longevity under deep shade in relation to seed size. *Journal of Ecology*, **84**, 681–689.

Seiwa, K. & Kikuzawa, K. (1991). Phenology of tree seedlings in relation to seed size. *Canadian Journal of Botany*, **69**, 532–538.

Skellam, J.G. (1951). Random dispersal in theoretical populations. *Biometrika*, **38**, 196–218.

Smith, C.C. & Fretwell, S.D. (1974). The optimal balance between size and number of offspring. *American Naturalist*, **108**, 499–506.

Sprent, J.I. & Thomas, R.J. (1984). Nitrogen nutrition of seedling grain legumes; some taxonomic, morphological and physiological constraints: opinion. *Plant, Cell and Environment*, **7**, 637–645.

Stiles, F.G. & Roselli, L. (1993). Consumption of fruits of the Melastomataceae by birds; how diffuse is coevolution? *Vegetatio*, **107/108**, 57–73.

Stocker, G.C. & Irvine, A.K. (1983). Seed dispersal by cassowaries (*Casuarius casuarius*) in north Queensland's rainforest. *Biotropica*, **15**, 170–176.

Swaine, M.D. & Whitmore, T.C. (1988). On the definition of ecological species groups in tropical rain forests. *Vegetatio*, **75**, 81–86.

Taylor, C.E. (1982). Reproductive biology and ecology of some tropical pioneer trees. PhD thesis, University of Aberdeen.
Thompson, K. & Rabinowitz, D. (1989). Do big plants have big seeds? *American Naturalist*, **133**, 722–728.
Vázquez-Yanes, C. (1976). Estudios sobre ecofisiologia de la germinacion en una zona calido-humeda de Mexico. In *Investigaciones sobre la regeneración de selvas altas en Veracruz, México* (Ed. by A. Gómez-Pompa, C. Vázquez-Yanes, S. del Amo Rodríguez & A. Butanda Cervera), pp. 279–387. Compania Editorial Continental, México D.F.
Vázquez-Yanes, C. & Orozco-Segovia, A. (1993). Patterns of seed longevity and germination in the tropical rainforest. *Annual Reviews of Ecology and Systematics*, **24**, 69–87.
Vázquez-Yanes, C. & Orozco-Segovia, A. (1996). Physiological ecology of seed dormancy and longevity. In *Tropical Forest Plant Ecophysiology* (Ed. by S.S. Mulkey, R.L. Chazdon & A.P. Smith), pp. 535–558. Chapman & Hall, New York.
Veenendaal, E.M., Swaine, M.D., Lecha, R.T., Walsh, M.F., Abebrese, I.K. & Owusu-Afriyie, K. (1996). Responses of West African forest tree seedlings to irradiance and soil fertility. *Functional Ecology*, **10**, 501–511.
Walters, M.B. & Field, C.B. (1987). Photosynthetic light acclimation in two rainforest *Piper* species with different ecological amplitudes. *Oecologia*, **72**, 449–456.
Westoby, M., Jurado, E. & Leishman, M. (1992). Comparative evolutionary ecology of seed size. *Trends in Ecology and Evolution*, **7**, 368–372.
Wheelwright, N.T., Haber, W.A., Murray, K.G. & Guindon, C. (1984). Tropical fruit-eating birds and their food plants: a survey of a Costa Rica lower montane forest. *Biotropica*, **16**, 173–191.
Zar, J.H. (1984). *Biostatistical Analysis*, 2nd edn. Prentice-Hall, Englewood Cliffs, NJ.

2. PATTERNS IN POST-DISPERSAL SEED REMOVAL BY NEOTROPICAL RODENTS AND SEED FATE IN RELATION TO SEED SIZE

P.-M. FORGET*, T. MILLERON† and F. FEER*

*Muséum National d'Histoire Naturelle, Laboratoire d'Ecologie Générale, CNRS URA 1183, 4 Avenue du Petit Château, F-91800 Brunoy, France; †Department of Rangeland Resources and The Ecology Center, Utah State University, Logan, Utah 84322, USA

SUMMARY

1 Animals are important agents of seed dispersal and plant recruitment in the tropics. Currently, it is generally assumed that seeds removed after primary dispersal are doomed by predation; the probability of secondary dispersal and further escape is seldom ascertained.

2 This chapter reports experiments that were carried out on Barro Colorado Island, Panama, to test the effect of seed size (>1 cm) and weight (>0.1 g) on the level of removal and fate for several tree species illustrating a variety of seed size and dispersal syndromes.

3 Seed fate was explored in the context of the habitat in which seeds land. Two habitats were recognized based on the frequency of agoutis observed along one transect across the forest: one agouti-rich area located close by the young forest, and one agouti-poor area further in the old forest.

4 Patterns of seed removal as well as the fate of small to large seeds, especially the probability of seed caching, changed between species and habitat depending on seed reward.

5 Small seeds (⩽1 g weight and c. 1 cm long) had a low probability of being missed by vertebrates, and there was no effect of the visitation rate of agoutis. Removal by terrestrial mammals mostly accounted for predation although the seeds of *Eugenia* (1.2 g) were cached by rodents more frequently (12%) than those of *Cupania* (0.2 g; 2%) and *Doliocarpus* (0.7 g; 0%).

6 Among medium seeds (c. 1–3 g and 1–2 cm), the habitat and the quality of seed nutrient reward in cotyledons differentially affected removal and fate. High removal in *Brosimum* seeds (2.2 g) with edible cotyledons corresponded to either predation or caching. On the contrary, removal in *Virola*

seeds (3 g), with toxic cotyledons, accounted for caching more frequently. Greater burying rates occurred in *Virola* where agouti sightings were lower.

7 Among large seeds (*c.* >3 g and >2 cm), fate was independent of habitat and month for the large nutrient-rewarding *Gustavia* seeds (12 g) only, seeds of which were more often buried than the less nutrient-rewarding *Licania* seeds (6 g). In the latter predation was dependent on habitat and month.

8 Similar patterns of seed removal mean contrasting seed fate pathways depending on seed size: a gradient of caching rate was observed from smaller to the larger seeds. Change in caching rate is assumed to be related to change in agonistic and caching behaviour among agouti populations which vary in density. A method of seed labelling is therefore essential to conclude whether rodents are effective seed dispersers.

INTRODUCTION

Seed dispersal by flying and arboreal frugivorous vertebrates dominates in neotropical rainforests (Janzen 1970, Howe & Smallwood 1982, Howe 1986). Seedfall patterns change depending on the spatial distribution of fruiting trees and seed attributes such as seed size that determine the guild of seed dispersers. The density of fruiting trees and the seasonal occurrence and abundance of fruits and alternative resources (insects and leaves) also affect how frugivores forage and behave, and likely determine where and how seeds will be deposited in the forest (Schupp, Howe, Augspurger & Levey 1989), scattered or clumped (Howe 1989). Post-dispersal removal of seeds often follows initial dissemination by animals and varies between species, on spatial and temporal scales as a function of habitat and forest types, food abundance and the community of predators (e.g. Schupp 1988a,b, Willson 1988, Osunkoya 1994) some of which may behave as dispersers (Janzen 1971a, Chambers & MacMahon 1994). The probability of seedling establishment will then be contingent on whether seeds are harvested or ignored, consumed or cached, and in this latter case retrieved or abandoned by either invertebrates, especially ants (e.g. Byrne & Levey 1993, Kaspari 1993, Levey & Byrne 1993) or vertebrates, mostly rodents (Price & Jenkins 1986, Forget & Milleron 1991, Crawley 1992, Terborgh, Losos, Riley & Bolanos Riley 1993, Terborgh & Wright 1994). Currently, the fate of those seeds that disappear is unknown and it is generally assumed that they are doomed to predation; the probability of secondary dispersal and further, of escape, is seldom ascertained. Seed fates need to be examined with reference to seed size and mode of primary dispersal, in the context also of the habitat in which the seeds land.

A preliminary approach to assessing rodent impact on the seedling establishment is to outline the possible pathways to different seed fates (Price & Jenkins 1986). Price and Jenkins's (1986) seed fate pathway diagram is useful for formulating hypotheses and then for designing experiments to test for the spatial and temporal effects of factors related to both seeds and rodents. Seed pathways are multiple, most of them leading to death but, in some instances, to seedling establishment. When seeds remain undispersed above ground they are often killed by vertebrate and invertebrate predators. Uncached seeds may germinate if their toxicity saves them from consumption by animals. Despite secondary dispersal (i.e. dispersal by terrestrial animals following primary dispersal by arboreal and/or flying animals), scatter- or larder-hoarded seeds (Morris 1962) may die in the caches due to pre-dispersal infestation, drought, pathogens or mammals that dig up the cotyledons before or after seedling establishment. Seeds may be retrieved quickly and recached further away by the same or another rodent (P.-M. Forget & T. Milleron, personal observation). Though poorly documented, such complex pathways may be more common than previously thought. Seedling establishment follows dispersal when seeds hidden in caches, and protected from post-dispersal predators, are abandoned or forgotten by cachers and not recovered by other predators.

When questioning the significance of seed removal by rodents, it is essential to know what internal and/or external factors trigger seed predation or dispersal (Vander Wall 1990). Extrinsic factors related to habitat, such as the availability of alternative seed resources (e.g. Forget 1992) and the animal community (for example, Osunkoya 1994), and intrinsic factors related to seed attributes, such as size, chemistry and nutrient content reward in cotyledons (e.g. Rankin 1978), control the level of seed harvesting by terrestrial animals (Price & Jenkins 1986, and references therein). Depending on habitat choice and rodent foraging activity within a given habitat, one might expect the level of seed removal and predation to change between and within species, as shown in several studies (Rankin 1978, Boucher 1981, De Steven & Putz 1984, Sork 1987, Schupp 1992, 1995, Osunkoya 1994). High seed removal in species that are strictly dependent on rodents for dispersal probably reflects high levels of scatter-hoarding (Forget 1992, 1996), whereas other abiotically dispersed species' seeds may be exclusively eaten (Rankin 1978, Forget 1989). Seasonal spatial distributions of resources may govern the probability that seeds will be harvested, consumed or dispersed by rodents (Hallwachs 1986, Forget 1993, Forget, Munoz & Leigh 1994). Seeds that are removed rapidly in the dry season, in general when food availability is low, will often be more predated than in the wet season (Rankin 1978, Hallwachs 1986).

Not only large-seeded species but also a wide range of species whose seeds offer small or intermediate nutrient rewards are removed and fall prey to small ground-dwelling vertebrates (Fleming 1974, Vandermeer 1979, Vandermeer, Stout & Risch 1979, Emmons 1982, Smythe, Glanz & Leigh 1982, Dirzo & Dominguez 1986, Larson & Howe 1987, Schupp 1988a,b, 1990, Chapman 1989, Schupp & Frost 1989, Zona & Henderson 1989, Estrada & Coates-Estrada 1991, Burkey 1993, 1994, Howe 1993, Hammond 1995). Small- and medium-sized seeds are cached by either small or large rodents, and this hoarding activity varies between rodent species, month, season and habitat (Fleming & Brown 1975, Forget 1991, 1993, Forget & Milleron 1991, Hallwachs 1994, Hammond 1995, Asquith, Wright & Clauss 1997). The smallest seeds, easy to handle and with the lowest energetic value will be most likely consumed by mice (which weigh <100 g), or discarded if they contain alkaloids toxic to mammals (Price & Jenkins 1986). Seeds with larger nutrient rewards will be either scatter-hoarded or eaten by spiny rats (250–500 g) and dasyproctid rodents (1–6 kg) (Emmons & Feer 1990), or discarded. Given a similar seed size, high-quality seeds should be harvested more often than low-quality seeds (Price & Jenkins 1986). The impact of small vertebrates on seed fate awaits further research, especially when the animals encounter and harvest seeds that have been dispersed already by other animal species (see Janzen & Martin 1982, Hallwachs 1986).

This chapter reports experiments that were carried out in Panama to test the effect of seed size on removal and fate for several tree species, illustrating a variety of dispersal syndromes. Several questions are addressed: how does seed fate change as a function of seed size? To what extent do some tree species depend on secondary seed dispersal by rodents? How does seed fate change across rodent population density gradients within a forest? Does the rate of caching change during the fruiting season?

STUDY SITE AND AREA

The study site is the tropical moist forest of Barro Colorado Island (BCI) (Leigh, Rand & Windsor 1982). BCI is covered with forest stands varying in age from relatively young (c. 100 years) to very old (c. 500 years) (Foster & Brokaw 1982, Piperno 1990). The climate of BCI is seasonal with a well-marked dry season of variable length and intensity, lasting from December or January to April, and a wet season for the rest of the year with a peak in rainfall in October–November (Rand & Rand 1982). There are two peaks in fruiting: in April–May (wind and animal dispersed species) and again in September–October (animal dispersed species) (Foster 1982). Because of habituation to humans, the BCI rodent community changes between forest habitats depending on the distance from the laboratory (Wright, Gompper & De Leon 1994). The results of Wright *et al.* (1994) led us to analyse spatial–temporal seed removal and fate by considering the number of diurnal rodent encounters along a trail within the study area. In the rodent community, an agouti (*Dasyprocta punctata*) ranks first in sighting counts, followed by a squirrel (*Sciurus granatensis*) (agoutis being about five times more sighted than squirrels) and then a paca (*Agouti paca*). Population levels of small rats such as spiny rats (*Proechimys semispinosus*) and mice (*Oryzomys* spp.) fluctuate widely from year to year in French Guiana, for instance (S. Ringuet personal communication, P.-M. Forget personal observation) as is also probably the case on BCI. Spiny rats were rarely observed by Wright *et al.* (1994) during nocturnal censuses while they may be much more abundant on BCI in other years (D.C. Tomblin & G.H. Adler, personal communication). In 1990, spiny rats were rarely seen on the study trail and seeds did not disappear at night during the first days of the experiments in the dry season (Forget 1993, P.-M. Forget personal observation) as well as in the wet season during this study (see Results). As a consequence, because of its large body size and abundance on BCI, as well as its food preference for the tested plant species (Smythe *et al.* 1982), agoutis are assumed to have been responsible for most of the caching in 1990.

The study was undertaken in June–August 1990 along the R.C. Shannon trail between the 100- and 700-m trail markers (sec Fig. 1 in Forget 1993). The study period coincided with the mid-wet season when food was abundant for rodents and preceeding the onset of food scarcity during August–March (Smythe 1970, 1978). The forest was classified as old but was close to a patch of young regrowth (Fig. 5 in Foster & Brokaw 1982, R. Foster personal communication). The diurnal rodent community was checked during monthly strip censuses ($n = 10$ per month) between November 1989 and May 1990. An animal was recorded when it was visible at a distance of *c.* 15–20 m on each side of the trail; distant alarming rodents were not counted if not within the range of visibility in the understorey. Agoutis ($n = 146$) accounted for more than 90% of all diurnal rodent sightings ($n = 160$) along the study trail. Agouti sightings changed from month to month, especially between February and April, presumably as a function of food availability in the area of the study trail (Forget 1993). Averaging the data for the entire 6-month period prior to experimentation, there was a decrease in the frequency of agoutis along the trail (Fig. 2.1), the greatest frequency being observed close to a patch of young tall forest surrounding the laboratory area, as shown by Wright *et al.* (1994). The decline in agouti sightings near the 250-m trail mark corresponds to a ravine where agoutis were rarely observed — one pair of agoutis lived on each side of the ravine. A

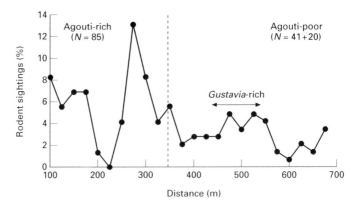

FIG. 2.1. Frequency of rodent sightings along the R.C. Shannon Trail on Barro Colorado Island, Panama, between November 1989 and May 1990. Definition of the two forest areas was based on the distance to trail origin, as an agouti-rich area at 100–325 m distance classes, and an agouti-poor area further away. One part of the trail (*Gustavia*-rich) was not used during the study of seed fate. Total number of agoutis encountered per transect section are shown in parenthesis; 20 animals were seen along the 100-m section at *Gustavia*-rich area, and 41 along the rest of the trail at the agouti-poor area.

greater density of key resources for agouti, such as *Dipteryx panamensis* and *Astrocaryum standleyanum* at the beginning of the Shannon Trail (Forget 1993; P.-M. Forget & T. Milleron personal observation) near the young forest, most probably accounted for the skewed rodent density versus the other part of the trail. Two forest areas were thus defined based on distance to origin of trail: an agouti-rich area at 100–325 m distance, and an agouti-poor area further away. A limit (>350 m) between the two zones was arbitrarily set to achieve equal numbers of seed piles ($n = 10$) per area for the experiments. Because another experiment was carried out in the area around the trail with an abundant population of *Gustavia superba* trees (the *Gustavia*-rich area in Forget 1992), as shown in Fig. 2.1, this part of the trail was not used for the present study.

STUDY SPECIES

The study species were the liana *Doliocarpus olivaceus* Sprague & L.O. Wms. ex Standl. (Dilleniaceae) and the trees *Cupania latifolia* H.B.K. (Sapindaceae), *Eugenia coloradensis* Standl. (Myrtaceae), *Brosimum alicastrum* Sw. subsp. *bolivarense* (Pitt.) C.C. Berg (Moraceae), *Virola nobilis* A.C. Smith (Myristicaceae), *Licania platypus* (Hemsl.) Fritsch (Chrysobalanaceae) and *Gustavia superba* (H.B.K.) Berg (Lecythidaceae)

TABLE 2.1. *Seed features, characteristics of the experimental design and mean (SD) percentage of seeds removed for each species per study month in 1990 on Shannon Trail at Barro Colorado Island, Panama.*

Species	Seed weight (g) ($n=30$)	Seed length (cm)	Study month	No. seeds per clump × no. sites	No. days per experiment	Removal (SD) %
Doliocarpus	0.21	1.1	August	5 × 20	1	91
Cupania	0.74	1.1	August	5 × 20	1	100
Eugenia	1.27	1.3	August	5 × 20	1	96
Brosimum	2.24*	1.8†	June	5 × 20	28	86
Virola	2.95	2	June	5 × 20	42	52
Licania	6.3	2.4	July	10 × 20	14	100
			August	10 × 19	1	99
Gustavia	12.6	1–3.5	June	10 × 20	28	88
			July	10 × 20	28	100

* S.J. Wright, personal communication ($n = 25$ seeds).
† Estrada & Coates-Estrada (1986).

(Table 2.1). Hereafter they are referred to by their genus in the text, table and figures. Nomenclature follows Croat (1978) except for *Virola* (Forget & Milleron 1991). In Table 2.1, these species are ranked by increasing seed weight. The first five species are likely to be dispersed by arboreal mammals, especially monkeys, and birds (Howe & Vander Kerkhove 1980, Oppenheimer 1982, Estrada & Coates-Estrada 1984). Peeled *Licania* seeds were observed beneath the palm fronds (P.-M. Forget personal observation) which are often used by bats as feeding roosts. Lacking large fauna to ingest the entire fruit (Janzen & Martin 1982, Howe 1986), *Gustavia* seeds are dispersed by scatter hoarding rodents (Forget 1992; see also Hallwachs 1986). All study genera are also consumed by spiny rats (Guillotin 1982, Adler 1995) which may also scatter the hoard seeds intensively (Forget 1991) when they occur at high densities on BCI (D.C. Tomblin & G.H. Adler personal communications).

METHODS

Experimental design

One clump of seeds cleaned of the arils or remains of pulp was placed at each of 20 sites within 2–5 m of the study trail, each seed set (pile) being separated by a 25-m interval; 10 clumps were set in the agouti-rich area and 10 clumps in the agouti-poor area. Clumps of 5–10 seeds were used in order to simulate clumped seed dispersal by frugivores (Howe 1989) beneath roost and perching trees. Clumps were made in sets of five seeds (areas of 50 × 50 cm; two sets 1 m apart in *Licania* and *Gustavia*) positioned each like a die-face number 5. To avoid further lateral movement, seeds were placed in a superficial depression in the ground, made with the thumb, and the location of each seed was marked with a wooden stake. Each seed was tagged with a 60-cm thread passing through it, which allowed the seed's fate to be followed, i.e. whether it was eaten, cached or lost (not found), after relocation of the thread with the intact cached seed or remains of the gnawed seeds (Forget 1990). In contrast to the spool-and-line method in which labelled seeds are attached to a fixed point (see Soné & Kohno 1996), rodents were allowed to move free thread-marked seeds long distances (>30 m). Seeds were retrieved, however, within 5 m for all species, except for *Licania* and *Gustavia* where the distance was up to 10 m. *Cupania*, *Doliocarpus* and *Eugenia* seeds were deposited mixed and simultaneously on 27 August 1990. Experiments with *Brosimum* and *Virola* seeds started on 14 and 29 June, respectively. *Licania* seeds were set on 27 July and 22 August, and *Gustavia* seeds on 21 June and 17 July. The number of days the experiment lasted differed between species depending on the time needed for all seeds to be removed, or for them to germinate or to rot (Table 2.1). Among *Cupania*, *Doliocarpus*, *Eugenia* and *Licania*, high seed removal occurred in the daytime within 12 h, and remained constant thereafter. The results only consider data collected after 1 day. Comparing medium-sized and large-sized seed species required consideration of the time needed for seeds to germinate (<4 weeks for *Brosimum* and *Gustavia*) or to rot (4–6 weeks for *Virola*) when not removed by animals. The data for *Virola* are from Forget and Milleron (1991) at the *Virola*-rich habitat (areas 1 & 2), and are presented here differently for comparison with *Brosimum*. The effect of month on seed fate was tested for *Licania* and *Gustavia* only.

Statistical analysis

The experiment conformed to a split-plot factorial design; the design had one between-block treatment (habitat: agouti-rich and agouti-poor areas; fate: eaten, cached and lost) and one within-block treatment (species) (see Kirk 1969, p. 247). Percentage of seeds removed and percentage of seeds in each fate category were based on the total number of seeds deposited per pile. Percentages were compared between species with similar seed size (small in *Doliocarpus*, *Cupania* and *Eugenia*; medium in *Brosimum* and *Virola*; large in *Licania* and *Gustavia*) and between months (June and July in *Gustavia*; July and August in *Licania*) using two- and three-way fully crossed

ANOVAs in PCSM-Plus V6.3 statistical package (PCSM 1994) after arcsin square-root transformation of the data (Sokal & Rohlf 1981). We wanted to address whether percentages in each fate category differed between habitats and species, and whether there were any interactions between these factors. Analysis allowed us to test for between-block variability due to the factor habitat and within-block variability due to the factor species, to the interactions of habitat and species, and of the two factors with error term different for each interaction (PCSM 1994).

RESULTS AND DISCUSSION

Change in seed fate: small-sized seeds

In August, seed removal was almost complete within 1 day (i.e. 12 h) for *Doliocarpus*, *Cupania* and *Eugenia* (Table 2.1). Neither species nor habitat (agouti-rich vs. agouti-poor area) or the interaction of species and habitat significantly affected the level of seed removal ($P > 0.19$). Similarly, neither habitat nor the interaction of habitat with species affected the percentage of seeds per fate category ($P > 0.69$). There was a significant effect of the species on seed fate (eaten: $F_{2,36} = 11.35$, $P = 0.0001$; cached: $F_{2,36} = 5.01$, $P = 0.012$; lost: $F_{2,36} = 3.62$, $P = 0.037$). Despite a similar seed size (c. 1 cm in their greatest dimension), the heavier seeds of *Eugenia* were observed cached more frequently (12%) than the lighter seeds of *Doliocarpus* (2%) and *Cupania* (0%) which were both essentially destroyed by the predators (Fig. 2.2).

Seeds as small (<1 g) as *Doliocarpus* and *Cupania* had a low probability not to be harvested by ground-dwelling vertebrates, and removal mostly meant predation by rodents. Because there was no effect on agouti density, removal and predation may have been due to small rodents such as spiny rats which may have been more abundant than previously thought on BCI (D.C. Tomblin & G.H. Adler personal communication). However, since most seeds disappeared at daytime (between 06:30 and 18:30), agouti and possibly squirrels were the main predators.

The experiment started in early August; that is, before the peak availability of small seeds 0.1–1 g, and when there were few large seeds in the forest (Foster 1982). By September–October, large rodents would be expected to be disinterested in such nutrient-poor seeds and be preferentially searching for more rewarding large seeds and caches (Smythe 1989, Forget 1992, Forget *et al.* 1994). One may postulate that removal of small seeds should decrease and fewer seeds should be consumed during September–October fruitpeak (see Foster 1982). At the same study area, with a comparable

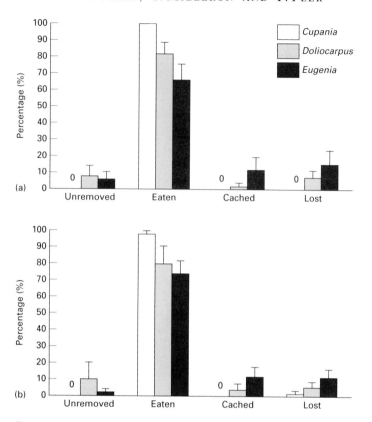

FIG. 2.2. Percentage of seeds (20 sites × 5 seeds) unremoved and the fate (eaten, cached or not retrieved within 5 m: lost) of removed seeds of *Doliocarpus*, *Cupania* and *Eugenia* in August 1990 at (a) the agouti-rich area and (b) the agouti-poor area along along the R.C. Shannon Trail on Barro Colorado Island, Panama. Bars = 1 SE of mean.

rodent community, Schupp (1988a,b, 1990) described the survival function of *Faramea occidentalis* (Rubiaceae) seeds that weighed 0.29 g on average. Within 4 weeks, seed removal ranged approximately 25–55% across four different years (Schupp 1990), and the survival probabilities of seeds were positively correlated with distance away from the parent tree in the understorey (Schupp 1988a,b). Because he carried out his experiment in January on BCI, that is, in the late wet to early dry season when overall food was scarce in the forest, Schupp (1988a) considered seed removal to be primarily due to seed predation. None the less, agoutis may occasionally bury *Faramea* seeds in November–December in the vicinity of parent trees (P.-M. Forget personal observation). Spiny rats which are known to consume *Faramea* seeds (Adler

1995) are also potential seed dispersers caching seeds below the litter and near logs (Forget 1991, Hoch & Adler 1997). Dirzo and Dominguez (1986) observed contrasting seed removal rates in *Trichilia martiana* (Meliaceae) between microhabitats in Los Tuxtlas, Mexico, and questioned the significance of these results given the possibility of secondary seed dispersal. Hammond (1995) found that on-site predation was less important than removal in *Erythrina goldmanii* (Fabaceae), *Spondias mombin* (Anacardiaceae) and *Bursera simaruba* (Burseraceae) weighing 0.21, 0.86 and 0.86 g, respectively. Local burial (<1 m) was evidenced in *Spondias*, and was of greater importance than predation. On the contrary, on-site predation was dominant in *Swietenia humilis* (Meliaceae) whose seeds weigh 0.80 g (Hammond 1995).

According to our results and Hammond's (1995) study, removal nearly equals predation by both small and large rodents when seeds weigh less than 1 g, especially when seeds are neither infested by insects after dispersal nor require burial for germination. Consequently, survival probability is exclusively dependent on patch suitability for seedling establishment and growth once seeds escape predators (Howe & Smallwood 1982, Schupp 1995). Seed survival rate through seedling establishment varies as a function of both spatial and temporal factors. On the one hand, survival may be related to either rodent density (Osunkoya 1994) or fruit tree density (Schupp 1992). On the other hand, selective pressures applied by different rodent species might be constrained by seasonal and year-to-year changes in either specific or plant community food availability depending on overall habitat richness. Regardless of the level of predation, the real impact of rodents as seed-eaters should be considered with caution for seeds nearly or greater than 1 g (as in *Eugenia*), which were secondarily scatter hoarded by rodents within 5 m (see also Hammond 1995).

Change in seed fate: medium-sized seeds

In June, seed removal rate was greater for *Brosimum* than for *Virola*, and there was a significant interaction of habitat and species on level of seed removal ($F_{1,18} = 18.47, P < 0.0001$). *Virola* seed removal rate was significantly greater ($F_{1,19} = 18.42, P < 0.001$) at the agouti-poor area (78%) than at the agouti-rich area (26%) while there was no difference ($P > 0.05$) for *Brosimum* (88 and 84%, respectively). The interaction of species and habitat was significant for each fate category (eaten: $F_{1,18} = 10.29, P = 0.0049$; cached: $F_{1,18} = 14.06, P = 0.0012$; lost: $F_{1,18} = 14.10, P = 0.0014$). In *Brosimum*, 58% of the seeds were recovered gnawed at the agouti-poor area whereas seeds were more frequently buried (26%) or lost (40%) at the agouti-rich area

(Fig. 2.3). Trends were different for *Virola*. Seeds were cached (42%) or lost (28%) at the agouti-poor area and remained largely unharvested at the other area (Fig. 2.3). In comparison to *Virola*, *Brosimum* seeds were a much preferred food item for rodents in general.

In Central America, *Brosimum* is an important primary source of fruit for howler monkeys (*Alouatta palliata*) that disperse seeds in clumps, on average up to several hundred metres away from parents (Estrada & Coates-Estrada 1984). Whether or not seeds are left in dung or fall beneath the parent tree does not affect germination rate but, based on seedling counts and survival under the tree crown, seeds that are dispersed are likely

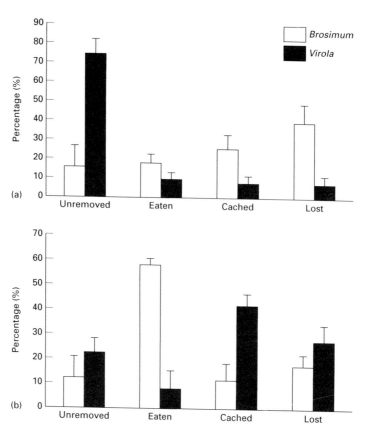

FIG. 2.3. Percentage of seeds (20 sites × 5 seeds) unremoved and the fate (eaten, cached or not retrieved within 5 m: lost) of removed seeds of *Brosimum* and *Virola* in June 1990 at (a) the agouti-rich area and (b) the agouti-poor area along along the R.C. Shannon Trail on Barro Colorado Island, Panama. Bars = 1 SE of mean.

to escape intense post-dispersal predation by mice (Estrada & Coates-Estrada 1991). Burkey (1994) found that *Brosimum* seed removal rate was negatively correlated with seed density but not at short distances (5–25 m) to the parent tree. He suggested that predator satiation was more important than dispersal distance in reducing the impact of seed-eaters. However, at Burkey's study site, close to a road that was human-disturbed and likely to have been in a hunted area, agoutis might have been less represented in comparison to small rodents such as Mexican deer mouse (*Peromyscus mexicanus*) and spiny pocket mice (*Heteromys desmarestianus*) (see Estrada, Coates-Estrada & Meritt 1994). In Los Tuxtlas, these two mice are the main consumers (Estrada & Coates-Estrada 1991) and potential scatter hoarders of *Brosimum* seeds (Fleming & Brown 1975, Emmons & Feer 1990). Therefore, it is probable that the contrast between low removal rate (40% after 20 days) reported by Burkey (1994) and higher values (100 and 88% after 10 and 2 days, respectively) in Estrada and Coates-Estrada (1986, 1991, respectively) and this study (86% after 28 days) was due to differences in rodent populations between habitats and this is likely to have been the result of the edge effect in a fragmented forest (Burkey 1993, Murcia 1995; see also Hammond 1995).

Seed removal by vertebrates in *Virola* is likely to be more influenced by tree neighbourhood, especially mean crop size (as in *Brosimum*, Burkey 1994), than by distance which instead affects seed damage by bruchids (Howe, Schupp & Westley 1985, Howe 1993). We observed that greater seed removal, associated with high caching rates in *Virola*, occurred where agouti sightings were lower. Change in agouti behaviour between areas may explain such a result. Korz, Heindrichs & Militzer (1996) observed that captive male agoutis (*D. punctata*) living in pairs were more often burying food and scraping the ground with their forefeet (to retrieve food and to bury) than other males living in large groups with several females. As suggested before for *Brosimum*, *Virola* seed fate is also likely to be altered by forest age. In both *Brosimum* and *Virola*, seed fate changes widely across forested habitats, possibly in relation to the ground-dwelling vertebrate community. Osunkoya (1994) showed that survival of diaspores was lower in a forest with the greatest abundance of rodents. Hoch and Adler (1997) found that the proportion of *Astrocaryum standleyanum* palm seed disappearing was dependent on the density of spiny rats on islands in Panama. The contrast in the proportion of seeds removed on the one hand, eaten or buried on the other hand, should then be correlated with the density of either small or large rodents across a diversity of habitats ranging from small and large islands to the mainland (Gliwicz 1984, Adler & Seamon 1991).

Although agoutis and other smaller rodents on BCI and mice in Los Tuxtlas may retrieve their hoards as well as seeds transported 2–5 cm deep by beetles (Estrada & Coates-Estrada 1991), so far it is impossible to characterize the true impact of rodents on seedling recruitment. Evidence of efficient secondary seed dispersal by agoutis on BCI was reported first by Forget and Milleron (1991) after creating artificial seed shadows. In contrast to *Virola*, lack of post-dispersal parasitism in *Brosimum* allows unburied seeds to germinate and develop as viable seedlings; nine out of 14 marked seeds had germinated within 4 weeks (P.-M. Forget & T. Milleron personal observation). Rodent species still play a potential role as secondary seed dispersers of insect-free seeds whose dispersal may be inefficient when buried seeds fail to establish.

This study shows that the removal as well as the fate of *Brosimum* and *Virola* seeds, especially the probability of seed caching, may change radically at a very small spatial scale (several hundred metres across a given forest) among habitats, apparently in relation to density of larger rodents. This suggests that either the same or different levels of seed removal, as a function of distance to a parent tree (Howe 1993, Burkey 1994) or habitat type (Burkey 1993, Osunkoya 1994, Asquith *et al.* 1997), overlap with contrasting seed-fate pathways. As shown for *Dipteryx panamensis* in the dry season (Forget 1993) and in a contrasting way to *Virola*, it is possible that the edible seed of *Brosimum* (and also *Licania*, see hereafter) is a preferred food item that has a high probability of being consumed by rodents when alternative food sources become rare. *Brosimum* may be hoarded more frequently when rodents have enough to eat, extra uncached food being eventually consumed by other vertebrates which usually compete with rodents for food (Smythe 1986). Habitat characteristics, such as the density of fruit trees (Schupp 1992, see also Forget 1992) and crop size (Howe 1993), and seed attributes possibly influence how agoutis would be triggered to eat and/or bury seeds. High seed removal in *Brosimum* should relate to either high predation or secondary dispersal whereas low or high seed removal in *Virola* should be a good indication, respectively, of weak or pronounced secondary seed dispersal by rodents.

Change in seed fate: large-sized seeds

In July, removal of both *Licania* and *Gustavia* seeds was almost 100% within 2 and 4 weeks, respectively. There was a significant effect of species on seed fate (eaten: $F_{1,18} = 36.68$, $P < 0.0001$; cached: $F_{1,18} = 19.68$, $P = 0.0003$; lost: $F_{1,18} = 6.65$, $P = 0.0189$). A greater proportion of *Licania* seeds were eaten at the agouti-rich area (69%) than at the agouti-poor area (40%), and conversely for the proportion of seeds cached (Fig. 2.4). No differences occurred in the proportion of *Gustavia* seeds within each seed fate category between habitats, caching being always dominant and predation lowest. The large nutrient-rich *Gustavia* seeds were more often buried than the less rewarding *Licania* seeds, for which the levels of predation were dependent on habitat and agouti density.

In contrast to the July experiment, 99% of *Licania* seeds deposited in early August were removed within 12 h. The effect of the interaction of habitat and month on seed fate was significant (eaten: $F_{1,36} = 10.69$, $P < 0.0024$; cached: $F_{1,36} = 6.41$, $P = 0.0158$; lost: $F_{1,36} = 9.46$, $P = 0.004$). Overall, less seeds were eaten in August (36%) than in July, and the proportion of

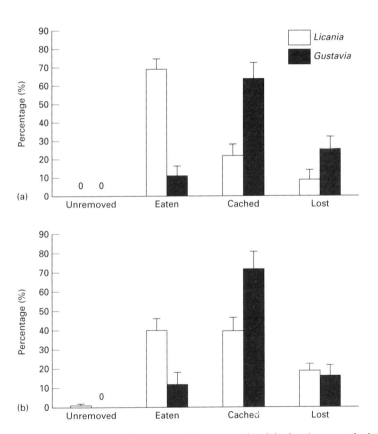

FIG. 2.4. Percentage of seeds (20 sites × 10 seeds) unremoved and the fate (eaten, cached or not retrieved within 10 m: lost) of removed seeds of *Licania* and *Gustavia* in July 1990 at (a) the agouti-rich area and (b) the agouti-poor area along along the R.C. Shannon Trail on Barro Colorado Island, Panama. Bars = 1 SE of mean.

cached and lost seeds changed between July (22 and 9%, respectively) and August (32 and 30%, respectively) at the agouti-rich area. Differences were not very marked at the agouti-poor area. Overall, a greater proportion of seeds were lost in August (33%) than in July (14%). A renewed interest of rodents for fallen nutrient-rich *Scheelea zonensis* seeds in late July–early August (Forget *et al.* 1994), when available food declined at the community level, can explain our results. Interpretation of differences in *Licania* seed fate, however, remains obscure without additional knowledge of the fate of unretrieved seeds that may have been transported greater distances (either eaten, larder- or scatter-hoarded) and without precise data of the alternative food resources in each habitat and period.

Removal of *Gustavia* seeds was marginally greater in July (100%) than in June (88%) ($F_{1,36} = 4.37$, $P = 0.043$). There was a significant effect of month on the level of seed predation ($F_{1,36} = 6.83$, $P = 0.013$). Overall, the proportion of seeds per fate category differed significantly ($F_{2,54} = 67.7$, $P < 0.001$). The proportion of seeds eaten was 10-fold greater in late July (12%) than in late June (1%). There was no effect of habitat on survival of cached seeds at agouti-rich and agouti-poor areas (14 and 11% in October 1990, respectively). There was a trend for greater survival of seeds cached in June (17.6) vs. July (8%), but the difference was not significant ($G = 3.47$, df = 1, $P > 0.05$).

As previously observed for smaller seeds, seed attributes may influence seed fate. The fresh *Licania* seeds were more often consumed than the non-germinated *Gustavia* seeds, the latter being almost exclusively cached. Difference of toxicity between seed species may explain such an outcome (see Forget 1992). In *Gustavia*, burying seeds may allow cotyledons to detoxify with, for instance, translocation of seed toxins to roots, stems and leaves. As with *Dipteryx panamensis* (Forget 1993), *Licania* falls within a bat-fruit dispersal syndrome and is also dispersed by agoutis following seed transportation by bats. Many other large-seeded bat-dispersed species could be also dependent on scatter-hoarding rodents for successful regeneration, especially when seeds are attacked by post-dispersal invertebrate predators as in the case of *Andira inermis* (Janzen, Miller, Hackforth-Jones *et al.* 1976). The lack of difference in *Gustavia* is consistent with the hoarding and survival rates of cached seeds shown for the *Gustavia*-rich area (Forget 1992). This suggests that when agoutis are present to disperse seeds, habitat features influence survival probability of cached and germinating seeds more than scatter-hoarding rate alone. Changing behaviour, especially agonistic behaviour between males and females (Smythe 1978, see also Korz *et al.* 1996), is likely to alter rates of seed caching between plant species and habitats from one month to another depending on overall food resources. In forested habitats with few or no dry season, key resources such as *D. pana-*

mensis seeds (De Steven & Putz 1984, Forget 1993) and previously buried palm nuts (Smythe 1989) supporting a persistent population of agoutis (Adler & Seamon 1991), a much greater proportion of seeds should be eaten instead of being cached immediately after seed fall.

PERSPECTIVES AND RECOMMENDATIONS

The morphological traits of seeds are important in shaping seed removal rate and loss (e.g. Willson 1988, Osunkoya 1994, see also Dirzo & Dominguez 1986) as well as seed fate. In this study, we provide evidence that caching rate tends to increase with seed weight (Fig. 2.5). A similar experiment by Hallwachs (1994) in a different neotropical forest showed the same relationship, implying that removal cannot be attributed to predation when seeds weigh more than 1 g. Indeed, Hallwachs (1994) brought to light new aspects of the scatter-hoarding behaviour of the agouti (*D. punctata*). She carried out several experiments across seasons to determine the food reward per piece of coconut transported, and consumed or cached by agoutis. The study clearly showed that (i) agoutis ate proportionately less of a food item as item weight increased; and (ii) the distance a 'seed' was carried and cached increased with weight of the food reward. 'Seeds' as small as 0.5 g were mostly cached within 5 m whereas mean distance for those cached items of 1–3 g was 6–7 m and much further for 6–30 g 'seeds'. A seasonal contrast was also highlighted by Hallwachs (1986) which agrees with Rankin's (1978) suggestion that rodents might eat proportionately fewer seeds and

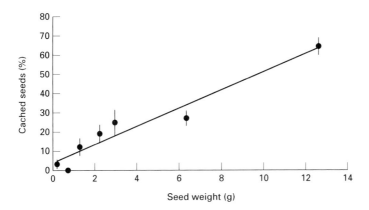

FIG. 2.5. Relationship between seed weight and mean (± SE) percentage of retrieved seeds cached within 5 m (seed < 3 g) and within 10 m (seed > 3 g) of the sites. Data for *Licania* (6.3 g) and *Gustavia* (12.3 g) are means for 2 month study periods. (Percentage of seeds cached = 4.85 × seed weight (g) + 3.47, $r^2 = 0.938$, $F_{1,5} = 75.7$, $P < 0.001$.)

cache more in the wet season than in the dry season. Not only does seasonal rainfall account for such a contrast, but caching may be low either in the dry season (Forget 1993) or in the late wet season (Forget et al. 1994; this study) depending on fruit availability (Smythe 1978).

Future studies should test the relationship between carrying distance and weight of food reward (Hallwachs 1994) considering animal home-range size and seasonal use of its foraging area. When food is scarce in the dry season, agouti may forage in another animal's area (thus increasing temporally the local rodent density, see Forget 1993) and carry fruits and seeds hundreds of metres (Hallwachs 1986) before returning to its home range. Conversely, they may transport seeds of a nearby food source (within several tens of metres) when food is locally plentiful in the wet season (Hallwachs 1986, Forget 1990, 1996). A better estimate of caching rates by reducing the lack of determination of seed fate (i.e. lost seeds) implies an adjustment of the searching radii dependent upon seed size. When there are several artificial seed piles per habitat, we recommend searching for marked seeds within 10 m for small ones c. 1–3 g, and at least 20 m for larger ones. If time is available, then it would be highly preferable to search at greater distances (e.g. 30–50 m). The category of lost seeds often comprises cached seeds that are not detected given the searching effort, but they may also be larder-hoarded by small rodents. If one may assume that caching rate and distance depend on the rodent species, then semi-permeable exclosures are recommended to discriminate the role of the small nocturnal species (Terborgh et al. 1993) especially when considering small- to medium-sized seeds. Given the wide range of seed size that nocturnal rodents consume (Guillotin 1982, Adler 1995) and maybe cache, checking marked seeds at dawn and dusk is also informative (Vandermeer et al. 1979, Forget 1991).

Post-dispersal seed fate is especially critical for those plant species whose successful seedling establishment rate is closely correlated with caching and incomplete cache recovery (Price & Jenkins 1986, Vander Wall 1990). A high percentage (nearly 100%) of post-dispersal predation is not surprising and should be viewed as a common picture given that rodents mostly rely upon these caches to survive when food is scarce; it may be the price of truly effective seed dispersal (i.e. dispersal giving rise to seedlings). On the other hand, a small percentage (1–10%, or less) of cached seeds escaping predators may match a large number of successfully established seedlings when the dispersed crop is large (>1000–10 000 seeds per tree). The most reliable way to validate the hypothesis that rodents are reliable seed dispersers is with labelled seed studies through the post-germination phase. A variety of techniques to trace removed seeds and fruits have been successfully used, including radioactive-labelling (Abbott & Quink 1970, Vander

Wall 1994, 1995), metal (Sork 1984) and magnet-tagging (Alverson & Diaz 1989), thread-tagging (Hallwachs 1986, Forget 1990, Iida 1996, Soné & Kohno 1996) and tracing with fluorescent pigments (Longland & Clements 1995). Ultra-light radio-transmitters (Soné & Kohno 1996) may also help to increase successful retrieval of cached seeds at distances far from their sources, but their current cost is prohibitive when large samples of seeds are required. The same methods as those used for following early seed fate may be used to estimate the rate of recovery of seeds by animals, and possibly recaching. All methods have the same disadvantage. Given the high rate of cache use by animals, the limited number of labelled seeds (several hundreds to thousands) prevents study of a large number of naturally buried seeds through to the seedling stage, when the marking itself does not prevent successful germination and establishment. Among these methods, the thread-marking method is cheap since it does not require special sophisticated equipment, is easy to apply using a drill, thread and good eyes to located cached seeds, and provides rapid recording of immediate seed removal and fate.

When natural caches are too scarce due to high post-dispersal predation, experimental but realistic burial must be used based on field observations of rodent behaviour; that is, spacing buried seeds around focal trees (phase I of seed dispersal *sensu* Chambers & MacMahon 1994), and at greater distances (to several hundred metres) simulating secondary seed dispersal (phase II). This last method presents a bias because the memory component of cache recovery is eliminated. Locating buried seeds at particular microsites (Willson 1988) where rodents hoard food (i.e. near logs, branches and at the base of palms and trees, see Kiltie 1981, Forget 1990), would be even better in order to simulate animal caching behaviour, and in this way reduce bias. Such an approach also has the advantage of testing the effect of burial on germination and seedling establishment and the effect of others predators (insects, peccaries) depending on microsite location such as gaps and shaded understorey (Schupp 1988a, Willson 1988, Schupp & Frost 1989, Osunkoya 1994).

Seed fate in the understorey may differ from that in gap openings. To understand the contrast in habitat choice between seed and seedling (Schupp 1995), one may assume that seeds removed in gap openings by rodents are reached within that environment. Thus, besides the degree of canopy openness, it is maybe the dense tangle of dead branches and logs (often associated with canopy disturbance), that favours survival of cached seeds which most likely establish and grow when gaps are sunny (see Schupp 1995). For instance, this is more probable for medium-sized seeds (*c.* 2.9 g) such as those of *Welfia georgii* (Schupp & Frost 1989) than for smaller-sized seeds (c. 0.3 g) such as those of *Faramea occidentalis* (Schupp 1988a).

Like morphology, plant phenology is another crucial parameter influencing the seed-fate pathway. Hoarding seems to be regulated by food abundance as suggested by experiments in the laboratory (Vandermeer 1979) or in the wild (Forget 1993, 1996). Caching appears to increase as satiation occurs, and cache recovery may be subject to the same ecological factors. Seasonal and spatial contrasts in seed fate are likely to be related to within-year changes of seed availability influencing the foraging, hoarding and feeding activities of rodents (Vandermeer *et al.* 1979, Forget 1993, Forget *et al.* 1994). Replicated experiments in space across the forest and in time throughout the fruiting season associated with standardized measures of resource availability at the community level, such as the number of fruiting trees and species and the average fruit/seed biomass falling on the ground, are thus essential. Because these parameters are likely to fluctuate annually, the net effect of rodents will therefore depend on the frequency of good and bad years; that is, years in which seed predators are or are not satiated (Janzen 1970). These considerations argue for more caution in drawing conclusions from short-term studies and for making longer periods of observations (i.e. 4 years for annual species, as in Schupp 1990; and ≥ 10 years for masting species).

Finally, the choice of habitats for experiments (natural or disturbed, entire or fragmented, with or without edge effects, hunted or not) largely condition the results. In Central America, for instance, long-term consequences of forest fragmentation on the populations of vertebrate predators (Estrada *et al.* 1994, Wright *et al.* 1994) have altered the level of seed removal and predation by rodents (De Steven & Putz 1984, Sork 1987, Forget *et al.* 1994, Hoch & Adler 1997) and by insects (Janzen 1971b, Janzen *et al.* 1976, Bradford & Smith 1977, Herrera 1989). Any change in the level of secondary seed dispersal as a function of habitat and rodent population density may have critical consequences on spatial distribution and survival of seeds and seedlings, and possibly on the maintenance of tree diversity (Leigh, Wright, Herre & Putz 1993).

ACKNOWLEDGEMENTS

Field study was made possible thanks to a post-doctoral Smithsonian Institution fellowship to P.M.F., and personal support by T.M. We thank the British Ecological Society for inviting us to participate in their annual meeting. Many thanks are extended to Judy Rankin and David Newbery who cordially accepted to comment and correct the text and to Gregory Adler and two anonymous reviewers for their criticism.

REFERENCES

Abbott, H.G. & Quink, T.F. (1970). Ecology of eastern white pine seed caches made by small forest mammals. *Ecology*, **51**, 271–278.

Adler, G.H. (1995). Fruit and seed exploitation by Central American spiny rats, *Proechimys semispinosus*. *Studies on Neotropical Fauna and Environment*, **30**, 237–244.

Adler, G.H. & Seamon, O.J. (1991). Distribution and abundance of a tropical rodent, the spiny rat, on islands in Panama. *Journal of Tropical Ecology*, **7**, 349–360.

Alverson, W.S. & Diaz, A.G. (1989). Measurement of the dispersal of large seeds and fruits with a magnetic locator. *Biotropica*, **21**, 61–63.

Asquith, N.M., Wright, S.J. & Clauss, M.J. (1997). Does mammal community composition control recruitment in neotropical forests? Evidence from Panama. *Ecology*, **78**, 941–946.

Boucher, D.H. (1981). Seed predation by mammals and forest dominance by *Quercus oleoides*, a tropical lowland oak. *Oecologia*, **49**, 409–414.

Bradford, D.F. & Smith, C.C. (1977). Seed predation and seed number in *Scheelea* palm fruits. *Ecology*, **58**, 667–673.

Burkey, T.V. (1993). Edge effect in seed and egg predation at two neotropical rainforest sites. *Biological Conservation*, **66**, 139–143.

Burkey, T.V. (1994). Tropical tree species diversity: a test of the Janzen–Connell model. *Oecologia*, **97**, 533–540.

Byrne, M.M. & Levey, D.J. (1993). Removal of seeds from frugivore defecation by ants in a Costa Rican rain forest. *Vegetatio*, **107–108**, 351–362.

Chambers, J.C. & MacMahon, J.A. (1994). A day in the life of a seed: movements and fates of seeds and their implications for natural and managed systems. *Annual Review of Ecology and Systematics*, **25**, 263–292.

Chapman, C.A. (1989). Primate seed dispersal: the fate of dispersed seeds. *Biotropica*, **21**, 148–154.

Crawley, M.J. (1992). Seed predators and plant population dynamics. In *Seeds* (Ed. by M. Fenner), pp. 157–191. CAB International, Oxon.

Croat, T.B. (1978). *Flora of Barro Colorado Island*. Stanford University Press, Stanford.

De Steven, D. & Putz, F.E. (1984). Impact of mammals on early recruitment of a tropical canopy tree, *Dipteryx panamensis*. *Oikos*, **43**, 207–216.

Dirzo, R. & Dominguez, C.A. (1986). Seed shadows, seed predation and the advantages of dispersal. In *Frugivores and Seed Dispersal* (Ed. by A. Estrada & T.H. Fleming), pp. 237–249. Junk, The Hague.

Emmons, L.H. (1982). Ecology of *Proechimys* (Rodentia, Echimyidae) in South-eastern Peru. *Tropical Ecology*, **23**, 280–290.

Emmons, L.H. & Feer, F. (1990). *Neotropical Rainforest Mammals. A Field Guide*. University of Chicago Press, Chicago.

Estrada, A. & Coates-Estrada, R. (1984). Fruit eating and seed dispersal by howling monkeys (*Alouatta palliata*) in a tropical rain forest of Los Tuxtlas, Mexico. *American Journal of Primatology*, **6**, 77–91.

Estrada, A. & Coates-Estrada, R. (1986). Frugivory by howling monkeys (*Alouatta palliata*) at Los Tuxtlas, Mexico: dispersal and fate of seeds. In *Frugivores and Seed Dispersal* (Ed. by A. Estrada & T.H. Fleming), pp. 93–104. Junk, The Hague.

Estrada, A. & Coates-Estrada, R. (1991). Howler monkeys (*Alouatta palliata*), dung beetles (Scarabaeideae) and seed dispersal: ecological interactions in the tropical rain forest of Los Tuxtlas, Mexico. *Journal of Tropical Ecology*, **7**, 459–474.

Estrada, A., Coates-Estrada, R. & Meritt, D. Jr (1994). Non-flying mammals and landscape changes in the tropical rain forest region of Los Tuxtlas, Mexico. *Ecography*, **17**, 229–241.

Fleming, T.H. (1974). The population ecology of two species of Costa Rican heteromyid rodents. *Ecology*, **55**, 493–510.
Fleming, T.H. & Brown, G.J. (1975). An experimental analysis of seed hoarding and burrowing behavior in two species of Costa Rican heteromyid rodents. *Journal of Mammalogy*, **56**, 301–315.
Forget, P.-M. (1989). La régénération naturelle d'une espèce autochore de la forêt guyanaise: *Eperua falcata* Aublet (Caesalpiniaceae). *Biotropica*, **21**, 115–125.
Forget, P.-M. (1990). Seed dispersal of *Vouacapoua americana* (Caesalpiniaceae) by caviomorph rodents in French Guiana. *Journal of Tropical Ecology*, **6**, 459–468.
Forget, P.-M. (1991). Scatterhoarding of *Astrocaryum paramaca* by *Proechimys* in French Guiana: comparison with *Myoprocta exilis*. *Tropical Ecology*, **32**, 155–167.
Forget, P.-M. (1992). Seed removal and seed fate in *Gustavia superba* (Lecythidaceae). *Biotropica*, **24**, 408–414.
Forget, P.-M. (1993). Post-dispersal predation and scatterhoarding of *Dipteryx panamensis* (Papilionaceae) seeds by rodents in Panama. *Oecologia*, **94**, 255–261.
Forget, P.-M. (1996). Removal of seeds of *Carapa procera* (Meliaceae) and their fate in rainforests in French Guiana. *Journal of Tropical Ecology*, **12**, 751–767.
Forget, P.-M. & Milleron, T. (1991). Evidence for secondary seed dispersal in Panama. *Oecologia*, **87**, 596–599.
Forget, P.-M., Munoz, E. & Leigh, E.G. Jr (1994). Predation by rodents and bruchid beetles on seeds of *Scheelea* palms on Barro Colorado Island, Panama. *Biotropica*, **26**, 420–426.
Foster, R.B. & Brokaw, N.V.L. (1982). Structure and history of the vegetation of Barro Colorado Island. In *The Ecology of a Tropical Forest: Seasonal Rythms and Long-term Changes* (Ed. by E.G. Leigh, Jr, A.S. Rand & D.M. Windsor), pp. 67–81. Smithsonian Institution Press, Washington DC.
Gliwicz, J. (1984). Population dynamics of the spiny rats *Proechimys semispinosus* on Orchid Island (Panama). *Biotropica*, **16**, 73–78.
Guillotin, M. (1982). Rythmes d'activité et régimes alimentaires de *Proechimys cuvieri* et d'*Oryzomys capito velutinus* (Rodentia) en forêt guyanaise. *Revue d'Ecologie* (*Terre et Vie*), **36**, 337–371.
Hallwachs, W. (1986). Agoutis *Dasyprocta punctata*: the inheritors of guapinol *Hymenaea courbaril* (Leguminosae). In *Frugivores and Seed Dispersal* (Ed. by A. Estrada & T.H. Fleming), pp. 119–135. Junk, The Hague.
Hallwachs, W. (1994). The clumsy dance between agoutis and plants: scatterhoarding by Costa Rican dry forest agoutis (*Dasyprocta punctata*: Dasyproctidae: Rodentia). PhD, thesis Cornell University, Ithaca, NY.
Hammond, D.S. (1995). Post-dispersal seed and seedling mortality of tropical dry forest trees after shifting agriculture, Chiapas, Mexico. *Journal of Tropical Ecology*, **11**, 295–313.
Herrera, C.M. (1989). Vertebrate frugivores and their interaction with invertebrate fruit predators: supporting evidence from a Costa Rican dry forest. *Oikos*, **54**, 185–188.
Hoch, G.A. & Adler, G.H. (1997). Removal of black palm (*Astrocaryum standleyanum*) seeds by spiny rats (*Proechimys semispinosus*). *Journal of Tropical Ecology*, **13**, 51–58.
Howe, H.F. (1986). Seed dispersal by fruit-eating birds and mammals. In *Seed Dispersal* (Ed. by D.R. Murray), pp. 123–189. Academic Press, Sydney.
Howe, H.F. (1989). Scatter- and clump-dispersal and seedling demography: hypothesis and implications. *Oecologia*, **79**, 417–426.
Howe H.F. (1993). Aspects of variation in a neotropical seed dispersal system. *Vegetatio*, **107/108**, 149–162.
Howe, H.F. & Smallwood, J.H. (1982). Ecology of seed dispersal. *Annual Review of Ecology and Systematics*, **13**, 201–228.

Howe, H.F. & Vander Kerckhove, G.A. (1980). Nutmeg dispersal by tropical birds. *Science*, 210, 925–927.

Howe, H.F., Schupp, E.W. & Westley, L.C. (1985). Early consequences of seed dispersal for a neotropical tree (*Virola surinamensis*). *Ecology*, 66, 781–791.

Iida, S. (1996). Quantitative analysis of acorn transportation by rodents using magnetic locator. *Vegetatio*, 124, 39–43.

Janzen, D.H. (1970). Herbivores and the number of species in tropical forests. *American Naturalist*, 104, 501–528.

Janzen, D.H. (1971a). Seed predation by animals. *Annual Review of Ecology and Systematics*, 2, 465–492.

Janzen, D.H. (1971b). The fate of *Scheelea rostrata* fruits beneath the parent tree: predispersal attack by bruchids. *Principes*, 15, 89–101.

Janzen, D.H. & Martin, P.S. (1982). Neotropical anachronisms: the fruits the gomphotheres ate. *Science*, 215, 19–27.

Janzen, D.H., Miller, G.A., Hackforth-Jones, J., Pond, C.M., Hooper, K. & Janos, D.P. (1976). Two Costa Rican bat-generated seed shadows of *Andira inermis* (Leguminosae). *Ecology*, 57, 1068–1075.

Kaspari, M. (1993). Removal of seeds from neotropical frugivore droppings: ant responses to seed number. *Oecologia*, 95, 81–88.

Kiltie, R.A. (1981). Distribution of palm fruit on a rain forest floor: why white-lipped peccaries forage near objects. *Biotropica*, 6, 69–103.

Kirk, R.E. (1969). *Experimental Design: Procedure for Behavioural Sciences*. Brooks Kole, Belmont.

Korz, V., Heindrichs, H. & Militzer, K. (1996). Behavioural and anatomical correlates of sympathetic arousal and stress in male Central American agoutis (*Dasyprocta punctata*). *Zeitschrift für Säugetierkunde*, 61, 112–125.

Larson, D. & Howe, H.F. (1987). Dispersal and destruction of *Virola surinamensis* seeds by agoutis: appearance and reality. *Journal of Mammalogy*, 68, 859–860.

Leigh, E.G. Jr, Rand, A.S. & Windsor, D.M. (1982). *The Ecology of a Tropical Forest: Seasonal Rhythms and Long-term Changes*. Smithsonian Institution Press, Washington DC.

Leigh, E.G. Jr, Wright, S.J., Herre, E.A. & Putz, F.E. (1993). The decline of tree diversity on newly isolated tropical islands: a test of a null hypothesis and some implications. *Evolutionary Ecology*, 7, 76–102.

Levey, D.J. & Byrne, M.M. (1993). Complex ant–plant interactions: rainforest ants as secondary dispersers and post-dispersal seed predators. *Ecology*, 74, 1802–1812.

Longland, W.S. & Clements, C. (1995). Use of fluorescent pigments in studies of seed caching by rodents. *Journal of Mammalogy*, 76, 1260–1266.

Morris, D. (1962). The behaviour of the green acouchi (*Myoprocta pratti*) with special reference to scatter hoarding. *Proceedings of the Zoological Society of London*, 139, 701–733.

Murcia, C. (1995). Edge effects in fragmented forests: implications for conservation. *Trends in Ecology and Evolution*, 10, 58–62.

Oppenheimer, J.R. (1982). *Cebus capucinus*: home-range, population dynamics, and interspecific relationships. In *The Ecology of a Tropical Forest: Seasonal Rhythms and Long-term Changes* (Ed. by E.G. Leigh, Jr, A.S. Rand & D.M. Windsor), pp. 253–272. Smithsonian Institution Press, Washington DC.

Osunkoya, O.O. (1994). Postdispersal survivorship of North Queensland rainforest seeds and fruits: effects of forest, habitat and species. *Australian Journal of Ecology*, 19, 52–64.

PCSM (1994). *Programme Conversationnel de Statistiques pour les Sciences et le Marketing*. Deltasoft, Meylan, France.

Piperno, D.R. (1990). Fitolitos, arqueologia y cambios de la vegetation en un lote de cincuentua hectareas de la isla de Barro Colorado. In *Ecologia de un Bosque Tropical* (Ed. by E.G.

Leigh, Jr, A.S. Rand & D.M. Windsor), pp. 141–152. Smithsonian Tropical Research Institute, Balboa.

Price, M.V. & Jenkins, S.H. (1986). Rodents as seed consumers and dispersers. In *Seed Dispersal* (Ed. by D.R. Murray), pp. 191–235. Academic Press, Sydney.

Rand, A.S. & Rand, W.M. (1982). Variation in rainfall on Barro Colorado Island. In *The Ecology of a Tropical forest: Seasonal Rhythms and Long-term Changes* (Ed. by E.G. Leigh, Jr, A.S. Rand & D.M. Windsor), pp. 47–59. Smithsonian Institution Press, Washington DC.

Rankin, J.M. (1978). The influence of seed predation and plant competition on tree species abundances in two adjacent tropical rain forests. PhD thesis. University of Michigan, Michigan.

Schupp, E.W. (1988a). Seed and early seedling predation in the forest understory and in treefall gaps. *Oikos*, 51, 71–78.

Schupp, E.W. (1988b). Factors affecting post-dispersal seed survival in a tropical forest. *Oecologia*, 76, 525–530.

Schupp, E.W. (1990). Annual variation in seedfall, postdispersal predation, and recruitment of a neotropical tree. *Ecology*, 71, 504–515.

Schupp, E.W. (1992). The Janzen–Connell model for tropical tree diversity: population implications and the importance of spatial scale. *American Naturalist*, 140, 526–530.

Schupp, E.W. (1995). Seed-seedling conflicts, habitat choice, and patterns of plant recruitment. *American Journal of Botany*, 82, 399–409.

Schupp, E.W. & Frost, E.J. (1989). Differential predation on *Welfia georgii* seeds in treefall gaps and the forest understory. *Biotropica*, 21, 200–203.

Schupp, E.W., Howe, H.F., Augspurger, C.K. & Levey, D.J. (1989). Arrival and survival in tropical treefall gaps. *Ecology*, 70, 562–564.

Smythe, N. (1970). Relationships between fruiting seasons and seed dispersal methods in a neotropical forest. *American Naturalist*, 104, 25–35.

Smythe, N. (1978). The natural history of the Central American agouti (*Dasyprocta punctata*). *Smithsonian Contribution to Zoology*, 257, 1–52.

Smythe, N. (1986). Competition and resource partitioning in the guild of neotropical terrestrial frugivorous mammals. *Annual Review of Ecology and Systematics*, 17, 169–188.

Smythe, N. (1989). Seed survival in the palm *Astrocaryum standleyanum*: evidence for dependence upon its seed dispersers. *Biotropica*, 21, 50–56.

Smythe, N., Glanz, W.E. & Leigh, E.G. Jr (1982). Population regulation in some terrestrial frugivores. In *The Ecology of a Tropical Forest: Seasonal Rhythms and Long-term Changes* (Ed. by E.G. Leigh, Jr, A.S. Rand & D.M. Windsor), pp. 227–238. Smithsonian Institution Press, Washington DC.

Sokal, R.R. & Rohlf, F.J. (1981). *Biometry*, 2nd edn. W.E. Freeman, New York.

Soné, K. & Kohno, A. (1996). Application of radiotelemetry to the survey of acorn dispersal by *Apodemus* mice. *Ecological Research*, 11, 187–192.

Sork, V.L. (1984). Examination of seed dispersal and survival in red oak, *Quercus rubra* (Fagaceae), using metal-tagged acorns. *Ecology*, 65, 1020–1022.

Sork, V.L. (1987). Effects of predation and light on seedling establishment in *Gustavia superba*. *Ecology*, 68, 1341–1350.

Terborgh, J. & Wright, S.J. (1994). Effects of mammalian herbivores on plant recruitment in two neotropical forests. *Ecology*, 75, 1829–1833.

Terborgh, J., Losos, E., Riley, M.P. & Bolanos Riley, M. (1993). Predation by vertebrates and invertebrates on the seeds of five canopy tree species of an Amazonian forest. *Vegetatio*, 107/108, 375–386.

Vander Wall, S.B. (1990). *Food Hoarding in Animals*. Chicago University Press, Chicago.

Vander Wall, S.B. (1994). Seed fate pathways of antelope bitterbrush: dispersal by seed caching yellow pine chipmunks. *Ecology*, 75, 1911–1926.

Vander Wall, S.B. (1995). Sequential patterns of scatter hoarding by yellow pine chipmunks (*Tamias amoenus*). *American Midland Naturalist*, **133**, 312–321.

Vandermeer, J.H. (1979). Hoarding behavior of captive *Heteromys desmarestianus*, (Rodentia) on the fruits of *Welfia georgii*, a rainforest dominant palm in Costa Rica. *Brenesia*, **16**, 107–116.

Vandermeer, J.H., Stout, J. & Risch, S. (1979). Seed dispersal of a common Costa Rican rain forest palm. *Tropical Ecology*, **20**, 17–26.

Willson, M.F. (1988). Spatial heterogeneity of post-dispersal survivorship of Queensland rainforest seeds. *Australian Journal of Ecology*, **13**, 137–145.

Wright, J.S., Gompper, M.E. & De Leon, B. (1994). Are large predators keystone species in Neotropical forests? The evidence from Barro Colorado Island. *Oikos*, **71**, 279–294.

Zona, S. & Henderson, A. (1989). A review of animal-mediated seed dispersal of palms. *Selbyana*, **11**, 6–21.

3. DISTURBANCE, PHENOLOGY AND LIFE-HISTORY CHARACTERISTICS: FACTORS INFLUENCING DISTANCE/DENSITY-DEPENDENT ATTACK ON TROPICAL SEEDS AND SEEDLINGS

D. S. HAMMOND*† and V. K. BROWN*

*International Institute of Entomology, 56 Queen's Gate, London SW7 5JR, UK; †Tropenbos Guyana Programme, 12E Garnett St, Campbellville, Georgetown, Guyana

SUMMARY

1 A large body of research has been dedicated to testing the predictions of the Janzen–Connell model of density/distance (or frequency)-dependent attack on seeds and seedlings over the past 25 years. Both theoretical and experimental field tests have been made but no consensus has been reached. A compilation of these studies suggests that invertebrate, but not vertebrate, attack generally conforms to the model despite varying forest types, locations and species. This disparity, linked to differences in search tactics and selection criteria, may explain in part why a consensus has been so elusive. More recently, the influence of spatial scale on predation effects has been emphasized.

2 A second aspect of seed and seedling availability which can influence frequency-dependent attack is the seasonal pattern (and scale) of reproduction in tropical trees and lianas. Large synchronous fruiting events within a population (or among populations of congeners), which can satiate predators, is a commonly accepted way of explaining low predation rates near trees. However, spatial and temporal scales of seed availability are not mutually exclusive influences; there are several different combinations of these two scales (reproductive adult density and individual adult crop size) which can produce the same number of seeds and seedlings available for attack.

3 The literature suggests that offspring 'escape' predators either through dispersal or by being unpredictably abundant in time. This *de facto* assumption of the Janzen–Connell model, along with the lack of attention paid to the important role that light availability might play in determining the

outcome of seed and seedling attack, have misled most attempts at predicting the effects of attack on population dynamics of tropical trees. An increase in light availability can 'buffer' the effects of attack by invertebrates, leading to establishment where normally under low light conditions there would be mortality. Light-dependent 'escape' may occur as a consequence of reducing seedling susceptibility to attack, by increasing the efficacy of compensatory growth in seedlings, or by marginalizing the impact of seed reserve on seedling growth (the redundancy model).

4   Disturbance of tropical forests occurs at different frequencies and magnitudes. This variation can affect the relative importance of light-dependent 'escape', since the level of disturbance determines the likelihood of an attacked seed or seedling being in a location exposed to a particular light environment.

INTRODUCTION

Most tropical trees stand a greater per capita chance of dying as seeds and seedlings than at any other time during pre-senescence. Invertebrate attack at these early developmental stages is often responsible for a large fraction of losses to a cohort (Janzen 1969, Janzen, Miller, Hackforth-Jones et al. 1976, Howe, Schupp & Westley 1985). This drastic reduction in offspring, especially near parent trees, prompted Janzen (1970) and Connell (1971) to suggest independently that seed and seedling predation might be related to the density and/or distance in which a cohort of offspring are spatially distributed around the parent tree. Connell (1971) explained how predation might influence survivorship and, ultimately, tropical tree diversity but it was Janzen (1970) who devised a graphical model (Fig. 3.1) describing the relationship between the dispersal of offspring away from adults, the density of seeds and seedlings and the concomitant advantage to be gained by 'escaping' (*sensu* Howe & Smallwood 1982) density or distance-dependent seed and seedling predators. The model predicted that disproportionate mortality beneath adults where density is highest, combined with a decreasing seed density away from the parent tree, should result in a minimum distance between conspecific adults where high seed density and probability of seed survival are simultaneously optimized. This 'minimum distance rule' (*sensu* Hubbell 1980) could explain the large numbers of species coexisting at low abundances since it would, theoretically, prevent any single or group of species becoming dominant. In (the more common) instance where parent trees were not found in singular isolation, Janzen (1970) suggested that the offspring recruitment pattern produced from overlapping seed shadows of conspecific adults would produce a more complex series of

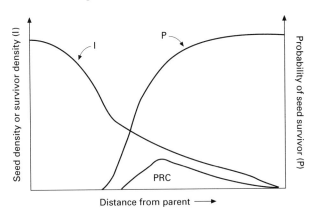

FIG. 3.1. Janzen's graphical model depicting a decline in seed or survivor density (I) and an increase in survival (P) as the distance from parent increases. The simultaneous optimization of these produces a peak in the population recruitment curve (PRC). (Adapted from Janzen 1970.)

recruitment 'depressions' and 'ridges' concomitant with the minimum spacing rule.

Janzen (1976, 1978) later proposed that seeds and seedlings could also escape their predators through changes in their temporal availability, as well as through spatial dispersion. By producing vastly larger numbers of fruits/seeds in synchronous supra-annual, or mast, events, tree populations would 'satiate' (*sensu* Lloyd & Dybas 1966) predators which would only be able to consume a part of the cohort. In addition, the intervening years of low seed production would prevent the predator population from expanding to a size capable of utilizing all of the resource available during a mast year. The survival of a larger percentage of offspring during masting years, combined with the regulation of predators during intermittent periods of little or no seed production, would presumably allow a species to recruit equally well over the entire area encompassed by the masting event (e.g. Hart 1995).

The popularity of Janzen's theories was bolstered by their apparent explanatory power and the implications they held for population dynamics. Yet, did Janzen really manage to encapsulate a fundamentally important mortality process, which effects a cohort of offspring at its most crucial stage of development? If so, why have theoretical and experimental studies over the past 25 years failed to reach a verdict on these seminal papers (Janzen (1970, 1976) and Connell (1971))? Are there factors or mechanisms that might confound or mask the outcome of density/distance-dependent attack processes?

THE JANZEN–CONNELL MODEL

The theoretical debate

Continual re-examination over the past 25 years, in the field and on paper, of Janzen's elegantly simple, but highly contentious, graphical model has led to neither solid acceptance or rejection of its predictions. Much of the debate has surrounded the spacing mechanism distilled from Janzen's spatial model and the implications of this spacing to the maintenance of tropical tree diversity. Hubbell (1980) evaluated the minimum distance rule proposed by Janzen (1970) in light of the observed spatial distributions of dry tropical forest trees in Costa Rica. Using a stochastic model, he argued that the spacing of adult conspecifics, as predicted by the Janzen–Connell (J–C) model, could not explain the high level of tree species richness observed. Furthermore, in direct contravention of Janzen's (1970) model, he suggested that the highest seed/seedling densities would be found beneath parent trees where the number of dispersed offspring is several orders of magnitude larger than the levels encountered within more distant annuli (concentric rings). As long as a small percentage of the larger number of offspring beneath trees survived, their post-establishment densities would still overshadow those of the fewer offspring that had been dispersed further away and experienced increased survivorship as a consequence.

Subsequently, Becker, Lee, Rothman & Hamilton (1985) and Clark and Clark (1984) re-evaluated the J–C model in light of Hubbell's arguments. Both concluded independently that Hubbell's analysis did not adequately refute the predictions of distance/density-dependent attack nor provide a viable alternative framework on which to explain the incongruity between early mortality processes and the spatial distributions of adult conspecifics in neotropical forests. Becker *et al.* (1985), and more recently Burkey (1994) argued succinctly that the outcome produced by modifying the J–C model as proposed by Hubbell (1980) is invalidated because it is the (relative) frequency of offspring (*sensu* Greenwood 1985), not the spacing between adults, and the accompanying decline in seed density that occurs between adults, which is important in evaluating seed and seedling mortality due to predation. Becker *et al.* (1985) further argued that the high recruitment below parent trees which Hubbell (1980) predicted was simply an artefact of the arbitrary survivorship probability assigned to offspring found beneath parent trees. Increasing mortality beneath trees by an order of magnitude caused a jump in peak recruitment from beneath the parent tree to over 100 m away. A more recent expansion of the J–C model (Armstrong 1989) concluded that the minimum distance rule could allow for the coexistence of a large number of species without necessarily leading to regular spacing as described by Hubbell (1980).

Empirical findings and field tests

Empirical studies examining various aspects of the J–C model have likewise yielded variable results (Table 3.1). Though attack upon seeds and seedlings of many species has proven to be density–distance responsive, a decrease in density or increase in distance from parent trees for others has shown only a weakly positive (e.g. Wright 1983) or even a negative effect (e.g. Schupp 1992, Burkey 1994, Hart 1995) upon the likelihood of avoiding different seed/seedling predators. The wide variation in the distance from the focal parent tree where predation declines and the effects of distance between conspecific adults on the probability of attack (e.g. Janzen 1972, Janzen *et al.* 1976, Wright 1983, Schupp 1988a, Burkey 1994) suggest that the spatial scale at which seed and seedling escape is density/distance-dependent varies considerably between species and sites.

TABLE 3.1. Comparison between vertebrate and insect attack on tropical seeds and seedlings based on their support of the Janzen–Connell density/distance-dependent effect. Studies that showed relatively high mortality below parent trees and/or a minimal distance between adult and offspring recruitment were considered to support the J–C model. Studies that showed predator satiation with an increase in conspecific adult density are considered to tentatively not support the model (denoted by '?'), since at smaller scales disproportionate attack may still occur beneath parent trees. Species that experienced nil or negligible vertebrate or invertebrate attack regardless of distance/density were not included.

Species	Forest type	Attack type	Seeds or seedlings or both?	Supports J–C model?	Source
Acacia farnesiana	Dry	Insect	Seeds	Yes	Traveset (1990)
Andira inermis	Dry	Insect	Seeds	Yes	Janzen *et al.* (1976)
Astrocaryum mac rocalyx	Wet	Insect	Seeds	Yes	Terborgh *et al.* (1993)
Carapa guianensis	Wet	Insect	Seeds	Yes	Hammond *et al.* (1994)
Chlorocardium rodiei	Wet	Insect	Seeds	Yes	Hammond (unpublished data)
Copaifera pubiflora	Moist	Insect	Seeds	Yes	Ramirez & Arroyo (1987)
Dioclea megacarpa	Dry	Insect	Both	Yes	Janzen (1971)
Gilbertiodendron dewevrei	Wet	Insect	Seeds	Yes	Hart (1995)
Julbernardia seretii	Wet	Insect	Seeds	Yes	Hart (1995)
Mora gongrijpii	Wet	Insect	Seeds	No	Hammond (unpublished data)
Macoubea guianensis	Wet	Insect	Seeds	No	Notman *et al.* (1996)
Normanbya normanbyi	Wet	Insect	Seeds	Yes	Lott *et al.* (1995)
Pouteria sp.	Wet	Insect	Seeds	No	Notman *et al.* (1996)
Scheelea zonensis	Moist	Insect	Seeds	Yes	Wright (1983)
Scheelea zonensis	Moist	Insect	Seeds	Yes	Wilson & Janzen (1972)
Sterculia apetala	Dry	Insect	Seeds	Yes	Janzen (1972)
Virola surinamensis	Moist	Insect	Seeds	Yes	Howe *et al.* (1985)

(Continued on p. 56.)

TABLE 3.1. Continued.

Species	Forest type	Attack type	Seeds or seedlings or both?	Supports J–C model?	Source
Virola michelii	Wet	Insect	Seeds	No	Forget (unpublished data)
Vouacapoua americana	Wet	Insect	Seeds	Yes	Forget (1994)
Astrocaryum macrocalyx	Wet	Vertebrate	Seeds	No	Terborgh et al. (1993)
Bertholletia excelsa	Wet	Vertebrate	Seeds	No	Terborgh et al. (1993)
Brosimum alicastrum	Wet	Vertebrate	Seeds	No	Burkey (1994)
Carapa guianensis	Wet	Vertebrate	Seeds	No	Hammond et al. (1994)
Chlorocardium rodiei	Wet	Vertebrate	Both	No	Hammond (unpublished data)
Dipteryx micrantha	Wet	Vertebrate	Seeds	No	Terborgh et al. (1993)
Dipteryx panamensis	Wet	Vertebrate	Seedlings	Yes	Clark & Clark (1984)
Dipteryx panamensis	Moist	Vertebrate	Seeds	No?	De Steven & Putz (1984)
Dipteryx panamensis	Moist	Vertebrate	Seeds	No	Forget (1993)
Eperua grandiflora	Wet	Vertebrate	Both	No	Forget (1992)
Eperua falcata	Wet	Vertebrate	Both	No	Forget (1989)
Faramea occidentalis	Moist	Vertebrate	Seeds	No?	Schupp (1992)
Faramea occidentalis	Moist	Vertebrate	Seeds	No	Schupp (1988a)
Gilbertiodendron dewevrei	Wet	Vertebrate	Both	No	Hart (1995)
Gustavia superba	Moist	Vertebrate	Seed	No?	Sork (1987)
Hymenaea courbaril	Wet	Vertebrate	Seed	No	Terborgh et al. (1993)
Julbernardia seretii	Wet	Vertebrate	Both	No	Hart (1995)
Macoubea guianensis	Wet	Vertebrate	Seeds	No	Notman et al. (1996)
Macrozamia communis	Moist	Vertebrate	Seeds	No	Ballardie & Whelan (1986)
Normanbya normanbyi	Wet	Vertebrate*	Seeds	No	Lott et al. (1995)
Pouteria sp.	Wet	Vertebrate	Seeds	No	Notman et al. (1996)
Scheelea zonensis	Moist	Vertebrate	Seeds	Yes	Forget et al. (1994)
Tachigalia versicolor	Moist	Vertebrate	Both	No	Kitajima & Augspurger (1989)
Virola michelii	Wet	Vertebrate	Seeds	No	Forget (unpublished data)
Virola surinamensis	Moist	Vertebrate	Both	No	Howe et al. (1985)
Insect			Yes		15 studies
			No		4
Vertebrate			Yes		2
			No		25

* Not including feral pigs.

Connell, Tracey & Webb (1984) and Condit, Hubbell & Foster (1992) looked at the recruitment of older juveniles (1–8 cm dbh) in permanent sample plots in North Queensland, Australia and on Barro Colorado Island (BCI), Panama, respectively. An analysis of juvenile recruitment as a function of distance from the nearest conspecific adult showed that only a minority of species exhibited distance-dependent mortality as described by Janzen (1970) at either of the two sites. However, the analyses were restricted to the most common species in the plot (Queensland: 32 out of 86 for large saplings, BCI: 80 out of 304). In line with Janzen's ideas, relative commonness (or even monodominance) in larger size classes could result when seed and seedling predators are entirely absent for some, but not all, species in a community due to factors that influence mortality rates, but are not driven by frequency-dependence (e.g. strong chemical defence (*sensu* Janzen 1974)). Moreover, though analysing established juveniles of 1–8 cm dbh provides a composite picture of recruitment, it cannot differentiate between the various processes (J–C mortality being one) which can alter the spatial distribution of recruits at much earlier stages in the recruitment process (Zagt & Werger, Chapter 8, this volume). Early mortality due to invertebrate/vertebrate/pathogenic attack sets the scene for later mortality processes, such as competition within and between populations of juvenile trees during their ascent to the canopy. Since it is unlikely that any mortality process could remain uniformly effective over time, testing these individual processes by comparison to the spatial distribution of juvenile size classes or intercensual mortality rates may not be wholly effective.

Influence of phenology and seed crop size on model predictions

Annual community fruiting is highly seasonal at most neotropical forest sites (Frankie, Baker & Opler 1974, Foster 1982, Sabatier 1985, ter Steege & Persaud 1991). Though there are many seed production patterns exhibited by individual neotropical woody plant species (e.g. *Tachigalia versicolor* Standl. & Wms. (monocarpic), *Ficus* spp. (polycarpic — subannual/continuous), *Unonopsis* spp. (polycarpic — continuous)), the contribution of canopy tree and liana species to these seasonal peaks in fruitfall commonly occurs through (polycarpic) supra-annual reproductive bouts of individual, conspecific adults (Janzen 1978). If this pattern is asynchronous within the population, then during any given fruiting period, only a fraction of the local population of any given species will contribute to the production, and thus to availability, of fruits and seeds. Synchronized supra-annual, or mast, fruiting bouts of (congeneric) polycarpic species in SE Asian rainforests are legendary (Ashton, Givinish & Appanah 1988), typify some dominants in central Africa (e.g. *Gilbertiodendron dewevrei*, Hart 1995) and may arguably occur in some neotropical taxa (e.g. *Eschweilera* spp., Mori & Prance (1987), *Andira inermis*, Janzen (1978), *Vouacapoua* spp. (Forget 1997, D. Hammond personal observation).

The 'escape' of seeds (and perhaps seedlings since mortality is more likely during early expansion of hypocotyl and primary leaves, though see below) due to (i) the relatively low, irregular availability accompanying asynchronous patterns of fruit production, or (ii) satiation of predators through synchronization of production, can have an effect on offspring recruitment similar to that resulting from the better known spatial effects of offspring distance from parent source and adult density on seed and seedling attack.

However, though most studies referring to J–C mortality have focused on seed and seedling survival as a product of either dispersal distance (spatial effect) or seed crop size (temporal effect) exclusively, the effects of temporal and spatial availability of seeds and seedlings on predation intensity are not mutually exclusive. Within many species populations, only certain 'groups' or 'patches' of trees synchronize reproduction, while neighbouring conspecific groups produce few or no seeds (e.g. Ballardie & Whelan 1986, Hart 1995). Alternatively, if reproduction is asynchronous, it may be that only a fraction of adults in a patch will reproduce in any given year; that seed crop size is relatively small compared to mast years (Janzen 1978); or that there is considerable variation in the size of individual adult tree's seed crop within a population in any given year (e.g. Hubbell 1980). In instances of both synchronous and asynchronous production, as the density of reproductive adults that are fruiting increases, so will the overall density of seeds within this aggregation. Thus, satiation could theoretically occur at any adult density, depending on the size of the seed crop produced by adults within a given patch. This point can be graphically illustrated by simultaneously plotting the effects of varying crop size and reproductive adult density against survivorship below a focal adult tree (Fig. 3.2).

According to Janzen's original model, survivorship below the parent should be infinitesimally small. More recently, Schupp (1992) showed that seed survivorship beneath parent trees of *Faramea occidentalis* in Panama increased with increasing adult density. Combining Janzen's and Schupp's ideas, we can create a survivorship curve which is sensitive to adult conspecific density (curve $C°$ in Fig. 3.2). Survivorship as a function of adult density is depicted here as a curvilinear function. It is likely that a predator population would, in the first instance of an increase in seed availability, be able to consume a proportionally larger number of seeds, which would keep survivorship fairly constant. However, as the number of seeds available continues to increase with higher adult densities, consumption should decline as a fraction of the total number available as predator attack rates fail to keep up with the increase in available seeds. Eventually, the number of seeds attacked will represent only an infinitesimally small fraction of the total available, and 'satiation' should occur ($D°$ on abscissa in Fig. 3.2). Naturally, the shape of the survivorship curve and the adult density at which satiation can occur depend on the number of individuals in a predatory species population and how efficient they can forage for and utilize the seeds of the target tree species.

In Fig. 3.2, curve $C°$ assumes that the total number of available seeds is

Tropical seed and seedling attack

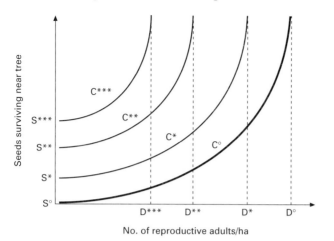

FIG. 3.2. The relationship between the spatial (reproductive adult density) and temporal (number of seeds per reproductive tree) abundance of seeds and the number of seeds surviving near the parent tree. Janzen's model prediction of survivorship is represented by the curve $C^°$. Curves $C^°$ to C^{***} represent increasing average seed crop sizes. Satiation should occur at the point along a curve where the change in survivorship as a function of adult density becomes asymptotic ($D^°$ to D^{***}).

simply a function of adult density; that is, average crop size per tree remains at some constant level. Increasing the average crop size per tree would push the seed survivorship curve up and to the left in Fig. 3.2 (example curves C^*–C^{***}). It is clear that in this construct, varying average crop size can alter both the level of survivorship beneath adult trees (S^*–S^{***} on the ordinate) as well as the density at which satiation should occur (D^*–D^{***} on the abscissa).

The potential confounding effects on seed survivorship patterns of simultaneously varying temporal and spatial availability of seeds suggest that the outcome of density/distance-dependent attack processes can be highly variable. Schupp's original findings for *Faramea* clearly show a strong explanatory relationship between adult density and survival beneath parent trees alone. However, *Faramea* is a subcanopy tree on BCI and typically fruits during the months when the majority of canopy trees and lianas do not (Foster 1982). According to Foster (1982), these species are more likely to exhibit reproductive synchrony, and thus Schupp's results are indeed more likely to reflect the distribution of adult conspecifics rather than the spatial variation in annual reproduction among these same adults. Most tree species (at BCI and elsewhere) do not produce seed during community depressions in fruiting, nor do they synchronize this event and Schupp's message concerning the effects of adult density needs to be duplicated for a wider spectrum of paired spatial and temporal fruitfall patterns in order to understand whether the 'escape hypothesis', as defined by both spatial and temporal avoidance of predators, can explain at least in part, patterns of tree species abundance and diversity in tropical forests.

INVERTEBRATE VS. VERTEBRATE ATTACK

Though Schupp (1992) clearly showed that adult density can be important in understanding the effects of seed and seedling predators, the evidence supporting a scale-dependent shift in the minimum distance effect is based solely on studies describing vertebrate attack (Schupp 1988a,b, Burkey 1994), which generally do not support the J–C model (Table 3.1), regardless of the scales considered. On the other hand, studies documenting cases where invertebrates are acting as the predators generally support an increase in attack as the density of seeds/seedings increases and/or the distance from parent decreases, though studies examining the role of scale in determining the efficacy of invertebrate attack are noticeably absent.

The disparity in support for the J–C model expressed by invertebrate and vertebrate studies may be explained by the different ways in which vertebrate and invertebrate predators search for and attack seeds and seedlings (crudely akin to the type I and III functional response curves of Holling (1959), respectively, but also see Spalinger and Hobbs (1992)) (Table 3.2).

Mammals frequently consume seeds and seedlings on a preferential basis, which is heavily influenced by the overall availability of resources (Schupp 1990, Forget 1993, D.S. Hammond unpublished data), the relative availability of resource types (Glanz, Thorington, Giacalone-Madden & Heaney 1982, Greenwood 1985), the relative nutritive value of prey (e.g. Smallwood & Peters 1986), the maintenance of home-ranges or territories (e.g. Smythe 1978), and the need to avoid their own predators (*sensu* Schupp 1988b, Bowers & Dooley 1993), as well as compete with other seed/seedling-eaters (Smythe 1986, but see August & Fleming 1984). Resource selection by vertebrate seed-eaters is rarely the product of long-standing, species-specific specialization. Neotropical rodents often cache seeds for later use when a particular resource is exceptionally abundant, but this action does not ultimately lead to an outcome consistent with Janzen's satiation process, since the final tally of surviving offspring depends on whether or not the hoarder returns to the cache (see Forget, Milleron & Feer, Chapter 2, this volume). The likelihood of cache revisitation can depend on the subsequent availability of other resources in the area (e.g. Forget 1993) and lead to frequent changes in the shape of the seedling shadow, contrary to the implicit assump-

TABLE 3.2. A comparison of general selection and search criteria for invertebrate and vertebrate seed and seedling-eaters in tropical forests.

Selection/searching parameter	Invertebrates	Vertebrates
Basis of selection	Reproduction and consumption	Consumption only
Selection range within resource type*	Narrow	Broad
Basis for shift between resource types	Inherent (life-cycle stage)	Opportunistic (seasonal or spatial scarcity)
Size relative to seed/seedling size	Small	Large
Number of individuals supported by single seed or seedling	≥1	<1
Type of chemical defence constraining resource use	Quality	Quantity
Basis of searching for resource	Specific host location Conspecific adult abundance (semiochemicals)	Relative prey value (energy intake/unit handling time)
Factors limiting search area	Mobility Host-plant patch size Host-plant patch heterogeneity	Territoriality Predator avoidance Mobility

* Resource types = fruits, seeds, leaves, bark, etc.

tion of Janzen's model that mortality depends predominately on the location of seeds by seed-eaters and that (secondary) dispersal increases the chance of seeds 'avoiding' their predators.

In contrast to vertebrate seed- and seedling-eaters (see Table 3.2), insect predators often restrict their efforts to one or several, often congeneric, species independent of community-wide resource availability (Janzen 1969). Specialization upon a limited number of hosts can release insects from many of the trade-offs which influence optimal foraging in vertebrates (e.g. searching and handling vs. predator avoidance). Fox example, adults searching for hosts may be attracted to a high density of seed or seedlings by semiochemical plumes (Fakas & Shorey 1972, Salom & McLean 1991), produced by a build up of conspecific adults already present at the site, rather than a need to maintain foraging efficiency. Invertebrates often can synchronize (seed) supply and (predator) demand by responding reproductively to the same seasonal cues which initiate seed and seedling production in their specific host plant rather than choose their prey based on trade-offs between the seasonal availability and relative nutritive quality of a much larger spectrum of seeds and seedlings. Many adult insects search for seeds as suitable oviposition sites for their offspring, which then act as post-dispersal seed predators during their juvenile growth stages, rather than as resources contributing to their own subsistence. Alteratively, some seed- or seedling-eating insects may simply descend, as juveniles or adults, from their residence in or near the crowns of parent trees (e.g. Janzen 1971, Fowler 1985), rather than search for resources in horizontal space like most vertebrate, post-dispersal seed predators.

Holling (1959) considered an animal's functional response as the change in feeding rate which takes place as a consequence of a change in resource adundance. The act of predation is broken down into components: (i) searching and locating; (ii) pursuit and attack; and (iii) handling and ingestion. Functional response curves types I and III (*sensu* Holling (1959)) could be crudely adapted to describe invertebrate and vertebrate feeding on seeds and seedlings in tropical forests, respectively. Holling's type I curve describes the situation where the prey (here, a seed) is larger than the predator (here, an insect). In this case, the pursuit and handling components are most important. However, since seeds are stationary, pursuit is inconsequential to insect predators and handling becomes the singular constraint on foraging success. If prey (seed) are smaller than their predators (most vertebrates) then searching efficiency will limit overall foraging efficiency because the large predator will need to consume many, rather than a single, prey item(s) (Table 3.2).

FIELD EXPERIMENTS IN GUYANA

Study site

Field experiments were carried out at the Wappu Reserve (900 ha), located 16 km SE of Mabura Hill (05°13′N; 58°48′W) in Central Guyana. The climate is tropical with an total annual rainfall (2400–3400 mm) falling mostly in December–January and from May–August. For further details of the geology, geography and climate at the site see ter Steege, Boot, Brouwer *et al.* (1996) and Gibbs and Barron (1993).

Experimental set-up

Several experiments were established in 1991–94 to test the effects of canopy openings of various sizes upon the outcome of insect and vertebrate attack on seeds and seedlings of a suite of Guyanan canopy trees and liana species. Experiments examined here tested the effects of: (i) canopy openness and distance from nearest conspecific adult upon the frequency of invertebrate attack and subsequent survival of *Chlorocardium rodiei* (Schomb.) Rohwer, Richter & van der Werff (Greenheart) (Lauraceae) seedlings; (ii) total defoliation upon the capacity of seedlings to recover in an understorey environment as a function of a species seed size; and (iii) gap size and partial cotyledon excision upon subsequent seedling survival and height in *Carapa guianensis* Aublet (Meliaceae).

The *Chlorocardium* experiment commenced in March 1991 in a 15-ha mixed forest plot on laterite (dystric leptosols). The plot was broken into a grid of 20 × 20 m cells and all adult *Chlorocardium* trees (>20 cm dbh) and all gaps (*sensu* Brokaw 1982) were measured and mapped onto the grid system. A total of 177 trees and 56 gaps were located within the plot in March 1991. Using this information, 10 trees (>20 cm dbh), 10 points on the grid system known to exceed 30 m distance from nearest *Chlorocardium* adult and 20 gaps (size range: 20–560 m^2) were selected at random as implantation sites. A block of eight seeds, *c.* 20 cm apart, were placed on the soil surface at each implantation site, encased in a mesh exclosure cage to prevent vertebrate attack. Additional sets of eight seeds were placed in adjacent blocks under insecticide treatments and a control, but only the results from the caged sets, which were used to examine the effects of invertebrate attack upon seedlings, are presented here.

The defoliation experiment commenced in early 1992, shortly after the peak in the fruiting season for that year. Thirty to forty seeds of each of 13 native species were placed in 2.5-l, perforated, polyethylene bags, one seed per bag. Species used were: *Anacardium giganteum* (Engl.) W. Hanc., *Carapa guianensis* Aublet, *Catostemma commune* Sandw., *Chlorocardium rodiei*, *Connarus perrottetii* (DC.) Planchon, *Elisabetha princeps* Benth., *Eschweilera sagotiana* Miers, *Mora gongrijpii* (Kleinhoonte) Sandw., *Pouteria speciosa* (Ducke) Baehni, *Pterocarpus officinalis* Jacq., *Renealmia floribunda* Schumann, *Securidaca* sp., and *Swartzia schomburgkii* Benth. Half of the seeds were placed in a gap (320 m^2, 89% canopy cover) and the remainder in a nearby understorey site (96% cover). Defoliation involved the complete removal of leaf area + terminal meristem after growth rate declined in each seedling. This method was employed in order to reduce the effects of varying rates of seedling growth between species. Survival and growth subsequent to defoliation was monitored up to 65 weeks from treatment application.

Linear least-squares regression models were employed to examine the relationship between the relative survivorship in understorey vs. gap sites (response variables) and seed size (predictor variable) for defoliated and intact groups. Residuals of each model were examined for violation of assumptions (normality, equal variances, independence) of regression analysis and the validity of the fitted models (leverage via Cook's distance) using the MGLH module of SYSTAT 5.1 (Wilkinson 1989).

The experiment examing the interaction between gap size and degree of cotyledon excision in *Carapa guianensis* commenced shortly after fruitfall in May 1993. Four hundred and fifteen *Carapa* seeds (11 trees) were collected and checked for damage. Of the remaining seeds, 320 were randomly assigned to one of four excision treatments (0 (control), 25, 50 and 75% of seed mass). Seeds were excised from the end opposite to the embryo. Excision scars were dipped in a bitumen-based tree wound dressing (Tree-Kote) in order to prevent pathogenic infection. Each seed was placed in a 2.5-l polyethylene bag and germinants randomly assigned to one of four light treatments:

1 Understorey (96% canopy cover);
2 Small gap ($320\,m^2$, 89%);
3 Intermediate gap ($730\,m^2$, 84%); and
4 Large gap ($3440\,m^2$, 78%).

Seedlings from each excision treatment were placed in a randomized block design within the centre of each gap and their growth monitored for 8 months thereafter.

ALTERNATIVE OUTCOMES TO INVERTEBRATE ATTACK ON SEEDS AND SEEDLINGS

Why escape may not mean avoidance: disturbance and light-dependent escape

A pair of commonly held assumptions underlie virtually all studies of density or distance-dependent seed and seedling attack in the neotropics: (i) that seeds and seedlings escape by avoiding being attacked by generalist or specialist predators; and (ii) that the light environment does not affect the outcome of the interaction between predator and prey, though light availability at the forest floor clearly must vary in some fashion with distance from parent tree. Studies evaluating density/distance-dependent attack rarely mention sublethal attack events or characterize the influence of a spatially varying light environment.

In retrospect, it is surprising that the debate surrounding the J–C model became as focused as it did, given the fact that it had already been well established that a change in light availability, through gap-phase dynamics, played a crucial role in tropical seedling establishment and later recruitment into larger juvenile size classes (Whitmore 1975). Becker *et al.* (1985) first formally recognized that, in order to evaluate the predictions of the J–C model, a plausible account of variation in light availability must be factored into the model. Clark and Clark (1984) warned that many other factors, especially gap formation, can contribute to later survivorship patterns of those seedlings that have 'escaped' J–C attack at earlier stages. More recently, Schupp (1988b) and Schupp and Frost (1989) presented results which suggest that vertebrate attack on seeds of two understorey specialists is disproportionately higher in gap habitats compared to understorey sites. They suggest that this difference might explain to some degree the disparity between J–C predictions and the observed distributions of larger size classes of trees. Schupp and Frost (1989) attributed the higher attack in gaps to an increase in vertebrate foraging activity at these sites, a direct consequence of the reduced predation hazard afforded by the dense tangle of secondary regrowth. However, this hypothesis is not wholly applicable to cases where invertebrates are the leading cause of seed and seedling loss, since the search tactics of these insects and their invertebrate predators are very different from those of vertebrates (see Table 3.2).

Recent research into the influence of the light environment on the efficacy of seed and seedling attack at Mabura Hill, Guyana, suggests that a shift from understorey to gap environment for many commonly abundant tree species can 'buffer' the effects of invertebrates which attack seeds and seedlings in a density/distance-dependent fashion. This is uniquely different from the argument that gap dynamics can influence spatial patterns of recruitment subsequent to J–C mortality processes (Clark & Clark 1984) or somehow alter the likelihood of attack through a change in the microsite conditions which makes the area no longer suitable to the predator (Hammond 1995) or even more suitable than understorey (Schupp 1995). Instead, it describes the process whereby individual offspring are able to survive an attack which, under much lower light conditions, would invariably lead to their death. Some of these individuals may go on to share a similar probability of being recruited as those conspecific sibs in gaps which avoid attack and are likely to experience a greater chance of recruitment than unattacked siblings dispersed to a low-light site in the understorey. The enhanced survival of attacked individuals once released from the low-light conditions in the understorey is akin, though inversely, to the notion that the application of a physical stress (here, low light) will have a greater impact on individuals previously existing at high densities (here, near parent trees) due to the stress caused by crowding (here, density-dependent attack) (Peterson & Black 1988).

How to escape without avoiding: life-history characteristics

The suitability of a host plant as food or for oviposition depends a great deal on the life-history characteristics of the species. Seed size, seed dormancy, maximum relative growth rate, leaf area ratio, the type and quantity of chemical defences, and mycorrhizal or rhizobial associations, among others, can contribute to the host-plant selection process and the degree to which seeds and seedlings suffer attack by animals (Janzen 1969, Coley 1983). However, life-history characteristics are typically the product of a long-standing series of trade-offs between many biotic and abiotic forces which influence their expression (Rees 1993) and need not be solely responsive to pressures exerted by predators alone, though this interaction may appear to be strongly selective. Character states which are able to convey an advantage to a species in a gap environment, for example high growth rate, may also indirectly benefit the plant in response to attack by invertebrate or vertebrate herbivores (*sensu* Coley 1983); that is, the response may be exaptive. Whether adaptive or exaptive, the basis of individual survival in juvenile plants is the ability to maintain a positive carbon balance; that is, the cost of maintaining tissue does not exceed the energy produced by it. This balance can be achieved in the face of invertebrate and vertebrate attack through several pathways, referred to here as three separate models.

The susceptibility model

One way of decreasing the likelihood of seedling mortality as a consequence of invertebrate attack is to minimize the amount of tissue which can be removed. This can be achieved by:
1 Reducing the palatability of tissue;
2 Inhibiting growth of the herbivore by suppressing nutrient assimilation from the tissue; or
3 Killing the herbivore via direct toxicity (Becker & Martin 1982).
The success of these strategies relies on the efficacy of secondary chemicals within the attacked stems, roots or leaves to reduce damage during any attack event. Most groups of phenolic secondary defence compounds increase in concentration within tropical plant leaves as light availability increases (Mole, Ross & Waterman 1988, Feibert & Langenheim 1988), though it is believed that the efficacy of this increase as a feeding deterrent may be linked to the relative nutrient value of the tissue (Mole & Waterman 1988), and nutrients also increase as light becomes more available (Chandler & Goosem 1982).

Phototoxic, or photosensitizing, phytochemicals, such as some furanocoumarins, express their full toxicities in the presence of light and are commonly found in several abundant neotropical families (e.g. Fabaceae, Moraceae) (Murray, Mendez & Brown 1982). They are often found in taxa that specialize in colonizing highly disturbed, high-light environments (e.g. *Ficus* spp., Asteraceae) (Towers 1984) so present a potentially important means in which to reduce susceptibility to attack which is unique to gap conditions. While the literature addressing the relationship between herbivores and plant defences is nearly as long-established as the J–C model (e.g. Feeny 1976, McKey, Waterman, Mbi *et al.* 1978, Coley, Bryant & Chapin 1985), much less is known about the abundance of phototoxins in neotropical seedlings in gaps and their putative role, as well as that of other secondary defence compounds, in reducing susceptibility to density/distance-dependent attack in high-light environments.

The compensatory growth model

The notion of compensatory growth, here defined as the replacement of tissue lost after seedling attack, is not new (see McNaughton 1983, Belsky 1986). However, this theory receives the bulk of its support from studies of grasslands and the herbivores that graze them (Owen & Wiegert 1981, Belsky 1986); very few studies were found which illustrate this model for tropical forest seedlings and none views compensatory growth as a means of light-dependent escape from density/distance-responsive predators. The anecdotal evidence available suggests that many large-seeded neotropical seedlings are able to compensate to some degree for the loss of tissue due to herbivory (Denslow 1980a, Forget 1992), though it is not certain if this recovery is only temporary if light availability is not high.

Evidence from our work in Guyana suggests that most large-seeded species undergo compensatory growth after severe defoliation (Figs 3.3a,b), but that extended survival after such an event is often dependent on the availability of light, not seed size (Fig. 3.4). In this experiment, seed size explained a significant proportion of the variation in the survival of seedlings in a gap compared to the understorey when seedling were not defoliated ($r^2 = 0.44$, $F_{1,11} = 8.75$, $P < 0.02$). Seed size could not account for any significant proportion of variation in survival after defoliation ($r^2 = 0.06$, $F_{1,11} = 0.75$, $P = 0.40$). Rejection of a statistical outlier (studentized residual >3, P set at 0.05) in the case of the defoliated set did not alter the outcome of the significance test ($F_{1,10} = 1.13$, $P = 0.30$).

Studies involving *Chlorocardium rodiei* over the past 4 years by us have shown that over 96% of living seeds in a cohort are eventually infested by the beetle *Sternobothrus* sp. (Scolytidae). A minimum of 40% of these

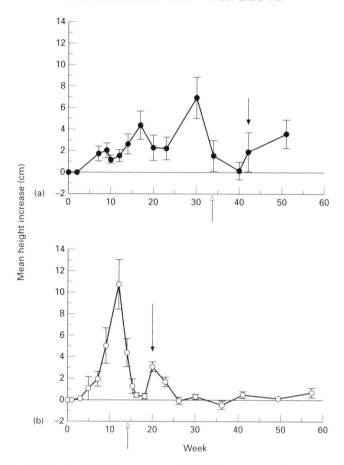

FIG. 3.3. Mean (± SE) shoot height increment in (a) *Elisabetha princeps* (Caesalpineaceae) ($n = 20$) and (b) *Swartzia schomburgkii*. (Caesalpiniaceae) ($n = 12$) after experimental defoliation of seedlings in an understorey environment. The open arrows indicate the time of defoliation and the closed arrows the ensuing compensatory growth.

individuals will later have their shoots attacked by the same species. The heavy infestation of the large seeds (mean = 63 g fresh weight) combined with later shoot-boring led to a faster, and ultimately higher, rate of mortality in seedlings confined to the understorey, especially where they were below parent trees (Fig. 3.5). However, distance from parent source (at constant density) is not a good predictor of mortality rate if seedlings in gaps are included in the model (Fig. 3.5). Since intraspecific variation in seed size was randomized prior to placement of seeds at understorey and gap locations, compensatory growth in response to the loss incurred during scolytid attack

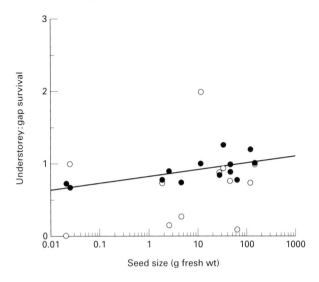

FIG. 3.4. The (untransformed) ratio of seedlings surviving in a gap to understorey as a function of average (log transformed) seed size for 13 Guyanian rainforest tree species under defoliation (open circles) and no defoliation (closed circles) treatments. Average seed sizes were taken from Hammond and Brown (1995).

would appear to be a plausible explanation for the difference in survival rates. Seedlings are typically entered from the base of the shoot and the adult beetle bores its way towards the apex until the taper decreases to a width where the shoot no longer presents sufficient pith for consumption. Since the shoot weight, height and the leaf area of *Chlorocardium* seedlings can increase substantially with an increase in light availability (Ter Steege, Bokdam, Boland *et al.* 1994), it is possible that seedlings heavily damaged by *Sternobothrus* can compensate for the loss of tissue through a higher growth rate in high light environments (*sensu* Coley 1983). Alternatively, seedlings in gaps may be less susceptible to fungal pathogens which many scolytid beetles are known to harbour (Rudinsky 1962) and *Sternobothrus* may be carrying (see Beaver 1973), since the conditions which support pathogen proliferation are more frequent in understorey than in gaps (*sensu* Augspurger 1983). In this case, light-dependent escape would be attributable to the susceptibility, rather than the compensatory growth, model.

The redundancy model

One strong advantage associated with having large seeds is the ability to persist as a seedling under light conditions which otherwise could not

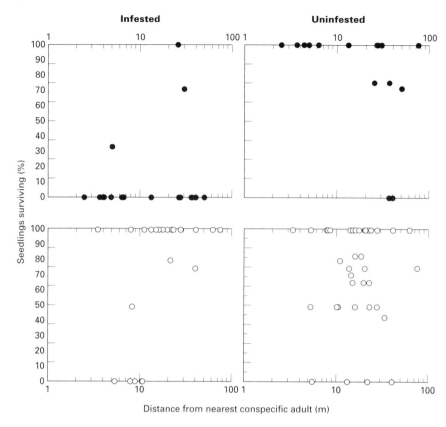

FIG. 3.5. The relationship between distance from nearest reproductive adult and survivorship of infested and uninfested *Chlorocardium rodiei* seedlings at 30 months after placement. Survivorship is given as the percentage of seedlings surviving. Only seedlings attaining a size that would allow infestation were included. Seedlings in gaps and understorey sites are represented by open and closed circles, respectively.

support growth (Foster 1986). Thus, tree species which are more commonly associated with large gaps tend to have smaller seeds, though gap preference can only explain a small fraction of the total variation in seed size in neotropical communities (Hammond & Brown 1995). Small-seeded 'pioneer' species also almost invariably have very high inherent growth rates, especially in comparison to larger-seeded, 'shade-tolerant' species, which tend to allocate more of their finite seed resource to chemical and structural defences (Kitajima 1994). In addition, the relative growth rate of large-seeded species tends to increase as light availability increases, though this is typically less than in 'pioneer' species (Boot 1994). Because a large seed reserve can be seen as a compensatory response strategy to damage inflicted by herbivores (or branchfalls) under low-light environments, it stands to reason that this reserve is less critical to the survival of offspring that are dispersed into high-light environments (gaps).

Seeds of many neotropical trees are still able to germinate after partial consumption of the cotyledons or endosperm as long as the embryo survives (Bonaccorso, Glanz & Sanford 1980, Hammond, Schouten, van Tienen *et al.* 1994). However, establishment of these seedlings is often dependent on light availability. Seeds and seedlings of *Carapa guianensis* Aublet (Meliaceae), a widespread neotropical tree, are commonly attacked in a density/distance-

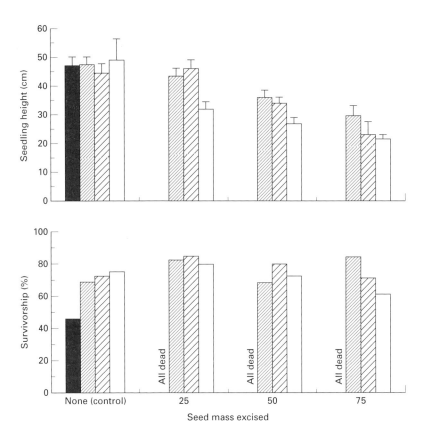

FIG. 3.6. The effects of light availability on the growth and survival of *Carapa guianensis* (Meliaceae) seedlings at 8 months after various levels of seed mass excision (X-axis). Seeds were placed near the centre of a large (□), intermediate (▨) and small gap (▨) and an understorey site (■). Initial sample size (n) equals 20 seedlings per excision per light treatment in all cases.

dependent fashion in Guyana by the pyralid moth *Hypsipyla* spp. (Hammond *et al.* 1994).

Seed excision trials in gap and understorey conditions suggested that survival of 9 month-old seedlings, germinating from seeds with mass reductions up to 75%, was significantly greater in gaps compared to the understorey (Monte-Carlo χ^2 contingency table analysis of 50000 trials with 4 columns (understorey + 3 gaps) × 4 rows (excision levels): $P = 0.0047 \pm 0.0008$ (the SE for P based on the number of trials performed). However, gap size alone did not have a significant effect on survival at the different levels of losses (Monte-Carlo χ^2 analysis of 50000 trials with 3 columns (3 gaps) × 4 rows (excision levels): $P = 0.9994 \pm 0.0001$) (Fig. 3.6). In cases like *Carapa*, the high light availability of the gap environment 'buffers' the effects of invertebrate attack because the increase in available, photosynthetically-active radiation amply compensates for the loss of the energy reserve contained in the seed; that is, the seed reserve is redundant to the seedling's requirements for survival, though is important in improving initial height extension (Fig. 3.6).

CONCLUSIONS

Twenty-five years after Janzen (1970) and Connell (1971) first proposed that seed and seedling attack might occur in a density/distance-dependent fashion and that this might maintain high tree species richness, several clear patterns are emerging. First, available evidence suggests that the predictions concerning attack patterns made by the J–C model are generally upheld by studies on invertebrate, but not vertebrate, attack. The way in which these two diverse groups of seed and seedling-eaters forage are very different, but this contrast has only recently been considered as an explanation of these divergent patterns (Schupp 1988b, Burkey 1994).

Second, recent studies have augmented the view of Janzen (1976, 1978) that heavy seed crops synchronized to mature as a single event could overwhelm potential predators by satiating their requirements (Ballardie & Whelan 1986, Hart 1995). It is further proposed here that spatial and temporal availability of seeds are inextricably linked and simultaneously influence the way in which attack will effect seedling recruitment. Reproductive adult trees clearly show varying spatial distributions and temporal fruitfall patterns at many tropical study sites (for example BCI – Foster (1982) and Hubbell & Foster (1990); La Selva – Lieberman & Lieberman (1994) and Newstrom, Frankie, Baker & Colwell (1994)) which could confound any attempts at discerning community scale effects of the classical J–C model of mortality. Future studies examining the role of density/distance dependence on seed and seedling mortality should attempt to address the relative influences of both spatial and temporal seed and seedling availability.

Third, the results presented here indicate that the means by which seeds or seedlings escape density/distance-dependent attack are not necessarily restricted to those processes that allow them to avoid attack (for example, dispersal, synchronized fruiting). The equivalence of escape with avoidance underpins current notions of J–C mortality processes. Light-dependent mechanisms that may provide a means of escaping without avoidance (presented here as three models) would alter the predictions of the J–C model substantially; further work is needed to elucidate whether this mode of escape is universally important, and what life-history characteristics, such as seed size, play a part in its occurrence.

Clearly, if offspring are able to escape density/distance-dependent mortality in gaps without avoiding attack, then the effect of this process on the number of established saplings available for later recruitment into the adult population would depend partly on the level of disturbance characterizing a particular area. The frequency and magnitude of disturbances can differ substantially between neotropical forest areas (Hammond & Brown 1995). Thus, it is likely that these differences could lead to biogeographic variability in the influence of light-dependent escape of seeds and seedlings from their invertebrate predators. Research from various tropical sites is needed in order to determine whether light-dependent escape can influence species recruitment, particularly after large and/or infrequent disturbance events.

The J–C model is only one of several theories that attempt to explain tropical forest tree diversity (Grubb 1977, Denslow 1980b, Hubbell & Foster 1986). It is likely that distance/density-dependent mortality processes predominately influence recruitment patterns in some populations of some species some of the time. Other modes of recruitment may explain a larger fraction of variation in recruitment when conditions alternatively emphasize their regulatory role. Thus, the effect of recruitment processes, such as that described by the J–C model, may wax and wane in their influence on population structures. Variability in temporal and spatial availability of seeds combined with light-dependent changes in the outcome of an attack event on seeds and seedlings are only two aspects of tropical forest dynamics which can alter the influence of J–C mortality on juvenile recruitment, providing an opportunity for other recruitment processes to have a greater influence on the tree community structure.

Janzen's model, as it was originally proposed, was an early attempt to explain patterns of high tropical tree diversity; Occam's razor had simply been oversharpened (*sensu* Grubb 1992). If the J–C model had to be viewed as a continual influence on seed and seedling attack events, it perhaps best reflects on the search patterns of invertebrate seed and seedling consumers, rather than the likelihood of seeds and seedlings escaping predation or the spacing between conspecifics trees within a population.

ACKNOWLEDGEMENTS

We thank K. Crandon, K. Lance, B. Cummings, D. Chapman, S. Roy, D. Angoy and the many other assistants who supported this work through their efforts in the field. The logistical support and assistance in Mabura Hill from the staff and management of Demerara Timbers Ltd is warmly acknowledged. This chapter profited greatly from the comments of Y. Basset, P.-M. Forget, D.M. Newbery, H. ter Steege, R. Zagt and two anonymous reviewers. Symposium participants provided additional useful insights. O. Cheesman's assistance in preparation of the manuscript was invaluable. Financial support for fieldwork was provided by the Overseas Development Administration (ODA–UK) and the Tropenbos Foundation (Netherlands). This chapter is dedicated to the forest ecologists and foresters whose work in the 1960s and 1970s provided a basis on which to begin rationally testing ideas concerning tropical forest diversity.

REFERENCES

Armstrong, R.A. (1989). Competition, seed predation, and species coexistence. *Journal of Theoretical Biology*, **141**, 191–195.

Ashton, P.S., Givinish, T.J. & Appanah, S. (1988). Staggered flowering in the Dipterocarpaceae: new insights into floral induction and the evolution of mast fruiting in the aseasonal tropics. *American Naturalist*, **132**, 44–66.

August, P.V. & Fleming, T.H. (1984). Competition in neotropical small mammals. *Acta Zoologica Fennica*, **172**, 33–36.

Augspurger, C.K. (1983). Seed dispersal of the tropical tree *Platypodium elegans*, and the escape of its seedlings from fungal pathogens. *Journal of Ecology*, **71**, 759–771.

Ballardie, R.T. & Whelan, R.J. (1986). Masting, seed dispersal and seed predation in the cycad *Macrozamia communis*. *Oecologia*, **70**, 100–105.

Beaver, R.A. (1973). Biological studies of Brazilian Scolytidae and Platypodidae (Coleoptera). II. The tribe Bothrosternini. *Papéis Avulsos de Zoologia*, **26**, 227–236.

Becker, P. & Martin, J.S. (1982). Protein-precipitating capacity of tannins in *Shorea* (Dipterocarpaceae) seedling leaves. *Journal of Chemical Ecology*, **8**, 1353–1367.

Becker, P., Lee, L.W., Rothman, E.D. & Hamilton, W.D. (1985). Seed predation and the coexistence of tree species: Hubbell's models revisited. *Oikos*, **44**, 382–390.

Belsky, A.J. (1986). Does herbivory benefit plants? A review of the evidence. *American Naturalist*, **127**, 870–892.

Bonaccorso, F.J., Glanz, W.E. & Sanford, C.M. (1980). Feeding assemblages of mammals at fruiting *Dipteryx panamensis* (Papilionaceae) trees in Panama: seed predation, dispersal, and parasitism. *Revista Biologia Tropical*, **28**, 61–72.

Boot, R.G.A. (1994). *Growth and Survival of Tropical Rain Forest Tree Seedlings in Forest Understorey and Canopy Openings*. Tropenbos Documents 6. Tropenbos Foundation, Wageningen, Netherlands.

Bowers, M.A. & Dooley, J.L. (1993). Predation hazard and seed removal by small mammals: microhabitat versus patch scale effects. *Oecologia*, **94**, 247–254.

Brokaw, N. (1982). Treefalls: frequency, timing and consequences. In *The Ecology of a Tropical Rainforest: Seasonal Rhythms and Long-term Changes* (Ed. by E. Leigh, Jr, A.S. Rand & D.M. Windsor), pp. 101–107. Smithsonian Institution Press, Washington, DC.

Burkey, T.V. (1994). Tropical tree species diversity: a test of the Janzen–Connell model. *Oecologia*, **96**, 533–540.

Chandler, G. & Goosem, S. (1982). Aspects of rainforest regeneration. III. The interaction of phenols, light and nutrients. *New Phytologist*, **92**, 369–380.

Clark, D.A. & Clark, D.B. (1984). Spacing dynamics of a tropical rain forest tree: evaluation of the Janzen–Connell model. *American Naturalist*, **124**, 769–788.

Coley, P.D. (1983). Herbivory and defensive characteristics of tree species in a lowland tropical forest. *Ecological Monographs*, **53**, 209–233.

Coley, P.D., Bryant, J.P. & Chapin, F.S. III. (1985). Resource availability and plant antiherbivore defense. *Science*, **230**, 895–899.

Condit, R., Hubbell, S.P. & Foster, R.B. (1992). Recruitment near conspecific adults and the maintenance of tree and shrub diversity in a neotropical forest. *American Naturalist*, **140**, 262–286.

Connell, J.H. (1971). On the role of natural enemies in preventing competitive exclusion in some marine animals and in rain forest trees. In *Dynamics of Populations* (Ed. by P.J. den Boer & G.R. Gradwell), pp. 298–310. Centre for Agricultural Publishing and Documentation, Wageningen.

Connell, J.H., Tracey, J.G. & Webb, L.J. (1984). Compensatory recruitment, growth, and mortality as factors maintaining rain forest tree diversity. *Ecological Monographs*, **54**, 141–164.

Denslow, J.S. (1980a). Notes on the seedling ecology of a large-seeded species of Bombacaceae. *Biotropica*, **12**, 220–222.

Denslow, J.S. (1980b). Gap partitioning among tropical rainforest trees. *Biotropica*, **12** (Suppl.), 47–55.

De Steven, D. & Putz, F.E. (1984). Impact of mammals on early recruitment of a tropical canopy tree, *Dipteryx panamensis*, in Panama. *Oikos*, **43**, 207–216.

Fakas, S.R. & Shorey, H.H. (1972). Chemical trail-following by flying insects: a mechanism for orientation to a distant odour source. *Science*, **178**, 67–68.

Feeny, P. (1976). Plant apparency and chemical defense. *Recent Advances in Phytochemistry*, **10**, 1–40.

Feibert, E.B. & Langenheim, J.H. (1988). Leaf resin variation in *Copaifera langsdorfi*: relation to irradiance and herbivory. *Phytochemistry*, **27**, 2527–2532.

Forget, P.-M. (1989). La régénération naturelle d'une espéce autochore de la forêt guyanaise: *Eperua falcata* Aublet (Caesalpiniaceae). *Biotropica*, **21**, 115–125.

Forget, P.-M. (1992). Regeneration ecology of *Eperua grandiflora* (Caesalpiniaceae), a large-seeded tree in French Guiana. *Biotropica*, **24**, 146–156.

Forget, P.-M. (1993). Post-dispersal predation and scatterhoarding of *Dipteryx panamensis* (Papilionaceae) seeds by rodents in Panama. *Oecologia*, **94**, 255–261.

Forget, P.-M. (1994). Recruitment pattern of *Vouacapoua americana* (Caesalpiniaceae), a rodent-dispersed tree species in French Guiana. *Biotropica*, **26**, 408–419.

Forget, P.-M., Munoz, E. & Leigh, E.G. Jr. (1994). Predation by rodents and bruchid beetles on seeds of *Scheelea* palms on Barro Colorado Island, Panama. *Biotropica*, **26**, 420–426.

Foster, R.B. (1982). The seasonal rhythm of fruitfall on Barro Colorado Island. In *The Ecology of a Tropical Forest: Seasonal Rhythms and Long Term Change* (Ed. by E.G. Leigh, Jr, A.S. Rand & D.M. Windsor), pp. 151–172. Smithsonian Institution, Washington DC.

Foster, S.A. (1986). On the adaptive value of large seeds for tropical moist forest trees: a review and synthesis. *Botanical Review*, **52**, 260–299.

Fowler, S.V. (1985). Difference in insect species richness and faunal composition of birch seedlings, saplings and trees: The importance of plant architecture. *Ecological Entomology*, **10**, 159–169.

Frankie, G.W., Baker, H.G. & Opler, P.A. (1974). Comparative phenological studies of trees in tropical wet and dry forests in the lowlands of Costa Rica. *Journal of Ecology*, **62**, 881–919.

Gibbs, A.K. & Barron, C.N. (1993). *The Geology of the Guianan Shield.* Oxford University Press. New York.

Glanz, W.E., Thorington, R.W., Giacalone-Madden, J. & Heaney, L.R. (1982). Seasonal food use and demographic trends in *Sciurus granatensis*. In *The Ecology of a Tropical Forest: Seasonal Rhythms and Long Term Change* (Ed. by E.G. Leigh, Jr, A.S. Rand & D.M. Windsor), pp. 239–252. Smithsonian Institution, Washington DC.

Greenwood, J.J.D. (1985). Frequency-dependent selection by seed-predators. *Oikos*, **44**, 195–210.

Grubb, P.J. (1977). The maintenance of species-richness in plant communities: the importance of the regeneration niche. *Biological Reviews*, **52**, 107–145.

Grubb, P.J. (1992). A positive distrust in simplicity—lessons from plant defences and from competition among plants and among animals. *Journal of Ecology*, **80**, 585–610.

Hammond, D.S. (1995). Post-dispersal seed and seedling mortality of tropical dry forest trees after shifting agriculture, Chiapas, Mexico. *Journal of Tropical Ecology*, **11**, 295–313.

Hammond, D.S. & Brown, V.K. (1995). Seed size of woody plants in relation to disturbance, dispersal, soil type in wet neotropical forests. *Ecology*, **76**, 2544–2561.

Hammond, D.S., Schouten, A., van Tienen, L., Weijerman, M. & Brown, V.K. (1994). The importance of being a forest animal: implications for Guyana's timber trees. *Proceedings of the 6th Annual NARI/CARDI Review Conference*, Georgetown, Guyana.

Hart, T.B. (1995). Seed, seedling and sub-canopy survival in monodominant and mixed forests of the Ituri Forest, Africa. *Journal of Tropical Ecology*, **11**, 443–459.

Holling, C.S. (1959). The components of predation as revealed by a study of small-mammal predation of the European pine sawfly. *Canadian Entomologist*, **91**, 292–320.

Howe, H.F. & Smallwood, J. (1982). Ecology of seed dispersal. *Annual Review of Ecology and Systematics*, **13**, 201–228.

Howe, H.F., Schupp, E.W. & Westley, L.C. (1985). Early consequences of seed dispersal for a neotropical tree (*Virola surinamensis*). *Ecology*, **66**, 781–791.

Hubbell, S.P. (1980). Seed predation and the coexistence of tree species in tropical forests. *Oikos*, **35**, 214–229.

Hubbell, S.P. & Foster, R.B. (1986). Biology, chance, and history and the structure of tropical rain forest tree communities. In *Community Ecology* (Ed. by J. Diamond & T.J. Case), pp. 314–329. Harper & Row, New York.

Hubbell, S.P. & Foster, R.B. (1990). Structure, dynamics and equilibrium status of old-growth forest on Barro Colorado Island. In *Four Neotropical Rainforests* (Ed. by A.H. Gentry), pp. 522–541. Yale University Press, New Haven, Connecticut.

Janzen, D.H. (1969). Seed-eaters versus seed size, number, toxicity and dispersal. *Evolution*, **23**, 1–27.

Janzen, D.H. (1970). Herbivores and the number of tree species in tropical forests. *American Naturalist*, **104**, 501–528.

Janzen, D.H. (1971). Escape of juvenile *Dioclea megacarpa* (Leguminosae) vines from predators in a deciduous tropical forest. *American Naturalist*, **105**, 97–112.

Janzen, D.H. (1972). Escape in space by *Sterculia apetala* seeds from the bug *Dysdercus fasciatus* in a Costa Rican deciduous forest. *Ecology*, 53, 350–361.
Janzen, D.H. (1974). Tropical blackwater rivers, animals, and mast fruiting by the Dipterocarpaceae. *Biotropica*, 6, 69–103.
Janzen, D.H. (1976). Why bamboos take so long to flower. *Annual Review of Ecology and Systematics*, 7, 347–391.
Janzen, D.H. (1978). Seeding patterns of tropical trees. In *Tropical Trees as Living Systems* (Ed. by P.B. Tomlinson & M.H. Zimmerman), pp. 83–128. Cambridge University Press, New York.
Janzen, D.H., Miller, G.A., Hackforth-Jones, J., Pond, C.M., Hooper, K. & Janos, D.P. (1976). Two Costa Rican bat-generated seed shadows of *Andira inermis* (Leguminosae). *Ecology*, 57, 1068–1075.
Kitajima, K. (1994). Relative importance of photosynthetic traits and allocation patterns as correlates of seedling shade tolerance of 13 tropical trees. *Oecologia*, 98, 419–428.
Kitajima, K. & Augspurger, C.K. (1989). Seed and seedling ecology of a monocarpic tropical tree, *Tachigalia versicolor*. *Ecology*, 70, 1102–1114.
Lieberman, M. & Lieberman, D. (1994). Patterns of density and dispersion of forest trees. In *La Selva: Ecology and Natural History of a Neotropical Rain Forest* (Ed. by L.A. McDade, K.S. Bawa, H.A. Hespenheide & G.S. Hartshorn), pp. 106–119. Chicago University Press, Chicago.
Lloyd, M. & Dybas, H.S. (1966). The periodical cicada problem. II. *Evolution*, 20, 466–505.
Lott, R.H., Harrington, G.N., Irvine, A.K. & McIntyre, S. (1995). Density-dependent seed predation and plant dispersion of the tropical palm *Normanbya normanbyi*. *Biotropica*, 27, 87–95.
McKey, D., Waterman, P.G., Mbi, C.N., Gartlan, J.S. & Struhsaker, T.T. (1978). Phenolic content of vegetation in two African rain forests: ecological implications. *Science*, 220, 61–64.
McNaughton, S.J. (1983). Compensatory plant growth as a response to herbivory. *Oikos*, 40, 329–336.
Mole, S. & Waterman, P.G. (1988). Light-induced variation in phenolic levels in foliage of rainforest plants. II. Potential significance to herbivores. *Journal of Chemical Ecology*, 14, 23–34.
Mole, S., Ross, J.A.M. & Waterman, P.G. (1988). Light-induced variation in phenolic levels in foliage of rain-forest plants. I. Chemical changes. *Journal of Chemical Ecology*, 14, 1–21.
Mori, S.A. & Prance, G.T. (1987). Chapter XI. Phenology. The Lecythidaceae of a lowland neotropical forest: La Fumée Mountain, French Guiana. *Memoirs of the New York Botanical Garden*, 44, 124–136.
Murray, R.D.H., Mendez, J. & Brown, S.A. (1982). *The Natural Coumarins: Occurrence, Chemistry and Biochemistry*. Wiley, New York.
Newstrom, L.E., Frankie, G.W., Baker, H.G. & Colwell, R.K. (1994). Diversity of long-term flowering patterns. In *La Selva: Ecology and Natural History of a Neotropical Rain Forest* (Ed. by L.A. McDade, K.S. Bawa, H.A. Hespenheide & G.S. Hartshorn), pp. 142–160. Chicago University Press, Chicago.
Notman, E., Gorchov, D.L. & Cornejo, F. (1996). Effect of distance, aggregation, and habitat on levels of seed predation for two mammal-dispersed neotropical rain forest tree species. *Oecologia*, 106, 221–227.
Owen, D.F. & Wiegert, R.G. (1981). Mutualism between grasses and grazers: an evolutionary hypothesis. *Oikos*, 36, 376–378.
Peterson, C.H. & Black, R. (1988). Density-dependent mortality caused by physical stress interacting with biotic history. *American Naturalist*, 131, 257–270.
Ramirez, N. & Arroyo, M.K. (1987). Variación espacial y temporal en la depradación de semillas

de *Copaifera pubiflora* Benth. (Leguminosae: Caesalpinioideae) en Venezuela. *Biotropica*, **19**, 32–39.
Rees, M. (1993). Trade-offs among dispersal strategies in British plants. *Nature*, **366**, 150–152.
Rudinsky, J.A. (1962). Ecology of Scolytidae. *Review of Applied Entomology*, **7**, 327–348.
Sabatier, D. (1985). Saisonnalité et déterminisme du pic de fructification en forêt guyanaise. *Revue d' Ecologie (Terre et Vie)*, **40**, 289–320.
Salom, S.M. & McLean, J.A. (1991). Flight behaviour of scolytid beetle in response to semiochemicals at different wind speeds. *Journal of Chemical Ecology*, **17**, 647–661.
Schupp, E.W. (1988a). Factors affecting post-dispersal seed survival in a tropical forest. *Oecologia*, **76**, 525–530.
Schupp, E.W. (1988b). Seed and early seedling predation in the forest understorey and in treefall gaps. *Oikos*, **51**, 71–78.
Schupp, E.W. (1990). Annual variation in seedfall, postdispersal predation, and recruitment of a neotropical tree. *Ecology*, **71**, 504–515.
Schupp, E.W. (1992). The Janzen–Connell model for tropical tree diversity: population implications and the importance of spatial scale. *American Naturalist*, **140**, 526–530.
Schupp, E.W. (1995). Seed–seedling conflicts, habitat choice, and patterns of plant recruitment. *American Journal of Botany*, **82**, 399–409.
Schupp, E.W. & Frost, E.J. (1989). Differential predation of *Welfia georgii* seeds in treefall gaps and the forest understorey. *Biotropica*, **21**, 200–203.
Smallwood, P.D. & Peters, W.D. (1986). Grey squirrel food preferences: the effects of tannin and fat concentration. *Ecology*, **67**, 168–174.
Smythe, N. (1978). The natural history of the Central American agouti (*Dasyprocta punctata*). *Smithsonian Contributions to Zoology*, **257**, 1–52.
Smythe, N. (1986). Competition and resource partitioning in the guild of neotropical terrestrial frugivorous mammals. *Annual Review of Ecology & Systematics*, **17**, 169–188.
Sork, V.L. (1987). Effects of predation and light on seedling establishment in *Gustavia superba*. *Ecology*, **68**, 1341–1350.
Spalinger, D.E. & Hobbs, N.T. (1992). Mechanisms of foraging in mammalian herbivores: new models of functional response. *American Naturalist*, **140**, 325–348.
Steege, H. ter & Persaud, C. (1991). The phenology of Guyanese timber species: a compilation of a century of observations. *Vegetatio*, **95**, 177–198.
Steege, H. ter, Bokdam, C., Boland, M., Dobbelsteen, J. & Verburg, I. (1994). The effects of man made gaps on germination, early survival, and morphology of *Chlorocardium rodiei* seedlings in Guyana. *Journal of Tropical Ecology*, **10**, 245–260.
Steege, H. ter, Boot, R.G.A., Brouwer, L.C., Caesar, J.C., Ek, R.C., Hammond, D.S. *et al.* (1996). *Ecology and Logging in a Tropical Rain Forest in Guyana. With Recommendations for Forest Management*. Tropenbos Series 14. The Tropenbos Foundation, Wageningen, The Netherlands.
Terborgh, J., Losos, E., Riley, M.P. & Bolaños-Riley, M. (1993). Predation by vertebrates and invertebrates on the seeds of five canopy tree species of an Amazonian forest. *Vegetatio*, **107/108**, 375–386.
Traveset, A. (1990). Post-dispersal predation of *Acacia farnesiana* seeds by *Stator vachelliae* (Bruchidae) in Central America. *Oecologia*, **84**, 506–512.
Towers, G.H.N. (1984). Interactions of light with phytochemicals in some natural and novel systems. *Canadian Journal of Botany*, **62**, 2900–2911.
Whitmore, T.C. (1975). *Tropical Rain Forests of the Far East*. Oxford University Press, Oxford.
Wilkinson, L. (1989). *SYSTAT: The System for Statistics*. SYSTAT, Inc., Evanston, IL.
Wilson, D. & Janzen, D.H. (1972). Predation on *Scheelea* palm seeds by bruchid beetles: seed density and distance from the parent palm. *Ecology*, **53**, 954–959.
Wright, S.J. (1983). The dispersion of eggs by a bruchid beetle among *Scheelea* palm seeds and the effect of distance to the parent palm. *Ecology*, **64**, 1016–1021.

4. GAP-SIZE NICHE DIFFERENTIATION BY TROPICAL RAINFOREST TREES: A TESTABLE HYPOTHESIS OR A BROKEN-DOWN BANDWAGON?

N. D. BROWN and S. JENNINGS
Oxford Forestry Institute, South Parks Road, Oxford, OX1 3RB, UK

SUMMARY

1 In this chapter, we review the literature for evidence that tropical rainforest trees are specialized for optimum growth in treefall gaps of different sizes. There is evidence of partitioning at an ecologically crude guild level (pioneers vs. climax species). However, as the overwhelming majority of tropical rainforest tree species are climax species, it is within this group that tests of the hypothesis are critical. Amongst climax species, whilst differences in ecology have been demonstrated, there is a paucity of evidence of niche differentiation on a gap-size gradient.

2 Reasons for the apparent lack of success in finding niche differentiation are explored. The difficulty of measuring forest gap sizes presents a significant practical problem in testing this hypothesis. The variability of microclimates on a variety of spatial and temporal scales undermines many of the simplistic assumptions about gap environments made in this theory. Both experimental and observational approaches to detecting differences in seedling responses to gaps suffer from numerous practical and theoretical shortcomings which are discussed.

3 We argue that the forest light climate is an inappropriate environmental gradient for niche differentiation of the majority of tropical rainforest trees. An excessive emphasis on canopy gaps has obscured the fact that processes occurring elsewhere in time and space are of crucial importance in determining community composition.

INTRODUCTION

A major theme of post-war ecological thinking has been the central role of competition in structuring communities. The powerful school of community ecology initiated by Robert MacArthur in the 1960s was based on the premise that the relationships between species were determined by their competitive interactions. This paradigm has proved to be inadequate to account for many aspects of variation in community and population structure. It has been clear since the important studies of insect populations by Andrewartha and Birch (1954) that stochastic perturbations in the environment play an important role in maintaining species coexistence. Under constant environmental conditions the outcome of competition between individuals will be determined by their relative competitive abilities. However, interspecific competition does not necessarily result in competitive exclusion under non-equilibrium conditions.

The relative importance of interspecific competition is still unresolved in the analysis of community structure in tropical rainforests. One of the most prevalent paradigms of forest dynamics, the gap-size niche differentiation hypothesis, places interspecific competition centre-stage in determining forest composition.

The importance of gaps in a forest canopy for the rapid growth of seedlings of many species into mature trees has been known to foresters for many centuries. Seedling growth is stimulated by release from shade suppression and perhaps by locally diminished root competition. The size of a canopy gap is the main determinant of both the amount and the duration of insolation that penetrates the forest. Gaps of different sizes are therefore assumed to create a major resource gradient. Richards (1952) discussed the importance of gap size in favouring species of different ecology. It is assumed that no one species is able to be competitively superior across a wide range of different gap sizes and that niche specialization is therefore an advantage (Sipe & Bazzaz 1995). Denslow (1980) has argued that a species may regenerate preferentially beneath a specific size of canopy gap that creates optimum conditions for its growth. This is because species' specializations bestow competitive superiority in one particular gap size but involve adaptive compromises that restrict success in gaps of differing size. As a consequence, it may be hypothesized that the size of a forest canopy gap exerts the most important control over the relative competitive status of seedlings (Brokaw 1982, Orians 1982, Brandani, Hartshorn & Orians 1988). The gap-size niche partitioning hypothesis suggests that a diversity of gap sizes promotes coexistence of differently adapted species and hence enhances community diversity.

A growing number of studies, however, have claimed that species interactions play little role in influencing community structure. Several authors have proposed that rather than species being highly specialized, most are generalists capable of adequate growth under a range of canopy conditions. Hubbell and Foster (1986a) propose that being in the right place at a propitious time might be more important to a tree's success in reaching the canopy than competitive superiority in that particular environment. Whilst competitive interactions will inevitably occur, they claim that the effect of these on community structure will be obliterated by stochastic events. Becker (1985), for example, reports of catastrophic mortality of seedlings of a fast-growing dipterocarp species harvested from a gap by a wild pig for construction of its nest.

In this chapter we examine research evidence for gap-size niche differentiation. We discuss some of the methodological and theoretical problems that have dogged this hypothesis and propose some ways in which it might be improved. We begin by examining the evidence that has been uncovered for gap-size niche differentiation, before discussing whether the methodologies used are adequate tests of this hypothesis.

EVIDENCE FOR GAP-SIZE NICHE DIFFERENTIATION

There seems little doubt that at an ecologically crude guild-level pioneer and climax species have different but overlapping distributions along the gap-size gradient (Brokaw 1987, Clark & Clark 1992, Clark, Clark & Rich 1993). This is explicable in terms of observed differences in germination requirements (Kennedy & Swaine 1992, Vásquez-Yanes & Orozco-Segovia 1993) growth response to light (Thompson, Kriedemann & Craig 1992a,b) and light energy dissipation (Scholes, Press & Zipperlen 1997). Severe damage was caused to the leaves of a climax species when moved into a high-light environment (Turner & Newton 1990); a pioneer species was apparently unaffected. Pioneer species have been shown to have substantially higher light saturated rates of photosynthesis than climax species (Oberbauer & Strain 1984, Press, Brown, Barker & Zipperlen 1996). Differences in architectural and physiological plasticity have been noted between pioneer and climax species (King 1991, Riddoch, Lehto & Grace 1991, Strauss-Debenedetti & Bazzaz 1991).

Within the pioneer guild Clark *et al.* (1993) showed microhabitat specialization between two pioneer *Cecropia* species. Brokaw (1987) concluded that three pioneer species had different but overlapping distributions across a gap-size gradient. However, Kennedy and Swaine (1992) have suggested that this result was a sampling artefact as rare species are more likely to occur, by chance, in larger gaps.

Obvious differences between guilds are of little value in confirming or refuting the gap-size niche differentiation hypothesis. Most tropical rainforest trees are climax species (Whitmore 1984). It is within this group that we need to find evidence for niche specificity if their coexistence is to be explained by this theory. Differences between climax species in seedling mortality and growth responses and physiological traits have been demonstrated experimentally on numerous occasions (e.g. Becker 1983, Brown & Whitmore 1992, Still 1996, Press et al. 1996). Small differences between species in photosynthesis at different levels of photosynthetically active radiation (PAR) have been demonstrated (Denslow, Schultz, Vitousek & Strain 1990; Press et al. 1996, Zipperlen & Press 1996). Differential ability to acclimate to new canopy conditions has also been reported (Popma & Bongers 1991). However, differences in physiological traits alone are insufficient evidence to confirm the existence of niche differentiation. It is necessary to demonstrate that there is a shift in the rank order of species' responses along a light gradient.

Hardly any evidence of gap-size specialization has been reported from observational studies of species distributions in natural forest. Brandani et al. (1988) found species differentiation within treefall gaps on different microsite types, as predicted by Orians (1982). However, these patterns appear to be due to differences in soil conditions and the distribution of treefall debris. Most observational studies in natural forest report no correlation between species composition and gap size or forest structure (Barton 1984, Hubbell & Foster 1986b, Raich & Gong 1990, Weldon, Hewett, Hubbell & Foster 1991, Brown & Whitmore 1992, Oberbauer et al. 1993) For example, Lieberman, Lieberman, Peralta and Hartshorn (1995) found that 86.5% of all species of tree ⩾10cm dbh in an 11-ha plot of the La Selva forest, Costa Rica, were randomly distributed with respect to levels of canopy openness. They concluded from this evidence that none of the species studied was restricted in its distribution to specific levels of canopy openness and that the vast majority of species were generalists in their response to light. Differences in the distribution of six climax tree species with respect to canopy structure were found by Clark and Clark (1992), but only at certain tree size classes. Furthermore, they noted a 'striking similarity' in performance of these species in different microsites.

PROBLEMS DETECTING NICHE DIFFERENTIATION

The measurement of gap size

Plants do not respond to gap size *per se*. Gap size is used implicitly in the niche differentiation hypothesis as a surrogate for microclimate. The use of geometric measures of gap size is therefore theoretically flawed, but could be acceptable in practical terms if they were found to correlate tightly with forest microclimate.

Forest microclimates are both laborious and expensive to monitor. It is not legitimate to make do with short-term measures of gap microclimate because they do not accurately sample the range of conditions which plants experience. Diurnal and seasonal changes in both the direction of solar radiation and in the types of weather, particularly cloudiness, mean that measurements made at one time of day or in one season will always be a biased sample.

In contrast, the use of some measure of gap size as a surrogate for microclimatic variables is methodologically convenient since most of the measures that have been used may be estimated quickly. Simple geometric measures such as those proposed by Brokaw (1982) and Runkle (1981) do not require expensive equipment.

Measurements of gap size are used because they are assumed to be correlated approximately with mean microclimate. This correlation is good at the centre of a gap (Whitmore, Brown, Swaine et al. 1993) but breaks down with increasing distance away from the centre. Peaks of PAR and soil moisture and extended periods of high temperatures and low humidity are found in gap centres with progressive decline into the surrounding closed forest (Becker, Rabenold, Idol & Smith 1988, Raich 1989, Brown 1993). A single estimate of gap size is a poor surrogate for the mean microclimate experienced by seedlings not growing at the centre of a gap. It does not provide an ecologically meaningful way of estimating the variation in microclimate from place to place across the forest floor. Geometric measures of gap size have been shown to give a poor estimate of the area over which seedling growth is affected by a gap (Popma et al. 1988). As Lieberman, Lieberman and Peralta (1989) have suggested and various authors demonstrated (Popma, Bongers, Martinez-Ramos & Veneklaas 1988, Brown 1993, 1996) the effects on both microclimate and seedling growth extend well beyond the physical limits of a gap.

The use of gap size as a surrogate for microclimate variables is further undermined by the discovery that gaps of the same size do not necessarily experience the same microclimate. Direct measurement of gap microclimates has revealed that gap shape, orientation and topography play an important role (Brown 1993). More significantly, Brown (1993) has shown that the microclimate found at the centre of a small gap also occurs over a much larger area around the periphery of a large gap. An important implication is that it is clearly nonsensical to claim that climax species may be specialized for growth in a specific size of gap (e.g. Bazzaz & Pickett 1980, Augspurger 1984, Denslow 1987, Thompson, Stocker & Kriedemann 1988, Osunkoya, Ash, Hopkins & Graham 1994) even if they grow optimally in a particular microclimate. Species that are highly specialized for growth in a particular microclimate will find appropriate conditions in a wide range of different sized gaps.

There is also considerable temporal variation in gap microclimates due to daily and seasonal changes in cloudiness and changes in the position of the solar track relative to a hole in the canopy. One result of this variation is that whilst gaps may differ significantly in their mean microclimate, for much of the time the microclimatic conditions in one gap will fall within the range of variation found in many others (Fig. 4.1).

Even the best geometric measures (see van der Meer, Bongers, Chatrou & Riéra 1994) fail to acknowledge that gap area will never be more than a crude, indirect approximation of the microenvironment of the seedlings affected by a gap. If the objective is to examine the relationship between seedling ecology and microclimate there is questionable value in measuring gap area at all. Niche differentiation along a light gradient can only be tested by measuring the pattern of irradiance on each seedling.

The frequent failure of observational studies to detect niche differentiation is unsurprising in view of the implausible use of gap size as a surrogate for microclimate. Seedlings may well exhibit niche differentiation

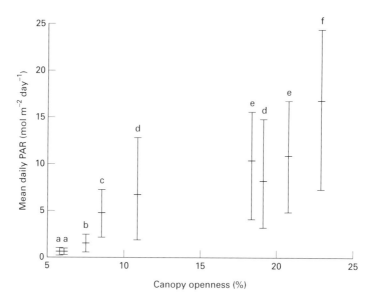

FIG. 4.1. Estimated mean daily photosynthetically active radiation (PAR) against canopy openness. Bars show the 5th and 95th percentile. Means with different letters are significantly different (two tailed t test, $P < 0.05$). PAR was measured for at least 65 days spread over 18 months at Revolta, Tapajós forest, Para, Brazil. Canopy openness was calculated from hemispherical photographs analysed with the program WINPHOT.

along a light gradient or some other microclimate gradient but this is unlikely to be demonstrated by comparison of relative growth in gaps of different size.

A superior method is to compare the level of light reaching individual seedlings with their growth performance (the 'dendrocentric' viewpoint of Lieberman *et al.* 1989). This too can either be monitored over long periods or estimated from forest canopy structure (Lieberman *et al.* 1989, Clark & Clark 1992).

Hemispherical canopy photography has offered a potential panacea for the problems of making quick, inexpensive and precise point estimates of the microclimate (Rich 1990). However, the errors involved in this method are unknown, difficult to estimate, and may well be substantial. It is likely that the errors made in estimating the light climate will be greatest in both absolute and relative terms when measuring beneath a closed forest canopy and in and around small gaps. The relationship between gap size and microclimate is not linear. There are larger increases in total daily PAR receipts for a unit increase in gap size when gaps are small (Brown 1993). Small gaps are significantly more abundant than large ones in unlogged tropical rainforest (Sandford, Braker & Hartshorn 1986, Brown 1990, Yavitt *et al.* 1995). Hemispherical canopy photography suffers from the major weakness that it is not able to give an accurate estimate of the light climate in that part of the range where the steepest gradient in light environments is known to occur.

The experimental approach

The most common method used to demonstrate gap-size niche differentiation among co-occurring species is an experimental approach. A number of studies have examined the responses of seedlings of different tree species in contrasting forest light environments (e.g. Turner 1990, Ashton, Gunatilleke & Gunatilleke 1995). However, such experiments present significant practical difficulties. No two treefall gaps are ever identical in size, shape or orientation and this creates the environmental heterogeneity necessary for niche differentiation. Even more problematic in forest-based experiments is the lack of control the scientist is able to exert over the multitude of other factors that influence growth responses through both direct and interaction. These may be imperfect covariates with gap size. For example, Whitmore and Brown (1996) report an increase in herbivory on dipterocarp seedlings growing in gaps of increasing size. The uniqueness of the environment in each treefall gap makes adequate experimental replication a serious practical problem (Ashton *et al.* 1995).

To avoid these difficulties ecologists have resorted to comparing seedling growth under more highly controlled artificial shade regimes. Most have attempted to simulate levels of light typically found in forest gaps but very few have adequately recreated either the temporal pattern or the spectral quality of forest light. The light regime in a forest and in forest gaps is characterized by periods of relatively low PAR (mostly from indirect sources) interspersed with bursts of high PAR (when direct sunlight penetrates the forest canopy). The majority of seedling light response studies have used shade cloth screens to reduce direct sunlight by a constant proportion. However, the efficiency of use by seedlings of fluctuating light is different to that of uniform illumination. Wayne and Bazzaz (1993) have shown that the daily pattern of light availability, independent of total PAR, significantly affected both seedling growth and plasticity in a seedling's sun–shade responses. They demonstrated that uniformly low PAR resulted in more growth than the same total PAR given in a shorter period at a high level. The use of uniform shade therefore appears likely to obscure rather than reveal species differences.

Furthermore, it is known that the spectral quality of light has a profound influence on seedling morphology and physiology. Most of the research on differential adaptation to shade has reduced PAR without any corresponding reduction in the ratio of red to far red light (R:FR). Such treatments will send ambiguous signals to plants which are likely to complicate both morphological and physiological responses. Zipperlen and Press (1996) concluded that seedling morphology is fundamentally interrelated with photosynthetic characteristics in determining the ecology of dipterocarp seedlings. The morphological response of a seedling to an artificial shade regime cannot therefore be ignored when seedling ecologies are compared. Figure 4.2 contrasts R:FR ratios recorded in a small gap in Sabah, Malaysia, with those recorded beneath spectrally neutral shade cloth at a similar level of PAR. R:FR ratios are always substantially lower in the forest than beneath the shade cloth. Lee, Baskaran, Mansor *et al.* (1996) have demonstrated the effects of reducing both PAR and R:FR on the growth of Malaysian tropical rainforest seedlings are substantial. However, all of the filters used cast uniform shade. Experiments are required such as those of Ashton (1995), which investigate the effects of fluctuations in both the quantity *and* quality of light.

Whilst the light in nursery shade-houses is more highly controlled other aspects of the environment differ from those found in a forest in many ways. Shaded nursery studies are typically conducted in the open away from the influence of forest shade. Here the relative humidity of the air may be lower and temperatures significantly higher than those found in the forest. Shade

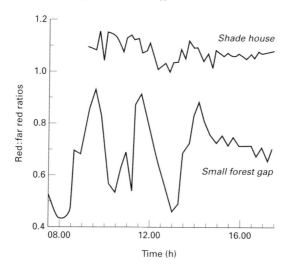

FIG. 4.2. The ratio of red (660 nm) to far red (730 nm) light recorded in the centre of a small gap and beneath spectrally neutral shade cloth at a similar level of total PAR (5–6 mol m^{-2} day^{-1}) over a period of 9 h in Sabah, Malaysia.

cloth severely restricts air movements over seedlings. Potted seedlings, free from root competition are commonly used in experiments, which makes the results difficult to interpret (discussed, for example, by Burslem, Turner & Grubb 1994).

The practical problems of experimental study of gap-size niche differentiation aside, such experiments face a number of theoretical difficulties. Few make explicit the hypothesis that they are attempting to test, which typically appears to be simple comparisons of seedling growth at contrasting gap sizes or light levels. Studies such as those of Augspurger (1984) and Turner and Newton (1990) remove competition by growing seedlings at wide spacings or singly in pots. Whilst this method will show the potential range of light levels over which a species may grow (or the light axis of the fundamental niche *sensu* Hutchinson 1957) it will reveal nothing of putative niche differentiation. It is often observed that the niche of a plant growing in the wild does not coincide with its physiologically optimum conditions for growth (e.g. Clymo & Reddaway 1972). Niche differentiation may only be demonstrated when plants are competing for a limiting resource and cannot be inferred from simple growth responses to different levels of that resource.

Another experimental method is to grow seedlings in mixed species arrays. This method has been used commonly for temperate species that grow in relatively species-poor communities. The realized niches that are revealed are not solely a property of the species themselves but a function of community composition. As a consequence, results are experiment-specific and difficult to generalize to a species-rich field situation. Even simple pairwise comparisons of competition between a very small number of species across a single environmental gradient would require impractically large numbers of trials.

The observational approach

An alternative approach to distinguishing niche specialization has been to examine the distribution in natural forest of all individuals of a species with respect to canopy structure (e.g. Whitmore 1974, Hubbell & Foster 1986b, Brokaw 1987, Weldon *et al*. 1991, Clark & Clark 1992).

With the exception of forests that have been devastated by extreme events (Whitmore 1974), such studies have by and large failed to find significant relationships between canopy gap size and plant community composition. However, they suffer from a common problem. Trees are very long-lived. Conditions that permitted the survival and growth from the juvenile stage may no longer pertain when the distribution of adults is assessed. When only juvenile stages are considered a different problem occurs. Only a small proportion of the juvenile individuals of any tree species will survive and grow to sexual maturity. Observational studies have no means by which the future champions may be distinguished from those that at some time in the future will find themselves at a competitive disadvantage. As Denslow and Hartshorn (1994) have pointed out we should therefore expect to find seedlings and saplings of shade-tolerant species randomly distributed among gaps and non-gap habitats. It should come as no surprise to discover the distribution of juveniles does not reflect niche-specific habitat conditions since most are doomed to die. Denslow and Hartshorn (1994) suggest that relative growth responses of different species to different canopy conditions may be a better indication of habitat requirements than the distribution of individuals. Even this approach suffers from the fact that very few individuals are likely to grow at a steady rate from seedling to mature adult and so present relative growth rates may give little indication of long-term performance. The relative positions of species in a competitive hierarchy may change over time. For example, short individuals of a fast-growing species may exhibit relatively slow growth whilst they overtake some slower-growing species. It is frequently proposed that some trees grow to sexual maturity in a series of repeated releases in response to a succession of small disturbances, rather than in a single uninterrupted burst of growth.

TENTATIVE SUGGESTIONS FOR AN IMPROVED MODEL

Competition for light is asymmetric. The superior competitor is not subject to the same resource scarcity as others lower in the competitive hierarchy. Acquisition of light is largely determined by relative plant size. The tallest plant intercepts the most light irrespective of whether or not it is able to respond with rapid growth. Plant size is an integration of its total growth history. It does not depend on whether conditions are currently optimum for growth nor on whether the plant has the highest rate of growth of those in competition for light. Pre-gap processes determine the size hierarchy at the time of gap creation. These processes may therefore be more important than the relative difference between species growth rates in gaps in determining the outcome of competitive interactions after gap formation. Brown and Whitmore (1992) demonstrated that seedlings of a slow growing, shade-tolerant dipterocarp species *Hopea nervosa* were able to persist in the understorey of lowland rain forest in Danum Valley, Sabah, longer than those of any other species studied. Consequently, when gaps formed *Hopea nervosa* seedlings had a significant height advantage over all other species. They retained this height advantage for over 3 years after gap creation. Many studies of tropical forest treefall gaps have discovered that the composition of the community growing in a gap merely reflects what was growing there before the gap was made rather than gap size (Uhl, Clark, Dezzeo & Maquirino 1988).

Tiny seedlings may be more vulnerable to a sudden change in microclimate (Brown & Whitmore 1992). Plant size may be significant in determining the ability of a plant to acclimate in a newly formed gap (Claussen 1996).

Pre-emption of light by plants which are already tall at the time that gaps are made will tend to reinforce the pre-existing size-hierarchy. The light intercepted by smaller seedlings or saplings will be determined primarily by their inferior position in that hierarchy and not by gap size. There is no necessary reason why there should be a strong relationship between growth and gap size for all but the tallest seedlings, even if species are narrowly specialized for optimum growth at particular levels of PAR. Rather than searching for species-specific growth responses to gap size it would be of greater ecological value to know what conditions (of species composition and vegetation structure) lead to the establishment of and subsequent changes to the initial size hierarchy. A study of how seedlings grow in the suboptimal light conditions beneath a taller individual will reveal more about how they are able to compete for a superior position in the size hierarchy.

We do not agree with Hubbell and Foster's (1986b) conclusion that stochastic events will obliterate the effects of differences in competitive ability. Analysis of plant distribution and relative competitive position must take full account of historical processes and not attempt to interpret it solely in terms of present canopy structure. It is because our conception of forest regeneration has been so gap-centred in both time and space that we have dismissed interactions taking place elsewhere as being chance historical events. However, it is clear that the nature of competitive interactions which take place in a gap are very often a function of pre-gap processes. Some tree species may use tolerance of shade conditions to pre-empt a leading position in the size hierarchy before gaps occur. Tilman (1982) has suggested that a trade-off may exist between the maximum rate at which a plant may exploit an abundant resource and its tolerance of resource scarcity. Rather than differing the level of resource they require for optimum growth (as proposed in the gap-size niche differentiation hypothesis) most plants will show better growth as resources become increasingly abundant. However, they will differ in the minimum amount of a resource they require to survive and in their maximum growth rate when resources are abundant. A shifting competitive hierarchy will therefore be established along a resource gradient (Keddy 1989). As resources become increasingly available species not able to tolerate resource scarcity are able to grow faster than those tolerant of scarcity. The pattern of growth of seedlings of three sympatric species of dipterocarp reported by Whitmore and Brown (1996) and Brown (1996) appears to confirm this model. All three species showed improved growth in larger gaps but they differed in their maximum growth rates and in the minimum level of canopy openness they required to survive.

We propose an added dimension to the shifting competitive hierarchy model which takes account of fluctuations in resource variability over time as well as space. Rather than restricting a species to resource-poor sites, scarcity-tolerance allows plants to use resource-poor periods to grow slowly to a size where they may pre-empt resources should they suddenly become available. Thus, they improve their competitive ability relative to other species that have higher relative growth rates at times when resources are abundant.

We do not believe that the forest light climate, on its own, forms an environmental gradient that is appropriate for niche differentiation of a large number of climax species. Variation occurs at different levels; large within-day and between-season variations will always be present, superimposed on variation caused by canopy changes (including gap creation and canopy closure) and that caused by the ontogenetic development of a seedling. This suggests that a species physiologically specialized for growth at a particular point on the PAR spectrum (i.e. optimum performance confined to a narrow range of PAR availability) will in fact encounter these conditions only fleetingly. In such a changeable environment a species that shows a growth response over a broad range of microclimatic conditions will be at a competitive advantage. Most of the time it will be in conditions allowing some growth. We argue that niche differentiation will only be found along predictable environmental gradients (e.g. water, edaphic and topographic gradients) such as that shown by Pemadasa and Gunatilleke (1981), Ashton *et al.* (1995) and Veenendaal and Swaine (Chapter 7, this volume). We speculate that the more predictable microclimate found in very large gaps and beneath closed forest may permit some degree of niche specialization.

Clark and Clark (1992) have called for studies of all size classes of target species rather than just seedlings. We agree that we should examine how individuals of different species cope with multiple periods of release and shade suppression at all sizes, and not assume that the most important processes occur at the seedling stage in gaps.

ACKNOWLEDGEMENTS

We are grateful to Malcolm Press, Doug Sheil, Tim Whitmore, Gil Vieira and Simon Zipperlen for having helped us develop our ideas.

REFERENCES

Andrewartha, H.C. & Birch, L.C. (1954). *The Distribution and Abundance of Animals.* University of Chicago Press, Chicago.

Ashton, P.M.S. (1995). Seedling growth of co-occurring *Shorea* species in the simulated light environments of a rain forest. *Forest Ecology and Management*, **72**, 1–12.

Ashton, P.M.S., Gunatilleke, C.V.S. & Gunatilleke, I.A.U.N. (1995). Seedling survival and growth of four *Shorea* species in a Sri Lankan rainforest. *Journal of Tropical Ecology*, **11**, 263–279.

Augspurger, C.K. (1984). Light requirements of neotropical tree seedlings: a comparative study of growth and survival. *Journal of Ecology*, **72**, 777–795.

Barton, A.M. (1984). Neotropical pioneer and shade tolerant tree species: do they partition treefall gaps? *Tropical Ecology*, **25**, 196–202.

Bazzaz, F.A. & Pickett, S.T.A. (1980). Physiological ecology of tropical succession: a comparative review. *Annual Review of Ecology and Systematics*, **11**, 287–310.

Becker, P. (1983). Effects of insect herbivory and artificial defoliation on survival of *Shorea* seedlings. In *Tropical Rain Forest Ecology and Management* (Ed. by S.L. Sutton, T.C. Whitmore & A.C. Chadwick), pp. 241–252. Blackwell Scientific Publications, Oxford.

Becker, P. (1985). Catastrophic mortality of *Shorea leprosula* juveniles in a small gap. *Malaysian Forester*, **48**, 263.

Becker, P., Rabenold, P.E., Idol, J.R. & Smith, A.P. (1988). Water potential gradients for gaps and slopes in a Panamanian tropical moist forest's dry season. *Journal of Tropical Ecology*, **4**, 173–184.

Brandani, A., Hartshorn, G.S. & Orians, G.H. (1988). Internal heterogeneity of gaps and species richness in Costa Rican tropical wet forest. *Journal of Tropical Ecology*, 4, 99–119.

Brokaw, N.V.L. (1982). The definition of treefall gap and its effect on measures of forest dynamics. *Biotropica*, 14, 158–160.

Brokaw, N.V.L. (1987). Gap-phase regeneration of three pioneer species in a tropical forest. *Journal of Ecology*, 75, 9–19.

Brown, N.D. (1990). *Dipterocarp regeneration in tropical rain forest gaps of different sizes.* DPhil thesis, University of Oxford.

Brown, N.D. (1993). The implications of climate and gap microclimate for seedling growth conditions in a Bornean lowland rain forest. *Journal of Tropical Ecology*, 9, 153–168.

Brown, N.D. (1996). A gradient of seedling growth from the centre of a tropical rain forest canopy gap. *Forest Ecology and Management*, 82, 239–244.

Brown, N.D. & Whitmore, T.C. (1992). Do dipterocarp seedlings really partition tropical rainforest gaps? *Philosophical Transactions of the Royal Society of London, Series B*, 335, 369–378.

Burslem, D.F.R.P., Turner, I.M. & Grubb, P.J. (1994). Mineral nutrient status of costal hill dipterocarp forest and adinandra belukar in Singapore: bioassays of nutrient limitation. *Journal of Tropical Ecology*, 10, 579–599.

Clark, D.A. & Clark, D.B. (1992). Life history diversity of canopy and emergent trees in a neotropical rainforest. *Ecological Monographs*, 62, 315–344.

Clark, D.B., Clark, D.A. & Rich, P.M. (1993). Comparative analysis of microhabitat utilization by saplings of nine tree species in Neotropical rain forest. *Biotropica*, 25, 397–407.

Claussen, J.W. (1996). Acclimation abilities of three tropical rainforest seedlings to an increase in light intensity. *Forest Ecology and Management*, 80, 245–255.

Clymo, R.S. & Reddaway, E.J.F. (1972). A tentative dry matter balance sheet for the wet blanket bog on Burnt Hill, Moor House NNR. *Moor House Occasional Paper*, 3. Nature Conservancy, London.

Denslow, J.S. (1980). Gap partitioning among tropical rainforest trees. *Biotropica*, 12 (Suppl.), 47–55.

Denslow, J.S. (1987). Tropical rain forest gaps and tree species diversity. *Annual Review of Ecology and Systematics*, 18, 431–451.

Denslow, J.S. & Hartshorn, G.S. (1994). Treefall gap environments and forest dynamic processes. In *La Selva: Ecology and Natural History of a Neotropical Rain Forest* (Ed. by L.A. McDade, K.S. Bawa, H.A. Hespenheide & G.S. Hartshorn), pp. 120–127. University of Chicago Press, Chicago.

Denslow, J.S., Schultz, J.C., Vitousek, P.M. & Strain, B.R. (1990). Growth responses of tropical shrubs to treefall gap environments. *Ecology*, 71, 165–179.

Hubbell, S.P. & Foster, R.B. (1986a). Biology, chance and history, and the structure of tropical tree communities. In *Community Ecology* (Ed. by J.M. Diamond & T.J. Case), pp. 314–324. Harper & Row, New York.

Hubbell, S.P. & Foster, R.B. (1986b). Canopy gaps and the dynamics of a Neotropical forest. In *Plant Ecology* (Ed. by M.J. Crawley), pp. 77–96. Blackwell Scientific Publications, Oxford.

Hutchinson, G.E. (1957). The multivariate niche. *Cold Spring Harbor Symposium on Quantitative Biology*, 22, 415–421.

Keddy, P.A. (1989). *Competition*. Chapman & Hall, London.

Kennedy, D.N. & Swaine, M.D. (1992). Germination and growth of colonising species in artificial gaps of different sizes in a dipterocarp rain forest. *Philosophical Transactions of the Royal Society of London, Series B*, 335, 357–368.

King, D.A. (1991). Correlations between biomass allocation, relative growth rate and light environment in tropical forest saplings. *Functional Ecology*, 5, 485–492.

Lee, D.W., Baskaran, K., Mansor, M., Mohamad, H. & Yap, S.K. (1996). Irradiance and spectral quality affect Asian tropical rain forest tree seedling development. *Ecology*, **77**, 568–580.

Lieberman, M., Lieberman, D. & Peralta, R. (1989). Forests are not just Swiss cheese: canopy stereogeometry of non-gaps in tropical forests. *Ecology*, **70**, 550–552.

Lieberman, M., Lieberman, D., Peralta, R. & Hartshorn, G.S. (1995). Canopy closure and the distribution of tropical forest tree species at La Selva, Costa Rica. *Journal of Tropical Ecology*, **11**, 161–178.

Oberbauer, S.F. & Strain, B.R. (1984). Photosynthesis and successional status of Costa Rican rain forest trees. *Photosynthesis Research*, **5**, 227–232.

Oberbauer, S.F., Clark, D.B., Clark, D.A., Rich, P.M. & Vega, G. (1993). Light environment, gas exchange and annual growth of saplings of three species of rain forest trees in Costa Rica. *Journal of Tropical Ecology*, **9**, 511–523.

Orians, G.H. (1982). The influence of treefalls on tropical forests on species richness. *Tropical Ecology*, **23**, 255–279.

Osunkoya, O.O., Ash, J.E., Hopkins, M.S. & Graham, A.W. (1994). Influence of seed size and seedling ecological attributes on shade-tolerance of rainforest tree species in northern Queensland. *Journal of Ecology*, **83**, 149–163.

Pemadasa, M.A. & Gunatilleke, C.V.S. (1981). Pattern in a rain forest in Sri Lanka. *Journal of Ecology*, **69**, 117–124.

Popma, J. & Bongers, F. (1991). Acclimation of seedlings of three Mexican tropical rain forest tree species to a change in light availability. *Journal of Tropical Ecology*, **7**, 85–97.

Popma, J., Bongers, F., Martinez-Ramos, M. & Veneklaas, E. (1988). Pioneer species distribution in Neotropical rain forest; a gap definition and its consequences. *Journal of Tropical Ecology*, **4**, 77–88.

Press, M.C., Brown, N.D., Barker, M.G. & Zipperlen, S.W. (1996). Photosynthetic responses to light in tropical rain forest tree seedlings. In *The Ecology of Tropical Forest Tree Seedlings* (Ed. by M.D. Swaine), pp. 41–58. *Man and the Biosphere Series*, Vol. 18. UNESCO/Parthenon, Paris/Carnforth UK.

Raich, J.W. (1989). Seasonal and spatial variation in the light environment in a tropical dipterocarp forest and gaps. *Biotropica*, **21**, 299–302.

Raich, J.W. & Gong, K.W. (1990). Effects of canopy openings on tree seed germination in a Malaysian dipterocarp forest. *Journal of Tropical Ecology*, **6**, 203–217.

Rich, P.M. (1990). Characterizing plant canopies with hemispherical photographs. *Remote Sensing Reviews*, **5**, 13–29.

Richards, P.W. (1952). *The Tropical Rain Forest*. Cambridge University Press, Cambridge.

Riddoch, I., Lehto, T. & Grace, J. (1991). Photosynthesis of tropical tree seedlings in relation to light and nutrient supply. *New Phytologist*, **119**, 137–147.

Runkle (1981). Gap regeneration in some old-growth forests of the eastern United States. *Ecology*, **62**, 1041–1051.

Sandford, R.L. Jr, Braker, H.E. & Hartshorn, G.S. (1986). Canopy openings in a primary neotropical lowland forest. *Journal of Tropical Ecology*, **2**, 277–282.

Scholes, J.D., Press, M.C. & Zipperlen, S.W. (1997). Differences in light energy utilisation and dissipation between dipterocarp rain forest tree seedlings. *Oecologia*, **109**, 41–48.

Sipe, T.W. & Bazzaz, F.A. (1995). Gap partitioning among maples (*Acer*) in central New England: shoot architecture and photosynthesis. *Ecology*, **75**, 2318–2332.

Still, M.J. (1996). Rates of mortality and growth in three groups of Dipterocarp seedlings in Sabah, Malaysia. In *The Ecology of Tropical Forest Tree Seedlings* (Ed. by M.D. Swaine), pp. 315–332. *Man and the Biosphere Series*, Vol. 18. UNESCO/Parthenon, Paris/Carnforth.

Strauss-Debenedetti, S. & Bazzaz, F.A. (1991). Plasticity and acclimation to light in tropical Moraceae of different successional positions. *Oecologia*, **87**, 377–387.

Thompson, W.A., Stocker, G.C. & Kriedemann, P.E. (1988). Growth and photosynthetic response to light and nutrients of *Flindersia brayleyana* F. Muell., a rainforest tree with a broad tolerance to sun and shade. *Australian Journal of Plant Physiology*, **15**, 299–315.

Thompson, W.A., Kriedemann, P.E. & Craig, I.E. (1992a). Photosynthetic response to light and nutrients in sun-tolerant and shade-tolerant rainforest trees. I. Growth, leaf anatomy and nutrient content. *Australian Journal of Plant Physiology*, **19**, 1–18.

Thompson, W.A., Kriedemann, P.E. & Craig, I.E. (1992b). Photosynthetic response to light and nutrients in sun-tolerant and shade-tolerant rainforest trees. II. Leaf gas exchange and component processes of photosynthesis. *Australian Journal of Plant Physiology*, **19**, 19–42.

Tilman, D. (1982). *Resource Competition and Community Structure*. Princeton University Press, Princeton.

Turner, I.M. (1990). The seedling survivorship and growth of three *Shorea* species in a Malaysian tropical rain forest. *Journal of Tropical Ecology*, **6**, 469–477.

Turner, I.M. & Newton, A.C. (1990). The initial responses of some tropical rain forest tree seedlings to a large gap environment. *Journal of Applied Ecology*, **27**, 605–608.

Uhl, C., Clark, K., Dezzeo, N. & Maquirino, P. (1988). Vegetation dynamics in Amazonian treefall gaps. *Ecology*, **69**, 763–781.

van der Meer, P.J., Bongers, F., Chatrou, L. & Riéra, B. (1994). Defining canopy gaps in a tropical rain forest: effects on gap size and turnover time. *Acta Oecologica*, **15**, 701–714.

Vásquez-Yanes, C. & Orozco-Segovia, A. (1993). Patterns of seed longevity and germination in the tropical rain forest. *Annual Review of Ecology and Systematics*, **24**, 69–87.

Wayne, P.M. & Bazzaz, F.A. (1993). Birch seedling responses to daily time course of light in experimental forest gaps and shadehouses. *Ecology*, **74**, 1500–1515.

Welden, C.W., Hewett, S.W., Hubbell, S.P. & Foster, R.B. (1991). Sapling survival, growth, and recruitment: relation to canopy height in a neotropical forest. *Ecology*, **72**, 35–50.

Whitmore, T.C. (1974). Change with time and the role of cyclones in tropical rain forest on Kolombangara, Solomon Islands. *Commonwealth Forestry Institute Paper*, **46**.

Whitmore, T.C. (1984). Gap size and species richness in tropical rain forests. *Biotropica*, **16**, 239.

Whitmore, T.C. & Brown, N.D. (1996). Dipterocarp seedling growth in rain forest canopy gaps during six and a half years. *Philosophical Transactions of the Royal Society, Series B*, **351**, 1195–1203.

Whitmore, T.C., Brown, N.D., Swaine, M.D., Kennedy, D., Goodwin-Bailey, C.I. & Gong, W.K. (1993). Use of hemispherical photographs in forest ecology: measurement of gap size and radiation totals in a Bornean tropical rain forest. *Journal of Tropical Ecology*, **9**, 131–151.

Yavitt, J.B., Battles, J.J., Lang, G.E. & Knight, D.H. (1995). The canopy gap regime in a secondary Neotropical forest in Panama. *Journal of Tropical Ecology*, **11**, 391–402.

Zipperlen, S.W. & Press, M.C. (1996). Photosynthesis in relation to growth and seedling ecology of two dipterocarp rain forest tree species. *Journal of Ecology*, **84**, 863–876.

5. DIFFERENTIAL EFFECTS OF SMALL-SCALE FISHING ON PREDATORY AND PREY FISHES ON FIJIAN REEFS

N.V.C. POLUNIN* and S. JENNINGS[†]
*Department of Marine Sciences & Coastal Management, University of Newcastle, Newcastle upon Tyne NE1 7RU, UK; † School of Biological Sciences, University of East Anglia, Norwich NR4 7TJ, UK

SUMMARY

1 The processes governing size of fish populations on coral reefs, and structure of the communities which they compose, are little known. The least known of these processes is predation and other forms of disturbance, which are reviewed here.

2 Seven traditional Fijian fishing-grounds, exhibiting a 60-fold range in fishing activity per unit of reef area, constituted a gradient of fishing pressure along which depletion of piscivores, as well as of other target species, was expected, and increase in some prey fish species, which were not fished, was predicted. The fishing involved was not substantially destructive of habitat.

3 Underwater visual census (UVC) point-counts of 7-m radius recorded 226 species of diurnally-active lagoonal reef fishes ⩾7 cm long, of which 98 were fished, while counts of 2-m radius recorded 85 species of fishes <7 cm long, and of these eight were juveniles of fishes caught at greater size. UVC biomass indices of fishes were calculated at the species level, and after trophic and taxonomic aggregation.

4 The 7-m radius samples in the least fished of the grounds had significantly higher biomasses of particular target fishes, as evinced by the coral trout *Plectropomus laevis*, which is commonly caught in the fishery, by groupers (Serranidae) and by piscivorous fishes generally.

5 There was no evidence that any unfished prey fishes had greater abundance at species, family or trophic levels in the heavily fished grounds relative to the lightly fished grounds, in either large (7 m) or small (2 m) UVC samples.

6 Fishing in the Fijian fishing-grounds has apparently depleted piscivores, but it was not possible to discern any increase in prey fish abundance. Understanding of fishing effects on reef populations, both of fishes and of invertebrates, and on the whole ecosystem, is reviewed.

INTRODUCTION

Much of the information on reef fish ecology relates to a select group of small-bodied and site-attached species, most of which are from a single family (Sale 1991a). Even where an ecological fact has been elucidated for a particular fish species or population in a particular location over a given time period, there remains substantial doubt as to its generality across a greater range of reef fish species, longer temporal scales and wider geographical areas (Jones 1991). Such uncertainty is not exclusive to reef fish population ecology. Population fluctuations in marine fish catches have puzzled fisherfolk for many centuries at least (Cushing 1988), while increase in the scientific capability to predict fish availability to fisheries has been the major purpose of fisheries science globally for more than a century (Smith 1994, Jennings & Lock 1996). However, particular doubt exists in the case of tropical reefs, with respect to the dynamics of large-bodied, long-lived and roving species, because of the limited ecological focus of previous and nearly all present work. The large species appear to be important to reef systems as predators, although there are few data on predation (Hixon 1991), and to the fisheries based on reefs as a resource which is susceptible to fishing and is appreciated by human consumers (Polunin, Roberts & Pauly 1996).

The pre-settlement life of tropical reef fishes is very poorly known compared to that of many temperate species (Leis 1991, Boehlert 1996, Roberts 1996, Sadovy 1996); in contrast, detail on the life of late juveniles and adults (Sale 1991a) of some reef species is probably greater than for many similar species at higher latitudes. Hypotheses available to explain how tropical reef fish population size and community structure are determined can be categorized in one of four types (Jones 1991): 'settlement-limitation' and 'lottery' hypotheses rely heavily on pre-settlement processes, while 'competition' and 'disturbance' hypotheses are dependent upon processes acting after larval settlement. Roles of pre-settlement processes in the context of settlement-limitation and lottery hypotheses of reef fish population size and community structure have been amply reviewed elsewhere (Sale 1980, Doherty & Williams 1988, Doherty 1991, Sale 1991b, Roberts 1996). In essence, settlement-limitation supposes that abundance of animals on a reef is determined prior to settlement by mortality in the plankton. Studies of populations which are considered limited in size by larval supply have focused on small species, and often merely on the juveniles of those species (for example, Meekan, Milicich & Doherty 1993), even though they may be quite long-lived (Polunin & Brothers 1989).

Lottery hypotheses postulate that while the size of an assemblage may be determined by competition, the actual composition of the assemblage varies, depending on the settlers that happen to be supplied from the plankton. Competition and disturbance as processes determining the size of populations and community structure of reef fishes have also been reviewed (e.g. Hixon 1991, Jones 1991). Essentially, competition hypotheses postulate that each piece of habitat should have its own particular mix of species and sizes of populations, given that resources determine which species become established and what the abundance of each is. However, where disturbance is important, and this may include predation, competition may not occur because populations are kept below a size at which resources can become limiting.

While competition hypotheses postulate that reef fish populations reach equilibria with the resources that limit their size, hypotheses relating to disturbance and predation rely on such equilibria not being attained (Doherty & Williams 1988). These hypotheses are not mutually exclusive overall (Jones 1991). If one particular population is kept below equilibrium by disturbance, or predation, competition between it and another population for limiting resources is clearly not occurring, but a role for competition is not excluded in another place and time, or for other populations at the same time and place. Much of the field work on competition and niche partitioning has focused on damselfishes (Pomacentridae) and in particular on those that are herbivorous and territorial. Much of the work has specifically examined aggressive behaviour. Thus, Ebersole (1977) showed that the frequency of aggressive acts against other fishes by a damselfish was related largely to diet overlap. Hourigan (1986) showed how experimental removal of a territorial damselfish affected the behaviour and distribution of other species which it attacked. Such work implies often that food may be a principal resource limiting populations of such fishes, but measuring its supply has been rarely attempted, and the results have been equivocal when it has been (Polunin & Klumpp 1992a). A good non-damselfish example of an ecological-partitioning study is that of Gladfelter and Johnson (1983), who showed that coexisting squirrelfishes (Holocentridae) could be distinguished from each other on the basis of diet, feeding periodicity and feeding habitat. Other resources, such as shelter from predators, may be important in many instances, but Robertson and Sheldon (1979) concluded that even though wrasse apparently fought over sleeping sites, aggression was not directly related to the density of these sites relative to that of fishes. In addition, Choat and Bellwood (1985) have shown how only very limited generalization may be possible from single site-specific studies of particular interspecies interactions. Behavioural mechanisms of competition are not as straightforward as might be imagined, and there is clearly little understanding of the role of competition in limiting population sizes and influencing community structures of fishes on reefs. What information exists is largely of an experimental nature.

Predators and prey

That reefs as a whole, and reef fish populations in particular, are subject to disturbance by physical events such as storms (Harmelin-Vivien 1995), and biological factors such as crown-of-thorns starfish outbreaks (Williams 1986) and sea-urchin die-backs (Hughes 1994), is obvious. There are, however, few general data on the impacts of such events (Harmelin-Vivien 1995). Predation is another form of disturbance, which may also reduce populations of prey below the levels at which they might otherwise compete. Since large fish predators tend to be targeted by reef fisheries, depletion of predators is thought widely to have occurred as a result of fishing (Jennings & Lock 1996). Although many fish stocks have been depleted, there have been very few attempts empirically to investigate the effects of such change (Beddington 1984). The expectation has commonly been that such depletion should have profound effects on prey organisms and on communities as a whole (Beddington 1984, Grigg, Polovina & Atkinson 1984, Munro & Williams 1985).

Understanding the role of predation in tropical reef communities relies on a number of sources of information, including experiments, correlative studies and modelling. Although experiments involving predator manipulation have been attempted several times, most have been unsuccessful, because total removal is difficult, even from small areas of reef (e.g. Shpigel & Fishelson 1991), while large-scale removal has so far proved impossible (Hixon 1991). Caley (1993), however, did find that when all known piscivores were successfully removed from small reef-flat areas, certain prey species did increase in abundance. In an example of a correlative study in the Florida Keys, Bohnsack (1982) compared fish communities on reefs from which piscivores had been depleted by fishing (an influence not destructive of the habitat), with those in a protected area where piscivore abundance remained high. Only one species of fish showed evidence of release from predation. Hixon and Beets (1993) showed that mean prey abundance was not negatively correlated with piscivore abundance, but that on artificial reefs, greatest prey abundance was found where there were fewer predators. On the basis of such studies, Hixon (1991) considered it likely that with so many piscivorous species on reefs, predation would be 'diffuse', in that the impacts of any one predator on a particular prey population would be small, because the predator would have many prey species from which to choose. Russ (1991) also concluded that fishing effects on non-target prey species were unlikely, given the variety of predators and their opportunism, and a dominant role of pre-settlement processes in determining size of fish populations established on reefs. If these suppositions are correct, then indirect effects of exploitation (Menge 1995) should be small, even if direct effects are large.

However, very few studies have estimated actual predation rates, and when Sudekum, Parrish, Radtke and Ralston (1991) did so, very high predation intensities by particular piscivores were found. Two species of jacks, *Caranx ignobilis* and *C. melampygus*, were estimated to ingest prey (Sudekum *et al.* 1991) at a rate that was greater than the maximum yields of Pacific reef fisheries (e.g. Dalzell, Adams & Polunin 1996). It has also been suggested, for example by Grigg *et al.* (1984) and Munro and Williams (1985), that depletion of large predators should be an optimal strategy for reef fishing, because this will increase the abundance of prey fishes and thus the potential yield. Nevertheless, the dynamics of marine communities are poorly known, and the possibility exists that predator depletion may have wider and unpredictable effects (Beddington 1984). Knowlton (1992) has argued that reefs have a number of characteristics which may predispose them to existing in more than one state. The stability or otherwise of such alternate states remains unknown, but there is mounting evidence from Jamaica (Hughes 1994) and Kenya (McClanahan 1992) that fishing may have effects on whole reef communities, impacts that may lead to what are being referred to as 'phase shifts' (Hughes 1994). Such phase shifts are indicative of 'top-down' influences in food-webs, where a consumer influences the abundance of the organisms on which it feeds. When repeated across more than one trophic linkage, such controls lead to 'cascade effects' in ecosystems (Paine 1980). If cascade effects occur in coral reef systems, then that would tend to be because there are relatively strong linkages between predators and prey, whereas, if such linkages were weak, predation effects would tend to be diffuse, and change in the rate of predation by a particular species, or group, would have small effects on prey populations. If the phase shifts that have occurred in Jamaica have been correctly interpreted as being initiated by the effects of fishing, then this would be because there are strong predator–prey linkages, such as a 'keystone predator' might support. Alternatively, predators may be generalists and fishing has been so intensive that there has been a wholesale depletion of them. Sea-urchins, which are thought to have mediated a phase shift on Jamaican reefs (Hughes 1994), may be subject to keystone predation in Kenya (McClanahan 1994), and there is circumstantial evidence that sea-urchins have widely increased as a result of fishing in the Caribbean (Hay 1984).

Another way in which predation rates might be inferred is from modelling studies, and attempts are being made increasingly to synthesize information on reefs into ecosystem models. These models have been either biomass-based (Polovina 1984, Arias Gonzalez 1993, Opitz 1993) or more process-oriented (McClanahan 1992, Polunin & Klumpp 1992b). The models do help to highlight gaps in knowledge but their power to predict particular processes such as predation will remain weak so long as the relevant empirical information is unavailable.

Jones (1991) considered that fluctuation in numbers is one of the few facts established by those investigating settlement limitation, yet the fact of settlement fluctuation is confined to a few small species and is no proof of settlement limitation of populations generally. The question remains as to what happens in the large species that are important in fisheries. Are they more, or less, likely to be settlement-limited? Is the structure of assemblages of these species dictated by planktonic processes, as many would have it? There is increasing evidence that many of the larger species are long-lived (e.g. Williams, Newman, Cappo & Doherty 1995, Choat & Axe 1997). Not only will many of these populations probably turn out to be more stable than is the case for the young of the small species, but there is substantial scope for these species to impose structure on other species in the community through predation, competition and other processes. In the meantime, analyses of large-scale changes in predator abundance are worthwhile for their potential in elucidating processes helping to structure whole communities.

Effects of reef fishing in Fiji

Fishing increases the mortality of exploited species, and, where predators are involved, should therefore reduce predation mortality of prey species and thus increase the abundances of the latter. Studies on predation and fishing effects have thus far suffered from poor replication, poor validation and limited scale. We carried out a study in Fiji at localities where fishing is not destructive of habitat and the reefs are permanently allocated to particular groups of rights users (e.g. Ruddle 1996) in exclusive grounds known as *qoliqoli*. There are over 400 such traditional fishing-grounds in Fiji, and we anticipated that fishing pressure in each *qoliqoli* would be quantifiable (Jennings & Polunin 1995a). Further, we expected that grounds would differ in fishing activity because the size of the rights-holding population varies among *qoliqoli* in relation to the area of available reef habitat (Jennings & Polunin 1995b). We set out to see whether we could detect differences in reef fish populations and communities in ecologically similar fishing-grounds subject to different levels of fishing effort, because such fishing in the Pacific widely targets large piscivorous species (Dalzell et al. 1996).

Our first specific objective was to test null hypotheses of no difference in fish biomass among the grounds. Where the null hypotheses were rejected, we sought to explain differences among grounds in relation to variations in fishing effort. There were two expectations for such cases. The first of these was that fishes targeted by the fishery should decline in abundance as fishing pressure increased. The second expectation was that fishes that constitute the prey of piscivorous target species should increase in abundance with increase in fishing pressure as their predators are removed. Any such inference as to effects of fishing would need to consider effects of habitat (Gosline 1965, Sale & Dybdahl 1975, Gladfelter & Gladfelter 1978) and recruitment differences among grounds, and the fishing history of the areas involved, independently of their current fishing intensity.

MATERIALS AND METHODS

Study areas

We selected seven grounds which are spread out from the vicinity of Suva, the capital of Fiji, in an east-southeasterly direction towards the Lau group of the Fijian islands (Fig. 5.1). We estimated fishing effort for six grounds (Nauluvatu, Ko Ono, Natusara, Yanuca, Cokovata and Moala; Fig. 5.1) through a study in which fishers participated by recording their own catches (Jennings & Polunin 1995a), and for the other ground (Suva) by direct observation (S. Jennings & N.V.C. Polunin unpublished data). Fishing for the multi-species reef-associated stock was rescaled to hours spent boat-based spear fishing over coral by day for sale, because significant differences in the catch data were found to be associated with each of these factors (Jennings & Polunin 1995b). Fishing activity was related to available reef area to provide an index of fishing pressure in reef habitats which could be compared among grounds. The fishing surveys offered abundant information on which species were commonly targeted. Although blast fishing does occur in Fiji, we obtained no evidence that it or other destructive fishing techniques have been employed in the seven grounds that we studied.

Underwater visual census

We conducted underwater visual censuses (UVC) using SCUBA in the lagoonal coralline habitat of each of the seven grounds during the period

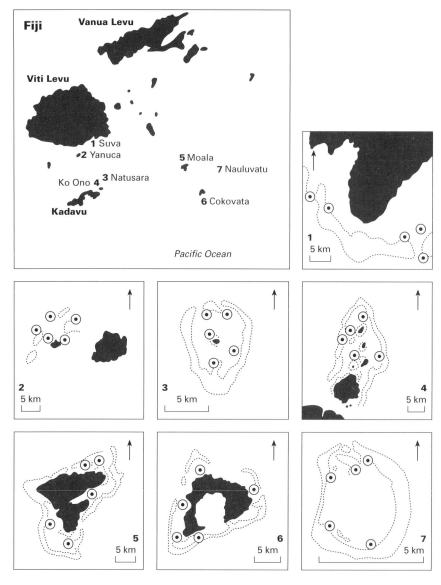

FIG. 5.1. Location in the Fijian archipelago and diagrammatic maps of the seven Fijian fishing grounds studied. *Qoliqoli*: 1, Suva; 2, Yanuca; 3, Natusara; 4, Ko Ono; 5, Moala; 6, Cokovata; 7, Nauluvatu. Scale bars are all 5 km, and the arrows point due north. *Qoliqoli* are exclusive fishing grounds. Dotted lines indicate approximate reef boundaries.

September 1992 to April 1993. Within each ground, five areas of lagoonal reef-edge habitat were selected on the basis that they were widely separated from each other. In each of the five areas, 16 circular UVC samples were taken by both of us from sites within an area up to 1 km across. The UVC method was based on those of Samoilys (1992) and Samoilys and Carlos (1992). The species in the UVC sample sites which were most wary of divers (e.g. groupers) were recorded from as far away as water clarity would allow, while small site-attached fishes (e.g. small surgeonfishes) were recorded close to. Body lengths of all fishes >7 cm in length were estimated in samples of 7-m radius (Samoilys & Carlos 1992) and numerical abundances of each species, identified using various works (Randall, Allen & Steene 1990, Allen 1975, Choat & Randall 1986) were recorded. Our UVC study was specifically focused on conspicuous, diurnally-active, benthic, reef-associated fishes. Accuracy of length estimation was determined for each of us in a swimming pool, and where necessary estimated lengths were adjusted. We visually estimated habitat variables (% of the bottom covered by rubble, sand, 'bare' rock, hard corals, and soft corals; and rugosity on a five-point scale; cf. Polunin & Roberts 1993); and we also measured the depth (m) at the centre, of each sample site. UVC data were also gathered in 2-m radius samples, on small fishes <7 cm in length, which were of insufficient size to be recruited to the fishery, and on habitat variables, as for the large samples.

Data analysis

We employed the UVC fish data together with fish weight/length information (Wright & Richards 1985, Kulbicki, Mou Tham, Thollot & Wantiez 1993; Smith & Dalzell 1993) to derive indices of biomass per unit sample area. Where the necessary data for a particular species were lacking, we employed the length/weight information for a species in the same genus, or otherwise of similar morphology. The species biomass data were aggregated into up to six trophic groups (omnivores, herbivores, corallivores (including colonial invertebrates other than corals), small-invertebrate feeders (hereafter 'invertebrate-feeders'), piscivores/large-invertebrate feeders (hereafter 'piscivores'), and planktivores) based on existing gut-contents data (principally Hiatt & Strasburg 1960, Vivien 1973, Hobson 1974, Sano, Shimizu & Nose 1984, Blaber, Milton & Rawlinson 1990) or up to 12 taxonomic groups (wrasses (Labridae), surgeonfishes (Acanthuridae), damselfishes (Pomacentridae), groupers (Serranidae), snappers (Lutjanidae), tetraodontiforms (Balistidae, Monacanthidae, Ostraciidae, Tetraodontidae and Diodontidae), parrotfishes (Scaridae), angelfishes (Pomacanthidae), butterflyfishes (Chaetodontidae), emperors (Lethrinidae), goatfishes (Mullidae) and rabbitfishes (Siganidae)). Parametric analysis of variance (ANOVA) was used on the means ($n = 16$) from each area ($n = 5$) in every ground ($n = 7$), to test null hypotheses of no differences among grounds in biomass and habitat data, for both small (2 m radius) and large (7 m radius) samples. Where series of tests were conducted, the experimentwise error rate, α', was calculated using the Dunn–Šidák method (Sokal & Rohlf 1995), to derive a very conservative estimate of those species showing biomass differences among grounds. Data were $\log_e(x + 1)$ transformed and multiple comparisons were made using Tukey's test the data presented in the paper are untransformed. A forward-selection multiple regression procedure (SAS version 6.07) was employed quantify the relative strength of association of fishing and habitat variables with significant variations in biomass observed among grounds. All habitat variables were subjected to an agglomerative hierarchical clustering procedure using the average linkage method (Sokal & Michener 1958).

RESULTS

Grounds and fishing effort

We expected a substantial range in fishing pressure and indeed found a 60-fold difference in reef fishing effort index among the grounds (Table 5.1). We also expected that fishing pressure would decline from a maximal level in the well-populated Suva area to a minimum in the uninhabited Nauluvatu. Although the fishing pressure at Nauluvatu was very low (72 h km^{-2} year^{-1}), the greatest fishing effort index was not in one of the grounds at or near the capital (739–2028 h km^{-2} year^{-1}) but rather at Moala and Cokovata (3582–4310 h km^{-2} year^{-1}), in the middle of the series (Fig. 5.1; Table 5.1).

Variations among grounds in abundance of large fishes (>7 cm length) in large samples (7-m radius)

Two hundred and twenty-six species of reef fishes were recorded by the underwater visual census (UVC) technique in the 7-m radius samples, of which 98 were considered target species (Table 5.2). Twenty-two (22%) of the target species showed differences in biomass among grounds at the 95% probability level (ANOVA, $P < 0.05$), though the large number of simultaneous comparisons meant a low experimentwise error rate ($\alpha' = 0.002$). Only two species showed statistically significant differences in biomass among grounds, with only one target species (the coral trout, *Plectropomus laevis*) exhibiting significantly greater biomass in the least fished ground than in the

TABLE 5.1. List of the traditional fishing-grounds studied, with data on areas of the fishing-ground (to 60 m depth) and reef (to 40 m depth), fishing effort and yield, and mean (±95% CL) depth, coral cover and rugosity of the large areas sampled using UVC (underwater visual census).

Ground	Total area of ground (km²)	Reef area within ground (km²)	Reef fishing effort index (h km⁻² year⁻¹)	Reef fishery yield (t km⁻² year⁻¹)	Depth of UVC sites (m)	Coral in UVC sites (% area)	Bottom rugosity in UVC sites
Suva	—	40.2	1509	—	5.8±2.5	21.5±11.3	2.6±0.3
Yanuca	85.4	16.8	2028	4.6	6.8±3.7	18.7±7.8	3.0±0.5
Natusara	42.9	15.5	739	1.9	5.9±0.9	21.4±12.9	2.9±0.9
Ko Ono	289.7	103.9	1121	2.6	7.7±1.9	14.8±22.1	2.7±1.5
Moala	130.1	42.6	4310	10.2	7.0±1.3	14.8±13.5	3.4±0.9
Cokovata	124.6	29.6	3582	8.2	5.9±1.3	15.8±13.4	2.7±0.9
Nauluvatu	19.4	10.7	72	0.3	8.3±7.2	14.2±9.7	2.6±0.9

others. Forty-three (33%) of the non-target species also exhibited differences (ANOVA, $P < 0.05$) in biomass among the seven grounds, but only one of these was statistically significant ($\alpha' = 0.002$) and a relationship with fishing effort was not clear in this case.

The fish data were aggregated into 12 taxonomic groupings (Table 5.2). Several of these groups showed variations in biomass among grounds (ANOVA, $P < 0.05$) but among target fishes only the data for groupers (Serranidae) were statistically significant (Table 5.3; Fig. 5.2a), although emperors (Lethrinidae) and surgeonfishes (Acanthuridae) also tended to exhibit lower biomass in grounds with higher fishing effort indices (Fig. 5.2b,c). Three of the families into which the species data were aggregated (butterflyfishes, angelfishes and damselfishes) are of species which are not exploited in these fisheries (Table 5.2), and of these the butterflyfishes showed a significant ($\alpha' = 0.0039$) overall difference among grounds (Table 5.3), but this was not readily attributable to variations in fishing effort (Fig. 5.2d).

Species recorded in the large UVC samples were allocated among six trophic groups (Table 5.2), three of which showed statistically significant ($\alpha' = 0.0073$) differences in biomass among grounds (Table 5.3). There were significant differences between the least-fished site and the remaining grounds for piscivores (Fig. 5.2e), while invertebrate-feeders showed a significant overall difference, with the least-fished grounds tending to have greater biomass (Fig. 5.2f). Out of the three trophic groups which were

TABLE 5.2. List of species recorded in the large and small UVC samples, taxonomic (Aca, Acanthuridae; Cha, Chaetodontidae; Lab, Labridae; Let, Lethrinidae; Lut, Lutjanidae; Mul, Mullidae; Poa, Pomacanthidae; Pom, Pomacentridae; Sca, Scaridae; Ser, Serranidae; Tet, Tetraodontiformes; —, unclassified; and trophic groups (Cor, corallivore; Her, herbivore; Inv, invertebrate-feeder; Omn, omnivore; Pis, piscivore; Pla, planktivore; —, unclassified); and fishery status (+, target species). Asterisks indicate species for which differences among grounds were statistically significant ($\alpha' = 0.0002$).

Fish species	Taxonomic group	Trophic group	Presence (+) in small samples	Presence (+) in large samples	Target species
Abudefduf sexfasciatus	Pom	Omn		+	+
Acanthurus lineatus	Aca	Her		+	
A. nigricauda	Aca	Her		+	+
A. nigrofuscus	Aca	Her		+	+
A. olivaceus	Aca	Her		+	+
A. pyroferus	Aca	Her		+	
A. sp.	Aca	Her		+	+
A. triostegus	Aca	Her		+	+
A. xanthopterus	Aca	Her		+	+
Amblyglyphidodon curacao	Pom	Omn		+	
A. leucogaster	Pom	Omn	+	+	
Amphiprion chrysopterus	Pom	Omn		+	
A. melanopus	Pom	Omn	+	+	
A. percula	Pom	Omn	+	+	
Anampses caeruleopunctatus	Lab	Inv		+	
A. meleagrides	Lab	Inv		+	
A. neoguinaicus	Lab	Inv	+	+	
A. twistii	Lab	Inv	+	+	
Anyperodon leucogrammicus	Ser	Pis		+	
Apogon kallopterus	—	Pla		+	
Aprion virescens	Lut	Pis		+	+
Arothron nigropunctatus	Tet	Cor		+	+
A. sp.	Tet	Omn		+	+
Aulostomus chinensis	—	Pis		+	
Balistapus undulatus	Tet	Inv		+	+
Blenny sp.	—	Her	+	+	
Bodianus axillaris	Lab	Inv		+	+
B. loxozonus	Lab	Inv		+	+
B. mesothorax	Lab	Inv		+	+
B. perditio	Lab	Inv		+	+
Calotomus sp.	Sca	Her		+	
Canthigaster janthinopterus	Tet	Om	+		
C. valentini	Tet	Her			
Cantherhines pardalis	Tet	Omn		+	
Centropyge bicolor	Poa	Her	+	+	
C. bispinosus	Poa	Her	+	+	
C. flavissimus	Poa	Her	+	+	

(Continued.)

TABLE 5.2. Continued.

Fish species	Taxonomic group	Trophic group	Presence (+) in small samples	Presence (+) in large samples	Target species
Cephalopholis argus	Ser	Pis		+	+
C. leopardus	Ser	Pis		+	+
C. urodeta	Ser	Pis		+	+
Cetoscarus bicolor	Sca	Her		+	+
Chaetodon auriga	Cha	Omn		+	
C. baronessa	Cha	Cor	+	+	
C. bennetti	Cha	Cor		+	
C. citrinellus	Cha	Cor	+	+	
C. ephippium	Cha	Omn		+	
C. kleinii	Cha	Omn		+	
C. lunula	Cha	Cor		+	
C. melannotus	Cha	Cor		+	
C. mertensii	Cha	Omn		+	
C. pelewensis	Cha	Omn	+	+	
C. plebeius	Cha	Cor		+	
C. rafflesi	Cha	Cor		+	
C. reticulatus	Cha	Cor		+	
C. trifascialis	Cha	Cor	+	+	
C. trifasciatus	Cha	Cor	+	+	
C. ulietensis	Cha	Cor		+	
C. unimaculatus	Cha	Cor		+	
C. vagabundus	Cha	Cor		+	
Cheilinus bimaculatus	Lab	Inv		+	
C. celebecus	Lab	Inv		+	
C. chlorourus	Lab	Inv		+	
C. digrammus	Lab	Inv	+	+*	+
C. fasciatus	Lab	Inv		+	+
C. orientalis	Lab	Inv		+	
C. oxycephalus	Lab	Inv		+	
C. sp.	Lab	Inv		+	
C. trilobatus	Lab	Inv	+	+	+
C. unifasciatus	Lab	Inv		+	+
Cheilodipterus quinquelineatus	—	Inv	+	+	
Choerodon jordani	Lab	Inv		+	
Chromis amboinensis	Pom	Pla		+	
C. atridorsalis	Pom	Pla	+	+	
C. brachyurus	Pom	Pla		+	
C. iomelas	Pom	Pla	+	+	
C. margaritifer	Pom	Pla	+	+	
C. reticulatus	Pom	Pla	+		
C. ternatensis	Pom	Pla	+	+	
C. vanderbilti	Pom	Pla	+		
C. viridis	Pom	Pla		+	

(Continued on p. 108.)

TABLE 5.2. *Continued.*

Fish species	Taxonomic group	Trophic group	Presence (+) in small samples	Presence (+) in large samples	Target species
C. weberi	Pom	Pla	+	+	
C. xanthurus	Pom	Pla		+	
Chrysiptera talboti	Pom	Her	+		
C. taupou	Pom	Her	+		
Cirrhitichthys falco	—	Pis	+	+	
Cirrhilabrus punctatus	Lab	Pla	+	+	
Coris aygula	Lab	Inv	+	+	+
C. gaimard	Lab	Inv	+	+	
Ctenochaetus binotatus	Aca	Her	+	+	+
C. strigosus	Aca	Her		+	+
C. striatus	Aca	Her	+	+	+
Dascyllus aruanus	Pom	Omn	+*	+	
D. reticulatus	Pom	Omn	+	+	
D. trimaculatus	Pom	Omn	+	+	
Dischistodus melanotus	Pom	Her	+*		
Epibulus insidiator	Lab	Inv		+	+
Epinephelus maculatus	Ser	Pis		+	+
E. merra	Ser	Pis	+	+	+
E. polyphekadion	Ser	Pis		+	+
Forcipiger flavissimus	Cha	Inv	+		
F. longirostris	Cha	Inv		+	
Gnathodentex aurolineatus	Let	Inv		+	+
Gnatholepis sp.	—	Inv	+		
Gomphosus varius	Lab	Inv		+	+
Gymnocranius sp.	Let	Inv		+	+
Halichoeres biocellatus	Lab	Inv	+	+	
H. hortulanus	Lab	Inv		+	
H. marginatus	Lab	Inv	+	+	
H. melanurus	Lab	Inv	+	+	
H. prosopeion	Lab	Inv	+	+	
H. trimaculatus	Lab	Inv	+	+	
Hemigymnus fasciatus	Lab	Inv		+	+
H. melapterus	Lab	Inv		+	+
Heniochus acuminatus	Cha	Inv		+	
H. chrysostomus	Cha	Inv		+	
H. monoceros	Cha	Inv		+	
H. varius	Cha	Inv		+	
Hipposcarus longiceps	Sca	Her		+	+
Labrichthys unilineatus	Lab	Cor	+	+	
Labroides bicolor	Lab	Pla	+	+	
L. dimidiatus	Lab	Pla	+	+	
Labropsis. australis	Lab	Inv	+	+	
L. xanthonota	Lab	Cor		+	

(Continued.)

TABLE 5.2. Continued.

Fish species	Taxonomic group	Trophic group	Presence (+) in small samples	Presence (+) in large samples	Target species
Lethrinus obsoletus	Let	Inv		+	+
L. olivaceus	Let	Inv		+	+
Lutjanus bohar	Lut	Pis		+	+
L. fulviflamma	Lut	Pis		+	+
L. fulvus	Lut	Pis		+	+
L. gibbus	Lut	Pis		+	+
L. kasmira	Lut	Pis		+	+
L. russelli	Lut	Pis		+	+
L. semicinctus	Lut	Pis		+	+
Macolor macularis	Lut	Pis		+	+
M. niger	Lut	Pis		+	+
Macropharyngodon meleagris	Lab	Inv		+	
Meiacanthus atrodorsalis	—	—	+	+	
Melichthys vidua	Tet	Her		+	+
Monotaxis grandoculis	Let	Inv		+	+
Mulloides flavolineatus	Mul	Inv		+	+
M. vanicolensis	Mul	Inv		+	+
Myripristis murdjan	—	Inv		+	+
M. violaceus	—	Inv		+	+
Naso brevirostris	Aca	Her		+	+
N. lituratus	Aca	Her		+	+
N. unicornis	Aca	Her		+	+
Neoniphon sammara	—	Inv		+	
Novaculichthys taeniourus	Lab	Inv		+	
Ostracion cubicus	Tet	Omn		+	
O. meleagris	Tet	Omn		+	
Oxymonacanthus longirostris	Tet	Cor	+	+	
Paracirrhites arcatus	—	Pis	+	+	
P. forsteri	—	Pis	+	+	
Paraluteres prionurus	Tet	Inv	+	+	
Parapercis hexophtalma	—	Pis		+	
Parupeneus barberinoides	Mul	Inv		+	+
P. barberinus	Mul	Inv		+	+
P. bifasciatus	Mul	Inv		+	+
P. ciliatus	Mul	Inv		+	+
P. cyclostomus	Mul	Inv		+	+
P. multifasciatus	Mul	Inv		+	+
P. pleurostigma	Mul	Inv		+	+
Plagiotremus laudandus	—	Pla	+	+	
P. rhinorhynchos	—	Pla	+	+	
Plectroglyphidodon dickii	Pom	Cor	+	+	
P. lacrymatus	Pom	Her	+	+	
P. johnstonianus	Pom	Cor	+	+	

(Continued on p. 110.)

TABLE 5.2. Continued.

Fish species	Taxonomic group	Trophic group	Presence (+) in small samples	Presence (+) in large samples	Target species
Plectropomus laevis	Ser	Pis		+*	+
P. leopardus	Ser	Pis		+	+
Plectrorhinchus chaetodonoides	—	Inv		+	+
Pomacentrus bankanensis	Pom	Her	+	+	
P. chrysurus	Pom	Her	+		
P. coelestis	Pom	Her	+	+	
P. grammorhynchus	Pom	Her	+	+	
P. imitator	Pom	Pla	+	+	
P. sp. 1	Pom	Her	+	+	
P. sp. 2	Pom	Her		+	
P. sp. 3	Pom	Pla	+	+	
P. sp. 4	Pom	Her	+	+	
P. sp. 5	Pom	Pla	+	+	
P. sp. 6	Pom	Pla	+		
P. vaiuli	Pom	Her	+	+*	
Pomacanthus imperator	Poa	Cor		+	
Pseudocheilinus evanidus	Lab	Inv	+	+	
P. hexataenia	Lab	Inv	+	+	
P. octotaenia	Lab	Inv	+	+	
P. sp.	Lab	Inv	+		
Pseudochromis sp.	Lab	Inv	+		
Ptereleotris evanidus	—	Pla		+	
P. tricolor	—	Pla		+	
Pygoplites diacanthus	Poa	Cor		+	
Sargocentron caudimaculatum	—	Inv		+	+
S. diadema	—	Inv		+	+
S. spiniferum	—	Inv		+	+
Scarus altivelis	Sca	Her		+	+
S. bleekeri	Sca	Her		+	+
S. chameleon	Sca	Her		+	+
S. dimidiatus	Sca	Her		+	+
S. flavipectoralis	Sca	Her		+	+
S. forsteri	Sca	Her		+	+
S. frenatus	Sca	Her		+	+
S. ghobban	Sca	Her		+	+
S. globiceps	Sca	Her		+	+
S. microrhinos	Sca	Her		+	+
S. niger	Sca	Her	+	+	+
S. oviceps	Sca	Her		+	+
S. psittacus	Sca	Her		+	+
S. rivulatus	Sca	Her		+	+
S. rubroviolaceus	Sca	Her		+	+
S. schlegeli	Sca	Her		+	+
S. sordidus	Sca	Her	+	+	+

(Continued.)

TABLE 5.2. Continued.

Fish species	Taxonomic group	Trophic group	Presence (+) in small samples	Presence (+) in large samples	Target species
S. sp.	Sca	Her		+	+
S. spinus	Sca	Her		+	+
Scolopsis bilineatus	—	Inv		+	+
S. lineatus	—	Inv		+	
Siganus argenteus	Sig	Her		+	+
S. doliatus	Sig	Her		+	+
S. puellus	Sig	Her		+	+
S. punctatus	Sig	Her		+	+
S. spinus	Sig	Her		+	+
S. vulpinus	Sig	Her		+	+
Stegastes fasciolatus	Pom	Her	+	+	
S. nigricans	Pom	Her	+	+	
S. sp. 1	Pom	Her		+	
Stethojulis bandanensis	Lab	Inv		+	
S. strigiventer	Lab	Inv		+	
Sufflamen bursa	Lab	Inv		+	
S. chrysopterus	Lab	Inv		+	
Synodus sp.	—	Pis		+	
Thalassoma amblycephalus	Lab	Pla	+	+	
T. hardwicke	Lab	Inv	+	+	
T. jansenii	Lab	Inv		+	
T. lunare	Lab	Inv	+	+	
T. lutescens	Lab	Inv	+	+	
T. quinquevittatum	Lab	Inv		+	
T. trilobatum	Lab	Inv		+	
Variola louti	Ser	Pis		+	+
Wrasse sp.	Lab	Inv		+	
Zanclus cornutus	—	Cor		+	
Zebrasoma scopas	Aca	Her	+	+	+

almost entirely made up of non-target fishes (corallivores, omnivores and planktivores; Table 5.2), only omnivores showed a significant difference among grounds in mean biomass ($F = 3.92, P = 0.0058$), although this was not directly associated with variations in fishing effort (e.g. Fig. 5.2g). The biomass of fishes targeted by the fishery varied significantly among grounds (ANOVA, $F = 3.86, P = 0.0062$), and a declining trend with increasing effort was evident. No such significant difference or trend was found for non-target fishes as a whole in the large (7-m radius) UVC samples.

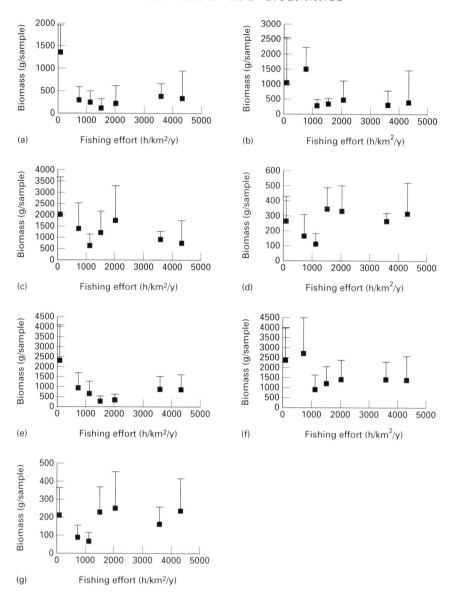

FIG. 5.2. Plots of mean (+95% CL) biomass index (g fresh weight per sample) against fishing effort in the ground (h km^{-2} year^{-1}) for groups of fishes in large (7-m radius) UVC samples: (a) Serranidae (one-way ANOVA among grounds, $F = 20.91$, $P = 0.0001$); (b) Lethrinidae (ANOVA $F = 3.37$, NS); (c) Acanthuridae (ANOVA $F = 2.76$, NS); (d) Chaetodontidae ANOVA $F = 5.00$, $P = 0.0014$); (e) piscivores (ANOVA $F = 6.60$, $P = 0.0022$); (f) invertebrate-feeders (ANOVA $F = 3.69$, $P = 0.0079$); (g) omnivores (ANOVA $F = 3.92$, $P = 0.0058$).

TABLE 5.3. *Taxonomic and trophic groups of fishes recorded in the large and small UVC samples, showing where (*) statistically significant differences (ANOVA, $\alpha' = 0.0039$ for taxonomic groups and $\alpha' = 0.0073$ for trophic groups) in biomass among grounds were found.*

Group	Large (7-m radius) samples F	Small (2-m radius) samples F
Taxonomic		
Surgeonfishes	2.76	0.86
Butterflyfishes	5.00*	1.38
Wrasses	1.32	0.78
Emperors	3.37	—
Snappers	1.51	—
Goatfishes	0.52	—
Angelfishes	2.74	0.90
Damselfishes	0.97	1.00
Parrotfishes	1.96	—
Groupers	20.91*	3.52
Rabbitfishes	0.52	—
Pufferfishes etc.	3.62	1.46
Trophic		
Corallivores	2.78	2.06
Herbivores	2.11	0.86
Invertebrate-feeders	3.69*	2.94
Omnivores	3.92*	0.84
Piscivores	6.60*	0.89
Planktivores	1.44	0.78

* $P \leq 0.05$.

Habitat variables in large and small samples

There were no significant differences (ANOVA, $P > 0.05$) among grounds in individual habitat variables for either small or large samples. Clustering of habitat data did not suggest overall differences among grounds which could be associated with variations in fishing effort (Fig. 5.3). Of the two groups of fishes which were not targeted by fishing but did show differences in biomass among grounds in the large UVC samples, there were significant regressions of biomass of omnivores on sand and rock cover (36% of variance explained), and of butterflyfishes on sand and soft coral cover (30% of variance explained). Fishing effort was not a significant variable in either multiple regression.

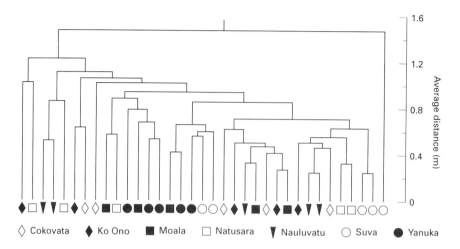

FIG. 5.3. Dendrogram of habitat data in large (7-m radius) samples by area ($n = 5$) showing groups formed by the hierarchical clustering procedure using the average linkage method. The symbols correspond to the grounds.

Variations among grounds in abundance of small fishes (<7 cm length) in small samples (2 m radius)

Of 85 species of fishes recorded by UVC in the small samples, eight of which were juveniles of fishery-targeted species (Table 5.2), 14 exhibited differences among grounds (ANOVA, $P < 0.05$), but only two of these (the damselfishes *Dascyllus aruanus* and *Dischistodus melanotus*) were statistically significant ($\alpha' = 0.0002$). In neither case, however, was there a trend in biomass that could be clearly associated with variation in fishing effort among grounds. The species data were aggregated into seven taxonomic groups (Table 5.2), where there were no significant differences ($\alpha' = 0.0063$) in biomass among grounds (Table 5.3). In no case were there obvious increases or decreases that could be explained simply by variations in fishing effort. The species data were aggregated into six trophic groups (Table 5.2); there were no statistically significant variations in biomass among grounds (Table 5.3) and no variations in biomass easily attributable to differences in fishing pressure.

DISCUSSION

Limitations of study and non-fishing effects

Fishes which are large, diurnally active, site-attached, abundant and conspicuous (e.g. parrotfishes, wrasses and surgeonfishes) can be well sampled by UVC, particularly if they are not extensively spear-fished, whilst others may not be well represented in UVC data, and this is so with some species and genera which are targeted by fisheries in Fiji and elsewhere in the South Pacific (Kulbicki 1988, Jennings & Polunin 1995c). Departures of UVC results from catch data are not fully understood, but in our case the aggregated data show trends that are similar to those exhibited by fishery data that were gathered independently of the UVC sampling (Jennings & Polunin 1995b).

We have no data specifically on levels of fishing activity prior to 1992 with which to support the supposition that the differences in fishing pressure among grounds have been maintained for several years. These variations should result primarily from differences in the density of resident populations. And these we understand have existed for a decade or more (Jennings & Polunin 1995b). In addition, absence of fish cold-storage facilities has restrained fisheries development of a commercial nature in the outlying grounds. So that most fish is still consumed within the grounds in those cases (Jennings & Polunin 1995b). This study was not able to measure recruitment and thus investigate the possibility that variations among grounds in the abundance of target species and groups were caused by, or contributed to, variations in this process. Another possibility for explaining those variations in biomass which we did find among grounds is that grounds differed in relevant habitat characteristics. We might not have discerned any such trends because we did not assess them in the UVC or because variations in them were undetectable amidst the substantial spatial variability. Such features might nevertheless have influenced the distribution, recruitment or survivorship of the fishes involved. We quantified a number of habitat variables that might have explained any differences in fish abundance ultimately found. Habitat characteristics such as rugosity and coral cover have been found elsewhere to be related to variations in fish abundance on reefs (e.g. Williams 1991). There was little indication of gross differences in such habitat features among grounds in Fiji, although variations in the abundance of butterflyfishes and omnivorous fishes as a group are as explicable in terms of such characteristics, as they are by the indirect effects of fishing.

There are many processes, variations in which might determine any spatial and temporal variability in recruitment, including mortality of eggs and larvae (Doherty & Williams 1988; Doherty 1991), and of juveniles after settlement from the plankton (Shulman & Ogden 1987, Jones 1991). While there has been reluctance to recognize fishing effects on the abundance even of fishery-target species (e.g. Russ 1991), empirical evidence for such effects has been mounting, particularly through studies of marine protected areas (e.g. Alcala & Russ 1990, Polunin & Roberts 1993, Watson & Ormond 1994, Bohnsack 1996). Such studies, however, tend to lack data on the actual status of the fisheries involved so that scientific understanding of fishing effects remains poorly developed (Jennings & Lock 1996).

Our study (Jennings & Polunin 1995a,b, 1996) is apparently the first to comprehensively quantify both fishing activity and reef fish abundances, to see whether simple predictions about effects of fishing at large spatial scales on these complex communities are borne out by the data (Russ 1991, Jennings & Lock 1996). The observed yields of reef fish per unit area of reef attain levels which are high in relation to those which reefs were once thought to support, but these Fiji data are consistent with what little is known on a long time-scale for other areas in the Pacific (Dalzell *et al.* 1996). The evidence is that the catches involved are sustainable, because yield does not evidently decrease at higher levels of fishing effort (Jennings & Polunin 1995b).

Piscivore depletion

In spite of difficulties inherent in assessing the abundance of mobile, infrequently-encountered and exploited fishes by underwater visual census (Jennings & Polunin 1995c), the data that we have gathered in Fiji support the prediction that some piscivorous species, and groups, have been depleted by fishing. The inferred depletion of target groups of such species, however, is not a steady process, with abundance inversely related in a linear fashion to fishing effort. Rather, the depletion occurs principally at low levels of fishing effort. Thus, for the lagoonal reef-edge habitat of the six most fished grounds, the grouper biomass was 19%, and biomass of all piscivores was 28% (Fig. 5.2a,e), of that at Nauluvatu, which was the least fished of the grounds studied (Table 5.1). This pattern is not peculiar to this UVC study; it is strongly reminiscent of the independently estimated catch per unit effort data for the very same grounds (Jennings & Polunin 1995b). Nor is this pattern particular to the small number of grounds that we happened to choose; we have observed a similar pattern in abundance of certain piscivorous target species of fishes, as well as of all piscivores, along a gradient of fishing pressure on reefs in the Seychelles, western Indian Ocean (Jennings, Grandcourt & Polunin 1995).

One explanation in relation to fishing is that slow-growing fish dominate the stock to begin with, that these suffer high mortality with the onset of fishing and are not replenished by recruitment of younger fish or movement of large fish of similar age into the grounds. Growth of fishing pressure might lead to a greater predominance of small fast-growing fish in the stock, and the yield from the stock would then rise with increasing effort. This faster growth would help explain how catch per unit effort, and biomass as indicated by UVC, might be maintained across a wide range of fishing efforts. Such a pattern would be consistent with what is thought generally to happen when a fishery stock is exploited (e.g. Sparre & Venema 1992) and such a change in age structure of a stock underlies some of the simplest fishery models used to assess reef fishery stocks (Appeldoorn 1996). There is evidence that average size and abundance of target fishes decline as a result of fishing on reefs as well (Alcala & Russ 1990) but, because growth rates and other aspects of the animals' biology remain poorly known, the consequences of these changes for reef fishery species have yet to be explored. Even some small species of reef fish are proving to be long-lived (e.g. Polunin & Brothers 1989, Choat & Axe 1996).

Another explanation for the pattern observed in piscivore abundance along the fishing gradient in Fiji and elsewhere is that fish are spatially unevenly distributed in unexploited grounds and that Fijian fishers are adept at locating such patches, for example by searching under water or by exploratory fishing. The patches are thus quickly depleted and increasing amounts of time will be required to locate unexploited patches. Such patchiness of distribution has been supposed in a modelling study of reef fishing and recruitment from marine reserves (Man, Law & Polunin 1995) but there is as yet little empirical evidence for it (Boehlert 1996). One form of such patchiness could be constituted by aggregations, such as exist for spawning purposes (e.g. Shapiro 1987). These might help to support high catch rates in the early stages of a reef fishery although they may be dissipated by fishing in later stages of development. The relatively invariable and low level of biomass, which it is inferred is maintained over a range of fishing pressures after an initial depletion, indicates that recruitment and growth are able to keep pace with exploitation in the fisheries.

The pattern of lesser abundance in grounds subject to more intense fishing was not evinced by all target groups. The snappers, for example, showed no evidence of decline in UVC biomass index in the grounds subject to higher fishing pressure. We did not quantify abundance of pelagic predatory fishes such as jacks because their occurrence was too sporadic, even though these are important at times in the fisheries involved (Jennings & Polunin 1995a) and could be important predators (Sudekum *et al.* 1991).

There are important implications of such an inferred pattern of predator decline in these Fijian fishing-grounds. Particular species, such as the coral trout *Plectropomus laevis*, may be good indicators of low-level fishing in Fiji, in that their abundance, as indicated by UVC, may decline rapidly as fishing develops from very low levels. Above these low levels of fishing pressure, it may be that such species change little further in abundance in relation to increase in fishing effort, although their productivity should increase to explain a growing catch. With greater replication along a fishing gradient than that which we have employed, however, differences other than those which we have found in the piscivores at low effort might also be found at higher fishing intensities.

The implications of such declines for the fisheries themselves are unclear, so long as the mechanisms underlying this 'sensitivity' to fishing effects are poorly known. The populations may become more productive at lower densities, if mean body growth rates are reduced in the unexploited fishery. It may also be that such depletion is only sustainable if recruitment continues to be supported from other reefs which are little fished. Spatial patterns of recruitment in reef fisheries are, however, very little understood (e.g. Man *et al.* 1995).

Effects on prey

The failure to find any strong indications of prey release at the species or group levels, in spite of substantial differences in mean biomass of piscivores, was unexpected. The analysis was conducted at two slightly different spatial scales on two different size ranges of fishes, albeit at the same sites in the same suite of grounds, with similar results. There are several possible explanations available as to why no increase in prey species was detected. To begin with, populations or groups may be resource-limited, as in competition hypotheses, with biomass kept at levels determined by the availability of particular resources (e.g. Beddington 1984). Alternatively, the populations or groups could be settlement-limited, their biomass levels determined by the supply of settlers from the plankton (e.g. Doherty & Williams 1988, Doherty 1991, Russ 1991).

A third general possibility is that the populations or groups are predation-limited, if certain major piscivore groups are not depleted by fishing. The latter may be the case for the snappers, which are expected to be major fish predators (Hiatt & Strasburg 1960, Hobson 1974, Sano *et al.* 1984,

Blaber, Milton & Rawlinson 1990) and for which we failed to detect any significant trends in abundance (Table 5.3). It could also be the case for pelagic species such as jacks (Carangidae), which are known to be important predators elsewhere (Sudekum et al. 1991) and do make a contribution to the reef-fish catch in the grounds that we studied (Jennings & Polunin 1995a).

At present we are unable to differentiate among such explanations. An important implication is that even where certain important piscivore groups are apparently depleted by fishing, effects on prey species of reef fishes should not necessarily be expected. Increase in abundance of prey fishes does occur as a result of predator depletion in lakes (e.g. Zaret & Paine 1973, Tonn, Paszkowski & Holopainen 1992). But on reefs the diversity of piscivores is greater, and the complexity of predator–prey interactions is considered greater than in fresh water (Jennings & Lock 1996). Amid growing evidence that fishing can have profound effects on reef ecosystems, particularly through the effects on invertebrate grazers (McClanahan 1992, Hughes 1994), we conclude that such effects will not invariably be found for reef fishes. If there is a widespread difference between fish and invertebrate prey, such as sea-urchins, in their susceptibility to fishing effects. Then, while finfish prey have apparently not responded to changes in large predators, sea-urchins could well have been doing so, on the Fijian reefs.

CONCLUSIONS

Quantification of fishing activity in seven traditional fishing-grounds in Fiji has made it possible for us to relate abundances of fishery-targeted and non-targeted species and groups of fishes to exploitation pressure. There are strong indications of fishing effects on a major family (groupers) and trophic group (large invertebrate-feeders and piscivores) which are targeted by fisheries, and on target fishes as a whole. There is no substantial evidence that abundance of non-target fishes or groups, many of which constitute the prey of the predators which are depleted by fishing, are indirectly affected by fishery exploitation. That prey abundance should not change in spite of predator depletion is predictable by the settlement-limitation hypothese, which suppose that abundance is largely determined in the plankton, prior to settlement. However, lack of prey release from predator depletion could also arise in other ways, for example from resource limitation, or indeed through predation, if predation is diffuse and fishing does not deplete all predators substantially. Top-down limitation of the abundance of fish prey by predators may exist on these reefs but indirect multi-species effects of fishing may occur only where predator–prey interactions are strong and probably specific, not diffuse, and fisheries substantially reduce the predators of particular prey. The latter appears to be the case with sea-urchins and their predators elsewhere in the tropics but in Fiji it does not apparently apply to reef fishes.

ACKNOWLEDGEMENTS

This study was funded by the British Government Overseas Development Administration through its Fish Management Science Programme. It would have been impossible without the co-operation of the fishing-rights holders of the *qoliqoli* involved and the support of the University of the South Pacific Marine Studies Programme and its Director, Professor G. Robin South.

REFERENCES

Alcala, A.C. & Russ, G.R. (1990). A direct test of the effects of protective management on the abundance and yield of tropical marine resources. *Journal du Conseil International pour l'Exploration de la Mer*, **46**, 40–47.

Allen, G.R. (1975). *Damselfishes of the South Seas*. TFH Publications, Neptune City, NJ.

Appeldoorn, R.S. (1996). Model and method in reef fishery assessment. In *Reef Fisheries* (Ed. by N.V.C. Polunin & C.M. Roberts), pp. 219–248. Chapman & Hall, London.

Arias Gonzalez, E. (1993). *Fonctionnement trophique d'un ecosystème récifal: secteur de Tiahura, Ile de Moorea, Polynésie Française*. PhD thesis, Ecole Pratique d'Hautes Etudes, University of Perpignan.

Beddington, J.R. (1984). The response of multispecies systems to perturbations. In *Exploitation of Marine Communities* (Ed. by R.M. May), pp. 209–225. Springer, Berlin.

Blaber, S.J.M., Milton, D.A. & Rawlinson, N.J.F. (1990). Diets of lagoon fishes of the Solomon Islands: predators of tuna baitfish and trophic effects of baitfishing on the subsistence fishery. *Fisheries Research*, **8**, 263–286.

Boehlert, G.W. (1996). Larval dispersal and survival in tropical reef fishes. In *Reef Fisheries* (Ed. by N.V.C. Polunin & C.M. Roberts), pp. 61–84. Chapman & Hall, London.

Bohnsack, J. (1982). Effects of piscivorous predator removal on coral reef fish community structure. In *Gutshop '81: Fish Food Habits Studies* (Ed. by G.M. Cailliet & C.A. Simenstad), pp. 258–267. Washington Sea Grant Publications, Seattle, WA.

Bohnsack, J. (1996). Maintenance and recovery of reef fishery productivity. In *Reef Fisheries* (Ed. by N.V.C. Polunin & C.M. Roberts), pp. 283–313. Chapman & Hall, London.

Caley, M.J. (1993). Predation, recruitment and the dynamics of communities of coral-reef fishes. *Marine Biology*, **117**, 33–44.

Choat, J.H. & Axe, L.M. (1996). Growth and longevity in acanthurid fishes according to otolith increments. *Marine Ecology—Progress Series*, **134**, 15–26.

Choat, J.H. & Bellwood, D.R. (1985). Interactions amongst herbivorous fishes on a coral reef: influence of spatial variation. *Marine Biology*, **89**, 221–234.

Choat, J.H. & Randall, R.E. (1986). A review of the parrotfishes (Family Scaridae) of the Great Barrier Reef of Australia with description of a new species. *Records of the Australian Museum*, **38**, 175–228.

Cushing, D.H. (1988). *The Provident Sea*. Cambridge University Press, Cambridge.

Dalzell, P., Adams, T.J.H. & Polunin, N.V.C. (1996). Coastal fisheries in the South Pacific islands. *Oceanography and Marine Biology Annual Review*, **34**, 395–531.
Doherty, P.J. (1991). Spatial and temporal patterns in recruitment. In *The Ecology of Fishes on Coral Reefs* (Ed. by P.F. Sale), pp. 261–293. Academic Press, San Diego.
Doherty, P.J. & Williams, D.M. (1988). The replenishment of coral reef fish populations. *Oceanography and Marine Biology an Annual Review*, **26**, 487–551.
Ebersole, J.P. (1977). The adaptive significance of interspecific territoriality in the reef fish *Eupomacentrus leucostictus*. *Ecology*, **58**, 914–920.
Gladfelter, W.B. & Gladfelter, E.H. (1978). Fish community structure as a function of habitat structure on West Indian patch reefs. *Revista Biologica Tropical*, **26** (Suppl. 1), 65–84.
Gladfelter, W.B. & Johnson, W.S. (1983). Feeding niche separation in a guild of tropical reef fishes (Holocentridae). *Ecology*, **64**, 552–563.
Gosline, W.A. (1965). Vertical zonation of inshore fishes in the upper water layers of the Hawaiian Islands. *Ecology*, **46**, 823–831.
Grigg, R.W., Polovina, J.J. & Atkinson, M.J. (1984). Model of a coral reef ecosystem III. Resource limitation, community regulation, fisheries yield and resource management. *Coral Reefs*, **3**, 23–27.
Harmelin-Vivien, M.L. (1995). The effects of storms and cyclones on coral reefs: a review. *Journal of Coastal Research*, Special Issue, **12**, 211–231.
Hay, M.E. (1984). Patterns of fish and urchin grazing on Caribbean coral reefs: are previous results typical? *Ecology*, **65**, 446–454.
Hiatt, R.W. & Strasburg, D.W. (1960). Ecological relationships of the fish fauna on coral reefs of the Marshall Islands. *Ecological Monographs*, **30**, 65–127.
Hixon, M.A. (1991). Predation as a process structuring coral reef fish communities. In *The Ecology of Fishes on Coral Reefs* (Ed. by P.F. Sale), pp. 475–508. Academic Press, San Diego.
Hixon, M.A. & Beets, J.P. (1993). Predation, prey refuges, and the structure of coral-reef fish assemblages. *Ecological Monographs*, **63**, 77–101.
Hobson, E.S. (1974). Feeding relationships of teleostean fishes on coral reefs in Kona, Hawaii. *Fishery Bulletin US*, **72**, 915–1031.
Hourigan, T.F. (1986). An experimental removal of a territorial pomacentrid: effects on the occurrence and behaviour of competitors. *Environmental Biology of Fishes*, **15**, 161–169.
Hughes, T.P. (1994). Catastrophes, phase shifts, and large-scale degradation of a Caribbean coral reef. *Science*, **265**, 1547–1551.
Jennings, S. & Lock, J.M. (1996). Population and ecosystem effects of reef fishing. In *Reef Fisheries* (Ed. by N.V.C. Polunin & C.M. Roberts), pp. 193–218. Chapman & Hall, London.
Jennings, S. & Polunin, N.V.C. (1995a). Comparative size and composition of yield from six Fijian reef fisheries. *Journal of Fish Biology*, **46**, 28–46.
Jennings, S. & Polunin, N.V.C. (1995b). Relationships between catch and effort in Fijian multi-species reef fisheries subject to different levels of exploitation. *Fisheries Management and Ecology*, **2**, 89–101.
Jennings, S. & Polunin, N.V.C. (1995c). Biased underwater visual census biomass estimates for target species in tropical reef fisheries. *Journal of Fish Biology*, **47**, 733–736.
Jennings, S. & Polunin, N.V.C. (1996). Effects of fishing effort and catch rate upon the structure and biomass of Fijian reef fish communities. *Journal of Applied Ecology*, **33**, 400–412.
Jennings, S., Grandcourt, E.M. & Polunin, N.V.C. (1995). The effects of fishing on the diversity, biomass and trophic structure of Seychelles' reef fish communities. *Coral Reefs*, **14**, 225–235.
Jones, G.P. (1991). Postrecruitment processes in the ecology of coral reef fish populations: a multifactorial perspective. In *The Ecology of Fishes on Coral Reefs* (Ed. by P.F. Sale), pp. 294–328. Academic Press, San Diego.

Knowlton, N. (1992). Thresholds and multiple stable states in coral reef community dynamics. *American Zoologist*, **32**, 674–682.

Kulbicki, M. (1988). Correlation between catch data from bottom longlines and fish censuses in the SW lagoon of New Caledonia. *Proceedings of the 6th International Coral Reef Symposium*, **2**, 305–312.

Kulbicki, M., Mou Tham, G., Thollot, P. & Wantiez, L. (1993). Length–weight relationships of fish from the lagoon of New Caledonia. *Naga*, **2+3**, 26–30.

Leis, J.M. (1991). The pelagic stage of reef fishes: the larval biology of coral reef fishes. In *The Ecology of Fishes on Coral Reefs* (Ed. by P.F. Sale), pp. 183–230. Academic Press, San Diego.

Man, A., Law, R. & Polunin, N.V.C. (1995). Role of marine reserves in recruitment to reef fisheries: a metapopulation model. *Biological Conservation*, **71**, 197–204.

McClanahan, T.R. (1992). Resource utilization, competition and predation: a model and example from coral reef grazers. *Ecological Modelling*, **61**, 195–215.

McClanahan, T.R. (1994). Kenyan coral reef lagoon fish—effects of fishing, substrate complexity, and sea-urchins. *Coral Reefs*, **13**, 231–241.

Meekan, M.G., Milicich, M.J. & Doherty, P.J. (1993). Larval production drives temporal patterns of larval supply and recruitment of a coral reef damselfish. *Marine Ecology—Progress Series*, **93**, 217–225.

Menge, B.A. (1995). Indirect effects in marine rocky intertidal interaction webs: patterns and importance. *Ecological Monographs*, **65**, 21–74.

Munro, J.L. & Williams, D.M. (1985). Assessment and management of coral reef fisheries: biological, environmental and socio-economic aspects. *Proceedings of the 5th International Coral Reef Congress*, **4**, 545–581.

Opitz, S. (1993). A quantitative model of the trophic interactions in a Caribbean coral reef ecosystem. In *Trophic Models of Aquatic Ecosystems* (Ed. by V. Christensen & D. Pauly), pp. 259–267. *ICLARM Conference Proceedings*, **26**. International Center for Living Aquatic Resources Management, Manila, Philippines.

Paine, R.T. (1980). Food webs: linkage, interaction strength and community infrastructure. *Journal of Animal Ecology*, **49**, 667–685.

Polovina, J.J. (1984). Model of a coral reef ecosystem. I. The ECOPATH model and its application to French Frigate Shoals. *Coral Reefs*, **3**, 1–11.

Polunin, N.V.C. & Brothers, E.B. (1989). Low efficiency of dietary carbon and nitrogen conversion to growth in an herbivorous coral-reef fish in the wild. *Journal of Fish Biology*, **35**, 869–879.

Polunin, N.V.C. & Klumpp, D.W. (1992a). Algal food supply and grazer demand in a very productive coral-reef zone. *Journal of Experimental Marine Biology and Ecology*, **164**, 1–15.

Polunin, N.V.C. & Klumpp, D.W. (1992b). A trophodynamic model of fish production on a windward reef tract. In *Plant–Animal Interactions in the Marine Benthos* (Ed. by D.M. John, S.J. Hawkins & J.H. Price), pp. 213–233. *Systematics Association Special Volume*, **46**. Clarendon Press, Oxford.

Polunin, N.V.C. & Roberts, C.M. (1993). Greater biomass and value of target coral-reef fishes in two small Caribbean marine reserves. *Marine Ecology—Progress Series*, **100**, 167–176.

Polunin, N.V.C., Roberts, C.M. & Pauly, D. (1996). Developments in tropical reef fisheries science and management. In *Reef Fisheries* (Ed. by N.V.C. Polunin & C.M. Roberts), pp. 361–377. Chapman & Hall, London.

Randall, J.E., Allen, G.R. & Steene, R.C. (1990). *Fishes of the Great Barrier Reef and Coral Sea.* Crawford, Bathurst, New South Wales.

Roberts, C.M. (1996). Settlement and beyond: population regulation and community structure. In *Reef Fisheries* (Ed. by N.V.C. Polunin & C.M. Roberts), pp. 85–112. Chapman & Hall, London.

Robertson, D.R. & Sheldon, J.M. (1979). Competitive interactions and the availability of sleeping sites for a diurnal coral reef fish. *Journal of Experimental Marine Biology and Ecology*, **40**, 285–298.

Ruddle, K. (1996). Geography and human ecology of reef fisheries. In *Reef Fisheries* (Ed. by N.V.C. Polunin & C.M. Roberts), pp. 137–160. Chapman & Hall, London.

Russ, G.R. (1991). Coral reef fisheries: effects and yields. In *The Ecology of Fishes on Coral Reefs* (Ed. by P.F. Sale), pp. 601–635. Academic Press, San Diego.

Sadovy, Y.J. (1996). Reproduction of reef fishery species. In *Reef Fisheries* (Ed. by N.V.C. Polunin & C.M. Roberts), pp. 15–60. Chapman & Hall, London.

Sale, P.F. (1980). The ecology of fishes on coral reefs. *Oceanography and Marine Biology an Annual Review*, **18**, 367–421.

Sale, P.F. (1991a). Introduction. In *The Ecology of Fishes on Coral Reefs* (Ed. by P.F. Sale), pp. 3–15. Academic Press, San Diego.

Sale, P.F. (1991b). Reef fish communities: open non-equilibrial systems. In *The Ecology of Fishes on Coral Reefs* (Ed. by P.F. Sale), pp. 564–598. Academic Press, San Diego.

Sale, P.F. & Dybdahl, R. (1975). Determinants of community structure for coral reef fishes in an experimental habitat. *Ecology*, **56**, 1343–1355.

Samoilys, M. (1992). Review of the underwater visual census method developed by the QDPI/ACIAR project: Visual assessment of reef fish stocks. Conference and Workshop Series QC92006. Department of Primary Industries, Brisbane.

Samoilys, M. & Carlos, G. (1992). Development of an underwater visual census method for assessing shallow water reef fish stocks in the South West Pacific. Unpublished report, Australian Centre for International Agricultural Research Project PN 8545, iii + 100pp. DPI Northern Fisheries Centre, Cairns, Q4870.

Sano, M., Shimizu, M. & Nose, Y. (1984). Food habits of teleostean reef fishes in Okinawa Island, southern Japan. *University of Tokyo, University Museum Bulletin*, **25**, 128pp.

Shapiro, D.Y. (1987). Reproduction in groupers. In *Tropical Snappers and Groupers: Biology and Fisheries Management* (Ed. by J.J. Polovina & S. Ralston), pp. 295–328. Westview, Boulder, CO.

Shpigel, M. & Fishelson, L. (1991). Experimental removal of piscivorous groupers of the genus *Cephalopholis* (Serranidae) from coral habitats in the Gulf of Aqaba (Red Sea). *Environmental Biology of Fishes*, **31**, 131–138.

Shulman, M.J. & Ogden, J.C. (1987). What controls tropical reef fish populations: recruitment or benthic mortality? *Marine Ecology—Progress Series*, **39**, 233–242.

Smith, A. & Dalzell, P. (1993). Fisheries resources and management investigations in Woleai Atoll, Yap State, Federated States of Micronesia. *South Pacific Commission Inshore Fisheries Research Project, Technical Document*, **12**, xiii + 64pp.

Smith, T.D. (1994). *Scaling Fisheries.* Cambridge University Press, Cambridge.

Sokal, R.R. & Michener, C.D. (1958). A statistical method for evaluating systematic relationships. *University of Kansas Science Bulletin*, **38**, 1409–1438.

Sokal, R.R. & Rohlf, F.J. (1995). *Biometry*, 3rd edn. Freeman, New York.

Sparre, P. & Venema, S.C. (1992). Introduction to tropical fish stock assessment. *FAO Fisheries Technical Paper*, **306/1**, xiii + 376pp.

Sudekum, A.E., Parrish, J.D., Radtke, R.L. & Ralston, S. (1991). Life history and ecology of large jacks in undisturbed, shallow, oceanic communities. *Fishery Bulletin US*, **89**, 493–513.

Tonn, W.M., Paszkowski, C.A. & Holopainen, I.J. (1992). Piscivory and recruitment: mechanisms structuring prey populations in small lakes. *Ecology*, **73**, 951–958.

Vivien, M.L. (1973). Contribution à la connaissance de l'éthologie alimentaire de l'ichtyofaune du platier interne des récifs coralliens de Tuléar (Madagascar). *Téthys*, **5** (Suppl.), 221–308.

Watson, M. & Ormond, R.F.G. (1994). Effects of an artisanal fishery on the fish and urchin populations of a Kenyan coral reef. *Marine Ecology—Progress Series*, **109**, 115–129.

Williams, D.M. (1986). Temporal variation in the structure of reef slope fish communities (central Great Barrier Reef): short-term effects of *Acanthaster planci* infestation. *Marine Ecology—Progress Series*, **28**, 157–164.

Williams, D.M. (1991). Patterns and processes in the distribution of coral reef fishes. In *The Ecology of Fishes on Coral Reefs* (Ed. by P.F. Sale), pp. 437–474. Academic Press, San Diego.

Williams, D.M., Newman, S.J., Cappo, M. & Doherty, P.J. (1995). Recent advances in the ageing of coral reef fishes. *South Pacific Commission Integrated Coastal Fisheries Management Project Technical Document*, **12**, 667–671.

Wright, A. & Richards, A. (1985). A multispecies fishery associated with local reefs in the Tigak Islands, Papua New Guinea. *Asian Marine Biology*, **2**, 69–84.

Zaret, T.M. & Paine, R.T. (1973). Species introduction in a tropical lake. *Science*, **182**, 449–455.

6. ARCHITECTURE AND DEVELOPMENT OF RAINFOREST TREES: RESPONSES TO LIGHT VARIATION

F. BONGERS and F. J. STERCK
Department of Forestry, Agricultural University, P.O. Box 342, 6700 AH Wageningen, The Netherlands

SUMMARY

1 Tropical rainforest trees are confronted with a variety of light levels during their life from seedling to mature tree. Their architectural changes during tree life are partly a response to light availability. Tree architecture determines to a large extent the ability to intercept light for carbon acquisition and the mechanical stability of a tree.

2 We studied the architectural changes during tree development as affected by tree height and light availability for two canopy tree species of a forest in French Guiana. Light environments, architectural tree models, trunk and crown allometry, mechanical design, and crown development are reported. The results are discussed as a section within this review.

3 Although light levels are highly variable among trees of the same height, in general they increase with height. Trees of shade-intolerant species receive higher light levels than trees of shade-tolerant species but they are sometimes found at low light levels in the forest understorey as well.

4 Height is linearly related with stem diameter when trees are young, but in later development stages height growth ceases while diameter growth still continues. Slenderness (height/stem diameter ratio) decreases with tree height. Juvenile trees of understorey species tend to be more slender than overstorey species of the same height, and this may be related to their wood properties. Later, trees of understorey species become less slender than trees of overstorey species because their height growth ceases at a smaller diameter while diameter growth continues. Height–diameter relationships have hardly been related to light availability.

5 Trees produce deep and narrow crowns (relative to their height) when they grow. For one of the two selected species, trees also produced deeper and narrower crowns when they received more light. No other studies on tropical trees that related responses in crown allometry to light could be found.

6 In their way to the canopy, trees take large risks with respect to mechanical stability. Our two study species showed the smallest margins in mechanical safety at a height of 20 m. For one of the species, the safety margins increased with light availability. Trees of shade-intolerant species tended to have lower safety margins than trees of shade-tolerant species.

7   Crown development (in terms of leaf area) refers to the demography, spatial arrangement, and structure of plant components at different levels within the crown hierarchy, from meristem to whole crown. At each of these levels crown development is affected by tree height and light level. Using path analysis, the influence of tree height and light availability on different components, and the consequences of these influences for one whole crown trait, total leaf area was analysed. Leaf area increased with both tree height and light availability, mainly because trees had produced more apical meristems.

8   Although light is a major factor influencing the architecture of a tree, studies taking this factor into account are scarce and mostly confined to seedlings. Partly, this is due to the difficulty in estimating light levels of trees larger than small saplings; partly, this is because the taller trees have a longer life history and their architecture is the result of a long process.

INTRODUCTION

Trees of tropical rainforest encounter many different light environments during their life from seedling to mature canopy tree, and may respond by various morphological and physiological traits to these light environments. Depending on the scale of light variation, such responses occur at different organizational levels; that is, the cell, leaf part, leaf, plant canopy, whole plant, plant population or community (Gross 1986, Chazdon 1988). In this chapter, the focus is on architectural responses from the leaf to the whole plant level.

Architectural responses have at least two important ecological components. First, they determine the display of leaves, the interception of light, and thus the ability to acquire and use carbon for future growth and survival. Second, they affect the balance between mechanical stress and (mechanical) resistance against this stress. Mechanical stress increases with biomass and wind forces, while mechanical resistance increases with growth in thickness of stem and branches. In the long run, the architectural responses affect the performance of trees in terms of growth, survival and reproduction.

When trees grow from seedling to canopy tree they undergo many changes in tree architecture. For a large part, the architectural changes may be typical to tree development irrespective of particular environmental conditions. These changes are further referred to as ontogenetic changes in tree architecture. Apart from this, there may be architectural responses to particular environmental conditions; that is, the availability of light. It may be hard to differentiate between these ontogenetic and environmentally induced effects (Coleman, McConnaughay & Ackerly 1994) because light environments change with ontogeny.

The present study concentrates on the architectural changes during ontogeny and on the architectural responses to light. Architecture is used as an overall term, encompassing a wide variety of morphological traits (Tomlinson 1987). We start with a case study for two canopy tree species of the tropical rainforest in French Guiana. Data are presented on the light environment, architectural tree model, trunk allometry, crown allometry, mechanical design, crown development and leaf display for trees of the selected species. Then, we continue with a review on the different topics as presented in the case study. For each of these topics, the focus is on ontogenetic changes and responses to light within a given species, as well as on the differences between species in these changes and responses. Species comparisons are usually made on the basis of known differences in shade tolerance (shade-tolerants vs. shade-intolerants) and, in some cases, on the basis of adult stature (overstorey vs. understorey) and life-span. General tendencies are drawn and gaps in knowledge are indicated.

TREE ARCHITECTURE AND LIGHT FOR TWO SPECIES AT LES NOURAGUES, FRENCH GUIANA

Light environment and tree architecture were studied for two canopy species of a tropical rainforest in French Guiana. The aim of this study was to describe and analyse the influence of the light environment on tree architecture for trees in different stages of life: from sapling to canopy tree.

Field work was carried out in a lowland tropical rainforest at the biological field station 'Les Nouragues' (4°05'N; 52°40'W), French Guiana. Two canopy tree species were selected, *Dicorynia guianensis* Amshoff and *Vouacapoua americana* Aubl. (both Caesalpiniaceae). With respect to the light environment, these species were compared with *Cecropia obtusa* Tréc. and *Pourouma bicolor* Mart. spp. *digitata* Tréc. (both Cecropiaceae). These species are from here on referred to only by their generic names. *Dicorynia*, *Pourouma* and *Vouacapoua* are considered shade-tolerant species, while *Cecropia* is considered a shade-intolerant species.

Dicorynia and *Vouacapoua* grow according to Troll's model (for *Dicorynia*, Oldeman 1974 for *Vouacapoua*, Hallé & Oldeman 1970, Hallé, Oldeman & Tomlinson 1978). According to this model, stem and branches develop as sympodial plagiotropic axes, and stem and branches are hard to distinguish within the crown. Drénou (1988), however, showed that more models were needed to describe the growth patterns of *Dicorynia*. In this case study, we did not explore the architectural models of the species further.

Populations of all four species were inventoried in October 1992, as part of a long-term study on the vegetative growth of these species. A 12-ha plot was searched for trees with stem diameters ≥10 cm at breast height (DBH). A central 1.5-ha plot was searched for trees with stem diameters <10 cm, but with a height above 50 cm. To study the variation in light levels on individual trees of these species the crown position index (CPI, Table 6.1) was estimated by two observers independent of each other, and mean values were calculated. Clearly, light levels (as inferred from the CPI values) increased with height for all species, although this was less conspicuous for *Cecropia* (Fig. 6.1). Within species, trees occurred at a wide variety of CPI values. Even trees of the shade-intolerant *Cecropia* were found at the lower CPI values indicative of the lowest light levels in the understorey.

For a subsample of *Dicorynia* and *Vouacapoua*, trees were measured annually for their light environment and architecture between 1992 and 1994. Trees shorter than 5 m were measured from the ground, while the taller trees (5–37 m) were measured by climbing neighbouring trees using spikes, alpine ropes or canopy walkways. The light environment of each individual was characterized with a hemispherical photograph. Photographs were made over individuals during a census in November 1993. A Canon Ti-70 body and Canon lens 7.5 mm/5.6 were fixed to a level, which was mounted on

TABLE 6.1. Crown position index (CPI) values and definitions. (Adapted from Clark & Clark 1992.)

CPI	Definition
5	Crown completely exposed (to vertical light and to lateral light within the 90°-inverted cone encompassing the crown)
4	Full overhead light (90° of the vertical projection of the crown exposed to vertical light; lateral light blocked within some or all of the 90°-inverted cone encompassing the crown)
3	Some overhead light (10–90% of the vertical projection of the crown exposed to vertical light)
2.5	High lateral light
2.0	Medium lateral light
1.5	Low lateral light
1	No direct light (crown not lit directly either vertically or laterally)

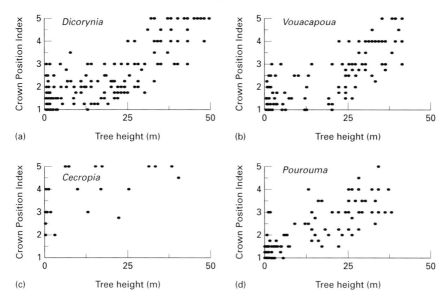

FIG. 6.1. Crown Position Index (CPI) values in populations of the four tropical rainforest species (a) *Dicorynia guianensis*, (b) *Vouacapoua americana*, (c) *Cecropia obtusa* and (d) *Pourouma bicolor* at Les Nouragues, French Guiana. ((a) and (b) from Sterck 1997.)

telescopic poles (up to 6 m long). Photographs were scanned and digitized with DeskScann II. Canopy openness was calculated as a percentage of the open sky, using PPFDCALC-2 (Ter Steege 1993). In this case study, canopy openness is used as an estimate of light availability.

Tree allometry was measured in terms of tree height, crown depth, crown width, DBH and ratios of these variables. Tree growth was assessed as the production of new leaves and growth units by counting leaves, leaf-scars and growth units for the same period. A growth unit (GU) was defined as the smallest repetitive woody unit produced by the meristems (Table 6.2, see also Sterck 1997). For *Dicorynia*, the growth unit is the metamer (White 1979, 1984, Bell 1991, Room, Maillette & Hanan 1994) which consists of an internode, a node, a compound leaf and an axillary bud. For *Vouacapoua*, the growth unit is the 'unit of extension' (Bell 1991) which consists of a sequence of metamers, each metamer (internode, node, axillary bud) with either a compound leaf or a scale leaf. Using this data set, light availability was related to trunk allometry, crown allometry, mechanical design, crown development and leaf display.

The relationship between DBH and height can be described well by the classical height–diameter equation (Kira 1978, Fig. 6.2a). Both species show a close-to-linear relationship between diameter and height when in the

TABLE 6.2. *Definitions for plant components in tree crowns. Plant components are in order of increasing size. (Partly adapted from Room et al. 1994.)*

Apical/axillary meristem	A cell or a group of cells, specialized for mitosis, initiating, or at the apex of a shoot. An axillary meristem becomes an apical meristem as soon as it starts to produce a metamer or shoot unit
Bud	An unextended, partly developed, shoot unit having at its summit the apical meristem which produced it; an unexpanded metamer or group of metamers (Bell 1991)
Leaf	Photosynthetic unit developed from a node
Metamer	An internode, the axillary bud(s) at its proximal end, and the leaf or leaves at its distal end, but not any shoots resulting from growth of axillary buds
Unit of extension	The vegetative axis, displayed by one period of meristem activity; the axis which has lengthened during the same flush. It consists of a sequence of metamers (Barthélémy 1991)
Sympodial unit	The product of one meristem; a set of metamers and/or units of extension originating from one meristem
Architectural unit	The complete set of axis types and their relative arrangements found in a species; the specific expression of the architectural model for a given species—cannot be seen until an individual is old enough to have expressed its architectural model in full (Barthélémy 1991)

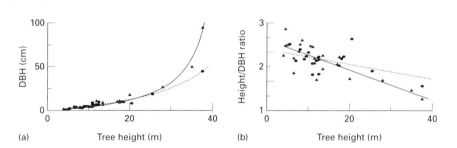

FIG. 6.2. Tree size and allometry of three rainforest species at Les Nouragues, French Guiana: (a) diameter (DBH) vs. height (H), (b) height/diameter ratio vs. height. *Dicorynia guianensis* is represented by the dotted, and *Vouacapoua americana* by the solid, line. For (a) a model derived from Kira (1978) was used: DBH = $(K \times H)/(H_{MAX} - H)$; for *Dicorynia* $H_{max} = 56.8$ and $K = 23.0$, $R^2 = 0.97$, $P < 0.001$; for *Vouacapoua* $H_{max} = 43.3$ and $K = 14.2$, $r^2 = 0.98$, $P < 0.001$. For (b) the model H/D ratio = $a + b \times H$ was used for *Dicorynia* $a = 2.10$, $b = -0.026$, $r^2 = 0.27$, $P < 0.05$; for *Vouacapoua* $a = 2.43$, $b = -0.052$, $r^2 = 0.66$, $P < 0.001$.

understorey, but the diameter sharply increases in trees taller than 25 m. Then height growth ceases while diameter growth still continues, leading to relatively thick trunks. The height/diameter ratio (also slenderness) decreases linearly with height for both species (Fig. 6.2b).

Crown allometry is expressed by the relative crown width (crown width/tree height ratio), the relative crown depth (crown depth/tree height ratio), and the crown shape (crown depth/crown width ratio). The scatter diagrams (Fig. 6.3) show that relative crown width decreased with height and

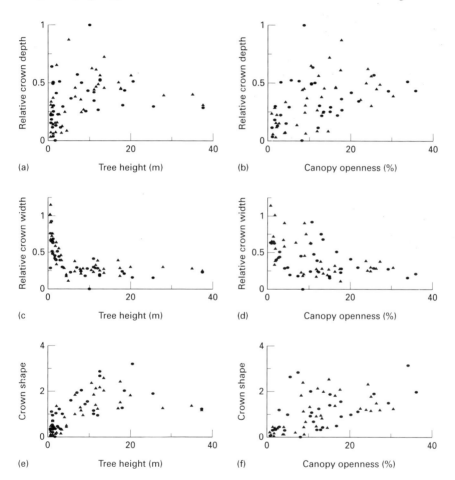

FIG. 6.3. Crown allometry of the tropical rainforest trees *Dicorynia guianensis* (circles) and *Vouacapoua americana* (triangles) in relation to tree height and canopy openness: (a,b) crown depth/tree height ratio, or relative crown depth; (c,d) crown width/tree height ratio, or relative crown width; and (e,f) crown depth/crown width ratio, or crown shape. Multiple regression parameters are given in Table 6.3.

with canopy openness and that relative crown depth and crown shape increased. The variability is very large, however. As large individuals that are in the canopy cannot be light-limited we analysed only the smaller ones. For the trees less than 22 m tall, which were still heading for the canopy, a multiple regression was performed for various allometric variables on tree height and canopy openness (Table 6.3). Nearly all of the regressions were significant and in most cases height was more important than canopy openness. For both species relative crown depth and crown shape increased with height while relative crown width decreased. For *Vouacapoua* only, these ratios changed in the same directions with light availability. Thus, trees of *Vouacapoua* produced narrower crowns at higher light levels and trees of *Dicorynia* did not.

Mechanical design refers to the stability of a tree and was expressed by the safety factor (critical buckling height/actual height, e.g. King 1981), using the elastic-stability model (Greenhill equation; see also McMahon 1973, Niklas 1994a,b) to calculate the critical buckling height, i.e. the height at which a tree is expected to buckle or snap under its own mass. Apart from

TABLE 6.3. Multiple regression parameters for the influence of height, canopy openness, and their interaction on different plant traits of *Dicorynia guianensis* and *Vouacapoua americana*. Only trees <22 m.

Dependent variable	Multiple regression		Coefficients			
	F	R^2	Constant	Height	Openness	Height × openness
Dicorynia						
Crown width[1]	163***	0.83	0.45***	0.187***		
Crown length[1]	115***	0.81	ns	0.347***		
Rel. crown width[2]	11***	0.56	0.60***	−0.031***	ns	ns
Rel. crown depth[2]	2.6(*)	0.23	0.18*	0.025*	ns	ns
Crown shape[3]	14***	0.62	ns	0.13***	ns	ns
Safety factor	4*	0.61	1.32**	ns	ns	ns
Vouacapoua						
Crown width[1]	211***	0.86	0.31***	0.265***		
Crown length[1]	263***	0.92	ns	0.415***		
Rel. crown width[2]	19***	0.63	0.85***	−0.059***	−0.029***	0.0028***
Rel. crown depth[2]	10***	0.53	ns	0.024*	0.019*	ns
Crown shape[3]	40***	0.79	ns	0.19***	0.056**	ns
Safety factor	6*	0.32	1.69*	−0.04*	0.016(*)	ns

[1] Linear regression on height only.
[2] Relative to tree height.
[3] Crown shape is the crown depth/crown width ratio.
Significance levels: ns, non-significant; (*) $P<0.1$; *$P<0.05$; **$P<0.01$; ***$P<0.001$.

Architecture of rainforest trees

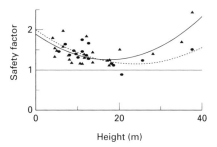

FIG. 6.4. Safety factors for *Dicorynia guianensis* (circles, dotted line) and *Vouacapoua americana* (triangles, solid line) in relation to tree height (H). Lines were fitted with a quadratic model: for *Dicorynia*, safety factor = $1.99 - 0.07 H + 0.0015 H^2$ ($r^2 = 0.41, P = 0.031$); for *Vouacapoua*, safety factor = $1.89 - 0.07 H + 0.0021 H^2$ ($r^2 = 0.53, P < 0.001$).

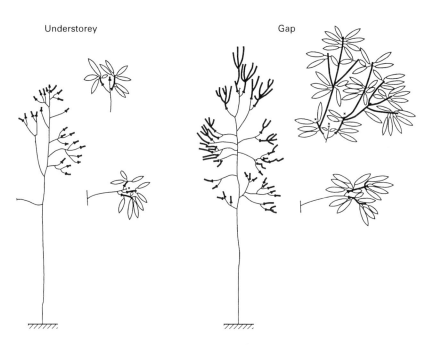

FIG. 6.5. Scale drawings showing crown development (1991–94) of *Dicorynia guianensis* in the understorey and in gaps Les Nouragues, French Guiana. Trees were 9 m high at the start of the measurements. For each tree the development at the top and halfway the crown is depicted in detail.

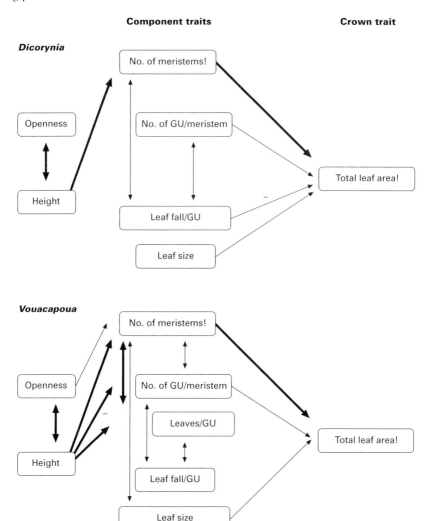

Fig. 6.6. Path-diagram presenting the effects of tree height and canopy openness on crown traits for (a) *Dicorynia guianensis* and (b) *Vouacapoua americana* at Les Nouragues, French Guiana. (Adapted from Sterck 1997.) The effects are split into two successive steps: (1) the effects of canopy openness and tree height on component traits; and (2) the effects of component traits on crown traits. The significant correlations and effects are indicated by double- and single-headed arrows, respectively. Thick-lined arrows indicate the stronger effects (path coefficient $P > 0.5$) and correlations (correlation coefficient $r > 0.5$), and thin-lined arrows the weaker effects ($P \leq 0.5$) and correlations ($r \leq 0.5$). Relationships were mostly positive, and in case of a negative relationship a – sign is given. The non-significant relationships are not shown; ! after a trait indicates that log-transformed values are used in the analysis.

height and diameter data, specific wood densities and stiffness values were used to calculate safety factors for each of the trees. The quadratic regression models (Fig. 6.4) indicate that safety factors decrease with height but later they start to increase again.

The most critical point is at around 20 m height. On the basis of multiple regression analysis for the shorter trees only (Table 6.3), it is shown that safety factors indeed decreased with height in trees shorter than 22 m, although this was not significant for *Dicorynia*. Interestingly, safety factors tended to increase with light availability for trees of *Vouacapoua*. These results suggest that trees of *Vouacapoua* have relatively high investments in their mechanical safety when they occur in more open conditions.

Development of tree crowns between 1992 and 1994 was studied and related to leaf display at the end of that period. As an illustration, two individuals of *Dicorynia* are depicted (Fig. 6.5) that were similar in 1992, one in the forest understorey, the other in a canopy gap. After 2 years the crown of the gap individual had developed much more than that of the understorey individual. The changes in crown development with tree height and canopy openness were related to one leaf display characteristic, total leaf area, using path analysis (Sokal & Rohlf 1981; Fig. 6.6). Clearly, the taller trees of both species had a larger total leaf area because they had more apical meristems. In addition, for *Vouacapoua*, taller trees increased their total leaf area by producing larger leaves, while they decreased their total leaf area by producing fewer GUs per meristem. Taller trees also produced more leaves per GU, but this did not significantly affect total leaf area. In *Dicorynia*, taller trees increased their total leaf area by dropping fewer leaves per GU. Interestingly, total leaf area also increased with canopy openness in *Vouacapoua*, mainly because canopy openness had a positive effect on the number of apical meristems. For *Dicorynia* this was not the case.

HOW VARIABLE ARE TREE LIGHT LEVELS?

On the basis of the CPI values, we have shown that light levels increase with height, although there is large variation in light levels among trees of the same height. Clearly, the overall increase is because trees are overgrown by fewer other trees when they become taller. Between species, trees of the only shade-intolerant species under investigation occurred more frequently at higher light levels than the three, more shade-tolerant, species (cf. Clark, Clark & Rich 1993). Trees of the shade-intolerant species are, however, not restricted to the most open sites and, like trees of more shade-tolerant species, they may occur with a wide variety of light availabilities (cf. Lieberman, Lieberman & Peralta 1989).

Other researchers have also found large variation in light levels among trees of the same species (e.g. Welden, Hewett, Hubbell & Foster 1991, Lieberman, Lieberman, Peralta & Hartshorn 1995) and a gradual increase in light with increasing tree size (Clark & Clark 1992). Only a few species (less than 10%) showed a preference for a particular light niche in the forest, while the majority of species did not (Welden *et al.* 1991, Lieberman *et al.* 1995). These studies were based on rough estimates of light, either by calculating a canopy closure index on the basis of distance and height relations with neighbouring trees (Lieberman *et al.* 1995) or by using two canopy height classes (Welden *et al.* 1991). Alternatively, hemispherical photographs have been used to estimate light levels over saplings (e.g. Clark *et al.* 1993). These may be more precise estimates than the methods used by Lieberman *et al.* (1995) and Welden *et al.* (1991). However, hemispherical photographs lack precision at light levels below 10% from the open sky (Whitmore *et al.* 1993). Photosynthetically active radiation (PAR)-sensors (e.g. Oberbauer, Clark, Clark & Quesada 1988, Pearcy 1989) are more precise but many data over a long time period are needed to accurately estimate light levels on individual trees. The use of hemispherical photographs (e.g. ter Steege 1993, Whitmore, Brown, Swaine *et al.* 1993, Van der Meer 1995) is a compromise in this respect. Clearly, both PAR-sensors and hemispherical photographs are difficult to apply on tall trees. We conclude that at present there are too little data available to draw a conclusion with respect to differentiation in light habitats (Denslow 1980, 1987) among the species in the forest, compared with the random occupation of light habitats (Hubbell & Foster 1986, Zagt & Werger, Chapter 8, this volume).

The information on temporal variation in light levels of individuals is even more scarce than the information on spatial variation in light. In fact, the differences in CPI between individuals of different heights (Fig. 6.1) indicate the overall trends in changes of light during tree life but they do not give an indication of the temporal change on individuals. Preliminary data show that CPI levels change on a scale of 1–2 years for the trees in our case study. On the basis of this and other long-term studies (for example, the data set of Clark & Clark 1992), the magnitude and frequency of change in light levels of individual trees may be assessed for trees of some species in the near future.

TREE ARCHITECTURAL MODELS

In our case study, we made reference to the architectural tree models of the tree species (see Hallé & Oldeman 1970, Hallé *et al.* 1978). These models or related descriptions of tree development may describe an important part of the ontogenetic changes in tree development. Each of the architectural models that has been described is defined by a set of growth rules for the trunk (monopodial vs. sympodial, orthotropic vs. plagiotropic), the branches (monopodial vs. sympodial, orthotropic vs. plagiotropic) and the inflorescenes (terminal vs. lateral, Hallé *et al.* 1978). As such, they describe particular developmental sequences of branching, from the seedling to the architectural unit. The architectural unit is the complete set of axis types and their relative arrangements found in a species (Barthélémy 1991). After forming a complete architectural unit trees may further expand by a process called reiteration (Oldeman 1974, 1990); that is, the model development pattern is repeated at more than one location in the crown.

Although originally species were thought to grow according to one model only, more detailed studies of the growth patterns of tree species have shown that this was not correct (e.g. Drénou 1988 for *Dicorynia guianensis*, Borchert & Tomlinson 1984 for *Tabebuia rosea*). For this reason, the recent architectural analyses as developed by the 'Montpellier school' refer to each of these growth rules separately instead of referring to one of the architectural models (e.g. Drénou 1994, Loubry 1994, Loup 1994). In 'standard' architectural analysis (Edelin 1984) all conditions are taken into account and the general ontogenetic changes in architecture and development are deduced and described (see also Hallé 1995). Frequently, the development of the young sapling is conforming to one of the models but more complex developmental patterns appear later in tree life. Unfortunately, few attempts have been made to relate architectural variability to the light environment. Most of the recent studies on tree architecture explain architectural variability by endogenous processes alone and implicitly neglect environmental factors that may influence tree architecture (e.g. Edelin 1991, Loubry 1994, Loup 1994).

Light and architectural models

For juvenile trees of species that grow according to one architectural model, the architecture and development has been related to light availability. Individuals of the same species that conform to the same model exhibit different crown shapes and leaf displays in different light habitats (Fisher & Hibbs 1982, Borchert & Tomlinson 1984). We do not know of studies that show changes in the architectural model, or in one of the major growth rules that form the basis of these models, in response to different light levels. Possibly, these growth rules are strongly determined genetically and do not change with light or other environmental factors (Hallé *et al*. 1978), but more factual evidence is needed to test this hypothesis. In addition, trees may respond to changes in the light environment by reiteration, the production of more model-like structures within the crown. Also on this topic, more factual evidence is needed by relating architectural response to light levels in the forest.

Tree species and their architectural models have been related to different levels of light availability. Shade-intolerants sometimes tended to conform to models that differ from those of shade-tolerants (e.g. Boojh & Ramakrishnan 1982). For example, for a rainforest in French Guiana, species conforming to the Roux's model (e.g. *Trema*, *Piper*) and Rauh's model (e.g. *Cecropia*) dominate in open sites, and species conforming to Troll's and Fagerlind's models dominate in the forest understorey (De Foresta 1983). For a rainforest in Colombia, however, species conforming to the models of Rauh, Roux and Troll dominate both in the open and closed forest sites (Vester & Saldarriaga 1993). In another study, trees of shade-intolerant and shade-tolerant species could hardly be distinguished on the basis of their architecture (Shukla & Ramakrishnan 1986). For shade-intolerants alone, Ackerly (1996) clearly showed that trees of different species conformed to many different architectural models and not to a very limited set of them.

These results show that architectural models cannot simply be related to functioning of trees in relation to ambient light levels. The value of the models is that they indicate certain constraints (growth rules) on tree development, and formalize the contribution of different crown components (basic construction units in a crown; Table 6.2) to the development of a tree from seedling to adult (e.g. Barthélémy 1991). Probably the capacity of a tree to respond opportunistically to environmental changes (by e.g. opportunistic reiteration) may have a much higher adaptive value than its inherent or deterministic blueprint as formalized in the architectural model. Comparative studies on the adaptive significance of architectural models, in which a large number of species and models are compared, together with studies of their opportunistic responses to environmental changes, will be rewarding in this respect.

On the other hand, it is possible that only some aspects of the models, and not whole models, are associated with particular light environments. For example, Leigh (1990) has shown that trees (and their models) with orthotropic stems were associated with the more open sites, while species with plagiotropic stems were associated with darker sites. Testing for associations between light environments and individual growth rules may be more rewarding than testing associations between light and whole models. (For some hypotheses concerning these last associations see Oldeman & van Dijk 1991.)

Tree architecture and virtual trees

Closely related to the above is the development of virtual trees that are produced using growth and development algorithms. De Reffeye, Dinouard and Barthélémy (1991) and Barthélémy (1991) simulated tree growth as represented by architectural tree models, including other growth rules as well. Their simulations can be used to study the deterministic nature of tree development. In addition, changes in the algorithms can then be used to simulate architectural responses to the environment (e.g. to light). In part, this work is comparable to the studies of Fisher, Honda and Borchert (Honda & Fisher 1978, Bell, Roberts & Smith 1979, Fisher & Honda 1979a,b, Honda 1982, Fisher & Hibbs 1982, Borchert & Honda 1984, Borchert & Tomlinson 1984, Fisher 1986, 1992). Focusing mainly on branching types and branching angles they were able to simulate virtual trees that largely mimic individuals in the field. Changing the algorithms gave them the possibilities to study the effects on crown form and leaf display. A third group is represented by Lindenmayer (1968) and Prusinkiewicz and Lindenmayer (1990). In the context of this chapter we will not treat this topic any further. For a general account on virtual plants in relation to the dynamics of crown components we refer to Room *et al.* (1994).

ALLOMETRY

Tree allometry refers to the relationships between different dimensions or sizes within a tree (e.g. height/diameter ratio). It results from the allocation of carbon to different tree parts minus the losses of these parts. (For a review of carbon allocation in trees see Cannell & Dewar 1994.) In Fig. 6.7 different types of allocation are related to different tree allometries. Losses are not considered in this model. We hypothesize that different allometries result from different growth strategies in relation to the environment:

1   Trees that still grow towards the canopy invest more in height and are relatively more slender.
2   Trees in the dark understorey invest more in leaf display without self-shading to capture as much light as possible and thus produce relatively wide crowns.
3   Trees that are water- or nutrient-limited invest more into their roots and produce an extensive root system.
4   Trees exposed to high wind stress or other forms of mechanical stress (for

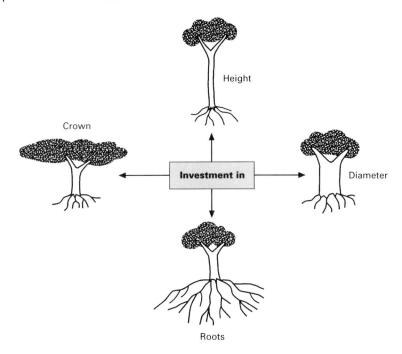

FIG. 6.7. A growth model for trees expressing four different ways of allocating (investing) carbon and its consequences for tree allometry.

example, liana or epiphyte loads) invest more in the thickness of supporting tissues, and produce relatively thick stems.

Clearly trees need to invest in each of their parts in order to function as a whole but the relative differences in investments among different parts may reflect responses to particular environmental conditions. In the following it is assumed that these different allometries occur among different trees of the same species but growing in contrasting environments, as well as among different species adapted to contrasting environments.

Although root systems have been described for tropical trees (Jenik 1978), little information on quantitative aspects of roots of individual trees is available, except for seedlings (e.g. Popma & Bongers 1988, Osunkoya, Ash, Graham & Hopkins 1993) and saplings (Becker & Castillo 1990, Sterck 1997). Only recently have detailed studies on the architecture of root systems of tropical trees become available (Atger & Edelin 1994a,b). We focus on the allometry of above-ground parts due to this lack of information on roots.

Trunks

Trunk allometry refers to the relationship between tree height and trunk diameter. As shown in our case study, this relationship changes with height. The relationship is close to linear when trees are young but later in the ontogenetic development height growth ceases while diameter growth continues. Several studies have shown that an asymptotically curved diameter—height relationship is general for trees (Kira 1978, Bongers, Popma, Maeve del Castillo & Carabias 1988, King 1991, 1996, Aiba & Kohyama 1996, Thomas 1996).

Many researchers used log-transformed diameter and height data in order to fit linear regressions (Hallé *et al*. 1978, Rich, Helenurm, Kearns *et al*. 1986, King 1991, Alvarez-Buylla & Martínez-Ramos 1992, Claussen & Maycock 1995, O'Brien, Hubbell, Spiro *et al*. 1995, O'Brien, Hubbell, Condit *et al*. 1997). Using this technique, tree species were compared for their slopes (Table 6.4). In general, the slopes of the regression lines were quite variable among species, with palm species having relatively low slopes because they lack secondary thickening of their stems. Among the trees of species with different adult statures, overstorey species had different slope values from understorey species (Kohyama & Hotta 1990, King 1996, O'Brien *et al*. 1997), indicating that the overstorey species invested more in height and less in diameter. It was argued that overstorey species need higher investments in height before they reach maturity than understorey species (see also Fig. 6.7). Within species, slopes changed with height (King 1991, 1996; Table 6.4) which is similar to the flattening of the height vs. diameter relationship in the asymptotic approach.

Interestingly, Thomas (1996) has shown that 28 out of 38 species had log-transformed height — diameter relationships which were significantly non-linear. Aiba and Kohyama (1996) came to the same conclusions for trees in a warm-temperate rainforest in Japan. Using the asymptotic height—diameter equation of Kira (1978), Thomas (1996) concluded that the overstorey species were even shorter at a given diameter than understorey species at small stature. In our model (Fig. 6.7), this would mean that the understorey species invested more in height than overstorey species. Thomas hypothesizes that this more slender habit of understorey species is related to their stiffer and denser wood (see Mechanical design below). This contrasts with the log height–log diameter regressions, probably because these latter do not account for the asymptotic nature of the height–diameter relationship.

Most allometric relationships are calculated for trees growing in the forest irrespective of their light environment. Height and diameter growth

TABLE 6.4. Compilation of data on the mechanical design of 23 tropical rainforest woody species: safety factors and diameter–height relationships.

Species	Type[1]	Safety factor[2]	Whole size range[3,4]		Slope for size classes[3,4]			Sources[5]
			Intercept	Slope	1–6 m	6–24 m	>24 m	
Dipterix panamensis	Emergent CT	4			0.74*	1.13*	3.22*	1,2
Lecythis ampla	Emergent CT				0.76*	1.34	2.61	2
Penthaclethra macroloba	CT	4			0.96*	1.37	4.57*	1,2
Penthaclethra macroloba	CT		−2.74	1.67*				6
Virola sebifera	CT	2.6			0.94*	1.44	2.35	1,2
Goethalsia meiantha	CT shade-intolerant				0.90*	1.62		2
Laetia procera	CT shade-intolerant				0.97*	1.10*	3.07	2
Ocotea hartshorniana	CT				0.86*	1.42		2
Pourouma aspera	CT				0.77*	1.68		2
Pourouma aspera	CT		−2.35	1.24*				6
Dicorynia guianensis	CT shade-tolerant	0.9–2.7	−0.43	1.19				3
Vouacapoua americana	CT shade-tolerant	1.1–2.6	−0.54	1.28*				3
Cecropia obtusifolia	CT shade-intolerant	1–20	−0.018	1.069*				5
Alphitonia petriei	CT shade-intolerant	1.1–5.5	−0.3	1.19*				4
Polyscias australiana	CT shade-intolerant	1.1–5.5	−0.3	1.23*				4
Cardwellia sublimis	CT shade-tolerant	1.3–6.0 (7.5)	−0.5	1.52				4
Syzygium papyraceum	CT shade-tolerant	1.3–6.0 (7.5)	−0.5	1.44				4
Casearia arborea	SubCT				0.99*	1.25*		2
Welfia georgii	P		−0.93	0.14*				6
Socratea durissima	P		−1.65	0.63*				6
Ocotea meziana	UT				1.08*	1.68		2
Rinorea deflexiflora	UT	5.5			1.16*	1.08		1,2
Anaxagorea crassipetala	UT	2.6			0.81*	1.60		1,2
Ocotea atirrensis	Treelet				0.70*			2
O. dendrodaphne	Treelet				0.80*			2

[1] CT canopy tree, P palm, UT understorey tree.
[2] Safety margins as buckling height/height ratios. Values in parentheses are outliers to stated range.
[3] Diameter–height relations: intercept and slopes of regression: log (diameter) = $a + b$ log (height).
[4] *Indicates that slopes are significantly ($P < 0.05$) different from 1.5, the slope indicating elastic stability.
[5] 1. King (1991); 2. King (1996); 3. Sterck (1997); 4. Claussen & Maycock (1995); 5. Alvarez-Buylla & Martínez-Ramos (1992); 6. Rich *et al.* (1986).

are cumulative processes in time, and height—diameter relationships may therefore result from the whole light history of the tree. Given the difficulty in quantifying light levels, and leaving aside the temporal changes in these light levels during tree life, the influence of the light environment on height—diameter relationships cannot be assessed properly. To be able to differentiate between light environments, studies need to be performed on trees growing under standardized open or closed conditions, for example in experimental plantations. Another possibility is long-term studies in which light environment, height and diameter are followed over time for large samples of trees (e.g. Clark & Clark 1992). For tropical trees such data are still hardly available.

Crowns

Crown allometry changes at different levels in the forest (Ashton 1978). Profile diagrams suggest that trees of lower stature are characterized by more elongated crowns (deep, narrow crowns), trees of intermediate stature by rounded crowns, and trees of the tallest stature by more broad and flat crowns (Richards 1952, 1983, Givnish 1984). It is unclear whether these differences are attributed to species (adult stature, age), ontogenetic growth phases, (light) environments, or combinations of these factors. Light environment is probably an important factor. Horn (1971, 1975) and also Brünig (1976) related tree and crown geometry to the ability to capture light. Wide, shallow crowns are advantageous over narrow, deep crowns when irradiation comes from above (but when it comes from the side this is reversed, cf. Kuuluvainen 1992). Narrow crowns require less material support and may be erected more quickly. In line with this, open-grown trees should have narrow crowns and shade-grown trees of the same species should have wider crowns. In addition, trees of shade-intolerant species should have narrow crowns and trees of shade-tolerant species should have wider crowns (Horn 1971, Kohyama & Hotta 1990). Factual data on crown allometry for tropical trees are scarce, however.

In our two study species, crowns became deeper and narrower with increasing height (Fig. 6.3, Table 6.3). In *Vouacapoua*, one of the two study species, crowns also became narrower with increasing light availability. This is probably due to the relatively high investments in vertical crown expansion, at the cost of horizontal crown expansion, at higher light availability. In *Dicorynia*, the second species, no such changes in crown allometry in response to canopy openness were observed. As far as we know, there are no other studies on such crown allometric responses to light availability for other tropical tree species and it remains unknown whether or not the response of *Vouacapoua* is common to other rainforest tree species.

Between species, crown allometry has been related to the adult tree stature, the preference for particular light habitats and the life-span of the species. King (1991, 1996) found that adults of understorey species bear larger crowns than juveniles of overstorey species of the same height. With other authors (Horn 1971, 1975, Shukla & Ramakrishnan 1986, Kohyama & Hotta 1990), King (1991) argued that trees of the shorter species increase their capacity to intercept light by extensive horizontal crown growth at the cost of height growth. However, O'Brien *et al.* (1997) could not show such a difference between understorey and overstorey species.

Parallel to these differences, Shukla and Ramakrishnan (1986) and Kohyama & Hotta (1990) found that trees of shade-intolerant species had narrower crowns than trees of shade-tolerant species. They used the same argument that King (1996) used for the difference between overstorey and understorey species: trees of shade-intolerant species invest more in height and may thus escape from shading by neighbours, while trees of shade-tolerant species invest in horizontal expansion to increase the probability of light interception (but at cost of height extension).

King (1996) also related crown allometry to the life-span of trees. At a given height, the longer-lived species had wider crowns than the shorter-lived species. High allocation to horizontal crown expansion in long-lived trees may allow them to exploit adjacent canopy gaps after the death of neighbouring trees. In contrast, high allocation to vertical crown extension may allow the shorter-lived trees to grow rapidly in height when openings occur (see also Kohyama & Hotta 1990). Thus, the responses in crown allometry to light are similar when we compare shade-intolerants with shade-tolerants, overstorey species with understorey species, and short-lived species with long-lived species. It can, therefore, be hypothesized that shade-(in)tolerance, adult tree stature and life-span are associated species traits.

MECHANICAL DESIGN

Mechanical tree design refers to a balance between mechanical stress due to self-loading or wind force on the one hand and the resistance against this stress on the other. Several models have been developed to quantify the mechanical design of trees. The elastic-stability model emphasizes stress due to static loading; that is, self-loading as well as snow or epiphyte loading (King & Loucks 1978, Waller 1986, Niklas 1992). This model is represented by the equation (after Greenhill 1881):

$$L_{cr} = c_1 \times (E/\rho)^{1/3} \times D^{2/3} \qquad (1)$$

The model expresses the height at which a tree starts to buckle under its own mass. This critical length (L_{cr}) or buckling height increases with a 2/3 power increase in trunk diameter D, and with a 1/3 power increase in the ratio of the elastic (or Young's) modulus E (a measure of stiffness of the wood) to wood density ρ. The constant c_1 is a measure of shape it has been modified by various researchers (e.g. McMahon 1973, King & Loucks 1978, King 1981, 1986, Holbrook & Putz 1989, Niklas 1994b).

The constant-stress model refers to stress due to dynamic wind forces (e.g. Niklas 1992). This model is represented by:

$$D = c_2 \times (S \times H)^{1/3} \qquad (2)$$

The model expresses the minimum diameter that a tree needs to resist wind forces. These forces are assumed to be proportional to 1/3 power increase in the product of total leaf area (or silhouette area S, Dean & Long 1986) and tree height H. The wind pressure and maximum wind stress are assumed to be constant, and are included in the constant c_2 (Niklas 1992). Wood properties such as the elastic modulus and wood density are not taken into account. A third model is the geometric-similarity model that predicts that the proportions of a tree remain constant during ontogeny (Niklas 1994a,b, O'Brien et al. 1995).

Safety factors

An important concept related to mechanical design is the biological safety factor. This factor has been defined in terms of the maximum load required to cause mechanical failure (Chazdon 1986, King 1987, Holbrook & Putz 1989), or in terms of the minimum structural dimensions required to prevent failure (McMahon 1973, Niklas 1992, 1994a,b). In general, a safety factor of $S = 1$ indicates that the strength of the tree has practically no reserves against self-loading and wind stresses. A small accident or extra load can cause tree fall or breakage. The larger the safety factor, the higher the mechanical safety margins of the tree.

In most cases safety factors are calculated (on the basis of the elastic-stability model) in terms of the critical buckling height or diameter (see McMahon 1973, King 1981, 1991, 1996, Dean & Long 1986, Rich et al. 1986, Alvarez-Buylla & Martínez-Ramos 1992, Niklas 1992, 1994a,b, Claussen & Maycock 1995, Sterck 1997):

$$\text{Safety factor} = L_{cr}/H \qquad (3)$$

This ratio of buckling height L_{cr} to actual tree height H quantifies the margins of safety against buckling under its own mass. Sometimes crown loads were also included in studies on temperate trees (Gere & Carter 1963, King & Loucks 1978, Holbrook & Putz 1989), and they were in one study of tropical trees (King 1987). Safety factors have also been calculated (for temperate trees only) by cutting well-defined windows into the stems (Mattheck 1991, 1995, Mattheck, Bethge & Schafer 1993).

Species do not remain 'elastically similar' (maintain a constant elasticity throughout their lengths, following the elastic-stability model) during ontogeny (Niklas 1992), nor do they maintain mechanical safety margins following from the constant-stress or geometric-similarity model (O'Brien et al. 1995). The lack of constant elastic similarity during ontogeny in shown by the data presented in Table 6.4. Most species have log diameter–log height slopes that are significantly different from 3/2 and are thus not elastically similar.

We have shown that safety factors of trees of two shade-tolerant species are smallest for intermediate-sized trees and larger for smallest and for large trees (Fig. 6.4; see also Claussen & Maycock 1995). In contrast, trees of some shade-intolerant species showed a decrease in safety factor from small to large trees (Alvarez-Buylla & Martínez-Ramos 1992, Claussen & Maycock 1995). Alvarez-Buylla and Martínez-Ramos (1992) hypothesized that selection for height growth rather than for strength and longevity (thus low safety factors) is to be expected in trees of shade-intolerant species. A clear reason for the extremely high safety factor in the young trees is not apparent, but it may be related to particular stem properties, for example the stem hollowness, wood density and wood stiffness. Although Niklas (1995) stated that wood characteristics may not have a large influence on mechanical tree design, their influence is so far unknown.

We show (Table 6.4) that trees generally remain well above the theoretical buckling limit. The palm *Socratea durissima*, which had smaller diameters than possible on the basis of the model, is the only exception. Juvenile palms, however, have a relatively thick stem (Rich et al. 1986). This implies that with an increase in height the margins of safety against mechanical failure decrease, or that stem tissue stiffness and strength increase, or that the crown becomes relatively narrow. In the other species, ontogenetic changes in wood properties may also affect the real safety margins of trees.

From our case study, it is inferred that trees of *Vouacapoua* over-compensate their more rapid height growth at higher light availability with stem thickness growth, and thus increase their mechanical safety. It may be speculated that such a response is of adaptive value in the more open and thus more wind-exposed, parts of the forest. As far as we know this is the only study which has directly related safety factors to ambient conditions in the forest. We think that the influence of the environment, especially light, should be taken into account in studies of the mechanical safety of tropical trees in the future.

CROWN DEVELOPMENT

Crown development in trees may refer to a hierarchy of plant traits, from meristem to whole crown (e.g. Oldeman & van Dijk 1991). The meristems produce components of increasing size and complexity; metamers, extension units, sympodial units and architectural units (White 1979, 1984, Barthélémy 1991, Bell 1991, Room et al. 1994; Table 6.2). The demography, spatial arrangement and structure of the components shape the crown and its characteristics, such as crown allometry, crown volume, total leaf mass, total leaf area, leaf area index (LAI), mantle area, effective crown length and bifurcation ratio (e.g. Fisher & Hibbs 1982, Borchert & Tomlinson 1984, Shukla & Ramakrishnan 1986, Canham 1988, Alvarez-Buylla & Martínez-Ramos 1992, Ackerly 1993, Sterck 1997). As these components accumulate during tree life they are at the basis of the increase in crown size and changes in crown allometry during tree development (see pages 143–44).

Crown components and leaf display

The demography of crown components influences the amount of leaf area and the spatial arrangement of leaves. This determines to a large extent the interception of light, and the amount of carbon that can be fixed and used for growth and survival. On the other hand, there are costs associated with leaf display as leaves have to be produced and supported by woody structures (Chazdon 1986, Küppers 1989). The demography of the different crown components has mainly been investigated for temperate trees (e.g. Maillette 1982a,b, 1987, Franco 1986, Jones & Harper 1987a,b). For tropical trees most work has been done on leaf demography. Leaves are faster produced but live shorter at high light availability (Shukla & Ramakrishnan 1984, Meave del Castillo 1987, Bongers & Popma 1990, Reich, Walters & Ellsworth 1992). Production rates of internodes (Sasaki & Mori 1981) and shoots (Fisher 1986) increase with light availability life-span was not assessed.

The spatial arrangement and structure of components was assessed for both temperate trees (e.g. Whitney 1976, Steingraeber & Waller 1986, Waller 1986, Canham 1988, Givnish 1988, Sipe & Bazzaz 1994) and tropical trees (Fisher & Hibbs 1982, Borchert & Tomlinson 1984, Chazdon 1985, Shukla &

Ramakrishnan 1986, Alvarez-Buylla & Martínez-Ramos 1992, Sterck 1997). Under high light levels, trees produce larger shoots (Fisher & Honda 1979a,b, Borchert & Tomlinson 1984, Fisher 1986) and larger internodes (Kohyama & Hotta 1990, King 1994, Sterck 1997). Leaves of trees may be smaller (Givnish 1984, Bongers & Popma 1988) or of the same size (Sterck 1997) but have a higher leaf mass per unit area (Bongers & Popma 1988) in high light.

The level of shade-tolerance of a species influences the type of components produced. Shade-intolerant species have other leaf characteristics than shade-tolerant species (Popma, Bongers & Werger 1992), and also the flexibility (shade/sun ratio) is different (Bongers & Popma 1988). King (1994) showed that 1–2-m tall saplings of shade-tolerant species produced shorter leaf supporting woody units under low light levels. Saplings of shade-intolerant species were not able to do so. The same patterns were found for seedlings (Popma & Bongers 1988) and taller trees (Sterck 1997). King (1994) argued that the saplings of shade-tolerant species were able to allocate more carbon to leaves by reducing their internode length in shade, and thus lower their light compensation point. This may permit trees of these species to survive shady conditions in the forest understorey. In contrast, the production of long leaf-supporting woody units may help trees of shade-intolerant species to grow quickly at high light levels but it reduces their ability to survive under persisting low light levels.

For 4–10-m tall juveniles of our shade-tolerant species *Dicorynia guianensis* and *Vouacapoua americana*, Sterck (1997) showed that the gap individuals produced more sympodial units (by branching), units of extension, metamers and leaves. Consequently, gap and understorey individuals produced different crown architectures within 3 years (see also Fig. 6.5). Gap individuals had larger total leaf area, LAI, and crown cover, and they achieved more height. In gaps, for individuals of both species, the leader grew faster than other branches within the crown (by producing more and longer metamers and units of extension), which might explain the more elongated crown shape at high light levels. For understorey individuals, however, there was no such difference between the leaders and other branches. Partly in contrast to this, Kohyama (1980) found that lateral branches of the temperate tree species *Abies mariessi* saplings grow better than the trunk, especially in the forest understorey. In the shade-tolerant *Acer saccharrum*, the leader grew larger relative to the lateral branches at lower light levels (Bonser & Aarssen 1994). Bonser and Aarssen suggested that this allowed trees to minimize the chances of being overgrown at low light levels and, conversely, to maximize light interception at higher light levels (see also Tucker, Hickley, Leverenz & Jiang 1987). It is unclear whether these contrasting growth habits reflect different growth strategies of the species, different growth phases or the different light environments of different (tropical vs. temperate) forests.

Apart from the amount and spatial arrangement of leaf area also the direction of the leaf display is important for light interception. Ackerly and Bazzaz (1995) showed that the direction of leaf display of saplings of four neotropical shade-intolerant tree species was related to the angular distribution of diffuse radiation and uncorrelated with the location of sunflecks. Ackerly (1993) suggested that during periods when the sky is overcast, availability of direct radiation is low. Plants then may respond primarily to diffuse radiation. This has to be taken into account in future studies since most growth studies on tropical tree seedlings have not distinguished between direct and diffuse light and have measured only total daily light levels (e.g. Fetcher, Strain & Oberbauer 1983, Popma & Bongers 1988, Turner 1991). This has an important implication: as diffuse light influences leaf display it also indirectly influences the capacity of the crown to capture direct light.

Path analysis

So far, this discussion on crown development referred to individual traits and to trees of one stage of life (seedling, sapling, juveniles). Few attempts have yet been made to study changes in different crown traits with ontogeny and light availability simultaneously. In our case study we attempted to study these processes simultaneously using path analysis (Wright 1934, Sokal & Rohlf 1981, Kingsolver & Schemske 1991).

In our example, we have illustrated (Fig. 6.6) how size and light affect crown development at various levels in the crown hierarchy, and how this may change a whole crown trait such as total leaf area. Clearly, taller trees produced a larger total leaf area as they had more apical meristems. For *Vouacapoua*, trees at higher light levels increased their total leaf area as they supported more apical meristems. Probably, trees at higher illumination levels have more apical meristems because they branch more frequently and are thus able to produce more total leaf area.

The results of path analysis have to be treated carefully. Apart from statistical aspects (Petraitis, Dunham & Niewiarowski 1996) it is important to consider the variation in environment (is the whole range of environments sampled?), the time of observation (development changes may take some time, e.g. total leaf area may take months (Popma & Bongers 1991, Ackerly 1997) or even years (Sterck 1997) to become established), and the fact that relationships are not necessarily linear.

The advantages of path analysis over regression analysis (of height and light on a given response variable) are clear. On the basis of a model (hypothesis), path analysis may show the relationships among different crown traits and their consequences for the whole crown. They may indicate different possible adaptive roles of architectural adjustments (in component traits) via their final effects on different (ecologically significant) whole crown traits (e.g. total leaf area). As such, path analysis can be a good tool for understanding processes of crown development.

Delay of responses

In general, individual trees are expected to be adapted to the environment in which they live; trees in the shade then should have a shade morphology and trees in the sun should have a sun morphology. Experimental studies on plant growth under different light levels show that (at least for seedlings) this indeed is the case (e.g. Fetcher *et al.* 1983, Popma & Bongers 1988, Ackerly 1997). However, as light levels change in time this morphology does not remain so clear (cf. Levins 1968). Plants may change their morphology in response to changes in light and the final plant morphology is the result of a mixture of shade- and sun-adapting processes. Plant structures with a long life-time (e.g. the stem) are therefore a mixture of sun and shade types while plant structures that live for a short time (such as leaves) can be changed relatively quickly and may be better adapted to the new situation.

Popma and Bongers (1991) showed that the leaves of seedlings of three shade-tolerant rainforest species were already different 3 months after a change in the environment. In contrast, stems needed a much longer time to be different from the stems under control situations. They also found that plants responded much faster after a change from low to high light than after canopy closure. Although high light levels can damage seedlings, the higher level of photosynthesis enables plants to produce enough assimilates for the production of new structures, better adjusted to the new situation (Popma & Bongers 1991).

Trees of shade-intolerant species have higher leaf turnover rates than those of shade-tolerant species and are thus thought to have faster responses to changes in light environment. Ackerly (1996) hypothesized that this also will be the case within the shade-intolerant species group, and found that the species with higher leaf turnover, *Heliocarpus appendiculates*, had a more rapid response in crown traits (e.g. specific leaf mass) and related traits (e.g. leaf area/plant mass ratio) than the species with lower leaf turnover, *Cecropia obtusifolia*. For trees taller and older than seedlings such data are hardly available, though it is very probable that the same mechanisms take place. In natural forest, however, it is hard to quantify the environmental fluctuations in light levels of individuals and their impact on crown development remains largely unexplored.

CONCLUSIONS

Architectural models and light

Architectural models and related descriptions of tree architecture may suitably describe part of the ontogenetic change in tree development but they fail to explain architectural responses to the light environment. Possibly, the growth rules that form the basis of the architectural models do not change in response to light. More evidence is needed to test this hypothesis and maybe additional growth rules are needed to be able to include light responses into the existing models. In addition, it has to be tested whether trees may respond to changes in light by reiteration of the model.

Between species, it was hard to relate different strategies for light, for example shade-intolerants vs. shade-tolerants, to particular architectural tree models. Possibly, only some of the basic growth rules of the models are related to the light environment, while others are not. For example, shade-intolerants tend to produce orthotropic stems while shade-tolerants tend to produce plagiotropic stems. In line with this, we think that tests of associations between individual growth rules and light environments will be more rewarding than tests of relations between whole architectural tree models and light environments. In a later stage, the architectural models may serve as frameworks for ordering the influences of light on separate (basic as well as additional) growth rules, and for modelling tree growth in relation to the light environment.

Allometry

The changes in trunk allometry with ontogeny are similar for different tree species: late in tree development, height growth ceases while stem diameter growth continues. Far fewer species have been studied for their ontogenetic changes in crown allometry. General changes in crown allometry with ontogeny cannot yet be depicted, except for the obvious increase in crown size.

Studies on responses in allometry of tropical trees to light were not found in the literature. For a large part this may be due to methodological problems: allometry is the result of growth and allocation processes over the whole life-span of a tree. Long-term data sets on allometry are scarce and long-term data on tree light levels are non-existent. In our case study, we have shown that crown allometry may change in response to light levels although we do not know whether the observed relationships are valid for other tropical rainforest tree species as well.

Both trunk and crown allometry differed between large stature and short stature species. Thomas (1996) convincingly demonstrated that small trees of short stature species have more slender stems than small trees of large-statured species. At a larger tree size this is reversed: for the small stature species the height growth ceases while their stem diameter still increases. For large stature species this process starts at a larger size.

For crown allometry, shade-intolerant species, tall-statured species and short-lived species had narrower crowns than shade-tolerant species, short-statured species and long-lived species, respectively. From this it is suggested that light preference, adult stature and life-span may be associated species traits, though evidence from more tree species is needed to test such a relationship.

The allometry of a tree is the result of a differential allocation and losses of carbon over time. The influence of light on this process has been little studied, with the exception of studies on seedlings. For larger individuals, the interpretation of allometry in terms of allocation of biomass to different parts is speculative so long as we are unable to quantify loss-rates of these parts. To link allometry to allocation and loss-rates, we need long-term studies on tree development and we cannot confine ourselves to studies on allometry alone.

Mechanical tree design

Mechanical tree design has been calculated on the basis of different models. Using these models, it appeared that trees do not remain constant in their mechanical safety margins during ontogeny. On the basis of the elastic-stability model, safety margins of canopy tree species decrease to a minimum at a tree height of 20 m, but increase when trees become taller. It is hypothesized that the low safety margins result from strong competition for light in the dark understorey; trees invest in height in order to escape from low light levels, but this is at the cost of mechanical safety (see also Kohyama & Hotta 1990).

Apart from our case study, we could not find studies which relate safety margins to different light environments (but see Holbrook & Putz 1989, for a temperate forest tree species). For *Vouacapoua*, the safety margins increased with increasing light availability. A similar result was found by

Holbrook and Putz (1989) for their tree species. This again indicates that trees make larger investments in height expansion at the expense of investments in mechanical safety at the lower light levels.

Among species, shade-intolerants tended to have lower safety margins than shade-tolerants, particularly in a late developmental stage. In fact, trees of the former group need to escape from suppression, while trees of the latter group need more stability to survive in the understorey as they have a higher chance of being hit by fallen debris (cf. Clark & Clark 1991). In addition, it has been suggested that the low safety margins are a physiological constraint of fast-growing shade-intolerant species (Alvarez-Buylla & Martínez-Ramos 1992). These hypotheses on interspecific differences can be checked for a large number of trees by combining data sets on allometry with data sets on wood properties.

Most empirical work so far is on the basis of the elastic-stability model, using data on the heights and diameters of individual trees and wood density and stiffness of the species. A start is made with more comprehensive studies which incorporate, for example, the dimensions and position of the crown. From the studies on gap dynamics and tree mortality (e.g. Van der Meer & Bongers 1996a,b) we know that several factors, including tree anchoring, determine whether large trees uproot or snap off. Putz, Coley, Lu et al. (1983) showed that wood properties were very important in this respect. Uprooted individuals tended to have denser, stiffer and stronger wood and to be shorter for a given trunk diameter. These factors should be included in the study of mechanical design of trees.

Models on mechanical stability need to be validated with field observations. The studies of King (1987), Holbrook and Putz (1989) and Niklas (1994a,b), in which mechanical stresses were applied to saplings and the effects studied, could serve as a starting point in this respect. For premature and mature trees these methods are impossible to use because of the large forces needed. A different option is the window-cutting method proposed by Mattheck et al. (1993). This method is (at least theoretically) also applicable to larger trees.

Multifactor analyses

Many factors, endogenous and exogenous, affect the architecture and development of trees. Although studies on simple factors are available, practically no work has been done combining factors. To understand development processes the factors should be studied simultaneously. In a recent review of canopy structure of tropical shade-intolerant tree species, Ackerly (1996) stresses the importance of integration of different growth processes in a tree crown in order to understand fully the dynamics of their constituent parts and the changes of canopy structure over time. Ackerly (1993) shows the importance of interdependence among individual plant traits related to overall leaf display. Leaf size, and changes in leaf size, must be evaluated in the context of variables such as branching pattern and internode length. In large-leaved *Cecropia* plants, for instance, excessive branching would strongly increase leaf overlap and thus self-shading, while this would be much less important for the small-leaved *Trema* plants. In the presented path analysis, we showed how different plant components change simultaneously with tree height and light availability, and how this may affect the whole crown. In addition, this method provides a means to distinguish between ontogenetic changes in architecture and architectural responses to light.

Variability in architecture

Species are thought to be variable in architecture to better cope with environmental variation and to have an advantage over species with less architectural variability. We hypothesize that shade-tolerant species are more flexible in their architecture than shade-intolerant species because the former are able to survive in a much wider range of light levels.

It is not easy to show whether or not variability in architecture leads to an advantage. Two questions should be answered first: what are the advantages of different architectures for the individual, and what are the advantages of variability in tree architecture for the population? Another important aspect is the problem of the time-lag: the time an individual needs to change its architecture in a way that it is better adapted to the new environment (Popma & Bongers 1991, Ackerly 1996).

Many advantages for individuals have been speculated (e.g. Givnish 1984, 1988, 1995): (i) trees capture light for photosynthesis more efficiently with a larger leaf area, or an enhanced leaf display; (ii) they better withstand damage by falling debris (Clark & Clark 1991); (iii) they have increased mechanical stability to be able to withstand load and/or windstress; and (iv) they are able to grow fast to have higher access to light and thus outcompete neighbours (for example, Kohyama & Hotta 1990, Niklas 1992, Givnish 1995, King 1996). Measurements of these advantages in the field are scarce, however. For questions such as: do the individuals with a specific architecture have a lower mortality, do they grow better (e.g. Sterck 1997), do they reach maturity earlier, do they reproduce better, and how do these questions relate to light environment; we could find no answers. Long-term data, as are available for some tropical forest plots (for example, Clark & Clark 1992, Condit 1995, Ashton, Chapter 18, this volume) could be analysed in these respects.

When advantages for individuals are not yet clear from field studies, we would expect a worse situation for populations. Do populations with a larger variability in architecture have greater survival than those with less variability? How is this related to light environment? This calls for demographic studies on tropical tree species where these aspects are explicitly taken into account.

We can conclude that, although light is a major factor influencing the architecture of a tree, studies taking this factor into account are scarce and mostly confined to seedlings. Partly, this is due to the difficulty in estimating light levels of trees larger than small saplings; partly it is due to the long lifespan of larger trees and the fact that the current architecture is the result of a long process. Studying trees of all sizes for a number of years and then linking the results could lead to a better understanding of tree development as affected by light environment.

ACKNOWLEDGEMENTS

This study is partly based on field work at the biological field station Les Nouragues in French Guiana. P. Charles-Dominique (Centre National de Recherche Scientifique, Brunoy) provided support and facilities for work. Field support was received from J.-W. Gunnink, R. van Heck, M. Karelse, K. Konings, P. van der Meer, L. Poorter, T. Rijkers and M. Smeenge. We are grateful to D. King, D. Newbery, B. Riera and an anonymous reviewer for constructive comments on the manuscript. This study is supported by grant W85-239 of the Netherlands Foundation for the Advancement of Tropical Research (WOTRO).

REFERENCES

Ackerly, D.D. (1993). *Phenotypic plasticity and the scale of environmental heterogeneity: studies of tropical pioneer trees in variable light environments.* PhD thesis, Harvard University, Cambridge, MA.

Ackerly, D.D. (1996). Canopy structure and dynamics: integration of growth processes in tropical pioneer trees. In *Tropical Forest Plant Ecophysiology* (Ed. by S.S. Mulkey, R.L. Chazdon & A.P. Smith), pp. 619–658. Chapman & Hall, London.

Ackerly, D.D. (1997). Allocation, leaf display and growth in fluctuating light environments. In *Plant Resource Allocation* (Ed. by F.A. Bazzaz & J. Grace), pp. 231–264. Academic Press, San Diego.

Ackerly, D.D. & Bazzaz, F.A. (1995). Seedling crown orientation and interception of diffuse radiation in tropical forest gaps. *Ecology*, **76**, 1134–1146.

Aiba, S.I. & Kohyama, T. (1996). Tree species stratification in relation to allometry and demography in a warm-temperate rain forest. *Journal of Ecology*, **84**, 207–218.

Alvarez-Buylla, E.R. & Martínez-Ramos, M. (1992). Demography and allometry of *Cecropia obtusifolia*, a neotropical pioneer tree: an evaluation of the climax–pioneer paradigm for tropical rain forests. *Journal of Ecology*, **80**, 275–290.

Ashton, P.S. (1978). Crown characteristics of tropical trees. In *Tropical Trees as Living Systems* (Ed. by P.B. Tomlinson & M.H. Zimmermann), pp. 591–615. Cambridge University Press, London.

Atger, C. & Edelin, C. (1994a). Premières données sur l'architecture comparée des systèmes racinaires et caulinaires. *Canadian Journal of Botany*, **72**, 963–975.

Atger, C. & Edelin, C. (1994b). Stratégies d'occupation du milieu souterrain par les systèmes racinaires des arbres. *Revue d'Ecologie (Terre et Vie)*, **49**, 343–356.

Barthélémy, D. (1991). Levels of organization and repetition phenomena in seed plants. *Acta Biotheoretica*, **39**, 309–323.

Becker, P. & Castillo, A. (1990). Root architecture of shrubs and saplings in the understory of a tropical moist forest in lowland Panama. *Biotropica*, **22**, 242–249.

Bell, A.D. (1991). *Plant Form: An Illustrated Guide to Flowering Plant Morphology.* Oxford University Press, Oxford.

Bell, A.D., Roberts D. & Smith A. (1979). Branching patterns: the simulation of plant architecture. *Journal of Theoretical Biology*, **81**, 351–375.

Bongers, F. & Popma, J. (1988). Is exposure-related variation in leaf characteristics of tropical rain forest species adaptive? In *Plant Form and Vegetation Structure* (Ed. by M.J.A. Werger, P.J.M. van der Aart, H.J. During & J.T.A. Verhoeven), pp. 191–200. SBP Academic Publishers, The Hague.

Bongers, F. & Popma, J. (1990). Leaf dynamics of seedlings of rain forest species in relation to canopy gaps. *Oecologia*, **82**, 122–127.

Bongers, F., Popma, J., Meave del Castillo, J. & Carabias, J. (1988). Structure and floristic composition of the lowland rain forest of Los Tuxtlas, Mexico. *Vegetatio*, **74**, 55–80.

Bonser, S.P. & Aarssen, L.W. (1994). Plastic allometry in young sugar maple (*Acer saccharum*): Adaptive responses to light availability. *American Journal of Botany*, **81**, 400–406.

Boojh, R. & Ramakrishnan, P.S. (1982). Growth strategy of trees related to successional status. I. Architecture and extension growth. *Forest Ecology and Management*, **4**, 359–374.

Borchert, R. & Honda, H. (1984). Control of development in the dichotomous branch system of *Tabebuia rosea*. A computer simulation. *Botanical Gazette*, **145**, 184–195.

Borchert, R. & Tomlinson, P.B. (1984). Architecture and crown geometry in *Tabebuia rosea* (Bignoniaceae). *American Journal of Botany*, **71**, 958–969.

Brünig, E.F. (1976). Treeforms and environmental conditions: an ecological viewpoint. In *Tree Physiology and Yield Improvement* (Ed. by M.G.R. Connell & F.T. Last), pp. 139–156. Academic Press, New York.

Canham, C.D. (1988). Growth and canopy architecture of shade tolerant trees: response to canopy gaps. *Ecology*, **69**, 786–795.

Cannell, M.G.R. & Dewar, R.C. (1994). Carbon allocation in trees: a review of concepts for modelling. *Advances in Ecological Research*, **25**, 59–104.

Chazdon, R.L. (1985). Leaf display, canopy structure, and light interception of two understorey palm species. *American Journal of Botany*, **72**, 1493–1502.

Chazdon, R.L. (1986). The costs of leaf support in understorey palms: economy versus safety. *American Naturalist*, **12**, 9–30.

Chazdon, R.L. (1988). Sunflecks and their importance to understorey plants. *Advances in Ecological Research*, **18**, 1–63.

Clark, D.B. & Clark, D.A. (1991). The impact of physical damage on canopy tree regeneration in tropical rain forest. *Journal of Ecology*, **79**, 447–457.

Clark, D.A. & Clark, D.B. (1992). Life history diversity of canopy and emergent trees in a Neotropical rain forest. *Ecological Monographs*, 62, 315–344.
Clark, D.B., Clark, D.A. & Rich, P.M. (1993). Comparative analysis of microhabitat utilization by saplings of nine tree species in a neotropical rain forest. *Biotropica*, 25, 397–407.
Claussen, J.W. & Maycock, C.R. (1995). Stem allometry in a North Queensland tropical rain forest. *Biotropica*, 4, 421–426.
Coleman, J.S., McConnaughay, K.D.M. & Ackerly, D. (1994). Interpreting phenotypic variation in plants. *Trends in Ecology and Evolution*, 9, 187–191.
Condit, R. (1995). Research in large, long-term tropical forest plots. *Trends in Ecology and Evolution*, 10, 18–22.
Dean, T.J. & Long, J.N. (1986). Validity of constant-stress and elastic-instability principles of stem formation in *Pinus contorta* and *Trifolium pratense*. *Annals of Botany*, 58, 833–840.
De Foresta, H. (1983). Le spectre architectural: Application a l'étude des relations entre architecture des arbres et écologie forestière. *Bulletin de Muséum Nationale Histoire Naturelle* 4, 5, 295–302.
Denslow, J.S. (1980). Gap partitioning among tropical rainforest trees. *Biotropica*, 12 (Suppl.), 47–55.
Denslow, J.S. (1987). Tropical rainforest gaps and tree species diversity. *Annual Review of Ecology and Systematics*, 18, 431–451.
De Reffeye, P., Edelin, C., Prevost, F., Jeager M. & Puech, C. (1988). Plant models faithful to botanical structure development. *Computer Graphics*, 22, 151–158.
De Reffeye, P., Lecoustre, R., Edelin, C. & Dinouard P. (1989). Modelling plant growth and architecture. In: *Cell to Cell Signalling: From Experimental to Theoretical Models* (Ed. by Goldbetter), pp. 237–246. Academic Press, London.
De Reffeye, P., Dinouard R. & Barthélémy, D. (1991). Modèlisation et simulation de l'architecture de l'orme du Japon *Zelkova serrata* (Thunb.) Makino (Ulmaceae): la notion d'axe de référence. In *L'Arbre, Biologie et Développement* (Ed. by C. Edelin), pp. 252–266. Naturalia Monspeliensia, Montpellier.
Drénou, C. (1988). Etude de l'architecture d'un arbre guyanais: l'angélique, *Dicorynia guianensis* Amshoff (Caesalpiniaceae). DEA thesis, Université Montpellier II, Montpellier.
Drénou, C. (1994). Approche architecturale de la senescence des arbres. Le cas de quelques angiospermes tempérées et tropicales. PhD thesis, Université Montpellier II, Montpellier.
Edelin, C. (1984). *L'architecture monopodiale: l'example de quelques arbres d'asie tropicale*. Thesis, Université Montpellier II, Montpellier.
Edelin, C. (1991). Nouvelles données sur l'architecture des arbres sympodiaux: le concept de plant d'organisation. In *L'Arbre: Biologie et Développement* (Ed. by C. Edelin), pp. 127–154. Naturalia Monspeliensa, Montpellier.
Fetcher, N., Strain, B.R. & Oberbauer, S.F. (1983). Effects of light regime on the growth, leaf morphology, and water relations of seedlings of two species of tropical trees. *Oecologia*, 58, 314–319.
Fisher, J.B. (1986). Branching patterns and angles in trees. In *On the Economy of Plant Form and Function* (Ed. by T.J. Givnish), pp. 493–523. Cambridge University Press. Cambridge.
Fisher, J.B. (1992). How predictive are computer simulations of tree architecture? *International Journal of Plant Science*, 153, 137–146.
Fisher, J.B. & Hibbs, D.E. (1982). Plasticity of tree architecture: specific and ecological variations found in Aubréville's model. *American Journal of Botany*, 69, 690–702.
Fisher, J.B. & Honda, H. (1979a). Branch geometry and effective leaf area: a study of *Terminalia*-branching pattern. 1. Theoretical trees. *American Journal of Botany*, 66, 633–644.
Fisher, J.B. & Honda, H. (1979b). Branch geometry and effective leaf area: a study of

Terminalia-branching pattern. 2. Survey of real trees. *American Journal of Botany*, **66**, 645–655.

Franco, M. (1986). The influence of neighbours on the growth of modular organisms with an example of trees. *Philosophical Transactions of the Royal Society, Series B*, **313**, 209–226.

Gere, G.M. & Carter, W.O. (1963). Critical buckling loads for tapered columns. *Transactions of the American Society of Civil Engineers*, **128**, 736–754.

Givnish, T.J. (1984). Leaf and canopy adaptations in tropical forests. In *Physiological Ecology of Plants of the Wet Tropics* (Ed. by E. Medina, H.A. Mooney & C. Vásquez-Yánes), pp. 51–84. Junk, The Hague.

Givnish, T.J. (1988). Adaptation to sun and shade: a whole plant perspective. *Australian Journal of Plant Physiology*, **15**, 63–92.

Givnish, T.J. (1995). Plant stems: biomechanical adaptation for energy capture and influence on species distributions. In *Plant Stems: Physiology and Functional Morphology* (Ed. by B.L. Gartner), pp. 3–49. Academic Press, San Diego.

Greenhill, G. (1881). Determination of the greatest height consistent with stability that a vertical pole or mast can be made, and the greatest height to which a tree of given proportions can grow. *Proceedings of the Cambridge Philosophical Society*, **4**, 65–73.

Gross, L.J. (1986). Photosynthetic dynamics and plant adaptation to environmental variability. *Lectures on Mathematics in the Life Sciences*, **19**, 135–170.

Hallé, F. (1995). Canopy architecture in tropical trees; a pictorial approach. In *Forest Canopies* (Ed. by M.D. Lowman & N.M. Nadkarni), pp. 27–44. Academic Press, San Diego.

Hallé, F. & Oldeman, R.A.A. (1970). *Essai sur l'Architecture et Dynamique de Croissance des Arbres Tropicaux*. Masson, Paris.

Hallé, F., Oldeman, R.A.A. & Tomlinson, P.B. (1978). *Tropical Trees and Forests: An Architectural Analysis*. Springer, Berlin.

Holbrook, N.M. & Putz, F.E. (1989). Influence of neighbours on tree form: effects of lateral shading and prevention of sway on the allometry of *Liquidambar styraciflua* (sweet gum). *American Journal of Botany*, **76**, 1740–1749.

Honda, H. (1982). Two geometric models of branching in botanical trees. *Annals of Botany*, **49**, 1–11.

Honda, H. & Fisher, J.B. (1978). Tree branch angle: maximizing effective leaf area. *Science*, **199**, 888–890.

Horn, H.S. (1971). *The Adaptive Geometry of Trees*. Princeton University Press, Princeton, NJ.

Horn, H.S. (1975). Forest succession. *Scientific American*, **232**, 90–98.

Hubbell, S.P. & Foster, R.B. (1986). Biology, chance and history and the structure of tropical rain forest tree communities. In *Community Ecology* (Ed. by J. Diamond & T.J. Case), pp. 314–329. Harper & Row, New York.

Jenik, J. (1978). Roots and root systems in tropical trees: morphologic and ecologic aspects. In *Tropical Trees as Living Systems* (Ed. by P.B. Tomlinson & M.H. Zimmermann), pp. 323–349, Cambridge University Press, Cambridge.

Jones, M. & Harper, J.L. (1987a). The influence of neighbours on the growth of trees. I. The demography of buds in *Betula pendula*. *Proceedings of the Royal Society, Series B*, **232**, 1–18.

Jones, M. & Harper, J.L. (1987b). The influence of neighbours on the growth of trees. II. The fate of buds on long and short shoots in *Betula pendula*. *Proceedings of the Royal Society, Series B*, **232**, 19–33.

King, D.A. (1981). Tree dimensions: maximizing the rate of height growth in dense stands. *Oecologia*, **51**, 351–356.

King, D.A. (1986). Tree form, tree height, and the susceptibility to wind damage in *Acer saccharum*. *Ecology*, **67**, 980–990.

King, D.A. (1987). Load bearing capacity of understorey treelets of a wet tropical forest. *Bulletin of the Torrey Botanical Club*, **114**, 419–428.

King, D.A. (1991). The allometry of trees in temperate and tropical forests. *Research and Exploration*, 7, 342–351.
King, D.A. (1994). Influence of light level on the growth and morphology of saplings in a Panamanian forest. *American Journal of Botany*, 81, 948–957.
King, D.A. (1996). Allometry and life history of tropical trees. *Journal of Tropical Ecology*, 12, 25–44.
King, D.A. & Loucks, O.L. (1978). The theory of tree bole and branch form. *Radiation and Environmental Biophysics*, 15, 141–165.
Kingsolver, J.G. & Schemske, D.W. (1991). Path analyses of selection. *Trends in Ecology and Evolution*, 6, 276–280.
Kira, T. (1978). Community architecture and organic matter dynamics in tropical lowland rain forests of Southeast Asia with special reference to Pasoh Forest, West Malaysia. In *Tropical Trees as Living Systems* (Ed. by P.B. Tomlinson & M.H. Zimmermann), pp. 561–590. Cambridge University Press, Cambridge.
Kohyama, T. (1980). Growth pattern of *Abies mariesii* saplings under conditions of open-growth and suppression. *Botanical Magazine Tokyo*, 93, 13–24.
Kohyama, T. & Hotta, M. (1990). Significance of allometry in tropical saplings. *Functional Ecology*, 4, 515–521.
Kuuluvainen, T. (1992). Tree architectures adapted to efficient light utilization: is there a basis for latitudinal gradients? *Oikos*, 65, 275–284.
Küppers, M. (1989). Ecological significance of above ground architectural patterns in woody plants: a question of cost–benefit relationships. *Trends in Ecology and Evolution*, 4, 375–379.
Leigh, E.G., Jr (1990). Tree shape and leaf arrangement: a quantitative comparison of montane forests, with emphasis on Malaysia and South India. In *Conservation in Developing Countries: Problems and Prospects* (Ed. by J.C. Daniel & J.S. Serrao), pp. 119–174. Bombay Natural History Society, Oxford University Press, Bombay.
Levins, R. (1968). *Evolution in Changing Environments: Some Theoretical Explorations*. Princeton University Press, Princeton, NJ.
Lieberman, M., Lieberman, D. & Peralta, R. (1989). Forests are not just Swiss cheese: canopy stereogeometry of non-gaps in tropical forests. *Ecology*, 70, 550–552.
Lieberman, M., Lieberman, D., Peralta, R. & Hartshorn, G.S. (1995). Canopy closure and the distribution of tropical forest tree species at La Selva, Costa Rica. *Journal of Tropical Ecology*, 11, 161–178.
Lindenmayer, A. (1968). Mathematical models for cellular interactions in development. *Journal of Theoretical Biology*, 18, 280–315.
Loubry, D. (1994). *Déterminisme du compartement phenologique des arbres en forêt tropicale humide de Guyane française*. PhD thesis, Université Paris 6, Paris.
Loup, C. (1994). *Essai sur le déterminisme de la variabilité architectectural des arbres*. PhD thesis, Université Montpellier II, Montpellier.
Maillette, L. (1982a). Structural dynamics of silver birch. I. The fates of buds. *Journal of Applied Ecology*, 19, 203–218.
Maillette, L. (1982b). Structural dynamics of silver birch. II. A matrix model of the bud population. *Journal of Applied Ecology*, 19, 219–238.
Maillette, L. (1987). Effects of bud demography and elongation patterns on *Betula cordifolia* near the tree line. *Ecology*, 68, 1251–1261.
Mattheck, C.I. (1991). *Trees: The Mechanical Design*. Springer Verlag, Berlin.
Mattheck, C.I. (1995). Biomechanical optimum in woody stems. In *Plant Stems: Physiology and Functional Morphology* (Ed. by B.L. Gartner), pp. 3–49. Academic Press, San Diego.
Mattheck, C.I., Bethge, K. & Schafer, J. (1993). Safety factors in trees. *Journal of Theoretical Biology*, 165, 185–189.

McMahon, T.A. (1973). Size and shape in biology: elastic criteria impose limits on biological proportions, and consequently on metabolic rates. *Science*, **179**, 1201–1204.

Meave del Castillo, J. (1987). *Longevidad de las hojas de tres especies de arboles perennifolios de selva tropical humeda*. MSc thesis, UNAM, Mexico D.F.

Niklas, K.J. (1992). *Plant Biomechanics: An Engineering Approach to Plant Form and Function*. University of Chicago Press, Chicago.

Niklas, K.L. (1994a). The allometry of safety factors for plant height. *American Journal of Botany*, **81**, 345–351.

Niklas, K.J. (1994b). Interspecific allometries of critical buckling height and actual plant height. *American Journal of Botany*, **81**, 1275–1279.

Niklas, K.J. (1995). Size-dependent allometry of tree height, diameter and trunk-taper. *Annals of Botany*, **75**, 217–227.

Oberbauer, S.F., Clark, D.B., Clark, D.A. & Quesada, M. (1988). Crown light environments of saplings of two species of rain forest emergent trees. *Oecologia*, **75**, 207–212.

O'Brien, S.T., Hubbell, S.P., Spiro, P., Condit, R. & Foster, R.B. (1995). Diameter, height, crown, and age size relationships in eight neotropical tree species. *Ecology*, **76**, 1927–1939.

O'Brien, S.T., Hubbell, S.P., Condit, R., Loo de Lao, S. & Foster, R. (1997). Height, crown size, and trunk diameter of 56 tree and shrub species in a neotropical forest. *Functional Ecology*, in press.

Oldeman, R.A.A. (1974). *L'Architecture de la Forêt Guyanaise*. Memoirs ORSTOM, **73**, 1–204.

Oldeman, R.A.A. (1990). *Forests: Elements of Silvology*. Springer, Berlin.

Oldeman, R.A.A. & van Dijk, J. (1991). Diagnosis of the temperament of rain forest trees. In *Rain Forest Regeneration and Management* (Ed. by Gómez-Pompa, T.C. Whitmore & M. Hadley), pp. 21–89. UNESCO, Paris.

Osunkoya, O.O., Ash, J.E., Graham, A.W. & Hopkins, M.S. (1993). Growth of tree seedlings in tropical rain forests of North Queensland, Australia. *Journal of Tropical Ecology*, **9**, 1–18.

Pearcy, R.W. (1989). Radiation and light measurements. In *Plant Physiological Ecology: Field Methods and Instrumentation* (Ed. by R.W. Pearcy, J. Ehleringer, H.A. Mooney & P.W. Rundel), pp. 97–116. Chapman & Hall, London.

Petraitis, P.S., Dunham, A.E. & Niewiarowski, P.H. (1996). Inferring multiple causality: the limitations of path analysis. *Functional Ecology*, **10**, 421–431.

Popma, J. & Bongers, F. (1988). The effect of canopy gaps on growth and morphology of seedlings of rain forest species. *Oecologia*, **75**, 625–632.

Popma, J. & Bongers, F. (1991). Acclimation of seedlings of three tropical rain forest species to changing light availability. *Journal of Tropical Ecology*, **7**, 85–97.

Popma, J. & Bongers, F. & Werger, M.J.A. (1992). Gap dependence and leaf characteristics of trees in a tropical lowland rain forest in Mexico. *Oikos*, **63**, 207–214.

Prusinkiewicz, P. & Lindenmayer, A. (1990). *The Algorithmic Beauty of Plants*. Springer, New York.

Putz, F.E. and Coley, P.D., Lu, K., Montalvo, A. & Aiello, A. (1983). Uprooting and snapping of trees: structural determinants and ecological consequences. *Canadian Journal of Forest Research*, **13**, 1011–1020.

Reich, P.B., Walters, M.B. & Ellsworth, D.S. (1992). Leaf life span in relation to leaf plant and stand characteristics among diverse ecosystems, *Ecological Monographs*, **62**, 365–392.

Rich, P.M., Helenurm, K., Kearns, D., Morse, S.R., Palmer, M.R. & Short, L. (1986). Height and stem diameter relationships for dicotyledonous trees and arborescent palms of Costa Rican tropical wet forest. *Bulletin of the Torrey Botanical Club*, **133**, 241–246.

Richards, P.W. (1952). *The Tropical Rain Forest. An Ecological Study*. Cambridge University Press, Cambridge.

Richards, P.W. (1983). The three-dimensional structure of the rainforest. In *Tropical Rain*

Forest: Ecology and Management, (Ed. by S.L. Sutton, T.C. Whitmore & A.C. Chadwick), pp. 3–10. Blackwell Scientific Publications, Oxford.

Room, P.M., Maillette, L. & Hanan, J.S. (1994). Modular and metamer dynamics and virtual plants. *Advances in Ecological Research*, 25, 105–157.

Sasaki, S. & Mori, T. (1981). Growth responses of dipterocarp seedlings to light. *Malaysian Forester*, 44, 319–345.

Shukla, R.P. & Ramakrishnan, P.S. (1984). Leaf dynamics of tropical trees related to successional status. *New Phytologist*, 97, 697–706.

Shukla, R.P. & Ramakrishnan, P.S. (1986). Architecture and growth strategies of tropical trees in relation to successional status. *Journal of Ecology*, 74, 33–46.

Sipe, T.W. & Bazzaz, F.A. (1994). Gap partitioning among maples (*Acer*) in Central New England: shoot architecture and photosynthesis. *Ecology*, 75, 2318–2332.

Sokal, R.R. & Rohlf, F. (1981). *Biometry: The Principles and Practice of Statistics in Biological Research*. W.H. Freeman, San Francisco.

Steingraeber, D.A. & Waller, D.M. (1986). Non-stationarity of tree branching patterns and bifurcation ratios. *Proceedings of the Royal Society, Series B*, 228, 187–194.

Sterck, F.J. (1997). *Trees and light: tree development and morphology in relation to light availability in a tropical rain forest in French Guiana*. PhD thesis, Wageningen Agricultural University, Wageningen.

Ter Steege, H. (1993). *Hemiphot, a Programme to Analyse Vegetation, Light, and Light Quality Indices from Hemispherical Photographs*. Tropenbos Documents, 3, Tropenbos Foundation, Wageningen.

Thomas, S.C. (1996). Asymptotic height as a predictor of growth and allometric characteristics in Malaysian rain forest trees. *American Journal of Botany*, 83, 556–566.

Tomlinson, P.B. (1987). Architecture of tropical plants. *Annual Review of Ecology and Systematics*, 18, 1–21.

Tucker, G.F., Hickley, Th.M., Leverenz, J. & Jiang, S.-M. (1987). Adjustments of foliar morphology in the acclimation of understorey Pacific silver fir following clearcutting. *Forest Ecology and Management*, 21, 249–268.

Turner, I.M. (1991). Effects of shade and fertilizer addition on the seedlings of two tropical woody pioneer species. *Tropical Ecology*, 32, 24–29.

Van der Meer, P.J. (1995). *Canopy dynamics of a tropical rain forest in French Guiana*. PhD thesis, Wageningen Agricultural University, Wageningen.

Van der Meer, P.J. & Bongers, F. (1996a). Patterns of treefalls and branchfalls in a neotropical rain forest in French Guiana. *Journal of Ecology*, 84, 19–30.

Van der Meer, P.J. & Bongers, F. (1996b). Formation and closure of canopy gaps in the rain forest at Nouragues, French Guiana. *Vegetatio*, 126, 167–179.

Vester, H.F.M. & Saldarriaga, J.G. (1993). Algunas características estructurales, arquitectónicas y florísticas de la sucesión secundaria sobre terrazas bajas en la region de Araracuara (Colombia). *Revista Facultad Nacional de Agronomía Medellin*, 46, 15–45.

Waller, D.M. (1986). The dynamics of growth and form. In *Plant Ecology* (Ed. by M.J. Crawley), pp. 291–320. Blackwell Scientific Publications, Oxford.

Welden, C.W., Hewett, S.W., Hubbell, S.P. & Foster, R.B. (1991). Sapling survival, growth and seedling establishment: relationship to canopy height in a neotropical forest. *Ecology*, 72, 85–96.

White, J. (1979). The plant as a metapopulation. *Annual Review of Ecology and Systematics*, 10, 109–145.

White, J. (1984). Plant metamerism. In *Perspectives on Plant Population Ecology* (Ed. by R. Dirzo and J. Sarukhan), pp. 15–47. Sinauer, Sunderland, MA.

Whitmore, T.C., Brown, N.D., Swaine, M.D., Kennedy, Y.D., Goodwin-Bailey, C.I. & Gong, W.K.

(1993). Use of hemispherical photographs in forest ecology: Measurement of gap size and radiation totals in a Bornean tropical rain forest. *Journal of Tropical Ecology*, **9**, 131–151.

Whitney, G.G. (1976). The bifurcation ratio as an indicator of adaptive strategy in woody plant species. *Bulletin of the Torrey Botanical Club*, **103**, 67–72.

Wright, S. (1934). The method of path coefficients. *Annals of Mathematics and Statistics*, **5**, 161–215.

7. LIMITS TO TREE SPECIES DISTRIBUTIONS IN LOWLAND TROPICAL RAINFOREST

E. M. VEENENDAAL and M. D. SWAINE

Department of Plant and Soil Science, University of Aberdeen, Cruickshank Building, St Machar Drive, Aberdeen AB24 3UU, UK

SUMMARY

1 The factors believed to determine the natural limits to species distribution in tropical forest are reviewed and the effects of changing climate noted. Most of the evidence for controlling influences is circumstantial with a dearth of experimental evidence.

2 The distributions of tropical forest tree species in West Africa in relation to environmental conditions are examined. Seasonal drought and soil fertility covary along a pronounced rainfall gradient and are the strongest candidates for determining the limits of species distributions.

3 Shadehouse experiments to test the growth responses of tree species seedlings to seasonal drought and soil type (acid, infertile soil taken from high rainfall forest and near-neutral, more fertile soil from moderate rainfall forest) are summarized and the results compared with transplants of seedlings between wet and moist forest.

4 In the drought experiment, a species with a wide range of natural distributions was subjected to a droughting regime in which soil matric potentials fell to values similar to those that occur annually in Moist Semi-deciduous forest. Seedling mortality varied greatly among species, exceeding 80% in species native to wet forest, and less than 5% for species whose distributions extend to lower rainfall areas.

5 In the soil-type experiment, a wide range of species were tested and included several expected from their natural distribution to show poor growth on infertile soil. Of 15 species tested, only three responded significantly to soil type. *Triplochiton scleroxylon* and *Mansonia altissima*, both deciduous Sterculiaceae absent from wet forest, showed significantly poorer growth on the infertile soil than on fertile soil at irradiances greater than about 10%. The drought-sensitive pioneer of wet forests, *Lophira alata*, showed the opposite response, growing better in the low fertility soil than in the more fertile.

6 Seedling transplants between wet and moist forest supported the differences in species drought responses recorded in the shadehouse experiments and showed that growth was faster for all species in the wet forest where seasonal drought is rare. The apparent sensitivity of *Mansonia altissima* to infertile soil was not evident in wet forest transplants, at least during the first 4 years.

7 Some preliminary results on ecophysiological differences among species are reported in a discussion of the factors likely to determine species distribution in these West African trees.

INTRODUCTION

It is widely taken as self-evident that species distributions are determined by environmental factors. Differences in species physiological tolerances and requirements are expressed in their differing abundances in relation to environmental variables or with composite environmental gradients (Austin 1985). Thus, species' natural distributions and the composition of vegetation reflect the environment and are widely used to predict it.

It is well known, however, that many species will grow outside their natural range if given the opportunity. Agriculturalists, foresters and horticulturalists have taken advantage of this plasticity, but mostly by trial and error. Some of the most successful and widely used tropical trees in plantations are exotics growing in environments similar to those of their provenance (e.g. *Tectona grandis* L.f., *Eucalyptus deglupta* Sm, *Gmelina arborea* Roxb.). However, species can also be grown in environments different from that of their natural range, in sites with lower rainfall, poorer soils or lower temperatures than might be expected. This is possible because man removes some limiting factors, most commonly by providing artificial dispersal, improving establishment and reducing competition. Cultivation thus reveals that the potential distributions of species are broader than their realized distributions. The natural distribution of a species is limited by more than physiology and is restricted by the actual opportunities for dispersal and establishment, by herbivore and pathogen pressure, climatic irregularities and by competition with other species. Such influences may account for outliers in a species' distribution: a conjunction of a rare dispersal event with timely arrival in a new forest gap may allow the establishment of a single plant or group of plants in unexpected outposts.

FACTORS DETERMINING SPECIES DISTRIBUTIONS

In Table 7.1 we enumerate some of the more important factors determining species distributions. The distinction between factors determining funda-

TABLE 7.1. Factors limiting tree species distribution.

Factor	For examples see:
Fundamental restrictions	
Temperature	Hogan *et al.* (1991)
Rainfall too little or soil too wet (flooding)	Hall & Swaine (1981), Wright (1992), Ter Steege (1994)
Nutrient supply inadequate or cation toxicity	Zech & Drechsel (1992), Grubb (1995)
Realized restrictions	
Mycorrhizal inoculum not present	Newbery *et al.* (1988), Alexander (1989), Bowman & Panton (1993), Lee & Alexander (1996)
Pathogen pressure	Augspurger (1983), Augspurger & Kelly (1984), Jarosz & Davelos (1995)
Dispersal ineffective or phenology mistimed	Veenendaal *et al.* (1996a), Veenendaal & Swaine (this chapter)
Large-scale disturbance (e.g. by fire)	Whitmore (1990), Swaine (1992), Hawthorne & Abu-Juam (1995)
Past climate fluctuations: species slow to spread from refugia	Adebisi Sowunmi (1986), Maley (1996)
Geographical limitations: species unable to transgress a natural boundary (e.g. Dahomey Gap)	Hall & Swaine (1981), Hawthorne (1996)
Competition with existing species	Richardson & Bond (1991)

mental and realized distributions is somewhat arbitrary; here the former is defined by the physical environment (Chapin, Rincón & Huante 1993).

In temperate regions, species distributions have often been related to temperature and water stress (Woodward 1987, Abrams 1994, Jeffree & Jeffree 1994, Sykes & Prentice 1995) and limits of different forest types appear closely related to minimum temperatures, for example −15°C for broadleaved evergreen temperate species or −40°C for some pines (Woodward 1987).

There is some evidence that tropical forest trees can respond physiologically and phenologically even to small changes in temperature, which in turn could influence distribution (Yanney Ewusie 1992). For instance, mast fruiting of dipterocarps has been linked with small changes in temperature (Ashton, Givnish & Appanah 1988). Hogan, Smith and Ziska (1991) suggested that tropical plants may well be more narrowly adapted to the prevailing temperature regime, because of the smaller fluctuations in the tropics, and thus be more sensitive to climate warming, but point out that physiological evidence is limited. Because of the lack of fluctuation, particularly in lowland tropical vegetation, water rather than temperature may have the most dominant influence on plant phenology. Temperature is, in any case, directly linked with the hydrological budget of forests through the process of evapotranspiration (Bruijnzeel 1989, Veenendaal et al. 1996a).

Most studies of lowland forest species distributions make links directly with rainfall or associated soil fertility gradients (Hall & Swaine 1976, Gentry & Emmons 1987, Newbery, Alexander, Thomas & Gartlan 1988, Russell-Smith 1991, Wright 1992, Oliveiro-Filho, Vilela, Carvalho & Gavilanes 1994). The relative importance with which authors view drought or soil fertility as key variables depends on their experience in different parts of the tropics. Where strong rainfall gradients exist seasonal drought is a more obvious candidate than in the 'ever wet' tropics. Subjective judgements such as these are sometimes the only basis for generating testable hypotheses about the factors limiting species distributions.

Effect of climate change on species distributions

To understand species distributions, we must consider not only current environmental conditions, but also their variation over time from the interannual scale to millennia. Historical pollen records in West Africa suggest large fluctuations in species and forest-type composition over the last 100 000 years associated with climatological variations (Maley 1996) implying changes in the distribution of individual species. Even recent climate fluctuations may have had a strong influence. Reynaud and Maley (1994) suggested, on the basis of pollen records for south western Cameroon, that occasional droughts over the last 200 years may have caused the dominance of *Lophira alata* (Nomenclature follows Hawthorne (1995) except where given) in some forest areas, while a present lack of regeneration of this species is attributed to a currently drier climate. This contrasts with earlier suggestions by Letouzey (1968) that past human disturbance would have been the cause for dominance of *L. alata*.

Recent severe droughts associated with the El Niño Southern Oscillation have had a marked impact on tree mortality in lowland tropical forests (Whitmore 1990) which could lead to changes in species distributions. In Sabah, the 1983 drought killed many large dipterocarps but the impact was enhanced by the associated fires which were particularly damaging in logged forests where there was up to 93% mortality of mature trees (Woods 1989, Goldammer & Siebert 1990). In West Africa, similar effects of fire associated with the 1983 drought were observed (Hawthorne & Abu-Juam 1995) where fire clearly limits the distribution of some species, causing significant differences in forest composition (Swaine 1992, Hawthorne

1994), accentuating the distinctive composition and expanding the area of 'fire-zone' forest recognized earlier (Hall & Swaine 1976, 1981). On Barro Colorado Island (Panama), overall mortality of tree species during the 1983 drought was elevated, but similar in both swamps and drier areas and affected both pioneers and non-pioneers. Some species were vulnerable to drought as adults while others, particularly pioneers, appeared vulnerable as small trees or seedlings (Howe 1990, Condit, Hubbell & Foster 1995).

Predictions of global climatic change suggest further changes will occur in species distribution. In the coming century, some parts of the world's precipitation may increase (for example, the Caribbean), but in West Africa the climate is expected to become drier (Bawa & Markham 1995). The manner in which species will respond to such climatological changes is difficult to predict as each species in a community may have a unique response to environmental and biotic variables (Chapin et al. 1993) and the plasticity of species response in relation to these factors for lowland tropical species is still largely unknown.

Spatial patterns in the environment and species distributions

Environmental limitation of species distributions operates at various scales. Water availability varies over a scale of kilometres with climate, and metres with soil type and topography (Webb 1959, 1968, Hall & Swaine 1981, Unwin & Kriedemann 1990, Furley, Proctor & Ratter 1992, Ter Steege, Jetten, Polak & Werger 1993, Bullock, Mooney & Medina 1995). As a result, the patterning of environmental conditions is not necessarily coherent so that spatial patterns of species occurrence may also lack coherence, particularly near geographical limits of species distributions. In West Africa, tree species largely confined to high rainfall forests may be found in forests with lower rainfall, but restricted to the lower (wetter) parts of the topography (Van Rompaey 1993, Hawthorne 1996). The scale of soil chemistry differences varies similarly, and differences may be abrupt when associated with lithological boundaries such as define ultramafic soils or those of heath forest (kerangas or caatinga on white sands). In such cases the association between species composition (including endemic species) and the unique environment is so strong that it appears unnecessary to test the implied hypotheses about the causes of species distributional limits.

The nature of the evidence for factors limiting species distributions

Most evidence for the factors controlling species distribution is circumstantial, based on associations with general environmental conditions or correlations between a species' abundance and a particular environmental variable. In fact, limiting factors are rarely specifically stated, but implied when floristic classifications are related to environment conditions (e.g. Hall & Swaine 1981). Even when classification is based on physiognomic characteristics such as the relative abundances of different life-forms (e.g. Webb 1959) these are expected to reflect limiting environmental conditions. If they do not, the classification has little benefit. The names applied to forest types also carry implied causes (e.g. wet, moist, dry forests) but many authors, cautious of such implications, have applied neutral descriptive terms commonly based on the leaf phenology of canopy trees (deciduous, evergreen, semi-deciduous, semi-evergreen; Richards 1952, Müller 1982, Walter & Breckle 1984, Whitmore 1990). Thus, there is a singular dearth of experimental evidence which tests hypotheses whether implied or explicitly stated.

We can argue that water and nutrient availability are the most ubiquitously variable factors in lowland tropical rainforest and thus likely to limit the distributions of many species. Since rainfall and soil fertility tend to covary (due to leaching under high rainfall) it is difficult to segregate their effects on individual species and an experimental approach is strongly indicated (Swaine 1996). In this review we focus primarily on the experimental evidence for seasonal drought and soil fertility as determinants of tree distribution, presenting our own evidence for West African forest species as case material.

SPECIES DISTRIBUTIONS IN WEST AFRICA

One of the best studied forest zones in terms of species distribution is the West African lowland tropical forest of Ghana (Hall & Swaine 1981, Hawthorne 1993, 1995, 1996). In addition, a wealth of information is becoming available on ecological aspects of fruiting and leaf phenology and ecological guild status (Hall & Swaine 1981, Agyeman 1994, Kyereh 1994, Hawthorne 1995, Swaine, Agyeman, Kyereh et al. 1997). This offers the opportunity to look in more detail at distribution patterns of individual species for which we have detailed ecological profiles and to test hypotheses for underlying physiological causes of species distribution.

The main forest types in Ghana were defined using floristic data (Wet and Moist Evergreen, Moist and Dry Semi-deciduous (Hall & Swaine 1976, 1981)) and are widespread in West Africa (Waterman, Meshal, Hall et al. 1978, Swaine & Hall 1986, Van Rompaey 1993). The forest types are associated with a strong gradient of annual rainfall which has had a marked effect on soil fertility. Soil pH varies from <4 in parts of the Wet Evergreen

Limits to tree species distributions

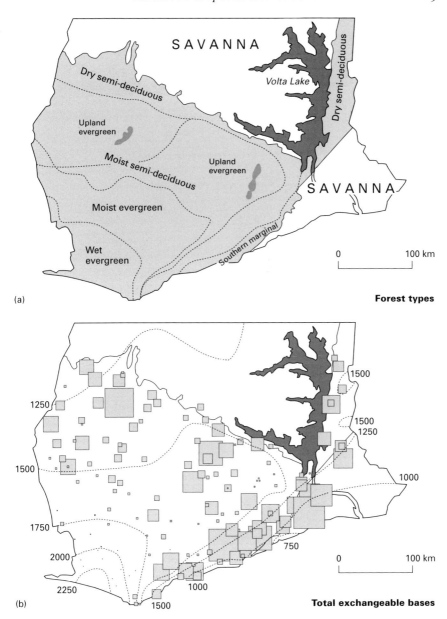

FIG. 7.1. Maps of (a) forest types within the forest zone (shaded) of Ghana, (b) the distribution of topsoil (0–15 cm) TEB (total exchangeable bases, symbol size proportional to concentration) and (c,d, continued on p. 170) the distribution of selected tree species in relation to approximate annual rainfall isohyets (dotted, mm). Species distributions are from Hawthorne (1993), total exchangeable bases from Swaine (1996).

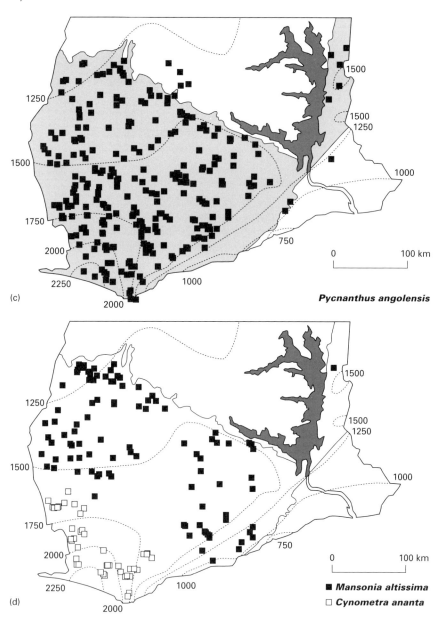

FIG. 7.1. *Continued.*

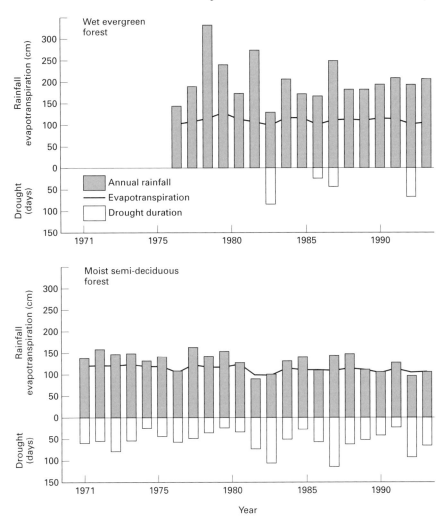

FIG. 7.2. Annual rainfall (solid bar), estimated actual evapotranspiration (line) and period (days year^{-1}) during which soil water deficit is >200 mm (shaded bar) for a Wet Evergreen forest (Draw River Forest Reserve; 5°12′ N, 2°20′ W) and a Moist Semi-deciduous forest (Tinte Bepo Forest Reserve; 7°04′ N, 2°06′ W) in Ghana. The evapotranspiration and soil water deficit were estimated from a water balance model using daily rainfall records for the sites in combination with monthly estimates of evapotranspiration from temperature data and periodic measurements of soil matric potential (see Veenendaal *et al.* 1996a).

forest to about pH 7 in Dry Semi-deciduous forests. Mineral nutrient concentrations in soil follow the same general trend (Fig. 7.1; Hall & Swaine 1976, 1981, Swaine 1996) but vary locally depending on parent material (highly weathered Precambrian formations, or younger acid granite intrusions or sandstone (Ahn 1970)). Most of these soils, however, are very ancient and deeply weathered, so that the effect of parent material is less marked than in younger soils. Local variation in parent material and topography may have more influence on hydrology.

The rainfall pattern results in differences in soil water regime between two contrasting forest types. Concurrent measurements of daily rainfall and soil matric potential in Wet Evergreen and Moist Semi-deciduous forest were combined with a temperature-dependent estimate of monthly evapotranspiration to develop a water balance model which was applied in both forest types (Veenendaal et al. 1996a). The model indicated that seasonal droughts, during which the soil matric potentials are reduced below permanent wilting point, occurred annually in the moist forest and only occasion-

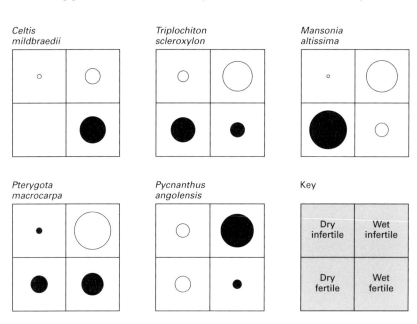

FIG. 7.3. Diagrammatic presentations of the occurrence of selected tree species amongst four categories of forest defined by differences in annual rainfall (greater or less than 1500 mm) and soil fertility (using a composite scalar including measurements of pH, total exchangeable bases (TEB), cation exchange capacity, C/N, P and bulk density). The frequency of a species occurrence in each cell is compared in a χ^2-test with the expected frequency based on all samples so that the areas of the circles in the cells are proportional to the squared deviations from expected; filled circles showing deviations greater than expected, open less (from Swaine 1996).

ally in wet forest. Figure 7.2 shows the variation in annual rainfall and calculated evapotranspiration over 2 years in a Wet Evergreen forest and in a Dry Semi-deciduous forest. Calculated periods when soil water deficit exceeded 200 mm are shown below the abscissa. The occasional droughts in the wetter forests are often associated with El Niño Southern Oscillation events (Diaz & Kiladis 1992).

An analysis of species occurrence in relation to soil fertility and annual rainfall for 51 common Ghanaian forest tree species (Swaine 1996) indicated that a large proportion of species show a preference, on a broad scale, for particular rainfall and soil fertility conditions, even when the distribution is apparently ubiquitous (cf. *Pycnanthus angolensis* in Figs 7.1 & 7.3). This suggests that both soil fertility and seasonal drought may limit the distribution of tree species in West Africa.

EXPERIMENTAL EVIDENCE FOR SEASONAL DROUGHT AS LIMITING FACTOR

Many tropical forest species from dry forest types regularly survive severe seasonal drought. Adaptations to drought may include drought avoidance through deciduousness, or drought resistance adaptations in leaf water relations, rooting patterns and stem anatomy (Table 7.2; and recent reviews: Wright 1992, Holbrook, Whitbeck & Mooney 1995, Mulkey & Wright 1996). Studies comparing wet and moist tropical forest species to assess the importance of these factors in determining spatial distribution patterns are few.

TABLE 7.2. Possible drought adaptations for lowland tropical trees. (From references in table and Holbrook *et al.* 1995.)

Adaptation	Source
Greater rooting depth of tap roots	Kummerow, Castellanos, Maass & Larigauderie (1990), Mulkey, Smith & Wright (1991), Huc *et al.* (1994)
Increased root volume to increase competition for available water	Murphy & Lugo (1986), Khalil & Grace (1992)
Deciduous or sclerophyllous leaves. Avoidance or resistance of drought associated with different stem anatomy to avoid cavitation and with different rooting patterns	Borchert (1994), Jackson & Grace (1994), Machado & Tyree (1994)
Changes in leaf water relations. Adaptations in osmotic potential or bulk modulus of elasticity	Sobrado (1986), Mulkey, Wright & Smith (1991), Khalil & Grace (1992), Fan *et al.* (1994)

Differences in leaf phenology and rooting pattern were associated with local distribution patterns in relation to water availability in dry Neotropical forest (Borchert 1994). Leaf water relations have been suggested to influence species distribution in warm temperate rainforest in Australia (Melick 1990) and along altitudinal gradients in Hawaii (Robichaux 1984). Genetically determined ecophysiological differentiation in leaf water relations and rooting pattern did not, however, lead to a difference in growth performance in *Psychotria* species from drier and wetter forest in Panama when the species were grown for 20 months at a common site on Barro Colorado Island (Hogan, Smith, Araus & Saavedra 1994).

Testing the hypothesis that the soil water regime accounts for interspecific differences in distribution should involve comparative experiments on growth and survival during dry periods in controlled experimental conditions as well as field transplant experiments. Such experiments should use seedlings in the first instance because young plants are likely to be more sensitive to drought stress than older trees (Condit, Hubbell & Foster 1995).

Drought avoidance and resistance strategies of individual species may be linked with successional status. For instance, pioneer trees may employ deciduousness to avoid cavitation and may have shallower rooting systems than non-pioneers (Sobrado 1993, Borchert 1994, Huc, Ferhi & Guehl 1994). Differences in plant water status have been observed in seedlings regenerating in the understorey of the forest compared to those regenerating in gaps, with the seedlings in the gap showing less negative leaf water potentials (Abrams & Mostoller 1995, Veenendaal, Swaine, Agyeman *et al.* 1996b). This suggests that differences in water availability within forests may confound species drought tolerance with shade tolerance.

We tested survival and recovery in a controlled droughting experiment and in a field transplant experiment in two contrasting sites for a number of West African tree species. In a simulated droughting trial, we used seedlings of species with differing natural distributions, regeneration niches and leaf phenology (Table 7.3). Drought-induced mortality >80% during the establishment phase (<1 year after germination) was demonstrated in the evergreen non-pioneer *Heritiera utilis* and the evergreen pioneer *Lophira alata*, species with a wet forest distribution (Fig. 7.4). In a third wet forest species, the evergreen non-pioneer *Cynometra ananta*, mortality was much less and more similar to the relatively drought-tolerant species from drier forest types. Surviving seedlings of this species, however, showed reduced regrowth and reduced leaf area ratio 6 weeks after rewatering more similar to the drought-sensitive species than to the drought-resistant species (Fig. 7.4; E.M. Veenendaal *et al.* unpublished data). Moist forest deciduous pioneer species *Pericopsis elata* and non-pioneers, *Mansonia altissima*, *Pterygota*

TABLE 7.3. *Ecological profiles of Ghanaian tree species used in a droughting experiment. Nomenclature and ecological guilds after Hawthorne (1995) and Veenendaal et al. (1996c). Leaf phenology after Hall and Swaine (1981) and personal observation's (authors). Abbreviations: P, pioneer; NPLD, non-pioneer light demander; NPSB, non-pioneer shade bearer; forest types are: WE, Wet Evergreen; UE, Upland Evergreen; MS, Moist Semi-deciduous; DS, Dry Semi-deciduous. Phenology: E, evergreen; D, deciduous.*

Species	Forest type distribution	Origin of seeds	Ecological guild	Leaf phenology	
				Seedling	Tree
Celtis mildbraedii	MS/DS	MS	NPSB/NPLD	E	E
Cynometra ananta	WE	WE	NPSB	E	E
Heritiera utilis	WE	WE	NPLD/NPSB	E	E
Lophira alata	WE/UE	WE	P	E	E
Mansonia altissima	MS/DS	MS	NPLD/P	E/D	D
Pericopsis elata	MS/DS	MS	NPLD/P	D	D
Pterygota macrocarpa	MS/DS	MS	NPLD/NPSB	E	D
Pycnanthus angolensis	Ubiquitous (wet sites)	MS	NPLD	E	E
Terminalia ivorensis	Ubiquitous	MS	P	D	D

macrocarpa and the evergreen *Celtis mildbraedii*, showed only a low mortality. Surprisingly, one of the two ubiquitous species, *Pycnanthus angolensis*, also showed high mortality under simulated drought (Fig. 7.4). This species occurs characteristically in disturbed forest and farmers' fields and has been associated with wetter parts of the topography (Fig. 7.3; Keay, Onochie & Stanfield 1964, Swaine 1996).

Seedlings exhibited differences in leaf retention during the trial, which were associated with their ecological guilds. The drought-tolerant pioneers, *P. elata* and *Terminalia ivorensis*, were deciduous losing their leaves within 2–3 weeks after permanent wilting. To a lesser extent this was also true in *M. altissima*. The drought-sensitive pioneer species, *L. alata*, kept green leaves until the seedlings died. Leaves of non-pioneer species remained green throughout, or until they died (E.M. Veenendaal *et al.* unpublished data). In this respect the phenology of young seedlings in terms of leaf phenology differs from larger trees of the same species. In this experiment, seedlings of *P. macrocarpa* retained their leaves but the species is deciduous as a large tree. The fact that young seedlings and saplings of tropical forest tree species differ in phenological behaviour and that this process may be influenced by their shade tolerance has been noted before (Bongers & Popma 1990, Yanney Ewusie 1992) but systematic observations relating plant size and phenology are lacking.

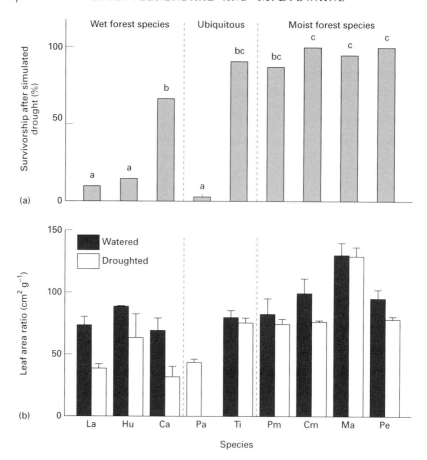

FIG. 7.4. Tree species responses to drought. (a) Percentage of tree seedlings surviving to form new leaves after being subjected to a simulated seasonal drought, and (b) leaf area ratio (bar shows +1 SE) 6 weeks after recovery in droughted (open bar) and in similarly-sized non-droughted seedlings (solid bar). The species are: La, *Lophira alata*; Hu, *Heritiera utilis*; Ca, *Cynometra ananta*; Pa, *Pycnanthus angolensis*; Ti, *Terminalia ivorensis*; Pm, *Pterygota macrocarpa*; Cm, *Celtis mildbraedii*, Ma, *Mansonia altissima*, Pe, *Pericopsis elata*. Other abbreviations as in Table 7.3. Columns sharing a letter in (a) are not significantly different at $P < 0.05$ with Fisher's exact test. (From E.M. Veenendaal *et al.* unpublished data.)

In a field transplant experiment in which wet forest species, *L. alata*, *P. angolensis*, *C. ananta* and the moist forest species, *M. altissima* were planted in forest gaps in a wet and in a moist forest site, mortality was highest in the moist forest site, corroborating the importance of drought over soil fertility for seedling mortality (Fig. 7.5). The field experiment thus supported the

Limits to tree species distributions

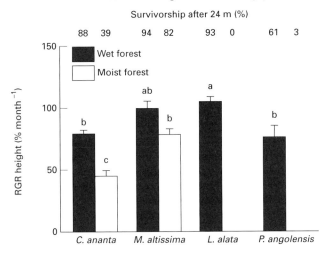

FIG. 7.5. Survival and relative growth rate (RGR) of height of saplings of selected tree species growing in gaps of wet (solid columns, Draw River) and moist (open columns, Tinte Bepo) forest for a period of 24 months. Abbreviations as in Fig. 7.4. Bars show +1 SE; columns sharing a letter do not differ significantly at $P < 0.005$ with Tukey's HSD test.

experimental results of drought sensitivity in *P. angolensis* and *L. alata*. A comparison of relative rate of height growth of seedlings in the field showed a non-significant reduction in the drier forest for *M. altissima* but a significant reduction in *C. ananta*, suggesting a higher sensitivity to drought in the latter species (Fig. 7.5).

If drought is a major factor limiting tree species distribution, then adaptive traits as suggested by Holbrook *et al.* (1995) for dry forest species (Table 7.4) should be present in our species. We have already noted differences in leaf retention between species as a means to avoid excessive desiccation, but differences in leaf water relations in species with an evergreen leaf strategy could also be expected. In a glasshouse experiment (12 h light, irradiance, photosynthetic proton flux density (PPFD) $<200\,\mu mol\,m^{-2}\,s^{-1}$), one wet and one moist forest species were grown in fertile topsoil and subjected to slow droughting up to permanent wilting point (6 weeks). The moist forest species, *P. macrocarpa*, was able to maintain turgor to a lower leaf water potential than the wet forest species, *H. utilis* (G. Burnett & E.M. Veenendaal, unpublished data). Differences in turgor and leaf water potential were significant between the two species, but not between the droughting treatments (two-way analysis of variance, $P > 0.05$) even though the data

TABLE 7.4. Characteristics of leaf water relations in droughted and non-droughted plants of *Pterygota macrocarpa* and *Heritiera utilis* (both Sterculiaceae). Abbreviations: tlp, turgor loss point; rwc, relative water content; π_{wilt}, osmotic potential at tlp; π_{sat}, osmotic potential at full turgor; ε_{max}, maximum bulk modulus of elasticity.

			Parameter			
Species	Treatment	n	tlp (1-rwc)	π_{wilt} (MPa)	π_{sat} (MPa)	ε_{max} (MPa)
P. macrocarpa	Watered	6	0.13	−2.14	−1.81	24
	Droughted	4	0.20	−2.27	−1.78	13
H. utilis	Watered	6	0.12	−1.59	−1.38	14
	Droughted	4	0.11	−1.73	−1.48	17

suggest that *P. macrocarpa* responded more to the droughting by adjustment of its cell wall elasticity thus increasing dehydration tolerance. Changes in cell wall elasticity are more often observed in trees subjected to mild drought stress (Fan, Blake & Blumwald 1994, Dias-Filho & Dawson 1995) and may help plants to maintain turgor pressure over a wide range of tissue water content. Generally, however, an increase in maximum elastic modulus can be expected, signifying more rigid cell walls that can support a more negative turgor pressure (Mulkey Wright & Smith 1991, Khalil & Grace 1992). Dias-Filho and Dawson (1995), however, showed the contrasting response in two co-occurring gap-invading species. Clearly, the way in which tropical tree species deal physiologically with drought and irradiance stress needs further investigation.

Deciduous species are likely to differ in their response from evergreen species in that their leaves are expendable and therefore less protected against desiccation (Sobrado 1986, Fanjul & Barradas 1987, Holbrook *et al.* 1995). Some evidence for this was provided from observations on *Terminalia superba*, a drought-resistant ubiquitous deciduous pioneer growing in moist semi-deciduous forest in Ghana. Osmotic potentials in leaves collected from drought-stressed seedlings in the forest during the dry season (Veenendaal *et al.* 1996b) were around −1.4 MPa in seedlings growing in the shade and −1.7 MPa in seedlings growing in the open (E.M. Veenendaal unpublished data). These findings are in line with differences found between deciduous and evergreen species in dry forest (Holbrook *et al.* 1995). Many questions remain as leaf water relations may change which leaf age, plant age and changes in irradiance, and drought tolerance may not be a constant feature throughout the life of a tree (Condit *et al.* 1995).

We conclude, however, that seedlings of species from wet forest show clear signs of higher sensitivity to drought and that seasonal droughts such as those observed in moist forest are an important limiting factor for species distribution.

EXPERIMENTAL EVIDENCE FOR SOIL FERTILITY AS A LIMITING FACTOR

Rainfall gradients are often closely related with soil fertility gradients because a long history of higher rainfall causes cumulative leaching of plant nutrients in wet forests. However, the way in which soil fertility influences tree distribution and particularly which nutrients are important is still very much an open question (for a recent review see Grubb 1995). There are many factors that confound the interpretation of soil fertility and nutrient availability in the field. For instance, soil nutrient availability can be highly heterogeneous, both spatially between gaps and closed forest or between lower and higher parts of the topography, and temporally, as in the transition from the dry to the wet season when there is a flush of nutrients due to accelerated decomposition of organic matter accumulated during the preceding dry period (Vitousek & Denslow 1986, Lodge, McDowell & McSwiney 1994, Silver, Scathena, Johnson et al. 1994, Sovan Roy & Singh 1994, Vogt, Vogt, Asbjorsen & Dahlgren 1995). This nutrient flush is similar to that in newly formed gaps, so that a capacity for rapid fine root growth in seedlings will permit rapid nutrient uptake (Grubb 1994) and may be advantageous in both situations.

Considerable emphasis has been placed on P as potentially the most important limiting nutrient in tropical forest (Vitousek & Sanford 1986, Silver 1994). However, P addition does not always lead to increased litter productivity or tree growth in forest ecosystems, possibly because underground productivity increases first (Denslow, Vitousek & Schultz 1987, Silver 1994, Grubb 1995). At the species level, several studies looking at species responses to increased nutrient supply have demonstrated an increase in growth when P is added. This response may be stronger in species native to more fertile soils and may be strongest in fast-growing pioneer species, which may use enhanced P uptake for growth compared to shade tolerant non-pioneers, which may store additional P (Tanner, Kapos, Freskos et al. 1990, Turner 1991, Raaimakers 1994, Burslem, Grubb & Turner 1995, Huante, Rincón & Chapin 1995, Rincón & Huante 1995, Raaimakers & Lambers 1996). Although a positive response is sometimes shown in species from tropical forests (Huante, Rincón & Acosta 1995), some experiments show no response by species to an increased P supply or point to other nutrients (e.g. Mg), or show different responses in different species (Turner, Brown & Newton 1993, Burslem *et al.* 1995, Burslem 1996).

Ancient soils in high rainfall areas typically have low nutrient concentrations and low pH. A low soil pH increases the solubility of some cations and may result in increased toxicity caused by H^+, Al and Mn and a reduced uptake of most other nutrients (Cuenca, Herrera & Medina 1990, Marshner 1991). Toxicity and cation deficiency are thus as likely as P deficiency to limit species.

Nutrient stresses in forestry plantations may provide an insight into potential nutrient limitations for growth in the planted species, even when the evidence (e.g. of foliar nutrient concentrations) is circumstantial and can be difficult to interpret because cause and effect cannot be explicitly distinguished. In West Africa, foliar nutrient studies have been reported for broadleaved indigenous trees in Liberia (Zech & Drechsel 1992). They suggest that both multiple nutrient deficiencies and Al toxicity limit tree growth of individual species on acid soils.

We designed an experiment to test growth performance of seedlings of Ghanaian tree species during the early establishment phase on contrasting soil types. For this we collected topsoil and subsoil from Wet Evergreen (Draw River) and Moist Semi-deciduous forest (Tinte Bepo); the former with lower pH and nutrient availability. These soils were fairly representative of their respective forest types, but somewhat richer in Ca and Mg (Table 7.5). Seedlings were grown in well-watered soil profiles reconstructed in pots at a common site. We expected species generally to grow

TABLE 7.5. Mean acidity and nutrient concentrations (mg kg^{-1} oven dry weight) in topsoil (0–10 cm) and subsoil (10–30 cm) from two contrasting forest sites ($n = 6$–9). Values in the same row sharing a superscript do not differ significantly at $P < 0.05$ according to Tukey's HSD method. Forest type means from Hall and Swaine (1976) are given for comparison.

Soil variable	Unit	Wet Evergreen forest			Moist Semi-deciduous forest		
		Topsoil	Subsoil	Type mean	Topsoil	Subsoil	Type mean
pH(H$_2$O)		4.87[a]	4.56[a]	4.2	6.67[b]	6.12[b]	5.6
N (total)	(mg g^{-1})	1.31[a]	0.87[a]	2.1	2.22[b]	0.91[a]	2.5
C (total)	(mg g^{-1})	18.6[a]	11.4[a]	22.4	23.5[b]	8.9[a]	22.2
C/N Ratio		14.3[a]	13.2[b]	—	10.7[c]	8.9[c]	—
P (avail.)	(mg kg^{-1})	6.9[a]	2.7[b]	6.6	13.6[c]	1.6[d]	12.2
Ca (avail.)	(mg kg^{-1})	1040[b]	302[a]	53	2451[c]	952[b]	1335
Mg (avail.)	(mg kg^{-1})	200[b]	59[a]	35	263[b]	151[b]	205

better in the more fertile soil (see for example, Huante Rincón & Acosta 1995) and pioneers to show a stronger response than shade-tolerant species because of a greater demand for soil nutrients imposed by higher relative growth rates (Escudero, Del Acro, Sanz & Ayala 1992, Thompson, Kriedemann & Craig 1992, Huante, Rincón & Chapin 1995, Raaimakers, Boot, Dijkstra & Pot 1995, Raaimakers & Lambers 1996). Species restricted in their natural distributions to more fertile soils were expected to grow poorly in the Wet Evergreen soil. As pioneers and shade-tolerant species can be expected to show maximum growth at different irradiances, the interaction between soil fertility and irradiance was included as a factor in the experiment.

We tested a total of 15 species (Table 7.6) differing in ecological guild (four pioneers, six non-pioneer light-demanders and five non-pioneer shade-bearers *sensu* Hawthorne 1995) and natural distribution, over a range of irradiances (Veenendaal, Swaine, Lecha *et al.* 1996c). Five of these 15, *Celtis mildbraedii, Mansonia altissima, Pericopsis elata, Pterygota macrocarpa* and *Triplochiton scleroxylon*, do not occur in the most infertile forest type.

TABLE 7.6. Species tested for soil type preference, their distributional bias (from Swaine 1996) and their preference for soil type in the pot experiment. The infertile soil was collected in Wet Evergreen forest, the fertile soil from Moist Semi-deciduous forest. Abbreviations are as in Table 7.3.

Species	Ecological guild	Family	Bias in soil type distribution	Soil type preference
Heritiera utilis	NPLD	Sterculiaceae	Infertile	None
Cynometra ananta	NPSB	Caesalpiniaceae	Infertile	None
Lophira alata	P	Ochnacaceae	Infertile	Infertile
Chrysophyllum pruniforme	NPSB	Sapotaceae	Infertile	None
Strombosia glaucescens	NPSB	Olacaceae	Infertile	None
Guarea cedrata	NPSB	Meliaceae	Infertile	None
Albizia zygia	NPLD	Mimosaceae	Ubiquitous	None
Blighia sapida	NPLD	Sapindaceae	Ubiquitous	None
Terminalia ivorensis	P	Combretaceae	Ubiquitous	None
Milicia excelsa	P	Moraceae	Ubiquitous	None
Celtis mildbraedii	NPSB	Ulmaceae	Fertile	None
Pterygota macrocarpa	NPLD	Sterculiaceae	Fertile	None
Triplochiton scleroxylon	P	Sterculiaceae	Fertile	Fertile
Mansonia altissima	NPLD/P	Sterculiaceae	Fertile	Fertile
Pericopsis elata	NPLD/P	Papilionaceae	Fertile	None

In our experiment only three species responded significantly in relative growth rate (RGR) to soil fertility at the seedling stage (Fig. 7.6). Of these, *L. alata* and *T. scleroxylon* are true pioneers, while *M. altissima* is somewhat more shade-tolerant. Each species responded differently to soil type. *T. scleroxylon* grew faster on more fertile soil from 6 to 8% of ambient irradiance upwards and the effect was highest at the highest irradiance level with up to 50% reduction in RGR in plants growing in the infertile soil. In *M. altissima* the effect was strong with 70% reduction in RGR relative to fertile soil and yellowing of the leaves above 16% of ambient irradiance (Fig. 7.6). In contrast, *L. alata* showed a 40% reduction in RGR in the fertile soil with intercostal chlorosis in seedlings growing at the highest irradiance. Swaine (1996) showed that both *M. altissima* and *T. scleroxylon* are biased in their natural distribution to more fertile soils (see Fig. 7.3).

Although soil affected RGR only in three of the species, others showed an influence of soil on foliar nutrient concentrations. In these species, the foliar nutrient concentrations of N (not shown), P and Mg were often higher in plants growing in the fertile soil (Fig. 7.7). Moist forest species in particular tended to have higher foliar nutrient concentrations in P and Mg than the wet forest species (Veenendaal *et al.* 1996c).

Manganese (not shown), which is more mobile in acid soils, was more often found in higher concentrations in plants growing in the acid soil, although the concentrations were too low to be toxic (>50 mg kg^{-1}; see also

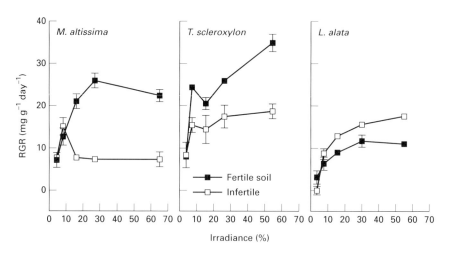

FIG. 7.6. Relative growth rate of three species in soil from Wet Evergreen (infertile, open symbol) and Moist Semi-deciduous (fertile, solid symbol) forests in relation to irradiance (% unshaded). (From Veenendaal *et al.* 1996c.)

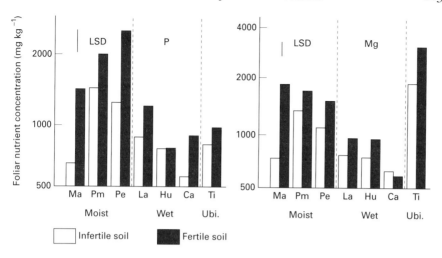

FIG. 7.7. Foliar nutrient concentrations of phosphorus and magnesium in forest tree seedlings (abbreviations as in Fig. 7.4) growing on infertile (open bar) and fertile (solid bar) forest soil. (From Veenendaal et al. 1996c.)

Zech & Drechsel 1992). In conclusion, the results from both our growth experiment and the foliar nutrient concentrations suggest that soil fertility may be a limiting factor at least for some species.

DISCUSSION

The results of our experiments have given clear indications about the factors limiting some species distributions, and revealed uncertainty in others. The restriction of *H. utilis* to wet forest is adequately explained by its intolerance of drought. For *L. alata*, its drought intolerance excludes it from dry forest except at low elevations on the topography where water is more available (see Van Rompaey 1993). Conversely, the drought tolerance of *T. ivorensis*, *P. macrocarpa*, *C. mildbraedii*, *M. altissima* and *P. elata* adequately explains their ubiquitous distribution or ability to grow in dry forest. However, to explain the exclusion of all except *T. ivorensis* from wet forest, we must invoke other factors. For *M. altissima* (and *T. scleroxylon*) their poor growth in pots of infertile soil (Fig. 7.6) suggests nutrient limitation (but see below for *M. altissima*), but for the remainder, inferior competitive abilities compared with wet forest specialists seems the better hypothesis.

Two exceptions are evident in Fig. 7.4. Why is *C. ananta* so restricted to the wettest part of the forest when our controlled and field experiments suggest that the species is intermediate in drought resistance? *C. ananta* seeds are dispersed explosively when the pods dry out during the Harmattan (the dry Saharan winds) at the end of December. The seeds are only viable for less than 4 weeks (personal observation by the authors) so that germination is presumably restricted to those forests with significant rainfall in the dry season, as occurs in Wet Evergreen forests (see Fig. 7.2). This combination of traits is likely to impose a severe limit on its ability to spread and establish in drier forest, where the driest part of the season lasts 2–3 months (Veenendaal *et al.* 1996a).

The other exception is the drought sensitivity observed in *P. angolensis*, a species with a very wide natural range in West Africa (Keay *et al.* 1964). Its establishment may be aided by local variability in soil water conditions in the field, but there may also have been experimental limitations. *P. angolensis* has a relative fleshy root system and probably forms a large tap root. This adaptation is not necessarily advantageous in a pot experiment, where rooting space is limited. However, *P. angolensis* also showed high mortality when transplanted into moist forest (Fig. 7.5) Tropical tree species are likely to differ in the ways they survive periodic water shortage in moist forests because of marked differences in shade tolerance and in leaf phenology (Table 7.3). Differences in leaf phenology and water status have been observed in seedlings grown in gaps compared to closed canopy (Bongers & Popma 1990, Veenendaal *et al.* 1996b). In the study of our West African species, we have not yet studied adaptations in root volume and depth to increasing drought which have been shown to be relevant in other studies (Mulkey, Smith & Wright, 1991, Khalil & Grace 1992). In ubiquitous species, we need also to test for possible genotypic differentiation.

Although *M. altissima* grew consistently poorly in a pot experiment on acid soil it is still (after 4 years—T. Baker personal communication) growing well in wet forest in a field experiment (compare Fig. 7.5 with Fig. 7.6). Factors other than soil fertility may thus play a role. A lack of mycorrhizal infection may dramatically reduce growth in establishing seedlings (Alexander 1989, Burslem *et al.* 1995), while the role of pathogens in tropical soils has been little researched, and may be significant (Augspurger 1983, Augspurger & Kelly 1984, Dobsen & Crawley 1994, Jarosz & Davelos 1995).

Finally, why do so few species respond to soil fertility, when uptake of nutrients appears to be enhanced in more fertile soils? Possibly the solution lies in changes in the rate of resource acquisition and its allocation within the plant as the tree grows larger. A particularly important aspect for the nutrient balance may be whether the species is deciduous or evergreen (Sands & Mulligan 1990). Evergreen species are particularly efficient in nutrient use, in that they reduce nutrient losses by prolonging leaf life-span (Sobrado 1991, Aerts 1995). Many species (e.g. *P. macrocarpa*, Table 7.3), are evergreen as young saplings and become deciduous later. There is currently a debate on whether the efficient resorption of nutrients before leaf fall is associated with low soil fertility. Although Pugnaire and Chapin (1993) did not find a difference in resorption capacity of species adapted to nutrient-poor sites in a Mediterranean climate, Grubb (1977, 1995) suggests that resorption capacity of trees may increase with increasing rainfall and thus with decreasing soil fertility. Our observations suggest that leaf nutrient concentrations in young fully expanded leaves in seedlings of tree species from moist forest types are often higher than those occurring in wet forest, even when they are grown in the same soil. If the resorption capacity of drier forest species is less than that of wet forest species (including the capacity of resorption of nutrients from roots as well as leaves) this could easily reduce the competitive ability of deciduous compared to evergreen species during advanced stages of sapling growth. Together with the effects of root growth and competition on nutrient uptake these matters deserve further attention.

CONCLUSIONS

From our work on West African tree seedlings, it is becoming clear that species display a wide range of tolerance to the seasonal droughts that occur in the drier parts of the forest zone. How they differ in their physiological adaptation is not yet known in detail but pointers for further research can be taken from the work on drier forest species (Holbrook *et al.* 1995). Soil fertility appears less often to be a determining factor in the distribution of tree species, at least during the early stages of life. Only three species out of 16 responded to soil type, despite a biased selection of species for testing to include those most likely to respond on the basis of natural distribution and associated soil conditions. In focusing on water and nutrients we must be careful to recognize that other factors may assume equal or greater importance in particular species. Little is known about the potentially important influences of plant pathogens and mycorrhizas which need to be studied in more detail before we can fully distinguish between species potential and realized distributions (Austin & Gaywood 1994).

The experiments summarized here highlight the need for experimental tests of the causal relationships implied by the correlation of species distribution with environmental conditions. The results from pot experiments, in which causes may be more explicitly tested, are somewhat unrealistic, largely because they lack the important influence of competition with other forest plants. Our early results from forest transplants lend support for some of the results from pot experiments but they also show that the expression of species differences may be significantly modified by forest conditions.

ACKNOWLEDGEMENTS

We are grateful to many colleagues in the Forestry Research Institute Ghana where this study was made, and to the Government of Ghana Forestry Department for permission to make manipulative experiments in their Forest Reserves. The experimental studies were funded by the ODA Forestry Research Programme (Oxford Forestry Institute ref. R4740). Useful comments were made on an earlier draft of this paper by P. Becker, D.B. Clark and P.J. Grubb.

REFERENCES

Abrams, M.D. (1994). Genotypic and phenotypic variation as stress adaptations in temperate tree species: a review of several case studies. *Tree Physiology*, **14**, 833–842.

Abrams, M.D. & Mostoller, S.A. (1995). Gas exchange leaf structure and nitrogen in contrasting successional tree species growing in open and understory sites during a drought. *Tree Physiology*, **15**, 361–371.

Adebisi Sowunmi, M. (1986). Change of vegetation with time. In *Plant Ecology in West Africa* (Ed. by G.W. Lawson), pp. 273–307. Wiley, Chichester.

Aerts, R. (1995). The advantages of being evergreen. *Trends in Ecology and Evolution*, **10**, 402–407.

Agyeman, V.K. (1994). *Environmental influences on tropical tree seedling growth*. PhD thesis, University of Aberdeen.

Ahn, P.M. (1970). *West African Soils*. Oxford University Press, Oxford.

Alexander, I. (1989). Mycorrhizas in tropical forests. In *Mineral Nutrients in Forest and Savanna Ecosystems* (Ed. by J. Proctor), pp. 169–188. Blackwell Scientific Publications, Oxford.

Ashton, P.S., Givnish, T.J. & Appanah, S. (1988). Staggered flowering in the Dipterocarpaceae: new insights in the evolution of mast fruiting in the aseasonal tropics. *American Naturalist*, **132**, 44–66.

Augspurger, C.K. (1983). Seed dispersal in the tropical tree *Platypodium elegans* and the escape of its seedlings from fungal pathogens. *Journal of Ecology*, **71**, 759–772.

Augspurger, C.K. & Kelly, C.K. (1984). Pathogen mortality of tropical tree seedlings: Experimental studies of the effects of dispersal distance, seedling density and light conditions. *Oecologia*, **65**, 211–217.

Austin, M.P. (1985). Continuum concept, ordination methods and and niche theory. *Annual Review of Ecology and Systematics*, **16**, 39–61.

Austin, M.P. & Gaywood, M.J. (1994). Current problems of environmental gradients and species response curves in relation to continuum theory. *Journal of Vegetation Science*, **5**, 473–482.

Bawa, K.S. & Markham, A. (1995). Climate change and tropical forests. *Trends in Ecology and Evolution*, **10**, 348–349.

Bongers, F. & Popma, J. (1990). Leaf dynamics of seedlings of rain forest species in relation to canopy gaps. *Oecologia*, **82**, 122–127.

Borchert, R. (1994). Soil and stem water determine phenology and distribution of tropical dry forest species. *Ecology*, **75**, 1437–1449.

Bowman, D.M.J.S. & Panton, W.J. (1993). Factors that control monsoon rain forest seedling establishment and growth in north Australian *Eucalyptus* savanna. *Journal of Ecology*, **81**, 297–304.

Bruijnzeel, L.A. (1989). Moist tropical nutrient cycling: the hydrological framework. In *Mineral Nutrients in Forest and Savanna Ecosystems* (Ed. by J. Proctor), pp. 383–416 Blackwell Scientific Publications, Oxford.

Bullock, S.H., Mooney, H.A. & Medina, E. (Eds) (1995). *Seasonally Dry Tropical Forests.* Cambridge University Press, Cambridge.

Burslem, D.F.R.P. (1996). Differential responses to nutrients, shade and drought among tree seedlings of lowland tropical forest in Singapore. In *Ecology of Tropical Forest Tree Seedlings* (Ed. by M.D. Swaine), pp. 211–244. UNESCO/Parthenon, Paris/Carnforth, UK.

Burslem, D.F.R.P., Grubb, P.J. & Turner, I.M. (1995). Responses to nutrient addition among shade tolerant tree seedlings in lowland tropical rain forest in Singapore. *Journal of Ecology*, **83**, 113–122.

Chapin, F.S., Rincón, E. & Huante, P. (1993). Environmental responses of plants and ecosystems as predictors of the impact of global change. *Journal of Biosciences*, **18**, 515–524.

Condit, R., Hubbell, S.P. & Foster, R.B. (1995). Mortality rates of 205 neotropical tree and shrub species and the impact of a severe drought. *Ecological Monographs*, **65**, 419–435.

Cuenca, G., Herrera, R. & Medina, E. (1990). Aluminium tolerance in trees of a tropical cloud forest. *Plant and Soil*, **125**, 169–175.

Denslow, J.S., Vitousek, P.M. & Schultz, J.C. (1987). Bioassays of nutrient limitation in a tropical rain forest soil. *Oecologia*, **74**, 370–376.

Dias-Filho, M.B. & Dawson, T.E. (1995). Physiological responses to soil moisture stress in two Amazonian gap-invader species. *Functional Ecology*, **9**, 213–221.

Diaz, H.F. & Kiladis, G.N. (1992). Atmospheric teleconnections associated with extreme phases of the Southern Oscillation. In *Historical and Paleoclimatic Aspects of the Southern Oscillation* (Ed. by H.F. Diaz & V. Markgraf), pp. 7–28. Cambridge University Press, Cambridge.

Dobsen, A. & Crawley, M. (1994). Pathogens and the structure of plant communities. *Trends in Ecology and Evolution*, **9**, 393–398.

Escudero, A., Del Acro, J.M., Sanz, J.C. & Ayala, J. (1992). The effects of leaf longevity and retranslocation efficiency on the retention time of nutrients in the leaf biomass of different woody species. *Oecologia*, **90**, 80–87.

Fan, S., Blake, T.J. & Blumwald, E. (1994). The relative contribution of elastic and osmotic adjustments to turgor maintenance of woody species. *Physiologia Plantarum*, **90**, 408–413.

Fanjul, L. & Barradas, V.L. (1987). Diurnal and seasonal variation in the water relations of some deciduous and evergreen trees of a deciduous dry forest of the western coast of Mexico. *Journal of Applied Ecology*, **24**, 289–303.

Furley, P.A., Proctor, J. & Ratter, J.A. (Eds) (1992). *Nature and Dynamics of Forest–Savanna Boundaries.* Chapman & Hall, London.

Gentry, A.H. & Emmons, L.H. (1987). Geographical variation in fertility, phenology and the composition of the understory in neotropical forests. *Biotropica*, **19**, 216–227.

Goldammer, J.G. & Siebert, B. (1990). The impact of droughts and forest fires on tropical lowland rain forest of East Kalimantan. In *Fire in the Tropical Biota* (Ed. by J.C. Goldammer), pp. 11–31 *Ecological Studies*, **84**. Springer, Berlin.

Grubb, P.J. (1977). Control of forest growth and distribution on wet tropical mountains. *Annual Review of Ecology and Systematics*, **8**, 83–107.

Grubb, P.J. (1994). Root competition in soils of different fertility, a paradox resolved, *Phytocoenologia*, **24**, 495–505.

Grubb, P.J. (1995). Mineral nutrition and soil fertility in tropical rain forests. In *Tropical Forests: Management and Ecology*, (Ed. by A.E. Lugo & C. Lowe), pp. 308–330. Springer, Berlin.

Hall, J.B. & Swaine, M.D. (1976). Classification and ecology of closed canopy forest in Ghana. *Journal of Ecology*, **64**, 913–951.

Hall, J.B. & Swaine, M.D. (1981). *Distribution and Ecology of Vascular Plants in a Tropical Rain Forest: Forest Vegetation in Ghana*. Junk, The Hague.

Hawthorne, W.D. (1993). *FROGGIE: Forest Resources of Ghana Graphical Information Exhibitor* (Manual and Computer program). IUCN/ODA, Cambridge.

Hawthorne, W.D. (1994). *Fire Damage and Forest Regeneration in Ghana. ODA Forestry Series*, **4**. NRI, Chatham.

Hawthorne, W.D. (1995). *Ecological Profiles of Ghanaian Forest Trees. OFI Tropical Forestry Papers*, **29**. Oxford Forestry Institute, Oxford.

Hawthorne, W.D. (1996). Holes and the sums of parts in Ghanaian forest: regeneration, scale and sustainable use. *Proceedings of the Royal Society of Edinburgh*, **104B**, 75–176.

Hawthorne, W.D. & Abu-Juam, M. (1995). *Forest Protection in Ghana*. IUCN/ODA, Cambridge.

Hogan, K.P., Smith, A.P. & Ziska, L.H. (1991). Potential effects of elevated CO_2 and changes in temperature on tropical plants. *Plant, Cell and Environment*, **14**, 763–778.

Hogan, K.P., Smith, A.P., Araus, J.L. & Saavedra, A. (1994). Ecotypic differentiation of gas exchange responses and leaf anatomy in a tropical forest understory shrub from areas of contrasting rainfall regimes. *Tree Physiology*, **14**, 819–831.

Holbrook, N.M., Whitbeck, J.L. & Mooney, H.A. (1995). Drought responses of Neotropical dry forest trees. In *Seasonally Dry Tropical Forests* (Ed. by S.H. Bullock, H.A. Mooney & E. Medina), pp. 243–276. Cambridge University Press, Cambridge.

Howe, H.E. (1990). Survival and growth of juvenile *Virola surinamensis* in Panama: effects of herbivory and canopy closure. *Journal of Tropical Ecology*, **6**, 259–280.

Huante, P., Rincón, E. & Acosta, I. (1995). Nutrient availability and growth rate of 34 woody species from a tropical deciduous forest in Mexico. *Functional Ecology*, **9**, 849–858.

Huante, P., Rincón, E. & Chapin, F.S. (1995). Responses to phosphorus of contrasting successional tree-seedling species from the tropical deciduous forest of Mexico. *Functional Ecology*, **9**, 760–766.

Huc, R., Ferhi, A. & Guehl, J.M. (1994). Pioneer and late stage tropical rain forest species (French Guyana) growing under common conditions differ in leaf gas exchange regulation, carbon isotope discrimination and leaf water potential, *Oecologia*, **99**, 297–305.

Jackson, G. & Grace, J. (1994). Cavitation and water transport in plants. *Endeavour*, **18**, 50–54.

Jarosz, A.M. & Davelos, A.L. (1995). Effects of disease in wild plant populations and the evolution of pathogen resistance. *New Phytologist*, **129**, 371–387.

Jeffree, E.P. & Jeffree, C.E. (1994). Temperature and the biogeographical distributions of species, *Functional Ecology*, **8**, 640–650.

Keay, R.W.J., Onochie, C.F.A. & Stanfield, D.P. (1964). *Nigerian Trees*. Nigerian National Press, Ibadan.

Khalil, Y.A.M. & Grace, J. (1992). Acclimation to drought in *Acer pseudoplatanus* L. (Sycamore) seedlings. *Journal of Experimental Botany*, **43**, 1591–1602.

Kummerow, J., Castellanos, J., Maass, M. & Larigauderie, A. (1990). Production of fine roots and the seasonality of their growth in a Mexican deciduous dry forest. *Vegetatio*, **90**, 73–80.

Kyereh, B. (1994). *Seed phenology and germination of Ghanaian forest trees*. PhD thesis, University of Aberdeen.

Lee Su See & Alexander, I.J. (1996). The dynamics of ectomycorhizal infection of *Shorea leprosula* seedlings in Malaysian rain forests. *New Phytologist*, **132**, 297–305.

Letouzey, R. (1968). *Étude Phytogéographique du Cameroun*. Lechevalier, Paris.

Lodge, D.J., McDowell, W.H. & McSwiney, C.P. (1994). The importance of nutrient pulses in tropical forest. *Trends in Ecology and Evolution*, **9**, 384–387.

Machado, J.L. & Tyree, M.T. (1994). Patterns of hydraulic architecture and water relations of two tropical canopy species with contrasting leaf phenologies: *Ochroma pyramidale* and *Pseudobombax septenatum*. *Tree Physiology*, **14**, 219–240.

Maley, J. (1996). The African rain forest – main characteristics of changes in vegetation and climate from the upper Cretaceous to the quaternary. *Proceedings of the Royal Society of Edinburgh*, **104B**, 31–73.

Marshner, H. (1991). Mechanisms of adaptation of plants to acid soils. *Plant and Soil*, **134**, 1–20.

Melick, D.R. (1990). Relative drought resistance of *Tristaniopsis laurina* and *Acmena smithii* from riparian warm temperate rain forest in Victoria. *Australian Journal of Botany*, **38**, 361–370.

Mulkey, S.S. & Wright, S.J. (1996). Influence of seasonal drought on the carbon balance of tropical forest plants. In *Tropical Forest Plant Ecophysiology* (Ed. by S.S., Mulkey, R.L., Chazdon, & A.P. Smith), pp. 187–216. Chapman & Hall, New York.

Mulkey, S.S., Smith, A.P. & Wright, S.J. (1991). Comparative life history and physiology of two understory species. *Oecologia*, **88**, 263–273.

Mulkey, S.S., Wright, S.J. & Smith, A.P. (1991). Drought acclimation of an understory shrub (*Psychotria limonenesis*; Rubiaceae) in a seasonally dry tropical forest in Panama. *American Journal of Botany*, **78**, 579–587.

Müller, M.J. (1982). *Selected Climatic Data for a Global Set of Standard Stations for Vegetation Science*. Junk, The Hague.

Murphy, P. & Lugo, A.E. (1986). Ecology of tropical dry forest. *Annual Review of Ecology and Systematics*, **17**, 67–88.

Newbery, D.M., Alexander, I.J., Thomas, D.W. & Gartlan, J.S. (1988). Ectomycorrhizal rainforest legumes and soil phosphorus in Korup National Park, Cameroon. *New Phytologist*, **109**, 433–450.

Oliveiro-Filho, A.T., Vilela, E.A., Carvalho, D.A. & Gavilanes, M.L. (1994). Effects of soils and topography on the distribution of tree species in a tropical riverine forest in south-eastern Brazil. *Journal of Tropical Ecology*, **10**, 483–508.

Pugnaire, F.I. & Chapin, F.S. (1993). Controls over nutrient resorption from leaves from evergreen Mediterranean species. *Ecology*, **74**, 124–129.

Raaimakers, D. (1994). *Growth of Tropical Rainforest Trees as Dependent on Phosphorus Supply*. Tropenbos series, **11**, Backhuys, Leiden.

Raaimakers, D. & Lambers, H. (1996). Response to phosphorus supply of tropical tree seedlings: a comparison between a pioneer species, *Tapipira obtusa* and a climax species, *Lecytis corrugata*. *New Phytologist*, **132**, 97–102.

Raaimakers, D., Boot, R.G.A., Dijkstra, P. & Pot, S. (1995). Photosynthetic rates in relation to leaf phosphorus content in pioneer versus climax forest species. *Oecologia*, **102**, 120–125.

Reynaud, I. & Maley, J. (1994). Histoire récente d'une formation forestière du Sud-Ouest Cameroun à partir de l'analyse pollinique. *Comptes Rendues de l'Academie des Sciences Serie iii Sciences de la Vie*, **317**, 575–580.

Richards, P.W. (1952). *The Tropical Rain Forest*. Cambridge University Press, Cambridge.

Richardson, D.M. & Bond, W.J. (1991). Determinants of plant distribution: evidence from pine invasions. *American Naturalist*, **137**, 639–668.

Rincón, E. & Huante, P. (1995). Influence of mineral nutrient availability on growth of tree seedlings from the tropical deciduous forest. *Trees, Structure and Function*, **9**, 93–97.

Robichaux, R.H. (1984). Variation in the tissue water relations of two sympatric Hawaiian *Dubautia* species and their natural hybrid. *Oecologia*, **65**, 75–81.

Russell-Smith, J. (1991). Classification, species richness, and environmental relations of monsoon rainforest in Northern Australia. *Journal of Vegetation Science*, **2**, 259–278.

Sands, R. & Mulligan, D.R. (1990). Water and nutrient dynamics and tree growth. *Forest Ecology and Management*, **30**, 91–111.

Silver, W.L. (1994). Is nutrient availability related to plant nutrient use in humid tropical forests? *Oecologia*, **98**, 336–343.

Silver, W.L., Scathena, F.N., Johnson, A.H., Siccama, T.G. & Sanchez, M.J. (1994). Nutrient availability in a montane wet tropical forest: spatial patterns and methodological considerations. *Plant and Soil*, **164**, 129–145.

Sobrado, M.A. (1986). Aspects of tissue water relations and seasonal changes of leaf water potential components of evergreen and deciduous species coexisting in tropical dry forests. *Oecologia*, **68**, 413–416.

Sobrado, M.A. (1991). Cost–benefit relationships in deciduous and evergreen leaves of tropical dry forest species. *Functional Ecology*, **5**, 608–616.

Sobrado, M.A. (1993). Trade-off between water transport efficiency and leaf life-span in a tropical dry forest. *Oecologia*, **96**, 19–23.

Sovan Roy & Singh, J.S. (1994). Consequences of habitat heterogeneity for availability of nutrients in a dry tropical forest. *Journal of Ecology*, **82**, 504–509.

Swaine, M.D. (1992). Characteristics of dry forest in West Africa and the influence of fire. *Journal of Vegetation Science*, **3**, 365–374

Swaine, M.D. (1996). Rainfall and soil fertility as factors limiting forest species distributions in Ghana. *Journal of Ecology*, **10**, 419–428.

Swaine, M.D. & Hall, J.B. (1986). Forest structure and dynamics. In *Plant Ecology in West Africa* (Ed. by G.W. Lawson), pp. 47–93. Wiley, Chichester.

Swaine, M.D., Agyeman, V.K., Kyereh, B., Orgle, T.K., Thompson, J. & Veenendaal, E.M. (1997). *Ecology of Forest Trees in Ghana. ODA Forestry Series*, **7**. University of Aberdeen, Aberdeen.

Sykes, M.T. & Prentice, I.C. (1995). Boreal forest futures: modelling the controls on tree species range limits and transient responses to climate change. *Water, Air, and Soil Pollution*, **82**, 415–428.

Tanner, E.V.J., Kapos, V., Freskos, S., Healey, J.R. & Theobald, A.M. (1990). Nitrogen and phosphorus fertilization of Jamaican montane rain forest trees. *Journal of Tropical Ecology*, **6**, 231–238.

Ter Steege, H. (1994). Flooding and drought tolerance in seeds and seedlings of two *Mora* species segregated along a soil hydrological gradient in the tropical rain forest of Guyana. *Oecologia*, **100**, 356–367.

Ter Steege, H., Jetten, V.G., Polak, A.M. & Werger, M.J.A. (1993). Tropical rain forest types and soil factors in a watershed area in Guyana. *Journal of Vegetation Science*, **4**, 705–716.

Thompson, W.A., Kriedemann, P.E. & Craig, I.E. (1992). Photosynthetic response to light and nutrients in sun tolerant and shade tolerant rain forest trees. I. Growth, leaf anatomy and nutrient content. *Australian Journal of Plant Physiology*, **19**, 1–18.

Turner, I.M. (1991). Effects of shade and fertiliser on two tropical woody pioneer species. *Tropical Ecology*, **32**, 24–29.

Turner, I.M., Brown, N.D. & Newton, A.C. (1993). The effect of fertiliser application on dipterocarp seedling growth and mycorrhizal infection. *Forest Ecology and Management*, **57**, 329–337.

Unwin, G.L. & Kriedemann, P.E. (1990). Drought tolerance and rainforest tree growth in a north Queensland rainfall gradient. *Forest Ecology and Management*, **30**, 113–123.

Van Rompaey, R.S.A.R. (1993). *Forest gradients in West Africa*. PhD thesis, Agricultural University, Wageningen.

Veenendaal, E.M., Swaine, M.D., Blay, D., Yelifari, N.B. & Mullins, C.E. (1996a). Seasonal and long term soil water regime in West African tropical forest. *Journal of Vegetation Science*, **7**, 473–482.

Veenendaal, E.M., Swaine, M.D., Agyeman, V.K., Blay, D., Abebrese, I.K. & Mullins, C.E. (1996b). Differences in plant and soil water relations in and around a forest gap in West Africa may influence seedling establishment and survival. *Journal of Ecology*, **84**, 83–90.

Veenendaal, E.M., Swaine, M.D., Lecha, R.T., Walsh, M.F. & Owusu-Afriyie, K. (1996c). Responses of West African forest tree seedlings to irradiance and soil fertility. *Functional Ecology*, **10**, 501–511.

Vitousek, P.M. & Denslow, J.S. (1986). Nitrogen and phosphorus availability in treefall gaps of a lowland tropical rainforest. *Journal of Ecology*, **74**, 1167–1178.

Vitousek, P.M. & Sanford, R.L. (1986). Nutrient cycling in moist tropical forest. *Annual Review of Ecology and Systematics*, **17**, 137–167.

Vogt, K.A., Vogt, D.J., Asbjorsen, H. & Dahlgren, R.A. (1995). Nutrient availability in a montane wet tropical forest: Roots, nutrients and their relationship to spatial patterns. *Plant and Soil*, **168/169**, 113–123.

Walter, H. & Breckle, S.W. (1984). *Ecological Systems of the Geobiosphere: 2. Tropical and Subtropical Zonobiomes*. Springer, Berlin.

Waterman, P.G., Meshal, I.A., Hall, J.B. & Swaine, M.D. (1978). Biochemical systematics and ecology of the Toddalioideae in the central part of the West African forest zone. *Biochemical Systematics and Ecology*, **6**, 239–245.

Webb, L.J. (1959). A physiognomic classification of Australian rain forests. *Journal of Ecology*, **17**, 137–167.

Webb, L.J. (1968). Environmental relationships of the structural types of Australian rain forest vegetation. *Ecology*, **49**, 296–311.

Whitmore, T.C. (1990). *An Introduction to Tropical Rain Forests*. Clarendon Press, Oxford.

Woods, P. (1989). Effects of logging, drought and fire on structure and composition of tropical forests in Sabah, Malaysia. *Biotropica*, **21**, 290–299.

Woodward, F.I. (1987). *Climate and Plant Distribution*. Cambridge University Press, Cambridge.

Wright, S.J. (1992). Seasonal drought, soil fertility and the species density of tropical forest plant communities. *Trends in Ecology and Evolution*, **7**, 260–263.

Yanney Ewusie, J. (1992). *Phenology in Tropical Ecology*, Ghana Universities Press, Accra.

Zech, W. & Drechsel, P. (1992). Multiple mineral deficiencies in forest plantations in Liberia. *Forest Ecology and Management*, **48**, 121–143.

8. COMMUNITY STRUCTURE AND THE DEMOGRAPHY OF PRIMARY SPECIES IN TROPICAL RAINFOREST

R. J. ZAGT*† and M. J. A. WERGER*

Department of Plant Ecology and Evolutionary Biology, University of Utrecht, PO Box 80.084, 3508 TB Utrecht, The Netherlands; † Tropenbos Guyana Programme, 12 E Garnett Street, Campbellville, Georgetown, Guyana

SUMMARY

1 Equilibrium and non-equilibrium theories have been proposed to explain the high species richness in tropical rainforests. Equilibrium theories suggest that species differ in their regeneration requirements and partition the available space over resource gradients. Non-equilibrium theories propose that most species have similar resource requirements and emphasize chance as the major factor determining the species composition of tropical rainforests.

2 Close inspection of the literature reveals that support for non-equilibrium theories comes from studies on large (diameter > 10 cm) or at least well-established trees, whereas support for equilibrium theories comes from studies on seedlings.

3 In interspecific comparisons, large trees are more likely than seedlings to be classified as generalists that do not show habitat specificity. This is because their growth and mortality responses are too slow and insensitive; the methods of measuring growth are too inaccurate; current methods are capable of measuring only part of the spectrum from very negative to very positive plant responses; and the abiotic environment cannot be quantified with sufficient precision as to detect small but significant differences between species. The growth and survival of seedlings are in a number of ways more easy to quantify with accuracy. However, large-scale seedling studies are scarce and often only quantify seedling numbers, not size.

4 The contributions of chance and determinism to successful regeneration are not constant during the different stages of the regeneration cycle. Processes occurring before gap formation partly determine the result of interspecific competition after a gap has formed. Even though the spatially and temporally stochastic availability of seedlings, and stochastic seedling mortality may dominate seedling dynamics in the understorey, interspecific

differences in growth and mortality and variation in resource availability may affect the size and condition of the seedlings at the onset of the gap phase. Therefore, seedling studies should not limit themselves to gaps, but also address the understorey.

5 Owing to a higher availability of resources, species-specific differences in growth are more pronounced in the gap phase than in the understorey, and the contribution of chance factors to regeneration is less.

6 Among generalist tree species, species that differ in competitive ability and in the abundance of seedlings and their spatial distribution may maintain themselves in the community at different population size distributions, which represent different pathways towards successful regeneration. Species that are weak competitors in gaps may locally recruit through chance by being common as seedlings.

7 In efforts to resolve the issue of species maintenance in rainforests, seedlings should be included in large-scale, long-term demographic studies, and the full variability of their biotic and abiotic environment should be addressed.

SPECIES RICHNESS AND GUILD DELIMITATION

Tropical rainforests are characterized by a remarkably high species richness (Whitmore 1984, Gentry 1988, Valencia, Balslev & Paz y Miño 1994). A number of theories have been advanced to explain the coexistence of surprisingly large numbers of species with apparently similar regeneration requirements in the same area. These theories are generally grouped as equilibrium and non-equilibrium theories (Connell 1978).

Equilibrium theories emphasize a differential adaptation of species to specific conditions of growth, survival or reproduction that occurs from time to time during the life-cycle of the species (Clark & Clark 1987). 'Gap partitioning', and 'compensatory mortality' are two applications of equilibrium theories. Gap partitioning occurs when species are niche-differentiated and, through different requirements of growth, survival and reproduction, regenerate optimally in gaps of different sizes (Hartshorn 1978, 1980, Whitmore 1978, 1984, Denslow 1980, Orians 1982, Brokaw 1985, 1987). Compensatory mortality in the form of frequency, density or distance-dependent mortality is seen as promoting species coexistence because it prevents self-replacement of species at a location. This mechanism is most often considered in the context of parent – offspring relationships as in the Janzen–Connell model (Janzen 1970, Connell 1971), but density-dependency of growth and mortality among established individuals fits also in this category (cf. Condit, Hubbell & Foster 1994). The evidence for the Janzen–Connell model is mixed (Condit, Hubbell & Foster 1992, Hammond & Brown, Chapter 3, this volume). Augspurger (1983a, 1984), Clark and Clark (1984), Connell, Tracey and Webb (1984), Howe, Schupp and Westley (1985) and Schupp (1988a,b) found either distance- or density-dependent depression of growth and/or survival among seedlings around adult conspecifics. However, in other cases the expected decrease of seedling mortality rates with increasing distance from the parent tree, or with decreasing seedling density, was not found (Hubbell 1979, Forget 1989, 1994, De Steven 1994, Hart 1995, Itoh, Yamakura, Ogino & Lee 1995), for example as a result of predator satiation. Negative density-dependent relationships among larger individuals have been demonstrated for some species (Hubbell & Foster 1987a, 1990, Barros-Henriques & Elias Girnos de Sousa 1989, Condit *et al.* 1992, 1994, Alvarez-Buylla 1994).

In contrast, non-equilibrium theories do not emphasize differential specialization in species; rather it is pointed out that many species show large similarities in their life-histories and functional responses to environmental conditions, and that chance factors and frequency rhythms of environmental disturbance largely explain species co-occurrence. Owing to the unpredictable identity of competing neighbours of an individual in species-rich communities, there is no constant selective pressure that fosters niche-differentiation. Instead, the neighbours exert a competitive pressure which is, at a population level, close to the spatial and temporal average of the community, thereby preventing the development of niche-specificity (Hubbell & Foster 1986b). The composition of the community is considered a 'random walk without a stable composition', determined by chance and historical effects (Hubbell & Foster 1986b).

In both views, gap dynamics play an important role in regulating the species composition of tropical rainforests. The sudden occurrence of an opening in the forest canopy, and its large variation in size, greatly changes microenvironmental conditions below the original canopy (Hartshorn 1978, Brokaw 1982). Furthermore, within each gap a spatial pattern in environmental conditions develops which is related to gap size, form and orientation. This allows, in principle, gap partitioning among colonizing species. Germination and the response of pre-established seedlings and saplings to the sudden opening of the forest canopy strongly determine their success in reaching reproductive size. Based on differences in response to these gaps, species are often grouped as long- and short-lived pioneers, gap-dependent, and gap-independent species (or equivalents; Aubréville 1938, Van Steenis 1956, Lieberman, Lieberman, Peralta & Hartshorn 1985, Clark & Clark 1987, Lieberman & Lieberman 1987, Bongers, Popma, Meave del Castillo & Carabias 1988, Swaine & Whitmore 1988). The extent to which the species show life-history differentiation within these groups is widely debated (Connell 1978, Hubbell & Foster 1986b).

A large body of evidence, gathered mainly at the 50-ha plot on Barro Colorado Island (BCI), Panama and at La Selva, Costa Rica in Central America (Hubbell & Foster 1990, Clark & Clark 1992, Lieberman, Lieberman, Peralta & Hartshorn 1995) over the past decade and a half, has led to the belief that the forest is not at equilibrium, and that most species possess a wide ecological amplitude with large interspecific overlap instead of narrow tolerance and high specificity (Lieberman *et al.* 1995). Only a few large and broad guilds are recognized, depending on the criteria used. The largest is usually a group of generalists or indifferents (Hubbell & Foster 1986a: 57 out of 81 species studied; Hubbell & Foster 1987b, Welden, Hewett, Hubbell & Foster 1991: 79 out of 104). Responses to light vary somewhat, but the variation is so large that only a limited number of species are shown to have a significant preference for lighter or darker sites (Hubbell & Foster 1986a: 24 out of 81 species studied; Welden *et al.* 1991: 9 out of 104; Lieberman *et al.* 1995: 14 out of 104). Virtually all species may occur under gap and non-gap conditions (Lieberman *et al.* 1995).

Direct evidence of gap partitioning is available for pioneer species which appear to respond to gap size by differential germination and establishment (Brokaw 1987, Raich & Gong 1990), but it is weaker for climax species. In experiments, the seedlings of a number of species have been demonstrated to respond differentially to high light conditions (Popma & Bongers 1988, Osunkoya, Ash, Graham & Hopkins 1993, Boot 1996). In the field many other factors, such as the variation in light availability within gaps and size differences, also determine success in a gap, and this may contribute to the absence of gap partitioning (Barton 1984, Brown & Whitmore 1992).

On the basis of these data, the current view seems to be that a small number of guilds can be distinguished among rainforest species (Lieberman *et al.* 1985, Hubbell & Foster 1986a, 1987b, Lieberman & Lieberman 1987, Bongers *et al.* 1988, Welden *et al.* 1991, Clark & Clark 1992; O'Brien, Hubbell, Spiro *et al.* 1995, Newbery, Campbell, Proctor & Still 1996), but that most species fall into a broad, generalist group.

From careful consideration of these studies, it seems that the support for theories that assume randomness to be the force organizing communities is usually derived from research on trees that are either large (as in Lieberman *et al.* 1995: dbh > 10 cm) or at least well established, as in the BCI studies (Hubbell & Foster 1990: dbh > 1 cm). Few of these studies include seedlings and saplings (Clark & Clark 1987, 1992: seedlings of 50 cm height and upwards).

In contrast, support for theories that assume species-specific differences in growth and survival frequently comes from studies on seedlings (Clark & Clark 1984, Popma & Bongers 1988, Osunkoya *et al.* 1993, 1994, Kitajima 1994). A striking illustration of this point is that *Quararibea asterolepis*, *Tetragastris panamensis* and *Trichilia tuberculata* are three of the many generalist, undifferentiated tree species of the 50-ha plot on BCI in terms of their established plant dynamics (Welden *et al.* 1991), whereas these species, in exactly the same plot, showed 'large differences in seedling and sapling recruitment', which were 'directional, rather than stochastic' (De Steven 1994, p. 380), when their smallest life-stages were considered.

In this chapter, we address the balance between species-specific (directional) and random processes in the determination of successful regeneration in the forest against the background of the methodology of studying trees and seedlings. In order to avoid drawing this discussion into an analysis of well-established differences between pioneers and non-pioneers, we focus on the large group of generalist species that germinate in the understorey, require gaps for regeneration, reach the canopy, and appear to coexist without apparent life-history differences or apparent habitat segregation. We pay special attention to the relation between plant performance and the light climate, even though we acknowledge that species may show differential responses to variation in water and nutrient availability, and that these factors thus contribute to the composition of communities. However, comparably few studies focus on these factors (but see recent papers by Huante, Rincón & Acosta (1995), Burslem (1996) and Veenendaal, Swaine, Lecha *et al.* (1996) for large comparative studies of the influence of nutrient availability on growth in tropical tree seedlings).

We use 'species richness' to mean the number of species per unit of homogeneous forest area, that is homogeneous in mesoclimate, hydrology and soil. We define a gap as any patch that is suitable for rapid plant growth, independent of the exact delineation of the canopy opening or the quantity of available resources. In regeneration we include the entire process of establishment of a seedling until the emergence of a reproducing individual (although we generally refer to the initial phase of that process). In the term 'seedling' we include saplings up to about 3 m height, and consider trees over 10 cm dbh as 'large trees'. Speaking about chance factors, we mean those factors that work in a probabilistic fashion, the occurrence of which cannot be influenced by the plant and the effect of which on regeneration processes is thus unpredictable. We choose a phytocentric view. So, for example, although many herbivores will find and attack plants according to a certain search strategy rather than chance, for individual plants it is unpredictable whether they will be attacked or not, and thus we consider herbivory a chance factor. Deterministic factors, as used here, encompass those characteristics in plants which lead to a predictably different response of different species which are subject to the same conditions.

WHY DO LARGE TREES ALWAYS APPEAR TO BE GENERALISTS?

The reason for the apparent discrepancy in results between seedling and 'large tree' studies is found in a number of methodological and conceptual limitations in the latter when they are compared with seedlings (Table 8.1).

Tree responses are slow and insensitive

Functional responses in large trees are often very small, or very slow. Changes in stem diameter in many species take a long time to become sufficiently large to be measurable. Death that is caused by potentially specific mortality agents, such as pathogens or drought, is often delayed until after a long period of decline. This hinders detection of this causal relationship. However, random responses, which are not very informative about species-specific differences, may be very strong, such as in the case of stochastic mortality due to wind-, tree- or branch-fall (Uhl, Clark, Dezzeo & Maquinero 1988, Yavitt, Battles, Lang & Knight 1995, Van der Meer & Bongers 1996). Furthermore, mortality among large trees is low (generally between 1 and 3% per year (Lieberman *et al.* 1985, Clark & Clark 1992,

TABLE 8.1. Aspects of the measurement of responses to environmental factors in large trees and seedling.

Feature	Large trees	Seedlings
Response of growth and survival to environmental factors	Limited, slow	Strong, rapid
Mortality rate	Low, largely stochastic	Potentially high, partly stochastic
Measurement of plant size	Difficult, inaccurate	Relatively easy, accurate
Measurement of negative response to unfavourable conditions	Nearly impossible	Possible
Light climate during the history of the individual	Usually highly variable	Usually of limited variability
Measurement of the light climate in the crown area	Difficult, inaccurate	Easier, often also inaccurate

Condit, Hubbell & Foster 1995), and this makes accurate determination of species-specific mortality rates difficult.

Seedlings, on the other hand, die quickly, in large numbers, and from a variety of causes. They show greater sensitivity to levels of pathogen infestation and herbivory (Augspurger 1984, Augspurger & Kelly 1984, Clark & Clark 1985, Osunkoya, Ash, Hopkins & Graham 1992, Itoh *et al.* 1995) than larger trees. Mortality rates of seedlings drop rapidly with increasing size (Hartshorn 1972, Clark & Clark 1992) and in most species survival of individuals above 1 cm diameter is already more than 97% (Welden *et al.* 1991; cf. Fig. 8.1). Seedling growth also responds strongly to variations in resource availability, such as light (Augspurger 1984, Popma & Bongers 1988, Osunkoya *et al.* 1993, Boot 1996).

Diameter measurements are inaccurate

It is hard to measure the size of a large tree accurately. Diameter at breast height (dbh) is the most common measure of tree size since it generally cor-

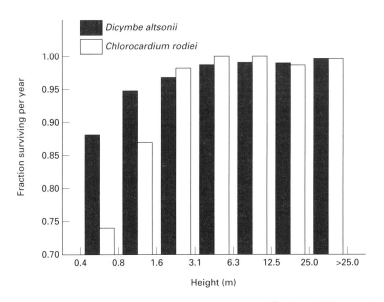

FIG. 8.1. Average yearly survival of the climax species *Dicymbe altsonii* and *Chlorocardium rodiei* in a tropical rainforest near Mabura Hill, Guyana. Data are from individually tagged individuals in two 1-ha demographic study plots monitored annually during 1991–95. Seedlings <1.3 m high were monitored in a 5% sample. The individuals were classified in categories of doubled height (for individuals below the canopy) and for individuals in the canopy (height >25 m). The population size per size class varied between 20 (larger trees) and 1015 (seedlings). (From Zagt 1997.)

relates well with other dimensions, such as height, crown diameter and crown area (Alvarez-Buylla & Martínez-Ramos 1992, Condit *et al.* 1995, O'Brien *et al.* 1995, King 1996) and biomass, which is the variable of real interest. However, the allometry of these dimensions to dbh is far from perfect, for example as a result of branch loss due to senescence or falling trees. The measurement error in dbh is of a comparable magnitude to the annual diameter growth in many tree species. Moreover, the dbh of large trees does not respond promptly and accurately to changes in biomass over the short time given to most demographic studies. Therefore, measurement of dbh increments is convenient, except that in long-term studies is not very suitable for the detection of small differences in growth between species. In seedlings and saplings it is easier to determine dimensions, including biomass (through allometric relationships which include more parameters), and hence responses to environmental variations.

Negative performance cannot be quantified

Height or diameter measurements are better for detecting positive than negative performance. Reductions in height or diameter are usually the result of damage and not a functional response. Generally, negative responses under adverse conditions are difficult to quantify in large trees. In seedlings, height or diameter measurements are also inadequate to register negative performance, but changes in biomass or leaf loss are easier to quantify.

The measurement of the light climate of a tree crown is difficult

The light climate is among the most variable environmental factors in the tropical rainforest and may be expected to cause clear plant responses. Its precise measurement is difficult, however. The present light climate in the crown of a large tree, which is sometimes used to demonstrate (the absence of) species-specific responses to crown light environments (Clark & Clark 1992, Lieberman *et al.* 1995), is not necessarily an accurate reflection of the crown light environments under which the tree grew. The physical environment around and above each individual is subject to continuous change as gaps close and new ones form. The older a tree, the weaker is the correlation between its present light climate and the light climate in which it grew. For seedlings this is less of a problem as their history is shorter. Moreover, seedlings are small and thus cover a short distance at the lower end of the vertical light gradient in the forest.

The larger a tree, the more difficult it is to quantify its light climate adequately. Many methods have been applied. There are indirect, qualitative methods such as the classification of sites in gaps (defined as areas with a canopy less than 10 m high) and non-gaps (canopy higher; Hubbell & Foster 1986a, Welden *et al.* 1991). Indirect but quantitative methods include the determination of the number of tree crowns above an individual (Clark & Clark 1987, 1992), the canopy closure index of Lieberman, Lieberman and Peralta (1989) and Lieberman *et al.* (1995), and gap size (Brokaw 1982). These methods have in common that they are based on the size or extent of vegetation around or above a plant, rather than the amount of light that the plant is receiving.

Other methods are more directly concerned with the amount of light received by the plant. The Dawkins crown position score (Dawkins & Field 1978, Clark & Clark 1992, Oberbauer, Clark, Clark *et al.* 1993) is a semi-quantitative measure, while hemispherical photographs (Chazdon & Field 1987, Whitmore, Brown Swaine *et al.* 1993, Ter Steege 1994), diffuse light (LAI meter) measurements (J.L. Machado, R.J. Zagt & P.B. Reich unpublished data) and radiation measurements with light sensors (Oberbauer, Clark, Clark & Quesada 1988, Brown 1993) are quantitative. Not all of these methods are equally easy to use in large trees, although most can be employed in some way (see e.g. Sterck (1997) for hemispherical photographs taken in tall trees).

Correlations between the scores obtained by these methods and measured irradiance are rarely determined (but see Chazdon & Field 1987, Clark & Clark 1992, Whitmore *et al.* 1993, ter Steege 1994), but these are likely to be low in the case of the qualitative and semi-quantitative methods as these have limited resolution and large inherent measurement errors. With some methods this hampers the detection of subtle interspecific differences in light requirements to such an extent that it is not likely that differences between species will be found.

It seems realistic to assume that there is a significant negative correlation between the size of the tree and the accuracy of the light measurement in its crown. Moreover, light availability has been shown to be highly variable within crowns (Oberbauer *et al.* 1988). Therefore, the value of single determinations of light is limited for large crowns. For seedlings, appropriate light measurements remain cumbersome and laborious, though less so than for large trees; and they are likely to achieve a higher accuracy than for large trees.

In conclusion, the problems associated with the study of large trees alone may have led to a failure to distinguish more than a few broad species groups and an over-emphasis of the importance of random processes governing regeneration. However, this is not to say that random processes are unimportant nor that a high degree of specialization occurs among climax species. The matter is simply not exhaustively explored by the most commonly applied research strategy of studying only large trees.

As shown, seedlings have clear advantages for demographical studies and are more suitable for evaluating the existence of species-specific regeneration strategies.

WHY HAVE SEEDLING STUDIES SO FAR NOT DEMONSTRATED SPECIES-SPECIFIC DIFFERENTIATION?

In spite of the advantages that seedlings have for the study of species richness and interspecific differences, these have not yet led to firm conclusions. Apart from practical problems of seedling identification and usually high seedling densities, three factors have contributed to this situation: (i) the lack of large-scale studies on seedling dynamics; (ii) the difficulties in quantifying dose (such as light) and effect (such as growth) for large populations of plants under field conditions; and (iii) the contribution of chance factors to regeneration.

Scarcity of seedling studies

For seedlings, there are hitherto few, if any, studies of the same scale and scope as the large plots in Central America for the study of large trees. The number of species studied has been much less and the time-span of the studies has usually been short (e.g. Howe *et al.* 1985, De Steven 1994; but see Li, Lieberman & Lieberman 1996). Also, many results were obtained under experimental conditions, in artificial treatments without competition from other species (e.g. Popma & Bongers 1988, Boot 1996).

Quantifying plant responses in seedlings

The methodological problems of quantifying environmental heterogeneity and plant responses are much easier to overcome in seedlings than in large trees. A suitable set of size parameters should ideally reflect as closely as possible the underlying mechanisms of plant performance, for example the parameters used for growth analysis (Lambers & Poorter 1992). Unfortunately, it is difficult to measure relative growth rate (RGR), leaf area ratio (LAR) and net assimilation rate (NAR) non-destructively (even for above-ground parts). For this reason, in many demographic studies only the number of seedlings and their stem length is determined (e.g. Brown & Whitmore 1992, De Steven 1994, Li *et al.* 1996), and the opportunities for more precise quantification of resource availability, plant size and plant responses are lost. The stem is hardly involved in the acquisition of resources; by itself is not very informative about the condition of the seedling and it is not necessarily correlated with plant growth (Whitmore & Brown 1996).

As leaves are the principal light-intercepting organs, it seems essential to determine the leaf area in demographic studies. Moreover, leaf area appears to be a highly responsive variable explaining interspecific differences in performance. Height growth was correlated with leaf area for *Dipteryx panamensis* and *Lecythis ampla* saplings, even in understorey conditions, at La Selva, Costa Rica (Oberbauer et al. 1988); for the same *D. panamensis* it was shown that seedling survival was correlated with leaf length, and with the number of leaves per seedling at 7 months after germination (Clark & Clark 1985). In *Pithecellobium elegans* and *L. ampla* at La Selva, height growth correlated with leaf area and the diffuse site factor (or ISF, Whitmore et al. 1993) in understorey conditions (Oberbauer et al. 1993). Also, in seedlings of 5 out of 10 pioneer species in light gaps on BCI, leaf area was correlated with survival (Garwood 1986).

In interspecific comparisons LAR correlates well with RGR as long as the light gradient is not too long (as may be assumed to be the case in the forest understorey; Kitajima 1994, Osunkoya et al. 1994, Huante, Rincón & Acosta 1995, Boot 1996). Although the amount of leaf-area per unit stem length is probably the best correlate for LAR, it is a time-consuming exercise to determine it on a large scale in the field. In species with a largely constant leaf size the number of leaves per unit stem length, or leaf density, can be used as an easy and satisfactory approximation of LAR that can easily be quantified in the field (Ashton 1990). In practice, however, this relation holds true for only a small fraction of the species (Fig. 8.2). Nevertheless, leaf density may prove to be a useful measure as it provides a performance index which can assume negative values, for example as a result of leaf abscission under unfavourable conditions such as deep shade, drought or sudden increases in light.

Chance-effects

Even if competitive exclusion among climax species in the rainforest affects regeneration, it may still be difficult to demonstrate it because of chance factors. Chance processes such as mortality from branch falls (Hartshorn 1972, Aide 1987, Clark & Clark 1991), and the occurrence of gaps in relation to the temporal and spatial constraints associated with seed rain and seedling persistence (Hubbell & Foster 1986a, Brokaw 1986, De Steven 1994, Van der Meer & Bongers 1996) may obscure the patterns resulting from species differentiation.

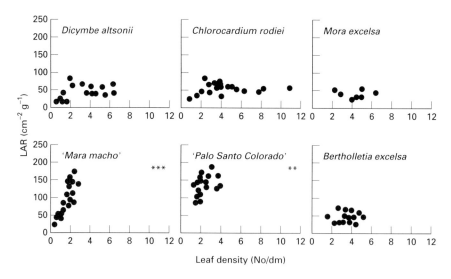

FIG. 8.2. The relationship between leaf density (the number of leaves per decimetre stem length) and the leaf area ratio (LAR, leaf area per unit plant weight; $cm^2 g^{-1}$) for six species of rainforest seedlings from Guyana and Bolivia. Not all species are identified with their scientific name. All seedlings were grown in pots under experimental conditions, in medium and low light (<12% of full sunlight). Asterisks denote significant correlations (**$P \leq 0.01$; ***$P \leq 0.001$); in the remaining species the correlations were not significant. (Sources: L. Poorter unpublished datas, Guyana (bottom row); R. Boot unpublished data, Bolivia (top row).)

The contributions of chance and determinism to successful regeneration are not constant during the different stages of the regeneration cycle. Therefore, in studies of seedling dynamics, the importance of chance and determinism must be evaluated for each of the various stages that contribute to the regeneration process.

FACTORS DETERMINING SUCCESSFUL REGENERATION IN CLIMAX SPECIES

The recruitment to maturity of a climax species is dependent on its success in two consecutive phases of regeneration. During the first phase, seedlings must become established and survive until a gap is formed. During the second phase, the seedlings must be successful after the gap has formed, given the initial species composition of that site. Hence, in the first phase the sites of eventual regeneration are unpredictable, so the relevant spatial context is the entire forest understorey (Burslem 1996). In the second phase the foci of regeneration are determined, so then the relevant spatial context is the gap. Only with the start of this phase will competition become an important determining factor.

These two phases are different in terms of resource availability. During the pre-gap phase the seedlings need to be tolerant of environmental conditions which are close to their physiological minimum requirements, whereas during the gap phase the major constraints to success are likely to arise from the presence of competing neighbours. These phases require different responses in terms of survival and growth from the seedlings. In the next two sections we argue that the presence of a species in a gap depends more on chance than determinism; and that the success of a seedling in a gap depends more on determinism than on chance.

FACTORS AFFECTING SEEDLING ABUNDANCE OF CLIMAX SPECIES BEFORE GAP FORMATION

The first phase, the presence of a species as a seedling in a patch at the moment a gap is created, depends on (see also Grubb 1977): (i) the quantity of available seeds in time and space, and (ii) on survival and growth of seedlings in the understorey. As a result the composition of the species pool at a given patch fluctuates in time. For a seedling it is important to be in the patch at the moment a gap is created: no gap can be colonized by species whose seedlings have not reached it.

Seed- and seedling distribution in the understorey

The quantity of seeds and seedlings varies in time (Frankie, Baker & Opler 1974, Garwood 1983, De Steven 1994, Li et al. 1996) and in space (Forget 1989, 1994, Bariteau 1992; Fig. 8.3). In the past, the dispersal of seeds into newly created gaps was considered an important feature (Denslow 1980, Connell 1989). This may be important for pioneers, but less so for climax species, as suitable conditions for their seedling establishment also occur outside the strict boundaries of gaps (cf. Popma, Bongers, Martínez-Ramos & Veneklaas 1988, Lieberman et al. 1989, 1995, Schupp, Howe, Augspurger & Levey 1989). In San Carlos (Venezuela), seedlings establishing prior to gap formation accounted for 83–97% of the vegetation at 4 years after gap formation, and their mortality in the gap was lower than the mortality of newly established seedlings (Uhl et al. 1988).

Nevertheless, as both the formation of gaps and the distribution of seeds in time and space are independent random processes, chance does play a dominant role in this aspect of regeneration. Shmida and Ellner (1984) have shown theoretically that non-uniform seed dispersal is an important

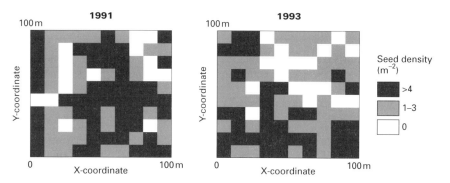

FIG. 8.3. Distribution of seeds of *Dicymbe altsonii* in one of the 1-ha demographic plots in a tropical rainforest area near Mabura Hill, Guyana (see Fig. 8.1). The distribution of freshly fallen seeds per 10 m × 10 m subplot is given for 1991 (left) and 1993 (right). Although the distribution of reproductive adults did not change between those years, the two seed distribution patterns differ considerably between the 2 years.

mechanism for the coexistence of (herbaceous) species with similar niches in species-rich communities.

Growth and survival in the understorey

Potential growth in the understorey differs between species. Interspecific and intraspecific differences in seed size (Foster & Janson 1985, Hammond & Brown 1995) contribute to differences in seedling size (Howe & Richter 1982, Osunkoya et al. 1993, 1994, Boot 1996). Relative growth rates, biomass allocation patterns and leaf dynamics vary between species and between individuals of the same species (Kohyama 1987, Popma & Bongers 1988, Bongers & Popma 1990, Osunkoya et al. 1993). Small differences in light availability may lead to differences in interspecific and intraspecific growth (Canham 1989): variations in growth rate at light levels below 2–4% of above-canopy light levels were shown by Howe et al. (1985), Oberbauer et al. (1988), Boot (1996) and Burslem (1996). Our own data from Guyana indicate that the growth rates of *Dicymbe altsonii* seedlings were correlated with the diffuse light environment in 22 understorey sites (Fig. 8.4). Similarly, survival may vary between species at these low light levels (Osunkoya et al. 1992, Li et al. 1996).

Owing to species-specific growth and survival in the understorey, patterns that form as a result of dispersal may be adjusted or even reversed, as was shown for seedlings of *Gilbertiodendron dewevrei* and *Julbernardia seretii* in Zaire (Hart 1995). *Gilbertiodendron* dominates the canopy whilst *Julbernardia* is less common. After seed dispersal more

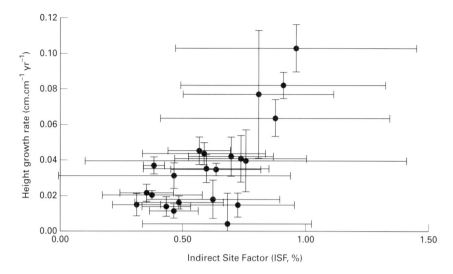

FIG. 8.4. The relationship between the light environment and the mean relative height growth rate of *Dicymbe altsonii* seedlings in 22 understorey plots of 5 m × 5 m (defined as plots with an average indirect site factor (ISF, Whitmore *et al.* 1993) less than 1.0%) near Mabura Hill, Guyana (see Fig. 8.1 for details). Light was measured with a LiCor LAI-2000 sensor, 10 samples per plot. Relative height growth rate is $(\ln h_1 - \ln h_0)/(t_1 - t_0)$, where h is height in cm and t time in days. Only seedlings with leaf densities of 0.5–1.0 leaves dm^{-1} that showed no signs of physical damage after 3 years are included in the analysis ($n = 264$). Seedling growth was correlated with ISF ($r = 0.60, P < 0.01$).

seedlings of *Julbernardia* became established because it was more successful than *Gilbertiodendron* in satiating seed predators. Later, however, *Gilbertiodendron* more than compensated for the initial disadvantage through a much higher 10-year survival in the understorey and it re-established its dominance by the time it reached a diameter of 2.5 cm (Hart 1995).

Therefore, although species-specific differences in growth and survival in the understorey may be small compared to gaps, they do contribute to predictable changes in the composition of seedling communities.

Stochastic aspects of survival in the understorey include mortality due to falling debris (Itoh *et al.* 1995, Clark & Clark 1991), disease (Augspurger 1983b) and herbivory (Coley 1983). Nevertheless, species may differ in their abilities to defend themselves against pathogens or herbivores (Coley 1983, Augspurger & Kelly 1984).

Variation in resource availability in the understorey

Resource availability varies on a small scale (Chazdon & Fetcher 1984) that may be relevant for seedling establishment and growth. Light availability is heterogeneous at scales of metres (J.L. Machado, R.J. Zagt & P.B. Reich, unpublished data; Becker & Smith 1990, Clark, Clark, Rich et al. 1996). Water infiltration rate and other soil hydrological properties vary at a scale of between 2 and 20m in Guyana (Jetten, Riezebos, Hoefsloot & van Rossum 1993). No tropical data are available, but in a temperate forest variation in soil pH, potassium and nitrate concentrations occurred over a range of 2m (Lechowicz & Bell 1991). This means that for most species appreciable differences in resource availability exist within the dispersal range of single individuals. Hence the conditions that seedlings encounter after germination in the understorey depend more on chance than on determinism, especially for species without directed dispersal mechanisms (Howe & Smallwood 1982).

Seedling dispersal, survival and growth in the understorey determine which seedlings are present at the moment a gap is formed. The distribution of seeds in space, the availability of resources at the site of germination and the formation of canopy openings are all stochastic factors which are likely to have a larger impact on seedling performance than species-specific differences.

FACTORS AFFECTING THE SUCCESS OF CLIMAX SPECIES IN GAPS

Soon after a gap is created, the composition of the pool of competing seedlings is determined; few seedlings that establish after the first few months can compete with the rapidly growing saplings that established earlier (Uhl et al. 1988). The success of a species in a gap then depends on: (i) the abundance of the species in the patch, (ii) the size hierarchy at the moment of gap formation, (iii) differences in growth responses and in mortality, and (iv) on the variability in within-patch resource availability. The presence of a species in the gap, or even eventually 'winning' a canopy position does not imply that this species is optimally adapted to the conditions in the gap, or that it is the best competitor: it means no more and no less than that it did not die. As long as a species is still present, it is in principle capable of capturing a canopy position.

Abundance of the species in the gap

The abundance of the species in the gap is a chance process, depending on the location and timing of gap formation relative to the pre-gap seed and seedling dynamics of the species at that site. Having a high abundance in a gap increases the probability of a species recruiting into the canopy through chance processes. One may argue that in forests where gaps are formed at a slow rate, seedling persistence in the understorey (which is deterministic) plays a more important role in the initial composition of a gap than in highly dynamic forests. In the first case it may take a long time before a gap is formed, so seedlings of persistent species have a higher chance than less-persistent species of surviving until a gap is formed. In the second case, the probability of gap formation soon after seed dispersal is higher and seedling persistence is of less importance.

Size-differences at the moment of gap formation

One determinant of successful regeneration is the hierarchy of size and vitality at the onset of competition (Connell 1989, Brown & Whitmore 1992, Garwood 1986 for pioneer species). Based on an allometric model, Kohyama (1991) emphasized height growth as the main criterion of sapling success and showed that 'the effort to maximize height growth on the dim forest floor has a critical meaning for the survival and competitive ability of saplings'. Seedling size at the moment of gap formation is partly a function of age and growth conditions in the understorey and partly of species-specific differences in seed size and seedling growth. Larger individuals have a head-start over smaller ones; also, at a given RGR, larger individuals gain biomass faster than smaller individuals (Hartshorn 1972, Uhl *et al*. 1988). Boot (1996) argued that it may take a long time before small individuals with a high potential RGR grow sufficiently to catch up with large individuals with a lower potential RGR. However, the result of competition between species differing in potential RGR is not only dependent on differences in size, but also on plant density in the competing stand and light availability (Whitmore & Brown 1996).

Growth and survival in gaps

Among similarly sized individuals of different species, the success of a seedling depends on differences in growth and survival in prevailing conditions of light, water and nutrients. Species have been shown to differ in RGR and allocation patterns at the different light levels that can be found in gaps of different sizes (Popma & Bongers 1988, Osunkoya *et al*. 1993, Boot 1996, Burslem 1996), and also in the capacity to acclimate to gap-related environmental conditions (Fetcher, Oberbauer, Rojas & Strain 1987, Popma & Bongers 1991). Therefore, given a similar size, and similar environmental conditions, predictable changes will occur in the size hierarchy between species, and eventually also in species composition due to competitive exclusion.

Mortality in gaps may be related to competitive exclusion or poor acclimation to the gap environment, yet mortality due to chance still occurs. As in the understorey, the factors include pathogens, herbivores and falling debris. The impact of minor damage due to these chance factors is likely to be less than in the understorey since the improved carbon balance of plants in gaps increases their recovery capacity.

Variation in resource availability within gaps

The success of a seedling in a gap depends also on the variability of resource availability within gaps as this determines its ability to compete. Within gaps, a root, bole and crown zone can be distinguished (Orians 1982, Vitousek & Denslow 1986, Brandani, Hartshorn & Orians 1988), each offering a specific combination of resource availability and damage patterns among seedlings. The light conditions within gaps (Clark & Clark 1987), and in their immediate surroundings (Popma *et al.* 1988, cf. Lieberman *et al.* 1989) may vary considerably, not only as a result of the gap geometry, but also because of remnant treelets, the fallen tree itself and developing vegetation. Unfortunately, this is rarely quantified (Canham, Denslow, Platt *et al.* 1990), in particular in relation to seedling performance (Van der Meer 1995). The relative position of a seedling in a regenerating patch is chance-dependent. On the other hand, competition occurs between direct neighbours, which experience relatively similar environmental conditions. Competition between neighbours is therefore still affected by directional processes. As the seedlings attain a larger size, competitive interactions stretch over an increasingly larger, potentially more variable, neighbourhood.

In conclusion, due to the greater availability of limiting resources, species-specific differences are more pronounced in gaps than in the understorey and more likely to determine the result of competition processes. Nevertheless, chance factors such as stochastic mortality and position within the gap contribute to variation in the result of competition.

REGENERATION STRATEGIES

Chance processes and species-specific responses can be expected to have greatly different effects on species recruitment, depending on variation in spatial patterns of seed and seedling abundance. Theoretically, species which are different in competitive vigour (the ability to be successful during the phases of regeneration that were described above), and in abundance at different spatial scales (patch vs. entire community) may maintain themselves in the community of different population size distributions, each representing different pathways towards successful regeneration. By capitalizing on stochastic processes, competitively inferior species may occasionally regenerate successfully at the expense of competitively more vigorous species, and thus maintain themselves in the community. Species characteristics that enhance success of inferior species are, for example, a long reproductive life, which is advantageous because it increases the number of seed dispersal events; concentration of offspring in space, which increases the probability of success in a limited number of sites; or, alternatively, a specialized dispersal mechanism which increases the probability of a seed establishing in a favourable microsite (directed dispersal, Howe & Smallwood 1982). Frequent gap formation leads to an increase in the number of opportunities for successful regeneration of inferior species. These considerations lead theoretically to at least four different classes of population size distributions, although each of them is subject to modification according to the specific characteristics of the species and site.

A species that is infrequent in a given patch by the time a canopy opening is formed above it, and is also a poor competitor, is likely to be excluded from this patch. If such a species is also infrequent in the entire community, then it is likely to be rare and will eventually disappear, unless its seeds and seedlings are concentrated in space, so that it is frequent in at least some patches. Such a situation was hypothesized by Condit *et al.* (1992) as a mechanism for inferior species to hold on to their canopy sites. We predict that such species are relatively long-lived and have a population structure that is dominated by adults, deficient in larger saplings and whose juvenile abundance is spatially variable (Fig. 8.5a).

A species that is more frequent in a given patch but competing poorly is more likely to be retained because there is a higher probability than in the previous case that an individual escapes stochastic mortality, and finds itself in suitable conditions for competition. Successful regeneration is only possible when the seedlings and saplings are very persistent under the unfavourable light conditions which develop at the forest floor during the rebuilding phase of gaps. Lacking competitive vigour, these species require a series of suitable regeneration events to reach the canopy. Mortality due to repeated gap formation and poor competitive ability allow only a few seedlings to reach larger size classes. If these species are also frequent at the community scale, we predict that their population size class distributions are bimodal, with a large number of adults, small seedlings and saplings, but little 'advanced regeneration'. Their frequency in the canopy is then explained by a low mortality rate in comparison with the previous 'pole' stage (Fig. 8.5b).

Strongly competing species that are infrequent in a given patch are in many cases retained by virtue of their competitive abilities. If these

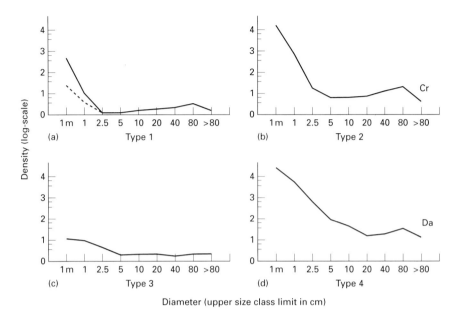

FIG. 8.5. Hypothetical population size distributions of species differing in competitive ability and abundance. (a) Type 1: infrequent and poorly competitive species; (b) type 2: frequent but poorly competitive species; (c) type 3: infrequent and highly competitive species and (d) frequent and highly competitive species. The size distribution of type 2 is based on actual population curves observed for *Chlorocardium rodiei* ('Cr') in two 1 ha demographic plots in tropical rain forest near Mabura Hill, Guyana; type 4 on the actual population curve observed for *Dicymbe altsonii* ('Da') in the same area. *C. rodiei* is a slow-growing species of low competitivity; *D. altsonii* is a fast-growing, competitive species. Both are common in the canopy. (From Zagt 1997.)

species are also infrequent at the community scale, they are predicted to have a relatively uniform population size distribution; that is, with an over-representation of adults and large saplings compared to normal 'reversed J' distributions. Possible reasons for their low abundance in the community are a short reproductive life-span, poor seedling survival in the forest understorey, small seedling stature or infrequent reproduction (Fig. 8.5c)

Species that are frequent in gaps and competitive are the most likely ones to gain access to the canopy. If they are also frequent in the community, we predict that these species have a well-balanced population structure in the form of a 'reversed J', with abundant large saplings. These species are likely to dominate the structure of the forest (Fig. 8.5d).

Other population size distributions have been published (Knight 1975, Bongers *et al.* 1988). It is difficult to interpret them without knowledge of the competitive ability of the species for which they were derived. Type I of

Bongers *et al.* (1988) and figs 6 and 9 of Knight (1975) represent 'reversed-J' population size distributions as in our Fig. 8.5d. Knight's fig. 10 is an uniform distribution reminiscent of our Fig. 8.5c. Species with irregular size distributions (Bongers type II, Knight fig. 7 & 8) or a distribution with few adults and saplings but a large seedling bank (Bongers type III) concern species with temporally irregular recruitment or growth rates. This pattern is indicative of poor competitiveness under adverse conditions, hence these might be variations of the size distributions presented in Fig. 8.5b. These population size distributions might show unimodality if adults are less persistent in the canopy than in our example or if the seedlings are very short-lived and thus have low average densities. Pioneers, not considered here, show comparable size distributions.

The species composition of the community and within the gaps influences the shape of the population size distribution of each species. In this respect it is necessary to keep in mind that competitive ability and abundance are both relative measures, and this implies that one species may have different size class distributions in communities with different species compositions.

CONCLUSIONS

Studies of the maintenance of species richness in tropical rainforests have concentrated on the response of large-sized trees to light availability. Canopy trees are the product of successful regeneration and are not subject to its critical demographical constraints. Their large size and complex form prevent the precise determination of growth and the environmental conditions to which they are subject. Because of this, the results of these studies are likely biased towards the detection of random processes governing forest regeneration.

Seedlings are more sensitive to variations in resource availability, and their responses in terms of growth and mortality are much larger than in large trees. This makes them more suitable for evaluation of mechanisms promoting species coexistence.

The contribution of chance and species-specific factors varies during the stages of seedling regeneration in the understorey and in gaps.

We argue that chance processes contribute highly to the initial species composition of gaps through the random seed density distributions, relative to the availability of resources in the understorey and to the occurrence of gaps. These chance processes are probably in most cases strong enough to even out the effect of species-specific growth and survival characteristics. Once the species composition of a patch is determined (soon after gap formation), regeneration will be dominated by predictable processes based on species-specific differences in growth and survival, and on their initial size. Chance effects such as chance mortality and position of the seedling within the gap will affect but not always cancel out these directional processes.

Seedling studies should be carried out on the scale of hectare plots rather than gaps, on the scale of years rather than months, and should encompass a large fraction of the available species pool. The full variability of biotic and abiotic environmental conditions should be addressed. As ecosystems may differ in important aspects, such as disturbance regime and diversity patterns of species, a comparable approach at different sites would increase our understanding of the relative contributions of chance and directionality to the coexistence of many species.

ACKNOWLEDGEMENTS

We thank René Boot and Laurens Poorter for providing unpublished data, and also along with Pieter Zuidema, Hans ter Steege, David Hammond and three anonymous reviewers, for comments on an earlier draft of this paper. R.J.Z. was funded in part by the Tropenbos Foundation, Wageningen.

REFERENCES

Aide, T.M. (1987). Limbfalls: a major cause of sapling mortality for tropical forest plants. *Biotropica*, **19**, 284–285.

Alvarez-Buylla, E.R. (1994). Density dependence and patch dynamics in tropical rain forests: matrix models and applications to a tree species. *American Naturalist*, **143**, 155–191.

Alvarez-Buylla, E.R. & Martínez-Ramos, M. (1992). Demography and allometry of *Cecropia obtusifolia*, a neotropical pioneer tree—an evaluation of the climax-pioneer paradigm for tropical rain forests. *Journal of Ecology*, **80**, 275–290.

Ashton, P.M.S. (1990). Method for the evaluation of advanced regeneration in forest types of south and southeast Asia. *Forest Ecology and Management*, **36**, 163–175.

Aubréville, A. (1938). La forêt coloniale: les forêts de l'Afrique occidentale française. *Annales Académie Sciences Coloniale*, **9**, 1–245.

Augspurger, C.K. (1983a). Offspring recruitment around tropical trees: changes in cohort distance with time. *Oikos*, **40**, 189–196.

Augspurger, C.K. (1983b). Seed dispersal of the tropical tree, *Platypodium elegans*, and the escape of its seedlings from fungal pathogens. *Journal of Ecology*, **71**, 759–771.

Augspurger, C.K. (1984). Seedling survival of tropical tree species: interactions of dispersal distance, light-gaps and pathogens. *Ecology*, **65**, 1705–1712.

Augspurger, C.K. & Kelly, C.K. (1984). Pathogen mortality of tropical seedlings: experimental studies of the effects of dispersal distance, seedling density, and light conditions. *Oecologia*, **61**, 211–217.

Bariteau, M. (1992). Régénération naturelle de la forêt tropicale humide de Guyane: étude de la répartition spatiale de *Qualea rosea* Aublet, *Eperua falcata* Aublet et *Symphonia globulifera* Linaeus f. *Annales des Sciènces Forestières*, **49**, 359–382.

Barros Henriques, R.P. & Elias Girnos de Sousa, E.C. (1989). Population structure, dispersion and microhabitat regeneration of *Carapa guianensis* in Northeastern Brazil. *Biotropica*, **21**, 204–209.

Barton, A.M. (1984). Neotropical pioneer and shade tolerant tree species: do they partition treefall gaps? *Tropical Ecology*, **25**, 196–202.

Becker, P. & Smith, A.P. (1990). Spatial autocorrelation of solar radiation in a tropical moist forest understory. *Agricultural and Forest Meteorology*, **52**, 373–379.

Bongers, F. & Popma, J. (1990). Leaf dynamics of rain forest species in relation to canopy gaps. *Oecologia*, **82**, 122–127.

Bongers, F., Popma, J., Meave del Castillo, J. & Carabias, J. (1988). Structure and floristic composition of the lowland rain forest of Los Tuxtlas, Mexico, *Vegetatio*, **74**, 55–80.

Boot, R.G.A. (1996). The significance of seedling size and growth rate of tropical rain forest tree seedlings for regeneration in canopy openings. In *The Ecology of Tropical Forest Tree Seedlings* (Ed. by M.D. Swaine), pp. 267–283 UNESCO, Paris and Parthenon, Carnforth UK.

Brandani, A., Hartshorn, G.S. & Orians, G.H. (1988). Internal heterogeneity of gaps and species richness in Costa Rican tropical wet forest. *Journal of Tropical Ecology*, **4**, 99–119.

Brokaw, N.V.L (1982). The definition of treefall gaps and its effect on measures of forest dynamics. *Biotropica*, **14**, 158–160.

Brokaw, N.V.L. (1985). Gap-phase regeneration in a tropical forest. *Ecology*, **66**, 682–687.

Brokaw, N.V.L. (1986). Seed dispersal, gap colonisation and the case of *Cecropia insignis*. In: *Frugivores and Seed Dispersal* (Ed. by A. Estrada & T.H. Fleming), pp. 323–331. Junk, Dordrecht.

Brokaw, N.V.L. (1987). Gap-phase regeneration of three pioneer tree species in a tropical forest. *Journal of Ecology*, **75**, 9–19.

Brown, N.D. (1993). The implications of climate and gap microclimate for seedling growth conditions in a Bornean lowland rain forest. *Journal of Tropical Ecology*, **9**, 153–168.

Brown, N.D. & Whitmore, T.C. (1992). Do dipterocarp seedlings really partition tropical rain forest gaps? *Philosophical Transactions of the Royal Society, Series B*, **335**, 369–378.

Burslem, D.F.R.P. (1996). Differential responses to nutrients, shade and drought among tree seedlings of lowland tropical forest in Singapore. In *The Ecology of Tropical Forest Tree Seedlings* (Ed. by M.D. Swaine), pp. 211–244. UNESCO, Paris and Parthenon, Carnforth, UK.

Canham, C.D. (1989). Different responses to gaps among shade tolerant tree species. *Ecology*, **70**, 548–550.

Canham, C.D., Denslow, J.S., Platt, W.J., Runkle, J.R., Spies, T.A. & White, P.S. (1990). Light regimes beneath closed canopies and tree-fall gaps in temperate and tropical forests. *Canadian Journal of Forest Research*, **20**, 620–631.

Chazdon, R.L. & Fetcher, N. (1984). Photosynthetic light environments in a lowland tropical rain forest in Costa Rica. *Journal of Ecology*, **72**, 553–564.

Chazdon, R.L. & Field, C.B. (1987). Photographic estimation of photosynthetically active radiation: evaluation of a computerized technique. *Oecologia*, **3**, 525–532.

Clark, D.A. & Clark, D.B. (1984). Spacing dynamics of a tropical rain forest tree: evaluation of the Janzen–Connell model. *American Naturalist*, **124**, 769–788.

Clark, D.B. & Clark, D.A. (1985). Seedling dynamics of a tropical tree: impacts of herbivory and meristem damage. *Ecology*, **66**, 1884–1892.

Clark, D.A. & Clark, D.B. (1987). Análisis de la regeneración de árboles del dosel en bosque muy húmedo tropical: aspectos teóricos y prácticos. *Revista de Biología Tropical*, **35** (Suppl. 1), 41–54.

Clark, D.B. & Clark, D.A. (1991). The impact of physical damage on canopy tree regeneration in tropical rain forest. *Journal of Ecology*, **79**, 447–457.

Clark, D.A. & Clark, D.B. (1992). Life history diversity of tropical and emergent trees in a neotropical rain forest. *Ecological Monographs*, **62**, 315–344.

Clark, D.B., Clark, D.A., Rich, P.M., Weiss, S. & Oberbauer, S.F. (1996). Landscape-scale evaluation of understory light and canopy structure: methods and application in a Neotropical lowland rain forest. *Canadian Journal of Forest Research*, **26**, 747–757.

Coley, P.D. (1983). Herbivory and defence characteristics of tree species in a lowland tropical forest. *Ecological Monographs*, **53**, 209–233.

Condit, R., Hubbell, S.P. & Foster, R.B. (1992). Recruitment near conspecific adults and the maintenance of tree and shrub diversity in a Neotropical forest. *American Naturalist*, **140**, 261–286.

Condit, R., Hubbell, S.P. & Foster, R.B. (1994). Density dependence in two understorey tree species in a neotropical forest. *Ecology*, **75**, 671–680.

Condit, R., Hubbell, S.P. & Foster, R.B. (1995). Mortality rates of 205 Neotropical tree and shrub species and the impact of a severe drought. *Ecological Monographs*, **65**, 419–439.

Connell, J.H. (1971). On the role of natural enemies in preventing competitive exclusion in some marine animals and in rain forest trees. In *Dynamics of Populations* (Ed. by P.J. den Boer & G.R. Gradwell), pp. 298–312. Pudoc, Wageningen.

Connell, J.H. (1978). Diversity in tropical rain forests and coral reefs. *Science*, **199**, 1302–1310.

Connell, J.H. (1989). Some processes affecting the species composition in forest gaps. *Ecology*, **70**, 560–562.

Connell, J.H., Tracey, J.G. & Webb, L.J. (1984). Compensatory recruitment, growth and mortality as factors maintaining rain forest tree diversity. *Ecological Monographs*, **54**, 141–164.

Dawkins, H.C. & Field, D.R. (1978). *A Long-term Surveillance System for British Woodland Vegetation. Occasional Paper*, **1**. Oxford University, Oxford.

De Steven, D. (1994). Tropical tree seedling dynamics: recruitment patterns and their population consequences for three canopy species in Panama. *Journal of Tropical Ecology*, **10**, 369–383.

Denslow, J.S. (1980). Gap partitioning among tropical rainforest trees. *Biotropica*, **12** (Suppl.), 47–55.

Fetcher, N., Oberbauer, S.F., Rojas, G. & Strain, B.R. (1987). Efectos del régimen de luz sobre la fotosíntesis y el crecimiento en plántulas de árboles de un bosque lluvioso tropical de Costa Rica. *Revista de Biologia Tropical*, **35** (Suppl. 1), 97–110.

Forget, P.-M. (1989). La régénération naturelle d'une espèce autochore de la forêt Guyanaise: *Eperua falcata* Aublet (Caesalpiniaceae). *Biotropica*, **21**, 115–125.

Forget, P.-M. (1994). Recruitment pattern of *Vouacapoua americana* (Caesalpiniaceae), a rodent-dispersed tree species in French Guiana. *Biotropica*, **26**, 408–419.

Foster, S.A. & Janson, C.H. (1985). The relationship between seed size and establishment conditions in tropical woody plants. *Ecology*, **66**, 773–780.

Frankie, G.W., Baker, H.G. & Opler, P.A. (1974). Comparative phenological studies of trees in tropical wet and dry forests in the lowlands of Costa Rica. *Journal of Ecology*, **62**, 881–919.

Garwood, N.C. (1983). Seed germination in a seasonal tropical forest in Panama: a community study. *Ecological Monographs*, **53**, 159–181.

Garwood, N.C. (1986). Constraints on the timing of seed germination in a tropical forest. In *Frugivores and Seed Dispersal* (Ed. by A. Estrada & T.H. Fleming), pp. 347–355. Junk, Dordrecht.

Gentry, A.H. (1988). Tree species richness of upper Amazonian forest. *Proceedings of the National Academy of Sciences USA*, **85**, 156–159.

Grubb, P.J. (1977). The maintenance of species-richness in plant communities: the importance of the regeneration niche. *Biological Reviews*, **52**, 107–145.

Hammond, D.S. & Brown, V.K. (1995). Seed size of woody plants in relation to disturbance, dispersal, soil type in wet Neotropical forests. *Ecology*, **76**, 2544–2561.

Hart, T.B. (1995). Seed, seedling and sub-canopy survival in monodominant and mixed forests of the Ituri Forest, Africa. *Journal of Tropical Ecology*, **11**, 443–459.

Hartshorn, G.S. (1972). *The ecological life history and population dynamics of* Pentaclethra macroloba, *a tropical wet forest dominant and* Stryphnodendron excelsum, *and occasional associate*. PhD thesis, University of Washington, Seattle.

Hartshorn, G.S. (1978). Tree falls and tropical forest dynamics. In *Tropical Trees as Living Systems* (Ed. by P.B. Tomlinson & M.H. Zimmermann), pp. 617–638. Cambridge University Press, Cambridge.

Hartshorn, G.S. (1980). Neotropical forest dynamics. *Biotropica*, **12** (Suppl.), 23–30.

Howe, H.F. & Richter, W.M. (1982). Effects of seed size on seedling size in *Virola surinamensis*; a within and between tree analysis. *Oecologia*, **53**, 347–351.

Howe, H.F. & Smallwood, J. (1982). Ecology of seed dispersal. *Annual Review of Ecology and Systematics*, **13**, 201–228.

Howe, H.F. Schupp, E.W., & Westley, L.C. (1985). Early consequences of seed dispersal for a neotropical tree (*Virola surinamensis*). *Ecology*, **66**, 781–791.

Huante, P., Rincón E. & Acosta, I. (1995). Nutrient availability and growth rate of 34 woody species from a tropical deciduous forest in Mexico. *Functional Ecology*, **9**, 849–858.

Hubbell, S.P. (1979). Tree dispersion, abundance and diversity in a tropical dry forest. *Science*, **203**, 1299–1309.

Hubbell, S.P. & Foster, R.B. (1986a). Canopy gaps and the dynamics of a neotropical forest. In *Plant Ecology* (Ed. by M.J. Crawley), pp. 77–98. Blackwell Scientific Publications, Oxford.

Hubbell, S.P. & Foster, R.B. (1986b). Biology, chance, and history and the structure of tropical rain forest tree communities. In *Community Ecology* (Ed. by J. Diamond & T.J. Case), pp. 314–329. Harper & Row, New York.

Hubbell, S.P. & Foster, R.B. (1987a). The spatial context of regeneration in a neotropical forest. In *Colonisation, Succession and Stability* (Ed. by A.J. Gray, M.J. Crawley & P.J. Edwards), pp. 395–412. Blackwell Scientific Publications, Oxford.

Hubbell, S.P. & Foster, R.B. (1987b). La estructura espacial en gran escala de un bosque neotropical. *Revista de Biología Tropical*, **35** (Suppl. 1), 7–22.

Hubbell, S.P. & Foster, R.B. (1990). Structure, dynamics, and equilibrium status of old-growth forest on Barro Colorado Island. In *Four Neotropical Rainforests* (Ed. by A.H. Gentry), pp. 522–541. Yale University Press, New Haven.

Itoh, A., Yamakura, T., Ogino, K. & Lee, H.S. (1995). Survivorship and growth of seedlings of four dipterocarp species in a tropical rainforest of Sarawak, East Malaysia. *Ecological Research*, **10**, 327–338.

Janzen, D.H. (1970). Herbivores and the number of tree species in tropical forests. *American Naturalist*, **104**, 501–528.

Jetten, V.G., Riezebos, H.Th., Hoefsloot, F. & Rossum, J. van (1993). Spatial variability of infiltration and related properties of tropical soils. *Earth Surface Processes and Landforms*, **18**, 477–488.

King, D.A. (1996). Allometry and life history of tropical trees. *Journal of Tropical Ecology*, **12**, 25–44.

Kitajima, K. (1994). Relative importance of photosynthetic traits and allocation patterns as correlates of seedling shade tolerance of 13 tropical trees. *Oecologia*, **98**, 419–428.

Knight, D.H. (1975). A phytosociological analysis of species-rich tropical forest on Barro Colorado Island, Panama. *Ecological Monographs*, **45**, 259–284.

Kohyama, T. (1987). Significance of architecture and allometry in saplings. *Functional Ecology*, **1**, 399–404.

Kohyama, T. (1991). A functional model describing sapling growth under a tropical forest canopy. *Functional Ecology*, **5**, 83–90.

Lambers, H. & Poorter, H. (1992). Inherent variation in growth rate between higher plants; a search for physiological causes and ecological consequences. *Advances in Ecological Research*, **23**, 187–261.

Lechowicz, M.J. & Bell, G. (1991). The ecology and genetics of fitness in forest plants. II. Microspatial heterogeneity of the edaphic environment. *Journal of Ecology*, **79**, 687–696.

Li, M., Lieberman, M. & Lieberman, D. (1996). Seedling demography in undisturbed tropical wet forest in Costa Rica. In *The Ecology of Tropical Forest Tree Seedlings* (Ed. by M.D. Swaine), pp. 285–314. UNESCO, Paris and Parthenon, Carnforth, UK.

Lieberman, D. & Lieberman, M. (1987). Forest tree growth and dynamics at La Selva, Costa Rica (1969–1982). *Journal of Tropical Ecology* **3**, 347–358.

Lieberman, D., Lieberman, M., Peralta, R. & Hartshorn, G.S. (1985). Mortality patterns and stand turnover rates in a wet tropical forest in Costa Rica. *Journal of Ecology*, **73**, 915–924.

Lieberman, D., Lieberman, M. & Peralta, R. (1989). Forests are not just Swiss cheese: canopy stereogeometry of non-gaps in tropical forests. *Ecology*, **70**, 550–552.

Lieberman, M., Lieberman, D., Peralta, R. & Hartshorn, G.S. (1995). Canopy closure and the distribution of tropical forest tree species at La Selva, Costa Rica. *Journal of Tropical Ecology*, **11**, 161–177.

Newbery, D. McC., Campbell, E.J.F., Proctor, J. & Still, M.J. (1996). Primary lowland dipterocarp forest at Danum Valley, Sabah, Malaysia. *Vegetatio*, **122**, 193–220.

Oberbauer, S.F., Clark, D.B., Clark, D.A. & Quesada, M. (1988). Crown light environments of saplings of two species of rain forest emergent trees. *Oecologia*, **75**, 207–212.

Oberbauer, S.F., Clark, D.B., Clark, D.A., Rich, P.M. & Vega, G. (1993). Light environment, gas exchange, and annual growth of saplings of three species of rain forest trees in Costa Rica. *Journal of Tropical Ecology*, **9**, 511–522.

O'Brien, S.T., Hubbell, S.P., Spiro, P., Condit, R. & Foster, R.B. (1995). Diameter, height, crown, and age relationships in eight neotropical tree species. *Ecology*, **76**, 1926–1939.

Orians, G. (1982). The influence of tree-falls in tropical forests on tree species richness. *Tropical Ecology*, **23**, 255–279.

Osunkoya, O., Ash, J.E., Hopkins, M.S. & Graham, A.W. (1992). Factors affecting survival of tree seedlings in North Queensland rainforests. *Oecologia*, **91**, 569–578.

Osunkoya, O.O., Ash, J.E., Graham, A.W. & Hopkins, M.S. (1993). Growth of tree seedlings in tropical rain forests of North Queensland, Australia. *Journal of Tropical Ecology*, **9**, 1–18.

Osunkoya, O., Ash, J.E., Hopkins, M.S. & Graham, A.W. (1994). Influence of seed size and seedling ecological attributes on shade-tolerance of rain-forest tree species in northern Queensland. *Journal of Ecology*, **82**, 149–163.

Popma, J. & Bongers, F. (1988). The effect of canopy gaps on growth and morphology of seedlings of rain forest species. *Oecologia*, **75**, 625–632.

Popma, J. & Bongers, F. (1991). Acclimation of seedlings of three tropical rain forest species to changing light availability. *Journal of Tropical Ecology*, **7**, 85–97.

Popma, J., Bongers, F., Martínez-Ramos, M. & Veneklaas, E. (1988). Pioneer species distribution in treefall gaps in Neotropical rain forest; a gap definition and its consequences. *Journal of Tropical Ecology*, **4**, 77–88.

Raich, J.W. & Gong, W.K. (1990). Effects of canopy openings on tree seed germination in a Malaysian dipterocarp forest. *Journal of Tropical Ecology*, **6**, 203–217.

Schupp, E.W. (1988a). Seed and early seedling predation in the forest understorey and in treefall gaps. *Oikos*, **51**, 71–78.
Schupp, E.W. (1988b). Factors affecting post-dispersal seed survival in a tropical forest. *Oecologia*, **76**, 525–530.
Schupp, E.W., Howe, H.F., Augspurger, C.K. & Levey, D.J. (1989). Arrival and survival in tropical treefall gaps. *Ecology*, **70**, 562–564.
Shmida, A. & Ellner, S. (1984). Coexistence of plant species with similar niches. *Vegetatio*, **58**, 29–55.
Sterck, F.J. (1997). *Trees and light. Tree development and morphology in relation to light availability in a tropical rain forest in French Guiana*. PhD Thesis Wageningen Agricultural University, Wageningen.
Swaine, M.D. & Whitmore, T.C. (1988). On the definition of ecological species groups in tropical forests. *Vegetatio*, **75**, 81–86.
Ter Steege, H. (1994). *Hemiphot, a Program to Analyze Vegetation Indices, Light and Light Quality from Hemispherical Photographs. Tropenbos Documents*, **3**. The Tropenbos Foundation, Wageningen.
Uhl, C., Clark, K., Dezzeo, N. & Maquinero, P. (1988). Vegetation dynamics in Amazonian treefall gaps. *Ecology*, **69**, 751–763.
Valencia, R., Balslev, H. & Paz y Miño, G. (1994). High tree alpha-diversity in Amazonian Ecuador. *Biodiversity and Conservation*, **3**, 21–28.
Van der Meer, P.J. (1995). Canopy dynamics of a tropical rain forest in French Guiana. PhD thesis, Wageningen Agricultural University, Wageningen.
Van der Meer, P.J. & Bongers, F. (1996). Patterns of treefalls and branchfalls in a neotropical rain forest in French Guiana. *Journal of Ecology*, **84**, 19–29.
Van Steenis, C.C.G.J. (1956). De biologische nomadentheorie, *Uakblad voor Biologen*, **36**, 165–172.
Veenendaal, E.M., Swaine, M.D., Lecha, R.T., Walsh, M.F., Abebrese, I.K. & Owusu-Afriyie, K. (1996). Responses of West African forest tree seedlings to irradiance and soil fertility. *Functional Ecology*, **10**, 501–511.
Vitousek, P.M. & Denslow, J.S. (1986). Nitrogen and phosphorus availability in treefall gaps in a tropical lowland rain-forest. *Journal of Ecology*, **74**, 1167–1178.
Welden, C.W., Hewett, S.W., Hubbell, S.P. & Foster, R.B. (1991). Sapling survival, growth and recruitment: relationships to canopy height in a neotropical forest. *Ecology*, **72**, 35–50.
Whitmore, T.C. (1978). Gaps in the forest canopy. In *Tropical Trees as Living Systems* (Ed. by P.B. Tomlinson & M.H. Zimmerman), pp. 639–655. Cambridge University Press, Cambridge.
Whitmore, T.C. (1984). *Tropical Rain Forests of the Far East*, 2nd edn. Clarendon Press, Oxford.
Whitmore, T.C. & Brown, N.D. (1996). Dipterocarp seedling growth in rain forest canopy gaps during six and a half years. *Philosophical Transactions of the Royal Society, Series B*, **351**, 1195–1204.
Whitmore, T.C., Brown, N.D., Swaine, M.D., Kennedy, D., Goodwin-Bailey, C.I. & Gong, W.-K. (1993). Use of hemispherical photographs in forest ecology: measurement of gap size and radiation totals in a Bornean tropical rain forest. *Journal of Tropical Ecology*, **9**, 131–151.
Yavitt, J.B., Battles, J.J., Lang, G.E. & Knight, D.H. (1995). The canopy gap regime in a secondary Neotropical forest in Panama. *Journal of Tropical Ecology*, **11**, 391–402.
Zagt, R.J. (1997). Tree demography in the tropical rain forest of Guyana. PhD Thesis, Utrecht University, Utrecht.

9. RISK-SPREADING AND RISK-REDUCING TACTICS OF WEST AFRICAN ANURANS IN AN UNPREDICTABLY CHANGING AND STRESSFUL ENVIRONMENT

K. E. LINSENMAIR

Theodor-Boveri-Institut für Biowissenschaften (Biozentrum), Lehrstuhl für Tierökologie und Tropenbiologie, Am Hubland, Universität Würzburg, D-97074 Würzburg, Germany

SUMMARY

1 Anurans as dwellers in African savannas must be able to cope with highly unpredictable conditions, due to irregular rainfall patterns, and survive climatically very harsh dry seasons. To secure reproductive success under such conditions appears to be difficult. The risk-reducing tactics and situation-dependent alternative strategies to improve the survival probabilities of tadpoles and metamorphosed froglets were investigated in three West African savanna anurans. In one species, special attention was also paid to its unusual mode of above-ground aestivation.

2 *Hoplobatrachus occipitalis* immediately oviposited in river-bed rock pools after the first rains. This species showed a very remarkable risk sensitivity *vis-à-vis* abiotic and biotic hazards when it distributed its eggs among different pools, by, among others tactics, measuring water-holding capacity of the pools. It usually also spreads its risk over time by repeatedly laying only a small proportion of its mature eggs during a single bout of spawning. In years with high rainfall, this risk-sensitive strategy was remarkably successful, but it resulted in a nearly complete loss of reproduction in very dry years. Later in the breeding season most individuals migrated to savanna ponds where they seemed to follow an alternative spawning strategy that, however, was much more risk averse because a small percentage of tadpoles reached metamorphosis every year.

3 *Bufo maculatus* mostly bred in shallow puddles, created by rising and falling of the Comoé river. These sites were extremely unpredictable in their prospects and most breeding attempts resulted in complete failures. No sophisticated risk-sensitive behaviour in adults could be detected. Tadpoles had to cope with the main risks of desiccation and predation. Some individuals were able to reduce the risk of desiccation by speeding up their development to a very limited extent. Tadpoles were more effective,

however, in coping with water-living predators by densely aggregating in the shallowest part of the puddle, although this was only attained by considerably increasing the risk of desiccation and by incurring additional energy costs.

4 *Hyperolius viridiflavus nitidulus* is a short-lived species, aestivating above ground, fully exposed to the harsh dry season climate. It is adapted in many ways to this very unusual aestivation habit. Since only half-grown subadults aestivate and need time to grow to adulthood, reproductive activity in *H. v. nitidulus* started relatively late in the season. Forming breeding aggregations and spawning synchronously may have helped in swamping the numerous tadpole predators then present in savanna ponds. Adults also experienced a very high predation pressure and thus rarely survived to the end of the potential breeding season, one that was considerably extended when compared to, for example, subterraneously aestivating species. In many years it should be of selective advantage for reed frogs to spawn still later, since the pressure exerted by predators and competitors seems then to be reduced. *H. v. nitidulus* has solved this problem by producing a rainy season generation, with a very fast post-metamorphic development showing profound modifications in many other behavioural and physiological traits.

INTRODUCTION

Anurans, having succeeded in invading xeric environments (see for example, Feder & Burggren 1992, Warburg 1997 for summaries), mostly lead fugitive lives due to some basic constraints (Shoemaker 1988, 1992), for example, low resistance against cutaneous evaporation. This also holds true for most of the 35 species (Schiøtz 1967, Rödel 1996) living in the study area in the West African Guinea savanna. However, a few of these species excel by being highly waterproof, aestivating above ground fully exposed to the harsh dry season climate.

Whether savanna anurans must seek the refuge of subterraneous aestivation sites or permanent bodies of water, or whether they can aestivate in exposed positions nearly anywhere in the savanna and opportunistically use short-term favourable weather conditions for foraging, should profoundly influence important life-history traits, foremost the temporal pattern of reproduction.

All fugitive species should try to reproduce early in the rainy season to accumulate sufficient energy (fat) and to attain a body size large enough for storing large water reserves to survive up to 5–6 months of subterraneous dormancy. On the other hand, those species that aestivate above ground have a considerably extended period for post-metamorphic development and are thus left with more time for breeding still late in the season.

The following account aims at demonstrating in an exemplary (and of course only fragmentary) manner:

1 How those species that lead fugitive lives, and therefore have to breed early, cope with unpredictable patterns of precipitation with untimely drying up of ephemeral ponds, with high predator pressure on their larvae and how they may elaborately spread the risk of reproductive failures under unpredictably changing conditions.

2 What mechanisms allow above-ground aestivating frogs successfully to sustain the stressful dry season climate and what consequences arise from this mode of aestivating for the population dynamics of such species.

MATERIALS AND METHODS

Study area and climate

Field data were collected in the southern part of the Comoé National Park ($8°5'-9°6'$N, $3°1'-4°4'$W) in the northeastern Ivory Coast. The study area belongs to the northern zone of the Guinea savanna. This tree–scrub savanna has a very pronounced seasonal climate with the first strong rains (>5 mm) falling extremely irregularly some time between the beginning of March and the end of April (Spieler & Linsenmair 1997a) and usually terminating between the end of September and mid-October. The mean annual precipitation amounted to 1100 mm during 1977–87 (Poilecot 1991) and has dropped to a mean of c. 950 mm (850–1070) during the last 6 years. Not only the date of the first rain differs greatly between years but also the amount and the small-scale spatial distribution of the rainfall events are highly variable and entirely unpredictable, creating an ever changing patchy environment with regard to water availability. The mean annual temperatures range between 25 and 28°C. The (usually rainless) core dry period comprises December to the beginning of March, but may also extend from the beginning of October to the end of April.

Study animals

Hoplobatrachus occipitalis is a large (adult body size 59–111 mm) long-lived, ranid frog, broadly distributed over West Africa (Lamotte 1967) and relatively tightly bound to open water. In the population studied, the dry season was spent at the large pools remaining in the bed of the Comoé River. Females glue their eggs singly to shallow ground in the breeding waters.

Tadpoles are carnivorous and cannibalistic and need a minimum developmental time of 25 days with average water temperature of 33°C. Between 1991 and 1994, 44 natural ponds, and during the last 2 years also 10 artificial ones, built with natural rocks and concrete, were checked daily for laid eggs. To study the effects of tadpole density and size on spawning behaviour, pools were emptied and refilled with water from unoccupied bodies of water. Tadpoles were kept in polyethylene containers with gauze-covered holes and these, put into the pools. For more details see Spieler and Linsenmair (1997a) and Spieler (1997).

Bufo maculatus is a medium-sized (adult body size 38–58 mm) and long-lived toad that is widely distributed over Africa south of the Sahara. To manipulate the behaviour of tadpoles experimentally, individual clutches were collected in the field and kept separately for 24 h in aquaria (100 × 50 × 30 cm). Thereupon, the behaviour of groups of tadpoles was video-taped in the presence and absence of predators as well as when they were presented with potential predatory cues (see Spieler 1997 for details).

Hyperolius viridiflavus nitidulus belongs to a group of very closely related species ('superspecies': Schiøtz 1971) of small frogs that very successfully thrive in many African savannas. They are very short-lived and are most striking in their above-ground aestivating behaviour. Three seasons, with three corresponding states of frogs, can be distinguished:

1 'wet season frogs' (WSF);
2 'transitional season frogs' (TSF); and
3 'dry season frogs' (DSF).

Under laboratory conditions young WSF were kept at a day–night regime of 30–24°C, 70–100% RH, with 13 h of light and 11 h of darkness. WSF were daily sprayed with water and fed, whereas TSF were kept at the same temperature and light regime, but were sprayed and fed three times a week. DSF were maintained at 35–25°C, 30–50% RH and 11 h of light and 13 h of darkness; they were given neither water nor food. Breeding groups were kept under wet season conditions. Regarding the wide spectrum of special methods applied in the studies on the aestivation physiology of reed frogs, of which only a short summary is given here, the reader is referred to the cited articles. Mean values ± 1 standard deviation were used as descriptive statistics.

RESULTS

From 1990 to 1996, the first rains in the study area never created savanna pools that held water for more than a few hours. Only after several stronger rains that saturated the savanna soil with water, these pools remained filled for some extended time. Rock pools along the river bank of the Comoé, on the contrary, filled during the first rains (of ≥5mm), offerring breeding opportunities from several weeks to occasionally 3 months earlier than the savanna pools. This breeding opportunity was regularly used by *Hoplobatrachus occipitalis*.

Risk-sensitive behaviour in H. occipitalis

Most rock pools offered very unpredictable breeding prospects; poor ones due to their high risk of drying out, and good ones because of the very low interspecific predation pressure in the majority of these pools. In the time from March to May, some 80% of the 54 monitored pools completely dried up, leading to the death of all tadpoles when rainless periods lasted for more than 10 days. Such dry spells occurred in most years, usually two or three times after the first 'effective' rain (i.e. a rain of at least 15 mm which always initiated a strong spawning bout). Another danger for the tadpoles was flooding of pools by the rising Comoé River. This allowed predatory fish to invade.

It should 'pay' the reproducing *H. occipitalis* to show some general as well as some particular bet-hedging and risk-spreading behaviours *vis-à-vis* the overall unpredictability of developmental conditions on the one hand, and the high variability in those parameters that result in differences in quality among pools on the other hand. As our studies (Spieler & Linsenmair 1997a,b) have shown, *Hoplobatrachus occipitalis* meets these expectations. During the 4 years (1991–94) different pools received different amounts of spawn at first ovipositions (Friedman's ANOVA for 41 regularly controlled pools with egg numbers between 0 and 850: $\chi^2 = 13.9$; df = 3, $P < 0.005$). Only 14 pools (34%) received eggs every year, while five (12%) pools never received a single egg. Most of the chosen pools received rather low numbers of eggs. For instance, after the first effective rain 3843 eggs were laid into 28 (of 54 controlled) pools with a median of only 64 eggs during the first night in 1995 (25%/75% quartiles: 19/119; range: 3–699). All amplectant pairs that could be directly observed visited more, and also laid in more, than one pool (Spieler 1997). Most females seemed to spend only a small fraction (mostly less than 100) of their mature eggs (five medium-sized females had 1590 ± 310 mature eggs in their gonads) during one laying bout.

Exceptions were only found in females laying during the very first spawning bout of the year into the three largest rock pools that were never seen to dry up completely (with 5–15 females, laying between 300 and 1600 eggs each). Whenever a first effective rain fell, even if it occurred as early as

February (as in 1996), even when separated by the next rain by 4 weeks or only by 3 days, or whether there were only two (as in 1992) or 11 (as in 1993) effective rains during March to May, spawning bouts immediately followed (with 78% of all eggs ($n = 39800$) laid on the first night, 20% on the second and 2% on the third night after the rain) as long as the rock pools were not flooded by the river.

What are the causes for the striking preference for some pools? Why are some pools avoided, although they seemed, when water-filled, very suitable to the human observer? To investigate this question, the number of eggs laid per pool was correlated with several abiotic pool parameters: surface area, maximal depth, volume, and water-holding capacity (WHC). A stepwise multiple regression showed that the best predictor of the preference of pools was the water-holding capacity (model-wide $r^2 = 0.256$, WHC partial $r^2 = 0.234$, $P < 0.001$, $n = 589$ spawning events; Spieler & Linsenmair 1997a). Differences in WHC between similar-sized pools are mainly caused by seepage through cracks and this could not be measured during a single visit at the time of spawning when all pools were completely filled with rainwater. Rather, this variable can only be judged after at least one repeat visit. The prediction that frogs need time to integrate information on WHC was shown by building 10 artificial pools and observing the subsequent spawning behaviour. Whilst the volume of these new pools during the first year was the best predictor of egg number (Fig. 9.1a), it became replaced by WHC during the first spawnings in the following year (Fig. 9.1b).

Hoplobatrachus occipitalis tadpoles are carnivorous and also cannibalistic and in most rock pools they were the only serious predators for (especially smaller) conspecific tadpoles (Spieler & Linsenmair 1997a). Given the predatory behaviour of conspecific tadpoles, spawning adult *H. occipitalis* should pay attention to the presence of conspecific tadpoles when selecting pools in a risk-sensitive manner. In comparing the egg distribution among occupied and unoccupied pools it became evident that tadpole-free pools were significantly preferred (Spieler & Linsenmair 1997a). Even eggs can produce a strong deterrent effect on oviposition. Further spawning into such pools was nearly entirely prevented at a density of 200 (16 to c. 24 h old) eggs per 100 l water. The same effect was achieved with 10 medium-sized tadpoles per 100 l. Quickly decaying chemical cues (remaining effective not more than 10 h) are responsible for the relative or absolute avoidance of occupied sites (Spieler & Linsenmair 1997a).

In favourable years, *H. occipitalis*, breeding in these rock pools, gained unusually high reproductive success because tadpole mortality was low (see Fig. 9.2). In addition, these tadpoles had a developmental advantage on average of nearly 3 weeks during the 5 years of this study (18 ± 23 days, range

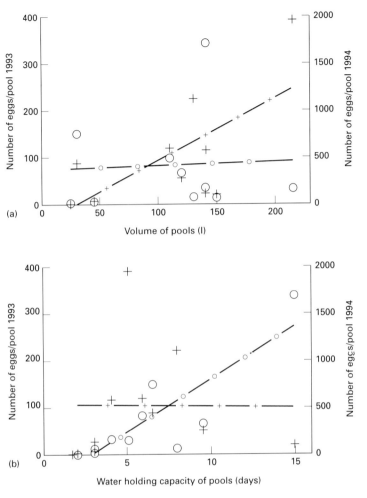

FIG. 9.1. Relationship between number of eggs oviposited by *Hoplobatrachus occipitalis* in 10 artificial pools and either (a) pool volume or (b) water-holding capacity in 1993 and 1994. The number of eggs/pool gives the sum of all eggs/pool within the first weeks of the rainy season comprising nine and eight oviposition events in 1993 and 1994, respectively. Crosses indicated data of 1993 and the line with crosses the corresponding regressions (a: $r = 0.60$, $P < 0.05$; b: $r = -0.01$, N.S., $P = 0.9$). Circles indicate data of 1994 and the line with circles the corresponding regressions (a: $r = 0.05$, N.S.; b: $r = 0.76$, $P = 0.001$).

7–63). In unfavourable years, however, between 80 and >90% of the breeding pools did not produce a single metamorphosed froglet.

When rock pools were flooded, most *H. occipitalis* left the river bed and migrated to savanna pools (visiting several of them and travelling distances

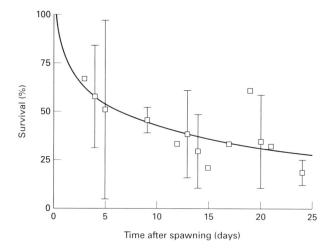

FIG. 9.2. Change in tadpole survival of *Hoplobatrachus occipitalis* over 13 different time intervals after spawning. Mean ± SD are given if several pools were investigated at one time. A negative exponential function best-fit line for the data ($r = 0.73$, $P < 0.01$). The sum of all eggs deposited in the 48 pools examined was 2395 (median: 26 eggs/pool, 25% quartile: 9.5, 75% quartile: 62.5).

of up to 6 km; see Spieler & Linsenmair 1997b) where they then spawned for some time. Indirect evidence suggests that *H. occipitalis* follows a different reproductive tactic there that corresponds to those of most of the savanna anurans: laying large amounts of eggs in single bouts that are intra- and interspecifically synchronized by strong rains. Thus, larger ponds receive hundreds of thousands of eggs within 2–4 days.

Systematic and quantitative sampling of communities and collections of metamorphosed froglets at drift fences provided strong evidence that predator pressure (and to some degree also competition) is extremely high leading to losses of more than 99% of potential progeny in most anuran species (M.-O. Rödel & R.E. Linsenmair unpublished data.). For example, out of 6350 newly hatched tadpoles comprising several species that were sampled from very small savanna pools where predation pressure was relatively low, only 90 individuals (1.4%) survived.

Bufo maculatus: *tadpoles coping with risks*

Bufo maculatus (and *Bufo regularis*) used shallow depressions on the margins of the river as their main breeding habitat. These depressions filled either by the rising or falling Comoé or by rainwater alone and could quickly and unpredictably change in water level. Therefore, they provided very unsafe habitats for developing tadpoles. Occasionally, they allowed very successful breeding but usually (>95% of more than 100 observed cases) complete reproductive failure occurred. When the level of the river water rose, the pools and depressions were invaded by swarms of predatory cichlids (and after strong rains partly also by *H. occipitalis* tadpoles). When the water level fell the probability of drying up increased strongly. Clutches of *B. maculatus* always seemed to comprise all the mature eggs a female possesses at the time of spawning. Contrary to *H. occipitalis*, *B. maculatus* was not as choosy in selecting breeding sites.

Within shallow river depressions, tadpoles are either very evenly distributed or highly aggregated. Two types of aggregations were observed. One actively moving about and one stationary type with the latter concentrated on small sections of the very margins of puddles in extremely shallow water (0.1–0.5 cm). These aggregations contained up to several thousand individuals concentrated on a patch with a diameter of not more than 10 cm. These aggregations often became cut off from open water and quickly dried up. What are the proximate cues used to elicit and maintain such risky aggregative behaviour and what are the ultimate benefits?

All 18 pools sampled in the field that had evenly distributed tadpoles did not contain water-living predators while predatory fish or *H. occipitalis* tadpoles were present where such stationary aggregations had been formed ($n = 35$ bodies of water). Both distributions differed significantly from a random distribution ($\chi^2 = 9.0, P < 0.01, n = 18; \chi^2 = 17.5, P < 0.001$). This behaviour was then experimentally investigated (Spieler 1996, 1997; M. Spieler & K.E. Linsenmair unpublished data). Briefly, we found that the stationary aggregation behaviour in *B. maculatus* tadpoles is proximately caused by a combination of two stimuli: (i) a species-specific or genus-specific chemical cue ('alarm substance') set free by injured tadpoles; and (ii) a mechanical stimulus caused by relatively fast movements of predators (or by similarly moving non-predatory tadpoles or fishes).

Since feeding opportunities should be reduced within a stationary aggregation, spending any extended time aggregated will lead to either a lengthened developmental time and/or will negatively affect the size at metamorphosis. Two groups of tadpoles, both living in their natural pools, one aggregated and the other evenly distributed, with an initially identical body mass, showed highly significant differences in mass gain already after 2 days. While aggregated tadpoles gained on average 4.0 mg within 2 days, the evenly distributed individuals showed an average increase of 6.4 mg (Mann–Whitney U-test: $z = -5.42, n = 50, P < 0.001$).

Cichlids usually leave the toads' breeding sites when falling water levels reach a critical limit, perhaps also after some time without opportunities to feed: Since costs accrue from remaining aggregated, tadpoles should dissolve their aggregations as soon as the danger is past. But how do they recognize the right moment? As long as all tadpoles remain aggregated predators have no chance to prey upon new tadpoles, thus no continued stimuli inform the prey of the predator's lasting presence. What then determines the duration of an aggregation?

To answer this question the following experiment was performed. Groups of 50 unaggregated sibling tadpoles were transferred to standard plastic boxes, where they always remained evenly distributed. Into the boxes of one set (A) of such groups a gauze container was placed containing one *H. occipitalis* tadpole and 10 *B. maculatus* tadpoles, which were always eaten up during the first 10 min. Into a second set (B) of identical boxes gauze containers with only the *H. occipitalis* tadpole were placed. According to our expectation tadpoles aggregated in A, but no aggregations formed in B. Within 1.5–5 h after the start of the experiment (placing the gauze container with the *H. occipitalis* tadpole into the plastic box was taken as the start) the tadpoles were cross-exchanged. In none of the five studied pairs of groups did B 'tadpoles' form an aggregation after the transfer to A. Obviously, alarm substances are either rapidly decaying or are highly volatile. Three of five groups originating from A and transferred to B, however, formed new aggregations (within 10–30 min) and remained aggregated for an additional 10–20 h. This demonstrated that, once aggregated, tadpoles remain in close association for a prolonged period without repeated stimulation by alarm pheromones. A once-repeated twin-stimulation (mechanical and chemical) at around the time when an aggregation started to dissolve, was sufficient to keep tadpoles closely together for another 24–48 h (Spieler 1997).

The population ecology of reed frogs

In the reed frog *Hyperolius viridiflavus nitidulus*, onset of breeding is highly variable and strongly dependent on the frequency of rainfall in spring. Only subadult frogs aestivate above ground and only they survive the dry season. Under favourable growth conditions, breeding may start as early as the end of April (observed in only 1 of 6 years: 1996) but can also be postponed until the second half of June and may last until mid- to even end- November. *H. v. nitidulus* breeds in savanna ponds of very variable size and shares these spawning sites with more than 25 other anuran species and with numerous predators as well. *H. v. nitidulus* always forms breeding aggregations. Clutches (of females belonging to the aestivated generation, laid in July and August (1992)) consisted on average of 485 ± 144 eggs ($n = 35$, range

210–910). Under laboratory conditions with *ad libitum* feeding, females produce new clutches in intervals of 15 ± 5 days. Contrary to the two aforementioned species, drying up of the ponds is not the foremost problems *H. v. nitidulus* has to cope with. Rather, high losses caused by predators during the aquatic phase seem to limit reproductive success. As in many other species, swamping the predators by repeatedly producing relatively large clutches and by synchronized breeding activity where rain triggers common spawning could be an important initial tactic helping to prevent total losses. In quantitatively sampling tadpole communities in savanna ponds ($n = 180$ samples of a water and ground column of $0.25\,m^2$) we obtained a tadpole to predator ratio of 6–10: 1 (M.-O. Rödel & K.E. Linsenmair unpublished data).

The number of eggs per population laid can be enormous. At one breeding pond (area: $220\,m^2$), with a relatively small chorus (maximally 82 males and 36 females present on one night) that was surveyed for 43 nights during July to the beginning of September, 171 amplectant pairs were observed. Certainly, some pairs were overlooked so that in reality 200 females or more may have spawned (on average 485 eggs/female, see above), amounting to an estimated 100 000 eggs laid in 43 nights into this pond. According to preliminary data on numbers of freshly metamorphosed froglets of *H. v. nitidulus*, collected on another pond that always harboured a larger breeding aggregation, less than 0.5% of the eggs laid by August led to successfully metamorphosing froglets. By chance, however, such a breeding association may attain much higher successes.

Although larval development is rather slow (taking *c.* 40–70 days in the laboratory at 28 ± 2°C: Schmuck, Geise & Linsenmair 1994), this does not hold true for the post-metamorphic development. A proportion of those froglets that are able to leave the water during the core rainy season show an extremely fast growth rate, attaining maturity within 7–8 weeks (according to laboratory results ($n \geq 500$ individuals raised)). These frogs form a very short-lived summer generation that reproduces, or tries to, during the same rainy seasons as their parents. In favourable years, where early breeding is possible, this second generation seems to attain higher abundances than the aestivated generation. This second generation then produces most if not all of the aestivating frogs. These fast-developing individuals are unable to aestivate and they die at the latest with the beginning of the core dry season. In many physiological and behavioural traits they differ fundamentally from those froglets leaving the water towards the end of the rainy season. Many questions about the special traits of these fast-developing individuals remain to be answered and are currently under investigation.

Aestivation in H. v. nitidulus

Hyperolius viridiflavus nitidulus, is an annual species (or even only an 'oligomensual' species) which is, contrary to the other two long-lived species, absolutely dependent on an annual reproduction. By producing a strongly modified, fast-developing summer form, *H. v. nitidulus* makes opportunistic use of the extended breeding phase, helping it to secure the required annual reproduction. However, for this species' population dynamics much depends on the survival rate of the aestivating juvenile froglets. Therefore, a very sketchy summary of previous results is given here.

Only half-grown subadults of 1.2–1.6 cm in length and c. 350–750 mg in mass survive an entire dry season. Aestivating frogs regularly select sites where they are directly exposed to solar radiation. They must be extremely resistant to heat and desiccation, since air temperatures in the shade may occasionally reach 45°C and RH may drop below 10%. Aestivating *H. v. nitidulus* are always brilliant white when temperatures exceed 37–39°C. If undisturbed, the frogs remain seemingly motionless on the same spot for weeks. They are, however, never torpid but instantaneously able to flee when threatened. Dehydrated frogs never feed. The critical threshold for switching from TSF to DSF status (refraining from foraging and other activities combined with a number of physiological and morphological changes) lies at approximately a mass loss of 10–14%, caused by dehydration (Geise & Linsenmair 1988, Schmuck & Linsenmair 1988, Schmuck, Kobelt & Linsenmair 1988, Kobelt & Linsenmair 1992, 1995). Body surface temperature of sun-exposed frogs does not differ significantly from air temperature (Kobelt & Linsenmair 1995). How do frogs maintain subcritical temperatures with radiation loads near 1000 W m^{-2}? The brilliant white colour suggests that skin pigments might hold the answer. Indeed, iridophores filled with guanine and hypoxanthine crystals (ratio 9:1) provide a protective shield by reflecting a large proportion of the incoming radiation (Kobelt & Linsenmair 1986, 1992). In the newly metamorphosed froglets a single iridophore layer functions as a multilayer structure (Land 1972, Kobelt & Linsenmair 1986, 1992). With increasing thickness toward the dry season, diffuse light scattering gains in importance and strongly broadens and strengthens the skin's reflectance. Dry-adapted *H. v. nitidulus* from the field reflected up to 72% of the incoming direct solar radiation (Kobelt & Linsenmair 1992). Besides radiation reflection, an optimized body form and behavioural tactics (e.g. in respect to choosing favourable aestivating sites) play an important role in protecting frogs against critical heating-up (for details see Kobelt & Linsenmair 1992, 1995).

Growth patterns in froglets

H. v. nitidulus froglets with a body weight of 550–650 mg and fat deposits of 100–150 mg are able to survive *c*. 100 days under the climatic conditions prevailing during the hottest last third of the dry season, without any food or any external water supply. In frogs weighing between 450 and 750 mg the amount of fat stored at the beginning of a dry period is the decisive variable determining the maximal survival time. In smaller frogs, water reserves limit survival. These size-dependent survival prospects should shape the growth of aestivating froglets and the pattern of energy allocation. Initially, body growth should be given absolute priority. With increasing body size, surface area allometrically decreases and so does evaporative water loss, whereas volume and, with it, water-storage capacity increase. When frogs reach 450 mg, growth should be slowed down and fat storage should receive priority. This prediction has been supported (Geise & Linsenmair 1988) but further data are needed before a complete picture can emerge. Since the danger of overheating increases with body mass, frogs should stop putting on weight at a relatively low body mass, at *c*. 750 mg (Kobelt & Linsenmair 1995), irrespective of their foraging opportunities. This latter expectation is clearly met in those froglets metamorphosing towards the end of the wet season.

DISCUSSION

Sympatrically living anurans in the investigated West African savanna community possess very different life-histories (see Rödel 1996 for a recent account). All anurans, however, are confronted with one major and multidimensional problem: the unpredictable variability of decisive ecological conditions, for instance the year-to-year rainfall pattern, in their temporal and spatial, small- and large-scale dimensions.

Under such conditions opportunistic behaviour is expected to be selected. Nearly all anurans in this Guinea savanna community depend on the rainy season for reproduction and for most an early reproduction should be very advantageous. Early metamorphosed froglets gain time for development, growth and fat accumulation, furthering their chances for successfully surviving a prolonged dry season. Even more important may be that early reproduction helps in avoiding all those predators, like predatory insect larvae, that need to grow until they become dangerous.

Among West African anurans, *H. occipitalis* seems to be the most opportunistic species that can already use the earliest rains (Spieler & Linsenmair 1997a). This is only possible since those individuals at least that spawn into rock pools spend the dry season near rivers (see Rödel 1996), where sufficient water is always available and, therefore, can become active at any moment (contrary to e.g. the subterraneously aestivating species: see Duellman & Trueb 1986). Already before the very first rains some of the females obviously possessed large amounts of mature eggs. The unerring precision and velocity with which *H. occipitalis* pairs found the preferred rock pools during the first spawning night make it very probable that the first breeders were mostly experienced individuals, well acquainted with the area and the quality of the pools. These results and other observations on individuals carrying radio-transmitters (Spieler & Linsenmair 1997a,b) and the fact that a variable that can only be assessed by more than one visit to the pools, namely WHC, as the predictor for the amount of eggs spawned into a unoccupied body of water, clearly point to remarkable memory capacities in *H. occipitalis*. It is to be expected that just homing in to the native rock pools (as in many other anuran species: Crump 1986, Sinsch 1992, Lüddecke 1996) or preferentially spawning into them cannot represent an adaptive tactic in such a unpredictably changing environment. By also paying attention to the presence of potential cannibalistic conspecific tadpoles in later spawning bouts, in a graded way, *H. occipitalis* achieves an excellent choice of breeding sites as far as predictable parameters of quality are concerned.

The best tactic to confront unpredictable conditions, such as the drying up or the flooding of pools, is spreading the risk by sparingly spending one's eggs and always retaining enough for a next breeding opportunity. Most probably, however, this tactic does not provide an advantage when breeding in savanna ponds in which laying large clutches may be the only way of occasionally 'drawing a winning number in a costly lottery'. In favourable years, *H. occipitalis* tadpoles have very good survival prospects in the rock pools and the ratio of eggs laid to metamorphosed froglets is most probably the highest for a non-brood caring species in the investigated savanna community. From the few, very strongly differing data available in the literature no clear picture for useful comparisons emerges (see summary in Duellman & Trueb 1986). All observations and systematic collections at savanna ponds substantiate that reproductive gains are much lower there, while the investment, in terms of eggs spawned, seems to be much higher (but this still has to be better quantified). In very unfavourable years, no *H. occipitalis* tadpole will reach metamorphosis in the rock pools and the same may, even more often, this holds true for the savanna ponds. However, in a long-living and very mobile species like *H. occipitalis* (our oldest laboratory-kept specimens have now attained 12 years without showing any signs of ageing), loss of individual reproductive success in one year can be buffered by success the following year and populations are unlikely to fluctuate strongly unless loss of recruitment occurs in a series of successive years and over a very extended area.

The breeding sites used predominantly and nearly exclusively by *B. maculatus* are even less predictable and still more unstable in their physical as well as their biological properties than even the smaller of the desiccation-prone rock pools used by *H. occipitalis*. Contrary to most other anurans in this community, *B. maculatus* shows no direct dependence on rainfall to elicit spawning and also lacks any other form of breeding synchronization in its riverine habitat. It is, most probably, the species with the most extended breeding season in the investigated community (Rödel 1996; K.E. Linsenmair personal observations).

B. maculatus lacks the kind of spatial and temporal bet-hedging behaviour shown by *H. occipitalis*. In contrast, this toad always puts all the mature eggs that it possesses at the moment of spawning, in to a 'single basket', which on occasions leads to high success. In much more than 90% of all trials, however, a complete failure results. Thus, the only form of risk-spreading shown by adult toads seems to be the successive production of several clutches as well as some spatial variation in breeding sites including the occasional use of savanna ponds.

How can toad tadpoles cope with the main problems of drying up and falling victim to predatory fish and cannibalistic tadpoles? Their possibilities to overcome these problems are rather limited. Speeding up development as soon as water level falls could be a strategy to reduce the danger of desiccating too early. This might occur despite the well-known trade-off between faster development and decreased body mass at metamorphosis (see e.g. Wilbur & Collins 1973), and although under-average body size reduces future survival prospects and leads to delayed and/or lowered reproduction (Semlitsch, Scott & Pechmann 1988, Breven 1990). Spieler (1997), however, found that the potential of tadpoles to speed up their development is very limited, since they already seem to take the fastest possible route to metamorphosis, which, however, is usually environmentally constrained. Under laboratory conditions, tadpoles kept singly and fed *ad libitum* showed the fastest development with a minimum time requirement of 17–20 days between hatching and metamorphosis. Groups of tadpoles, both in the laboratory and the field, always needed significantly more time until metamorphosis (in the field from at least 28–35 days to >60 days at different group sizes). That density of conspecific tadpoles may influence growth and development is also known from many other species (see e.g. Wilbur 1977, Smith-Gill & Gill 1978, Newman 1988, 1989). By reducing the water volume (from 320 to 160 cm^3) the average development time in groups of *B. maculatus* tadpoles increased significantly; it also increased, however, the temporal variance with the consequence that the very earliest tadpoles to metamorphose were considerably faster under these conditions (Spieler 1997).

Toad tadpoles have developed highly effective mechanisms against water-living predators that represent the main threats in riverine breeding habitats. Fleeing to the shallowest margins of the puddles prevents the considerably larger predators from following. Staying there, however, is risky and costly. Tadpoles can easily become cut off from the open water and feeding is hampered. This prolongs developmental time and thus increases both the risk of desiccation and of being preyed upon. Tadpoles, therefore, should show this escape behaviour only when a corresponding predator is present. By reacting only to a combination of two stimuli, an alarm substance from injured tadpoles and a mechanical stimulus from the swimming predator, tadpoles should in most cases reliably prevent aggregating and thus incur opportunity costs or, even worse, make them more vulnerable to terrestrial predators. By remaining up to 24–48h in an aggregation without further adequate stimulation, once the behaviour is elicited, tadpoles might strike a best compromise between a still impending predation risk, on the one hand, and costs accruing from reduced energy intake on the other. This, however, has not yet been experimentally investigated.

Whereas it is evident why *B. maculatus* tadpoles seek refuge in the shallowest parts of the pools where predatory fish cannot follow, it is, however, not immediately self-evident why they usually try to form dense aggregations. Aggregation behaviour is widespread in tadpoles and apart from feeding aggregations (e.g. Lescure 1968, Wassersug 1973, Blaustein 1988) and aggregations induced by favourable abiotic conditions (Duellman & Trueb 1986), it is usually discussed as a means of reducing the individual predation risk of schooling tadpoles (the selfish herd concept: Hamilton 1971, Bertram 1978). Such behaviour is effective through, among other factors, the dilution (Milinski 1977a,b, Kenward 1978) and the confusion effect (Heller & Milinski 1979, Milinski 1979, Landau & Terborgh 1986). While it is easy to see how these effects may protect tadpoles in aggregations that move within bodies of water (see Caldwell 1989, Tyler 1989, Kehr & Schnack 1991), it is not evident, how one of these effects should increase the survival probabilities of the *B. maculatus* tadpoles. If birds were to prey on these tadpoles, the aggregation behaviour could be advantageous within this context. This, however, was never observed. Thus, these aggregations may serve completely different functions. They could reduce the danger of drying up by diminishing the surface area that is exposed to evaporation through very close body contact and/or by creating, through common movements, shallow depressions that help them remain within the water film when the water level is critically sinking.

Among those *H. occipitalis* that breed in rock pools there were always several females (5–15) that spawned into the large pools that offered the best survival prospects and thus achieved reproductive success. Though small pools were often used only by a single female, this never led to an outstandingly large success for that female, since either none or several pools produced froglets. In *B. maculatus*, however, always only very few females within a very large population (comprising many hundreds of individuals along a stretch of about *c.* 1 km of river during the late dry to early rainy season) were reproductively successful because all tadpoles from single clutches are likely to perish whereas a high percentage of tadpoles from very few clutches may survive to metamorphosis. In other words, the variance in reproductive success was very high within one year. Even life-time reproductive success of individuals in a population is likely to remain high. Thus, a large percentage of individuals within a population are likely to be descendants of very few pairings which may lead to some interesting genetic consequences.

Hyperolius viridiflavus nitidulus differs from the aforementioned two species in a number of very important traits of its life history. It is a short-lived species that breeds exclusively in savanna ponds, and it must succeed in breeding every year. Furthermore, it is a species that neither retreats to permanent bodies of water during the dry season nor aestivates subterraneously. Instead, individuals remain above ground, exposing themselves fully to the harsh climate prevailing (for more details than given in the results see Kobelt & Linsenmair 1986, 1992, Geise & Linsenmair 1988, Schmuck & Linsenmair 1988, Schmuck *et al.* 1988, 1994). The only point that is discussed here is: what are the likely consequences of the unusual mode of aestivation on the population ecology of this species?

Since only subadult frogs are able to survive entire dry seasons, *H. v. nitidulus* is never mature when first rains fall. At that time individuals still need to grow considerably (at least doubling to quadrupling their weight before attaining minimal adult weight (K.E. Linsenmair, unpublished observations)). After such first rains, reed frogs only become active when they have been able to fully rehydrate and will stop any activity as soon as they lose again 10–14% of their body weight through transpiration, whereupon they resume aestivation behaviour and physiology (Geise & Linsenmair 1986, 1988, Schmuck & Linsenmair 1997, K.E. Linsenmair unpublished data unpubl. results). Usually, the frequency of rain is low at the beginning of the rainy season and then it may take *H. v. nitidulus* up to 3 or even 4 months after the first rain (e.g. in March) to reach adulthood and accumulate the energy reserves required for reproduction. If, however, the frequency of rain is high, with soil saturation soon attained, resulting in more or less regular night-time dew fall for some successive nights, development will be speeded up and (as was the case in 1996) maturity reached as early as the end of April/beginning of May. However, reed frogs are never among the very early breeders in the investigated community. Consequently, they encounter high predation within the savanna ponds. This problem is aggravated by the rather long time this species requires for larval development. One way of overcoming this severe predation pressure is by producing as many progeny as possible (reed frog tadpoles are palatable to all of the predators I have observed to date). According to results from the laboratory, *H. v. nitidulus* females (and two closely related species, *H. v. ommatostictus* and *H. marmoratus taeniatus* (Grafe, Schmuck & Linsenmair 1992)) invest very heavily in egg production (and the males in calling) and produce successive clutches under *ad libitum* conditions in the shortest possible intervals. Our field data are not yet sufficient to allow us to give any reliable figure on spawning intervals under natural conditions. Telford and Dyson (1990) reported an average interval between clutches of 57 days in *H. marmoratus marmoratus* in the field. Whatever the figure for the West African *H. v. nitidulus*, it seems justified to assume that these frogs will not only try to maximize their reproductive output in the laboratory but also in the field. As a consequence, high mortality must be expected (of 104 males marked in choruses only 16 could be recaptured once or several times with probably most others preyed upon). Also, many direct observations corroborate this prediction, with several snake species being the most effective predators of adult frogs, especially at breeding sites.

When reed frogs are able to start reproducing early after the dry period, one should expect them not to survive until the end of their breeding phase (with spawning still observed in October) due to the just-mentioned strong predation pressure on an energy-maximizing species, and this seems to hold true. The aestivated generation becomes replaced by a post-metamorphically very fast developing generation (the fastest post-metamorphic development known; Duellman & Trueb 1986) that differs profoundly in its behaviour and physiology from the aestivated generation. To date, we do not yet fully understand what proximate factors are responsible for the choice of one of the alternative pathways: either accelerate development to maturity with no capability to aestivate, or slow down growth and development and start the trajectory immediately that leads, already at metamorphosis, to extraordinary resistance against high temperature, strong direct solar radiation load and desiccation. Water quality deterioration is responsible for the latter alternative, at least in part, in the form of accumulation of nitrogenous waste products (Schmuck *et al.* 1994).

Those individuals that accelerate development and are able to reproduce in time (i.e. towards the end of the rainy season), may have, rather often, good prospects of achieving a lower cost to benefit ratio of spawned eggs to metamorphosed tadpoles. Late in the season predator pressure seems to be considerably reduced and interspecific competitors are less abundant. If, however, this oligomensual generation develops too late, becoming only mature well after the end of the rainy season, it will in average years have no chance at all to reproduce. We do not yet know whether, under conditions of a delayed start of reproduction in the aestivated generation, the generation of accelerated developers may be entirely skipped. Besides a delayed growth at the end of the dry season through an unfavourable rainfall pattern, still another constellation may prevent producing a second generation, namely low numbers of dry season survivors. Owing to especially long and severe dry seasons, but perhaps also to other factors, *H. v. nitidulus* shows very pronounced fluctuations in abundance. The number of aestivating juveniles that survived to the beginning of the rainy season fluctuated by at least two orders of magnitude within 10 years of sampling (within an overall time span of 21 years) (K.E. Linsenmair unpublished data). When numbers are very low, but growth conditions are very favourable, a few individuals might be unable to overcome the high predation pressure in the savanna ponds when they are still spawning during the first two-thirds of the rainy season. Under such circumstances it could pay the adults to postpone their reproduction (e.g. by using size of the calling aggregation of males as a proximate cue) until later in the season when conditions are more favourable and not to produce a non-aestivating generation. There are some hints pointing in this direction; they need, however, to be backed up by more data and experimental results. Whatever the outcome of this particular question will be, *H. v. nitidulus* demonstrates that it is no easy-to-solve task to survive in an unpredictably changing and temporally very harsh environment as a very short-lived and small frog that does not follow the fugitive but the resisting tactic.

ACKNOWLEDGEMENTS

The work reported here was supported by the Deutsche Forschungsgemeinschaft (SFB 251 TP B3) and by the Volkswagen-Stiftung (AZ I/64 102). I thank the Ministre des Eaux et Forêts and the Ministre de la Recherche Scientifique, Republique de Côté d'Ivoire for granting the research permit for conducting field work in the Parc National de la Comoé. Dr Ulmar Grafe's help in improving the English and Dr B. Fiala's various contributions to the editing of the manuscript are gratefully acknowledged as are their critical comments. I thank my respective former and present PhD students for their most valuable contributions to our studies in general and this report in particular, especially Mark-Oliver Rödel and Marko Spieler.

REFERENCES

Bertram, B.J.R. (1978). Living in groups: predators and prey. In *Behavioral Ecology: An Evolutionary Approach* (Ed. by J.R. Krebs & N.B. Davies), pp. 64–96. Sinauer, Sunderland, MA.

Blaustein, A.R. (1988). Ecological correlates and potential function of kin recognition and kin association in anuran larvae. *Behavioral Genetics*, **18**, 449–464.

Breven, K.A. (1990). Factors affecting population fluctuations in larval and adult stages of the wood frog (*Rana sylvatica*). *Ecology*, **71**, 1599–1608.

Caldwell, J.P. (1989). Structure and behavior of *Hyla geographica* tadpole schools, with comments on classification of group behavior in tadpoles. *Copeia*, **1989**, 938–950.

Crump, M.L. (1986). Homing and site fidelity in a neotropical frog, *Atelopus varius* (Bufonidae). *Copeia*, **1986**, 438–444.

Duellman, W. & Trueb, L. (1986). *Biology of Amphibians.* McGraw-Hill, New York.

Feder, M.E. & Burggren, W.W. (Eds) (1992). *Environmental Physiology of the Amphibians.* University of Chicago Press, Chicago.

Geise, W. & Linsenmair, K.E. (1986). Adaptations of the reed frog *Hyperolius viridiflavus* (Amphibia, Anura, Hyperoliidae) to its arid environment. II. Some aspects of the water economy of *Hyperolius viridiflavus nitidulus* in wet and dry season conditions. *Oecologia*, **68**, 533–541.

Geise, W. & Linsenmair, K.E. (1988). Adaptations of the reed frog *Hyperolius viridiflavus* (Amphibia, Anura, Hyperoliidae) to its arid environment. IV. Ecological significance of water economy with comments on thermoregulation and energy allocation. *Oecologia*, **77**, 327–338.

Grafe, T.U., Schmuck, R. & Linsenmair, K.E. (1992). Reproductive energetics of the African reed frogs, *Hyperolius viridiflavus* and *H. marmoratus. Physiological Zoology*, **65**, 153–171.

Hamilton, W.D. (1971). Geometry for the selfish herd. *Journal of Theoretical Biology*, **31**, 295–311.

Heller, R. & Milinski, M. (1979). Optimal foraging of sticklebacks on swarming prey. *Animal Behavior*, **27**, 1127–1141.

Kehr, A.I. & Schnack, J.A. (1991). Predator–prey relationship between giant water bugs (*Belostoma oxyurum*) and larval anurans (*Bufo arenarum*). *Alytes*, **9**, 61–69.

Kenward, R.E. (1978). Hawks and doves: factors affecting success and selection in goshawk on woodpigeons. *Journal of Animal Ecology*, **47**, 449–460.

Kobelt, F. & Linsenmair, K.E. (1986). Adaptations of the reed frog *Hyperolius viridiflavus* (Amphibia, Anura, Hyperoliidae) to its arid environment: I. The skin of *Hyperolius viridiflavus nitidulus* in wet and dry season conditions. *Oecologia*, **68**, 533–541.

Kobelt, F. & Linsenmair, K.E. (1992). Adaptations of the reed frog *Hyperolius viridiflavus* (Amphibia, Anura, Hyperoliidae) to its arid environment. VI. The iridophores in the skin of *Hyperolius viridiflavus taeniatus* as radiation reflectors. *Journal of Comparative Physiology B*, **162**, 314–326.

Kobelt, F. & Linsenmair, K.E. (1995). Adaptations of the reed frog *Hyperolius viridiflavus* (Amphibia, Anura, Hyperoliidae) to its arid environment. VII. The heat budget of

Hyperolius viridiflavus nitidulus and the evolution of an optimised body shape. *Journal of Comparative Physiology B*, **165**, 110–124.

Lamotte, M. (1967). Le Parc National du Niokolo-Koba, Fascicule III; XXX. Amphibiens. *Memoires de l'Institut Fondamentale de l'Afrique Noire*, **84**, 420–426.

Land, M.F. (1972). The physics and biology of animal reflectors. *Progress in Biophysical Molecular Biology*, **24**, 75–106.

Landeau, L. & Terborgh, J. (1986). Oddity and the 'confusion effect' in predation. *Animal Behavior*, **34**, 1372–1380.

Lescure, J. (1968). Le comportement social des batraciens. *Revue Comportement Animales*, **2**, 1–33.

Lüddecke, H. (1996). Side fidelity and homing ability in *Hyla labialis* (Anura, Hylidae). *Alytes*, **13**, 167–178.

Milinski, M. (1977a). Experiments on the selection by predators against spatial oddity of their prey. *Zeitschrift für Tierpsycholgie*, **43**, 311–325.

Milinski M. (1977b). Do all members of a swarm suffer the same predation? *Zeitschrift für Tierpsychologie*, **45**, 373–388.

Milinski, M. (1979). Can an experienced predator overcome the confusion of swarming prey more easily? *Animal Behavior*, **27**, 1122–1126.

Newman, R.A. (1988). Adaptive plasticity in development of *Scaphiopus couchii* tadpoles in desert ponds. *Evolution*, **42**, 774–783.

Newman, R.A. (1989). Developmental plasticity of *Scaphiopus couchii* tadpoles in an unpredictable environment. *Ecology*, **70**, 1775–1787.

Poilecot, P. (1991). *Un Écosystème de Savane Soudanienne: Le Parc National de la Comoé (Côte d'Ivoire)*. Unesco, Paris.

Rödel, M.-O. (1996). *Amphibien der Westafrikanischen Savanne*. Edition Chimaira, Frankfurt/Main.

Schiøtz, A. (1967). The treefrogs (Rhacophoridae) of West Africa. *Spolia Zoologica Musei Hauniensis*, **25**.

Schiøtz, A. (1971). The superspecies *Hyperolius viridiflavus* (Anura). *Vidensabeilge Meddedelser fra Dansk Naturhistorisk Forening*, **134**, 21–76.

Schmuck, R. & Linsenmair, K.E. (1988). Adaptations of the reed frog *Hyperolius viridiflavus* (Amphibia, Anura, Hyperoliidae) to its arid environment. III. Aspects of nitrogen metabolism and in the reed frog *Hyperolius viridiflavus taeniatus*, with special reference to the role of iridophores. *Oecologia*, **75**, 354–361.

Schmuck, R. & Linsenmair, K.E. (1997). Regulation of body water balance in reedfrogs (superspecies *Hyperolius viridiflavus* and *H. marmoratus*: Amphibia, Anura, Hyperoliidae) living in unpredictably varying savannah environments. *Journal of Comparative Biochemistry and Physiology*, in press.

Schmuck, R., Kobelt, F. & Linsenmair, K.E. (1988). Adaptations of the reed frog *Hyperolius viridiflavus* (Amphibia, Anura, Hyperoliidae) to its arid environment. V. Iridophores and nitrogen metabolism. *Journal of Comparative Physiology B*, **158**, 537–546.

Schmuck, R., Geise, W. & Linsenmair, K.E. (1994). Life cycle strategies and physiological adjustments of reedfrog tadpoles (Amphibia, Anura, Hyperoliidae) in relation to environmental conditions. *Copeia*, **1994**, 996–1007.

Semlitsch, R.D., Scott, D.E. & Pechmann, J.H.K. (1988). Time and size at metamorphosis related to adult fitness in *Ambystoma talpoideum*. *Ecology*, **69**, 184–192.

Shoemaker, V.H. (1988). Physiological ecology of amphibians in arid environments. *Journal of Arid Environments*, **2**, 145–153.

Shoemaker, V.H. (1992). Exchange of water, ions, and respiratory gases in terrestrial amphibians. In *Environmental Physiology of the Amphibians* (Ed. by M.E. Feder & W.W. Burggren), pp 125–150. University of Chicago Press, Chicago.

Sinsch, U. (1992). Sex-biased site fidelity and orientation behaviour in reproductive natterjack toads (*Bufo calamita*). *Ethology, Ecology and Evolution*, **4**, 15–32.

Smith-Gill, S.J. & Gill, D.E. (1978). Curvilinearities in the competition equations: an experiment with ranid tadpoles. *American Naturalist*, **112**, 557–570.

Spieler, M. (1996). Strategies of minimizing stress in West African anurans. *Verhandlungen der Deutschen Zoologischen Gesellschaft*, **89**(1), 260.

Spieler, M. (1997). *Anpassungen westafrikanischer Anuren an Austrocknungsrisiko und Räuberdruck in einem saisonalen Lebensraum*. PhD thesis, University of Würzburg.

Spieler, M. & Linsenmair, K.E. (1997a). Choice of optimal oviposition sites by *Hoplobatrachus occipitalis* (Anura: Ranidae) in an unpredictable and patchy environment. *Oecologia*, **109**, 184–199.

Spieler, M. & Linsenmair, K.E. (1997b). Migration and diurnal shelter in a ranid frog from a West African savannah: a telemetric study. *Amphibia–Reptilia*, in press.

Telford, S.R. & Dyson, M.L. (1990). The effect of rainfall on interclutch interval in painted reed frogs (*Hyperolius marmoratus*). *Copeia*, **1990**, 644–648.

Tyler, M.J. (1989). *Australian Frogs*. Viking O'Neil, Victoria.

Warburg, M. (1997). *Ecophysiology of Xeric-inhabiting Amphibians*. Springer, New York.

Wassersug, R.J. (1973). Aspects of social behavior in anuran larvae. In *Evolutionary Biology of the Anurans, Contemporary Research on Major Problems* (Ed. by J.L. Vial), pp 273–297. University Missouri Press, Missouri.

Wilbur, H.M. (1977). Interactions of food level and population density in *Rana sylvatica*. *Ecology*, **58**, 206–209.

Wilbur, H.M. & Collins, J.P. (1973). Ecological aspects of amphibian metamorphosis. *Science*, **182**, 1305–1314.

10. LIMITS TO EXPLOITATION OF SERENGETI WILDEBEEST AND IMPLICATIONS FOR ITS MANAGEMENT

S. MDUMA,[†] R. HILBORN[‡] and A. R. E. SINCLAIR[*]

[*]Centre for Biodiversity Research, Department of Zoology, 6270 University Boulevard, University of British Columbia, Vancouver, B.C., Canada;
[†]Serengeti Wildlife Research Centre, P.O. Box 3134, Arusha, Tanzania;
[‡]School of Fisheries WH-10, University of Washington, Seattle WA 98195, USA

SUMMARY

1 Many large tropical mammals are currently harvested, either through government-run culling programmes or through unregulated and frequently illegal activities. Pressure for exploitation of these species will grow as populations increase and many species that are currently unmanaged will likely come under some form of manipulation. Principle challenges in managing large tropical mammals include:
  (i) methods for estimating abundance;
  (ii) understanding the interaction between environmental conditions, density, and reproductive success and survival; and
  (iii) estimation and regulation of harvest.
We present the history of understanding the potential harvest of Serengeti wildebeest as a case study in exploiting large tropical mammals.

2 A population dynamics model was developed that incorporates:
  (i) intermittent sampling of the demographic parameters (population size, births, recruitment and dry season adult mortality) including their respective sampling errors; and
  (ii) the influence of dry season food availability.
Parameters not incorporated in the model include wet-season adult mortality, and human off-take. The model attempts to fit a curve that is most consistent with the census data by estimating the unknown human off-take.

3 The best estimates of human off-take range between 15 000 and 40 000 wildebeest per year, consistent with independent estimates of legal and illegal harvesting (20 000–40 000 wildebeest) given by Mduma (1996).

4 There is an opportunity for expanding the legal harvest if the illegal harvest is reduced. By providing a legalized harvest, the local population

could benefit from the protected areas and the incentives for illegal harvest could be reduced, but considerable caution is required.

5 The long-term conservation of the wildebeest population while exploiting its biological potential must meet the following conditions:
  (i)   appropriate economics of harvesting;
  (ii) implementation of appropriate harvesting strategies that recognize the importance of environmental fluctuations on biological potential; and
  (iii) an adequate monitoring and evaluation system that monitors both the population and harvest.

INTRODUCTION

Throughout the world, natural habitats are being lost at an accelerating rate as a direct result of human population increase (Sinclair & Wells 1989, Soulé 1991, Geer 1992, Tolba El-Kholy, El-Hinnawi *et al*. 1992, McNeely 1994, Sinclair, Hik, Schmitz *et al*. 1995). In the developing countries of the tropics human population expansion is increasing the exploitation of wild populations of indigenous species, particularly of large mammals. Such exploitation applies to both herbivores for their meat and carnivores for their skins, claws, teeth and other parts. Clearly, increasing consumption, if unregulated, will lead to over-exploitation and possible extinction of such species (Caughley, Dublin & Parker 1990, Leader-Williams, Albon & Berry 1990).

In Africa, Dasmann and Mossman (1960), Dasmann (1964) and Mossman (1975) proposed that conservation of African ungulates would be ensured if animals could be harvested economically. In particular, if the meat could be sold at a profit to local people at prices that they could afford, this would provide the incentive for those people to conserve their heritage (Talbot & Swift 1966). This idea has been expounded at length elsewhere (Field 1979, Walker 1979, Eltringham 1984). It is favoured particularly in southern Africa (Child & Child 1987, Child 1990, Bothma 1989) and requires commercial 'game cropping' operations using modern equipment, transport and hygiene regulations. However, these early studies misunderstood the population dynamics of indigenous flora and fauna (Walker 1976, 1979, Walker, Emslie, Owen-Smith & Scholes 1987, Macnab 1991) as well as the economics of harvesting (Clark 1973 a,b, 1990, Caughley 1993). In eastern Africa, these schemes have all met with economic failure because overheads were too high and domestic species were more productive (Parker 1984, 1987, Macnab 1991).

Exploitation of wildlife in the Serengeti ecosystem, Tanzania, is an exception to the above generalization. Local peoples have been hunting in this area, probably for centuries. Although most of the harvesting is illegal after the Park was established in the 1950s, its existence is indisputable (Turner 1987, Arcese, Hando & Campbell 1995, Campbell & Hofer 1995). Instead of expensive modern equipment, hunters use traditional methods such as snares, pitfalls and poison-tipped arrows. The meat is dried and sold locally at prices affordable to the local community. Transport costs are minimal because the migrant herbivores, largely wildebeest (*Connochaetes taurinus*), leave the Serengeti National Park and literally walk up to the village gates. During the wet season the animals return to the Serengeti plains as part of their annual migration. On the plains they are too far away for hunters to reach them, not because of the presence of the National Park, but because the intervening area has always been unsuitable for habitation for biological reasons, at least in historical times. Thus, there was a natural 'closed season', and the system was probably in equilibrium until recently. However, as elsewhere in Africa, the human population has been increasing in the surrounding areas by 3% per year, and in some areas by as much as 15% per year, through immigration and reproduction (Campbell & Hofer 1995). The concern of the Tanzanian authorities is that the off-take, unregulated and illegal, could exceed biological capacity and result in a collapse of the wildebeest population and cause a major ecological change in the Serengeti ecosystem (Sinclair & Arcese 1995a).

Our studies, therefore, investigated the current size of the hunting off-take and estimated the harvest of wildebeest that could be tolerated under a regulated cropping regime. Previous harvesting schemes on free-living wild populations in eastern Africa have had only the most rudimentary census data with which to estimate off-take, with the exception of some elephant-culling operations (Laws, Parker & Johnstone 1975, Parker 1984). This study used 30 years of census, recruitment and mortality data to estimate harvest levels. It represents a case study, which could be employed as the basis for sustainable harvest of other systems employing traditional hunting methods.

Hilborn and Sinclair (1979) explored models of the wildebeest and their predators to examine the likely impacts of different rainfall regimes. Neither legal nor illegal harvesting were being considered at that time, but the models used were essentially identical to those which can be used for determining the potential harvest. Hilborn and Sinclair (1979) argued that the data available at the time, which covered the period of exponential growth, combined with basic energetic knowledge of large ungulates, suggested that the wildebeest population would be food limited when the population size reached 1.5 million. This prediction was subsequently born out as the population did level off around 1.4 million animals (Sinclair 1995).

Pascual and Hilborn (1995) explored harvesting strategies under environmental fluctuations using a non-age-structured model. They found that sustainable yields (i.e. those that did not lead to extinction) were on the order of 60 000–70 000 animals per year and that, to achieve maximum sustainable yield, there would have to be significant year-to-year changes in harvest; in years of low rainfall the harvest would have to be much lower than in years of high rainfall.

In this chapter we consider the interaction between legal and illegal harvests. We also add age structure and consider in more detail the processes of birth, calf survival and adult mortality. Finally, we consider the implementation and impediments to achieving a successful legalized harvesting programme.

METHODS

Data sources

Analysis of the population dynamics and the potential harvest in the Serengeti wildebeest population used data collected since 1961. These data are given in Table 10.1 with their sources in Table 10.2. The data are (i) dry season rainfall measured from numerous stations throughout the Serengeti over the entire study period; (ii) censuses of total wildebeest population conducted 15 times since 1961; (iii) estimates of yearling/adult ratio conducted 24 times since 1962; and (iv) estimates of dry season adult mortality rate conducted nine times since 1968. Pregnancy rates were estimated from autopsy and analysis of hormone levels in faecal samples (Mduma 1996). In this analysis we have assumed a 0.85 pregancy rate in all years, since variation was small. The relationship between rainfall and dry-season grass production (taken from Sinclair 1979, Fig. 4.4) is given in the model specifications below.

The population dynamics model

For convenience, all numbers of animals refer to an arbitrary date of March 1, on which we assume that all calves are born. The following terms are defined as:

G_y, the amount of green grass growing per month in kg ha^{-1} in the dry season in year y;

TABLE 10.1. *Data used in the model of wildebeest populations in Serengeti National Park. July to November rainfall values were the averages for gauges at Kogatende, Bologonja, Klein's Camp (original), Lobo and Togoro.*

Year	Dry season rainfall (mm)	Wildebeest abundance (× 1000)	Standard error	12 months calves as proportion of population	Adult mortality (%)
1961	38	263	—		
1962	100			0.230	
1963	104	357	—	0.160	
1964	167			0.170	
1965	167	439	—	0.093	
1966	165			0.110	
1967	79	483	—	0.100	
1968	91			0.130	1.7
1969	77				1.4
1970	134			0.110	
1971	192	693	28.8	0.139	0.8
1972	235	773	76.7	0.139	0.5
1973	159			0.097	
1974	211				
1975	257				
1976	204			0.140	
1977	300	1444	200	0.146	
1978	187	1248	355	0.114	
1979	84				
1980	99	1337	80	0.094	
1981	163				
1982	97	1208	272	0.145	2.7
1983	228			0.103	2.1
1984	208	1337	138	0.111	
1985	83				
1986	44	1146	134	0.109	
1987	112				
1988	191				
1989	202	1407	109	0.156	
1990	127			0.135	
1991	254	1221	177		
1992	153			0.173	4.0
1993	19			0.062	7.9
1994	227	917	173	0.073	1.0
n	33	14	11	24	9

F_y, the amount of food available per wildebeest in kg month^{-1} in the dry season in year y;

$N_{y,i,0}$, the number of calves born in year y, sex i (1 = females, 2 = males);

$N_{y,i,1}$, the number of yearlings in year y, sex i;

TABLE 10.2. Sources of the data used in the model as shown in Table 10.1.

Year	Wildebeest abundance	12-month calves as % of population	Adult mortality
1961	Sinclair (1973)		
1962		Watson (1967)	
1963	Sinclair (1973)	Watson (1967)	
1964		Watson (1967)	
1965	Sinclair (1973)	Watson (1967)	
1966		Watson (1967)	
1967	Sinclair (1973)	Sinclair (1979)	
1968		Sinclair (1979)	Sinclair (1979)
1969			Sinclair (1979)
1970		Sinclair (1979)	
1971	Sinclair (1973)	Sinclair (1979)	Sinclair (1979)
1972	Sinclair & Norton-Griffiths (1982)	Sinclair (1979)	Sinclair (1979)
1973	Norton-Griffiths (1973)	Sinclair (1979)	
1976		Sinclair (1979)	
1977	Sinclair & Norton-Griffiths (1982)	Sinclair (1979)	
1978	Sinclair & Norton-Griffiths (1982)	Sinclair (unpubl.)	
1980	Sinclair & Norton-Griffiths (1982)	Sinclair (unpubl.)	
1982	Sinclair et al. (1985)	Sinclair (unpubl.)	Sinclair et al. (1985)
1983		Sinclair (unpubl.)	Sinclair et al. (1985)
1984	Borner, Fitzgibbon, Borner et al. (1987)	Sinclair (unpubl.)	
1986	Sinclair (unpubl.)	Sinclair (unpubl.)	
1989	Campbell & Borner (1995)	Sinclair (unpubl.)	
1990		Sinclair (unpubl.)	
1991	Campbell & Borner (1995)		
1992		Mduma (unpubl.)	Mduma (unpubl.)
1993		Mduma (unpubl.)	Mduma (unpubl.)
1994	Farm & Woodworth (1994)	Mduma (unpubl.)	Mduma (unpubl.)

$N_{y,i,2}$, the number of animals in year y, sex i age 2;

$N_{y,i,3}$, the number of animals in year y, sex i age 3 and older;

Y_y, the proportion of the total population that is yearlings;

T_y, the total wildebeest population summed over all ages and sexes;

V_y, the total wildebeest population weighted by relative vulnerability to snares;

$v_{i,j}$, the relative vulnerability of individuals of sex i and age j to snares;

$s_{y,i,j}$, the survival rate from natural mortality in year y, sex i, age j;

m_y, the mortality rate per month of animals 1 year or older;

$u_{y,i,j}$, the survival rate from harvesting in year y, sex i, age j;

R_y, the dry season (July–October) rainfall in millimetres averaged over the Serengeti woodlands;

P_j, the pregnancy rate of age-j females;

a, b, c, d, parameters relating survival to food per animal;
C_y, the harvest in year y of all ages and sexes.

The amount of green grass growth is proportional to dry-season rainfall. Using fig. 4.4 of Sinclair (1979):

$$G_y = 1.25 R_y \tag{1}$$

where G_y is the amount of grass produced measured in kilograms per hectare per month (kg ha^{-1} month^{-1}); and R_y the dry-season (July–October) rainfall measured in millimetres and averaged over the northern Serengeti region. Green food supply in the dry season is considered the limiting resource for wildebeest (Sinclair, Dublin & Borner 1985).

Dry-season rainfall was given in Sinclair (1979) but many of the gauges used in that sequence were discontinued, so we used gauges that had complete histories to generate the rainfall shown in Table 10.1.

The amount of food per animal is the total grass grown per month per hectare, times the number of hectares utilized in the dry season ($c.\ 0.5 \times 10^6$), divided by the total number of wildebeest:

$$F_y = \frac{G_y\, 0.5 \times 10^6}{T_y}. \tag{2}$$

The number of births is the pregnancy rate of 3 year and older females times the number of 3 year and older females. It was assumed that 2-year-old females have no successful reproduction, and that the pregnancy rates assumed were 0.85 for 3-year-olds and older so that the number of births is:

$$N_{y,i,0} = p_3 N_{y-1,1,3}. \tag{3}$$

The survival rate of calves from birth to their first birthday is assumed to depend on grass per animal with survival reaching an asymptote:

$$s_{y,i} = \frac{aF_y}{b + F_y}. \tag{4}$$

This equation is the simplest form in which survival is zero when there is no food and at a maximum when food is unlimited. The dry-season survival rate of individuals 1 year or older is assumed to be proportional to grass per animal, which is converted to a monthly mortality rate. This assumption is based on the empirical measures of mortality since the 1960s (Sinclair 1979, Sinclair et al. 1985, Sinclair & Arcese 1995b). We implicitly assume that all adult mortality takes place during the 4-month dry season based on seasonal measures (Mduma 1996). The same mortality is assumed for all ages older than 1 year and all sexes:

$$S_{y,i} = \frac{cF_y}{d+F_y} \qquad (5)$$

$$m_y = 1 - (s_{y,i})^{0.25} \qquad (6)$$

where $s_{y,i}$ is the survival over the 4 months of the dry season, and the 0.25 exponent converts to monthly survival. Two of the constants used in the previous calculations, the 1.25 in the rain-to-grass relationship and the 500 000 ha of available dry-season habitat, are only important when relating survival to a functional analysis of food availability.

We allow for age/sex specific vulnerability to harvesting (v). Given a particular total harvest and age/sex specific vulnerability to harvest, the age/sex specific exploitation rates can be calculated by first computing the total population size weighted by age/sex specific vulnerability (V) and then computing the age/sex specific exploitation rates (u):

$$V_y = \sum_{i=1,2} \sum_{j=1,3} N_{y,i,j} v_{i,j} \qquad (7)$$

$$u_{y,i,j} = \frac{v_{i,j} C_y}{V_y}. \qquad (8)$$

The number of yearlings is the number of calves born the previous year times the calf survival rate:

$$N_{y,i,1} = N_{y-1,i,0} s_{y-1,0} (1 - u_{y-1,i,0}). \qquad (9)$$

The number of individuals age 2 years is the number of yearlings the previous year times the adult survival rate:

$$N_{y,i,2} = N_{y-1,i,1} s_{y-1,i,1} (1 - u_{y-1,i,1}). \qquad (10)$$

The number of individuals age 3 years and older is the sum of the number of age 3 years and older and the number of 2-year-olds alive the previous year times the adult survival rates:

$$N_{y,i,3} = (N_{y-1,i,3} + N_{y-1,i,2}) s_{y-1,2} (1 - u_{y-1,i,2}). \qquad (11)$$

Estimating harvest

There are few reliable data on the total harvest of wildebeest. The only published figures are those of Campbell and Hofer (1995) of 87 000 and 118 922 per year, estimated in the late 1980s. Our model for estimating sustainable harvest required a measure of actual harvest each year. Although snaring of wildebeest and other animals has been going on for many years in the Serengeti, three recent changes may have affected the number of wildebeest taken by snares. First, the wildebeest population has increased five-fold since the 1960s, and the number taken by snares would have increased because of higher abundance. Second, the human population adjacent to the Park has been growing at about 3% per year (Campbell & Hofer 1995), and this additional population pressure has certainly led to more snaring. Third, while National Parks and the Tanzanian Wildlife Department have consistently had anti-poaching enforcement programmes, anti-poaching effort was at a minimum from 1977 to about 1984, and ranger posts were often totally without vehicles, fuel and even staff (Sinclair 1995, Arcese, Hando & Campbell 1995). We presume there was a concomitant increase in numbers of wildebeest killed by snares since 1977.

We have examined three possible ways to model the harvest of wildebeest. The first method (harvest model I) assumes that the harvest has been proportional to the human population size and the size of the wildebeest herd. If we assume arbitrarily that the human population size in 1960 was 1.0, and grew at 3% per year (Campbell & Hofer 1995), then the human population size could be described as:

$$H_{y+1} = 1.03 H_y. \quad (12)$$

The harvest of wildebeest is then:

$$C_y = V_y H_y q \quad (13)$$

where q is a parameter representing the proportion of the vulnerable wildebeest population taken per unit of human population, and we estimate q. This approach accounts for the first two 'facts' about poaching, namely that it has increased as human population has increased, and as the wildebeest herd has increased.

The second approach (harvest model II) assumes that poaching was insignificant until 1977, and has been constant since then. This is modelled by assuming that $C_y = 0$ prior to 1977 and that C_y is a constant to be estimated after 1977.

The third approach (harvest model III) is to modify the first approach by a year-specific factor reflecting the level of anti-poaching activity (E_y) so that:

$$C_y = V_y H_y q E_y \quad (14)$$

where we assume that E_y is 1 for all years before 1977 and after 1984 but is either a fixed constant or a parameter to be estimated between 1977 and 1984.

Fitting the model to the data

We used three sources of data in fitting the model (Table 10.1): the aerial census of wildebeest numbers; the proportion of the population on the plains that is made up of yearlings; and estimates of adult dry-season mortality.

There are four free parameters relating calf and adult survival to food available per animal (a, b, c and d). Then, depending upon which harvest model is used, there are one or two additional parameters. Our approach is to find the values of these parameters that provide the best fit to the three available data sets.

Initial conditions

We assumed the population was exactly equal to the census in 1961 (e.g. 263 000 animals) and that 10% were yearlings, 10% were 2-year-olds and 80% were 3 years or older. We assumed that the sex ratio was 1:1.

Fitting method

We fitted the model by simulating a trajectory of numbers and survivals from the starting conditions, known rainfall, and parameters a, b, c and d, and the harvest parameter(s). For any predicted values, a total goodness-of-fit was calculated by assuming that each observed type of data was log-normally distributed. As is normal with the maximum likelihood method we used negative log-likelihoods, the three components of which were calculated as follows:

$$L_T = \sum_y \frac{\left(\ln(T_{obs}) - \ln(T_{pre})\right)^2}{2\sigma_T^2} \qquad (15)$$

$$L_{s_0} = \sum_y \frac{\left(\ln(Y_{y,obs}) - \ln(Y_{y,pre})\right)^2}{2\sigma_{Y_y}^2} \qquad (16)$$

$$L_m = \sum_y \frac{\left(\ln(m_{obs}) - \ln(m_{pre})\right)^2}{2\sigma_m^2}. \qquad (17)$$

where the three Ls are the likelihoods of the census of total abundance, the likelihood of the estimate of yearling percentage, and the likelihood of the estimate of adult dry-season mortality. We have omitted the first term of the log-normal distribution in each of these equations because it is constant. The σs are estimated as follows. There are published estimates of the standard error of censuses since the 1970s. These estimates are derived from standard distribution theory and the coefficient of variation (*cv*) of almost all estimates is about 0.2. We prefer the log normal model because most modern abundance estimates produce log-normal rather than normal estimates. We conducted runs of the model using the year-by-year *cv*s and a normal distribution and there was no appreciable difference in the results. No standard errors have been published for the data on proportion of yearlings, although they are based upon very large samples so that estimates of *cv*s would be very small. We used 0.2 to allow for more uncertainty than the sample sizes would suggest. Sinclair *et al.* (1985) have published some variance estimates of the standard deviations of the mortality data for 1982 and 1983. The standard errors for these data are also quite low, with *cv*s less than 0.2. These *cv*s were checked against the goodness-of-fit to the model, and the models fitted the data reasonably well with *cv*s of less than 0.2 for all types of data. However, these *cv*s are probably too low because mortality rates were estimated in only a few places and for a few years. Thus, we consider a conservative *cv* of 0.6 is more realistic for this analysis.

Total negative log likelihood is therefore:

$$\left[\ln(L_T) + \ln(L_Y) + \ln(L_m)\right].$$

RESULTS

Fitting the model

A baseline fit in which no harvest is assumed depicts the data during the growth of the population in the 1960s and 1970s, the stability in the 1980s, and the large decline in 1993 (Fig. 10.1a). Although the individual census estimates from 1977 to 1991 are not followed with any reliability, the major changes in population abundance are well represented. The data, and the fits to the data for components of the model, are shown in Fig. 10.1 (b,c) for percentage of yearlings and adult survival. Figure 10.2 presents the calf and adult survival as a function of food availability.

Harvest model I was then fitted by assuming that the total harvest was proportional to the human population size and the wildebeest herd size. The best fit to the data was obtained when vulnerable proportions of the population q was very close to zero, indicating that the best estimate of harvest was c. 15 000 per year. Part of the reason for this is that from 1963 to 1977 the herd

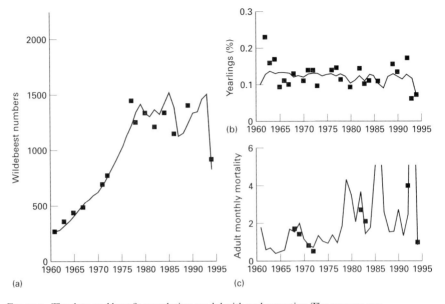

FIG. 10.1. The data and best fit population model with no harvesting. The squares are observations and the solid lines the model fits. (a) Is the total wildebeest population size, (b) is the percentage of the population that is yearlings, and (c) is the adult mortality rate per month of the dry season.

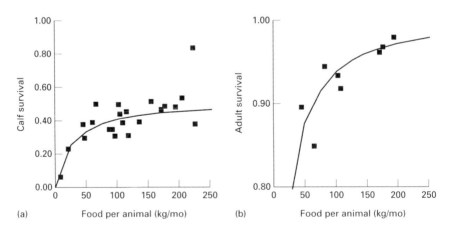

FIG. 10.2. Survival of (a) calves, and (b) adults plotted against estimated food per animal per month. The solid squares are the observed survival values and the lines are the functional relationships fitted by the model.

grew at 10% per year. Since yearlings during this period averaged 13% of the population and there was an additional 2–4% adult mortality, any significant snaring mortality in the 1960s and 1970s makes it very difficult for the model to fit the observed rate of increase. The total negative log-likelihood for this model was 21.6.

Harvest model II produced almost identical estimates to those of the first; the negative log-likelihood was 21.5. The estimated harvest from 1977 onwards was 15 000 animals per year. Harvest method III, assuming an anti-poaching factor of 5, produced a slightly better fit with the total negative log-likelihood at 21.2. The P level for a change in negative log-likelihood from 21.6 to 21.2 is 0.37. Thus, we conclude these models all fit the data equally well. The value of 5 was chosen arbitrarily to represent the major increase in poaching after 1977. The parameter q was chosen by the fitting procedure to maximize the likelihood. This fit estimated that the harvest was 30 000–40 000 from 1978 to 1984, but declined to a few thousand after 1984.

The goodness-of-fit to the model was measured by the negative log-likelihood (Hilborn & Mangel 1997) and none of these fits was significantly different from the others. We used the likelihood ratio test, and the Akaike information criterion to test for the best model. Both suggested that the no harvest alternative was the preferred model. This is because the no-harvest model has one less parameter, and the best fitting model (model III) did not fit the data significantly better. However, the no-harvest model could be ignored because we know there was a harvest. Therefore, we must incorporate some harvest. Recent analyses of poaching (Arcese *et al.* 1995, Mduma 1996) present no evidence that the illegal harvest has changed in the past few years. Thus, an assumption of a constant poaching off-take is best represented by harvest model II.

Confidence bounds on current harvest

We used the likelihood ratio test to calculate confidence bounds on the harvest assuming harvest model II. This was done using the method of likelihood profile, in which the post-1977 harvest was set at different levels, and assessed how good a fit to the data could be obtained by searching over the other parameters, a, b, c and d. The 95% confidence interval for the parameter of interest, harvest, occurred when the negative log-likelihood was less than 1.96 units greater than its minimum estimate. Since the minimum estimate is 21.54, the 95% confidence interval for harvest is any value of harvest with a negative log-likelihood under 23.52. Figure 10.3 (a) shows the results for this analysis. The negative log-likelihood is quite flat from the

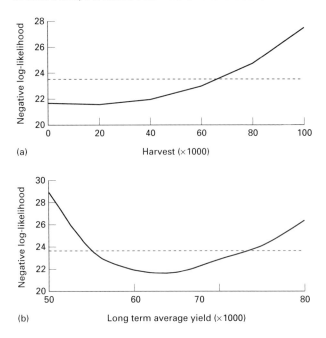

FIG. 10.3. (a) The likelihood profile of total annual harvest in years since 1977; (b) the potential harvest with 150 mm of dry season rainfall. The solid line is the negative log-likelihood, and the thin horizontal line represents the 0.05 confidence level.

region 0 to 40000 animals per year, and the 95% confidence interval is roughly from 0 to 67000.

Level of sustainable harvest

The sustainable harvest depends on the values of the parameters a, b, c and d, as well as the rainfall. The best estimate of the sustainable yield with a dry-season rainfall of 150 mm (the average during dry years) is 66000. Figure 10.3(b) shows the likelihood profile of the long-term average yield. The thin horizontal line is the 0.05 level so the 95% confidence interval is approximately 56000–74000. If the rainfall averages 200 mm (the average during wet years), the sustainable yield would be 80000 (95% CL 70000–118000). In this analysis we use maximum sustainable yield (MSY) purely as a benchmark and a full analysis should include variability in rainfall as was carried out by Pascual and Hilborn (1995). They explored a wide range of harvest strategies when dealing with fluctuating rainfall, using a similar model that did not include much of the recent data.

DISCUSSION

Wildebeest population dynamics

The wildebeest population in the Serengeti ecosystem increased from c. 0.25 million in the early 1960s to c. 1.4 million in the late 1970s and it has remained at around 1.3 million until the drought of 1993. Our evidence indicates that since the late 1970s the population is limited primarily by intraspecific competition for dry-season food (Sinclair et al. 1985, Sinclair & Arcese 1995b).

MSY is used here as a convenient benchmark for potential yield in a constant environment and using it does not imply that it is the level of exploitation that can be sustained each year. The subject of sustained yield has been explored extensively by Hilborn and Walters (1992), and effect of environmental variability on potential yield in the context of Serengeti wildebeest is discussed in Pascual and Hilborn (1995). The confidence intervals for the present estimates of MSY reflect the uncertainty in the parameters of the model.

Our estimates of the potential yield and the consequences of sustained harvesting depend on present environmental conditions continuing in the future. The most likely disruption is a reappearance of the exotic viral disease, rinderpest. In such an event, and any other shift in population dynamics, our present calculations would be invalid. Our estimates also depend on the assumption that the level of illegal harvest has been near constant for the past 15 years. If harvesting has expanded in the past 5 years (as may be expected from an expanding human population although we do not yet have evidence for it), then the current level of illegal harvest could be greater than the sustainable level.

A second likely event could be a major long-term drought. The drought of 1993 was the most severe since rainfall records began in 1938, with dry-season rainfall being the lowest yet recorded. Our model predicted that the population level should have dropped to 850000 animals and the census in 1994 counted 917000 animals (SE = ±18%) (Farm & Woodworth 1994), giving some support for our model.

Current estimates of illegal harvest

The above estimates of illegal harvest (0–40000 wildebeest) are much lower than those of Campbell and Hofer (1995). However, our estimates from model II are consistent with the independent estimate of 29 298 wildebeest per year (23 113–39 714, 95% CL) harvested both legally and illegally (Mduma 1996). The 95% upper limit of model II was 67 000 wildebeest harvested per year (Fig. 10.3a).

While the model is complex, the basic calculations come from knowing only population size, calf recruitment and adult mortality. Given these figures the probable removals by harvesting can be estimated, and appear to be low. The likelihood profiles show that removals of up to 40 000 animals per year fit the data well. However, removals of over 75 000 per year (as suggested by Campbell & Hofer (1995)), are incompatible with the data. The census data and the yearling recruitment data are probably reliable. If there is a problem with the data it must lie in the estimates of adult mortality rate. The most likely bias, however, is that adult mortality is actually higher than estimated and, therefore, that recent harvests are lower. Adult mortality could be higher for two reasons. First, we assumed no wet-season mortality, although we know it accounts for 2–5% loss per year (Mduma 1996). Second, we assumed that the transect method used to estimate mortality rates detected all of the carcasses in the surveyed area. However, our mortality counts are almost certainly an underestimate.

Sex ratio

An unpublished report (Georgiadis 1988 in Campbell & Hofer 1995) suggested that the current harvest of wildebeest is highly skewed towards males (88–98% of his sample of wildebeest caught in snares were males). If the illegal harvest was so biased towards males, then the current sex ratio of adults would be greatly in favour of females. A random sample of the age/sex structure in 1980 showed 4290 females and 4530 males, and in 1990 showed 6597 females and 5969 males (Mduma 1996). Both of these results are inconsistent with a significant harvest of predominantly males. Further, in an analysis of ranger reports on carcasses found in 1992 and 1993, 28 female and 23 male wildbeest were reported. We conclude, therefore, that the current harvest is not highly sex-specific. Since males and females tend to segregate into same-sex groups in the dry season when snaring occurs, it is likely that some small samples of snared animals would by chance be all of one sex.

The potential for legalized harvest

The model indicates that the current estimated level of illegal harvest is roughly half of the potential harvest at average levels of rainfall. Thus, additional harvesting is apparently available. However, any attempt to set a legal harvest above 40 000 would require a reduction in the illegal harvest. The calculations show that the population size that provides a maximum sustainable yield when dry-season rainfall averages 150 mm is c. 800 000 wildebeest.

The potential for a long-term legal harvest depends on the ability of management agencies to reduce the illegal harvest. The legal and illegal harvests are related. A legalized harvest could either aggravate or reduce the illegal harvest. We assume here that the illegal harvest could be reduced, and discuss how a legal harvest might be implemented.

Impediments to implementing a legal harvest

While a sustained harvest of wildebeest is theoretically possible, there are a number of concerns regarding the sustainability of a *legal* harvesting programme. These concerns include:

1 *A growing illegal harvest.* Roughly half of the sustainable harvest of wildebeest is now taken in the illegal harvest. If the assumption of constant harvest over 15 years is wrong the current illegal harvest could be exceeding the sustainable yield. Also, if the 1993 drought reduced the population as much as our model predicts, then there is little room to add a legal harvest to the illegal harvest. There is only room for a substantial legal harvest if the existing illegal harvest is reduced. However, it is possible that the addition of a legal harvest might aggravate, rather than reduce, the illegal harvest. By permitting legal possession of wildebeest meat and implements of harvesting (rifles, bow and arrow, snares), prevention of poaching may become considerably more difficult. If human population density on the perimeter of the Park continues to grow, we expect that poaching pressure will increase from the present value and the potential for legal harvest would be very small indeed.

2 *Economic inefficiency of harvesting.* Macnab (1991) has reviewed attempts at game cropping around the world and particularly in Africa. There are very few examples where game cropping has proved to be both cost effective and controlled. A major impediment to any legalized wildebeest harvest is finding a technology that is both cost effective and controllable. No large-scale legal harvesting should be undertaken until an economically viable method is shown to be effective.

3 *Inability to monitor.* Any harvesting strategy requires monitoring the levels of catch. This has proved to be difficult in most wildlife and fisheries harvesting schemes where the agency infrastructure and budgets are much larger than those available on the perimeter of Serengeti National Park.

Once legal harvesting is in place, it could prove difficult to determine and regulate.

4 *Inability to reduce harvesting as necessary.* Most of the harvesting strategies that might be adopted would require periodic adjustments in the quota. Such adjustments would from time to time mean reduction in quota. Experience in wildlife and fisheries around the world indicates that user groups are often resistant to reductions in quota. It is quite possible that the government might be unable to implement a harvesting strategy that required harvest reductions.

Counteracting the economic benefits of poaching

A sustained legal harvest of wildebeest requires a major reduction in the current illegal harvest. The benefits of poaching are the acquisition of meat and/or money from the sale of meat. The costs of poaching are the time lost from other activities, any money spent on poaching equipment such as snares, and the expected probability of capture, arrest, and possible fine or imprisonment. Counteracting poaching requires a combination of both reduced benefit and increased cost to the poachers (Leader-Williams *et al.* 1990, Milner-Gulland & Leader-Williams 1992, Wells, Brandon & Hannah 1992, Newmark Leonard, Sariko & Gamassa 1993). The most cost-effective way to reduce poaching could be by providing legal meat at a low price rather than by increasing the costs of poaching through anti-poaching patrols. Legal meat could come from legalized harvest of wildebeest, or from subsidized imports of other sources of protein (e.g. beef) from other regions of Tanzania.

However, wildebeest harvesting alone is not a long-term way to provide substantial benefits to the local population near the Serengeti. Any harvest poses the threat of population collapse if the harvest does not decline during times of reduced abundance. If the level of any form of harvest, legal or illegal, were to remain constant and the herd were to decline due to a drought or an outbreak of disease, such harvesting would accelerate the decline and potentially cause a population crash.

Whether development assistance is through legalized culling of wildebeest or direct financial assistance to villages it is essential that such assistance is tied to reduction in poaching. Wells *et al.* (1992) have discussed many integrated development and conservation programmes and shown that they almost always fail to tie development to conservation. This mistake must not be repeated in the Serengeti.

Monitoring and research needed to support a harvesting programme

Two types of data are essential for a legalized harvesting programme: (i) estimates of wildebeest abundance; and (ii) estimates of both the total legal and illegal catches. The monitoring of abundance is by far the most important, and is currently in place as part of the Tanzanian Wildlife Conservation Monitoring (TWCM). There is no systematic programme in place to estimate the illegal harvest, yet such a programme would be necessary both to determine the impact of harvesting on the wildebeest population in the future and to adjust local benefits in response to changes in poaching effort. A programme to monitor the illegal harvest should consist of systematic searches for snares to estimate the total amount of snaring effort, and measurement of snaring success. Such a systematic survey could be integrated with a programme to measure reductions in poaching effort as part of a benefits-for-conservation programme.

If a legalized harvest were implemented, it is essential that it be reliably monitored. If the legal harvest is determined on a village-by-village basis, monitoring the harvest could be as difficult as monitoring the illegal harvest. If quotas are allocated to villages, there will be economic pressure to exceed the quota illegally.

CONCLUSIONS

The long-term viability of the wildebeest population in the Serengeti, the viability of the entire ecosystem, and indeed of most natural areas in tropical regions is primarily threatened by encroachment of human settlement and by the illegal harvest of wildlife. The problem of encroaching human settlement is beyond the scope of this chapter, but poaching and legalized harvesting are closely related. One of the principal problems faced in management of African National Parks and other natural areas is to allow the local population to benefit from the existence of protected areas. It is often hoped that by providing a legalized harvest the local population would both benefit from the protected areas and the incentives for illegal harvest would be reduced. We think that considerable caution is merited on both fronts.

Wildebeest harvesting alone is not a long-term way to provide substantial benefits to the local population near the Serengeti. First, the cost of harvesting and administration of harvest regulations would probably exceed the value of the harvested wildebeest. Second, any benefits of wildebeest harvesting would be diminished by human population growth. We believe that wildebeest harvesting can perhaps be a useful supplement to the regional economy, but not a major one. The Serengeti Regional

Conservation Strategy (Mbano, Malpas, Maige *et al.* 1995) has explored the economic feasibility of harvesting methods, and found serious impediments using either fire-arms or bow and arrows. The most cost-effective method is the use of snares but, because they are inhumane and they are the tools of poachers, there is a strong reluctance to use this method.

Given that growing human population and continued human encroachment on the Park are expected, it may be dangerous for the local population to think of the wild animals in the Park as a potential food resource. Even if poaching were reduced to allow for a legalized harvest, a breakdown of anti-poaching programmes could result in a major outbreak of poaching at some time in the future.

Therefore, we end with the caution that any harvest poses the threat of population collapse if the harvest does not decline during times of reduced abundance. The biological potential does exist, but making use of the potential without threatening the long-term conservation of the resource faces many difficulties. In addition to biological potential the following conditions must be met: (i) the economics of harvesting are appropriate; (ii) an appropriate harvest strategy can be implemented that recognizes the importance of environmental fluctuations on biological potential; and (iii) an adequate monitoring and evaluation system is implemented that monitors the population and the harvest.

ACKNOWLEDGEMENTS

This work was commissioned by the Serengeti Regional Conservation Strategy, which provided the authors with considerable support. Mr B. Mbano and Mr W. Mapunda, the two directors of SRCS during the term of this study deserve special thanks. Mr M. Maige and Dr B. Farm provided useful discussion while the report was being prepared. Tanzanian National Parks, and Serengeti National Park, the Serengeti Wildlife Research Institute and the Tanzanian Commission for Science and Technology were all extremely helpful in facilitating this work. Drs Heribert Hofer, Ken Campbell and Andre Punt provided useful comments on the preliminary draft of this manuscript. Funding for S.M. and A.R.E.S. came from the Canadian Natural Sciences and Engineering Research Council (NSERC), the New York Zoological Society, and SRCS; funding for R.H. was provided by the University of Washington, SRCS and NSERC.

REFERENCES

Arcese, P., Hando, J. & Campbell, K. (1995). Historical and present day anti-poaching efforts in Serengeti. In *Serengeti II: Research, Management and Conservation of an Ecosystem* (Ed. by A.R.E. Sinclair & P. Arcese), pp. 506–533. University of Chicago Press, Chicago.

Borner, M., Fitzgibbon, C.D., Borner, M., Caro, T.M., Lindsay, W.K., Collins, D.A. et al. (1987). The decline of the Serengeti Thomson's gazelle population. *Oecologia*, **73**, 32–40.

Bothma, du P.J. (1989). *Game Ranch Management*. J.L. Van Schaik, Edms. Bpk., Pretoria.

Campbell, K. & Borner, M. (1995). Population trends and the distribution of Serengeti herbivores: implications for management. In *Serengeti II Research, Management and Conservation of an Ecosystem* (Ed by A.R.E. Sinclair & P. Arcese), pp. 117–145. University of Chicago Press, Chicago.

Campbell, K. & Hofer, H. (1995). People and wildlife: spatial dynamics and zones of interaction. In *Serengeti II: Research, Management and Conservation of an Ecosystem* (Ed. by A.R.E. Sinclair & P. Arcese), pp. 543–570. University of Chicago Press, Chicago.

Caughley, G. (1993). Elephants and economics. *Conservation Biology*, **7**, 943–945.

Caughley, G., Dublin, H. & Parker, I. (1990). Projected decline of the African elephant. *Biological Conservation*, **54**, 157–164.

Child, B. (1990). Assessment of wildlife utilization as a land use option in the semiarid rangeland of southern Africa. In *Living with Wildlife* (Ed. by A. Kiss), pp. 155–176. *World Bank Technical Paper*, **130**. Africa Technical Dept Series. The World Bank, Washington, DC.

Child, G. & Child, B. (1987). Economic characteristics of the wildlife resource. In *Wildlife Management in Sub-Saharan Africa*, pp. 163–178. Foundation Internationale pour la Sauvegarde du Gibier, Paris.

Clark, C.W. (1973a). The economics of over-exploitation. *Science*, **181**, 630–634.

Clark, C.W. (1973b). Profit maximization and the extinction of species. *Journal of Political Economics*, **81**, 950–961.

Clark, C.W. (1990). *Mathematical Bioeconomics: The Optimal Management of Renewable Resources*, 2nd edn. Wiley, New York.

Dasmann, R.F. (1964). *African Game Ranching*. Pergamon Press, Oxford.

Dasmann, R.F. & Mossman, A.S. (1960). The economic value of Rhodesian game. *Rhodesian Farmer*, **30**, 17–20.

Eltringham, S.K. (1984). *Wildlife Resources and Economic Development*. Wiley, New York.

Farm, B.P. & Woodworth, B.L. (1994). *Status and Trends of Wildebeest in the Serengeti Ecosystem*. Frankfurt Zoological Society, Arusha, Tanzania.

Field, C.R. (1979). Game ranching in Africa. In *Applied Biology*, Vol. 4 (Ed. by T.H. Coaker), pp. 63–101. Academic Press, New York.

Geer, S. (1992). Alarming soil degradation around the world revealed by latest data. *Environmental Conservation*, **19**, 268–270.

Hilborn, R. & Mangel, M. (1997). *The Ecological Detective: Confronting Models with Data*. Princeton University Press, Princeton, NJ.

Hilborn, R. & Sinclair, A.R.E. (1979). A Simulation of the wildebeest population, other ungulates, and their predators. In *Serengeti: Dynamics of an Ecosystem* (Ed by A.R.E. Sinclair & M. Norton-Griffiths), pp. 287–309. University of Chicago Press, Chicago.

Hilborn, R. & Walters, C.J. (1992). *Quantitative Fisheries Stock Assessment: Choice, Dynamics and Uncertainty*. Chapman & Hall, New York.

Laws, R.M., Parker, I.S.C. & Johnstone, R.C.B. (1975). *Elephants and their Habitats.* Clarendon Press, Oxford.

Leader-Williams, N., Albon, S.D. & Berry, P.S.M. (1990). Illegal exploitation of black rhinoceros and elephant populations: Patterns of decline, law enforcement and patrol effort in Luangwa Valley, Zambia. *Journal of Applied Ecology*, **27**, 1055–1078.

Mbano, B.N.N., Malpas, R.C., Maige, M.K.S., Symonds, P.A.K. & Thompson, D.M. (1995). The Serengeti Regional Conservation Strategy. In *Serengeti II: Research, Management and Conservation of an Ecosystem* (Ed. by A.R.E. Sinclair & P. Arcese), pp. 605–616. University of Chicago Press, Chicago.

Macnab, J. (1991). Does game cropping serve conservation? A reexamination of the African data. *Canadian Journal of Zoology*, **69**, 2283–2290.

McNeely, J.A. (1994). Lessons from the past: forests and biodiversity. *Biodiversity and Conservation*, **3**, 3–20.

Mduma, S.A.R. (1996). Serengeti wildebeest population dynamics: regulation, limitation and implications for harvesting. PhD thesis, University of British Columbia.

Milner-Gulland, E.J. & Leader-Williams, N. (1992). A model of incentives for the illegal exploitation of black rhinos and elephants: poaching pays in Luangwa Valley, Zambia. *Journal of Applied Ecology*, **29**, 388–401.

Mossman, A.S. (1975). International game ranching programs. *Journal of Animal Science*, **40**, 993–999.

Newmark, W.D., Leonard, N.L., Sariko, H.I. & Gamassa, D.-G.M. (1993). Conservation attitudes of local people living adjacent to five protected areas in Tanzania. *Biological Conservation*, **63**, 177–183.

Norton-Griffiths, M. (1973). Counting the Serengeti migratory wildebeest using two-stage sampling. *East African Wildlife Journal*, **11**, 135–149.

Parker, I.S.C. (1984). Perspectives on wildlife cropping or culling. In *Conservation and Wildlife Management in Africa* (Ed. by R.H.V. Bell & E. McShane-Caluzi), pp. 233–253. US Peace Corps, Office of Training and Program Support Forestry and Natural Resources Sector, Washington DC.

Parker, I.S.C. (1987). Game cropping in the Serengeti region. *Parks*, **112**, 12–13.

Pascual, M.A. & Hilborn, R. (1995). Conservation of harvested populations in fluctuating environments: the case of the Serengeti wildebeest. *Journal of Applied Ecology*, **32**, 468–480.

Sinclair, A.R.E. (1973). Population increases of the buffalo and wildebeest in the Serengeti. *East African Wildlife Journal*, **11**, 93–107.

Sinclair, A.R.E. (1979). The eruption of the ruminants. In *Serengeti: Dynamics of an Ecosystem* (Ed. by A.R.E. Sinclair & M. Norton-Griffiths), pp. 82–103. University of Chicago Press, Chicago.

Sinclair, A.R.E. (1995). Serengeti past and present. In *Serengeti II: Research, Management and Conservation of an Ecosystem* (Ed. by A.R.E. Sinclair & P. Arcese), pp. 3–30. University of Chicago Press, Chicago.

Sinclair, A.R.E. & Arcese, P. (1995a). Serengeti in the context of worldwide conservation efforts. In *Serengeti II: Research, Management and Conservation of an Ecosystem* (Ed. by A.R.E. Sinclair & P. Arcese), pp. 31–46. University of Chicago Press, Chicago.

Sinclair, A.R.E. & Arcese, P. (1995b). Population consequences of predation-sensitive foraging: the Serengeti wildebeest. *Ecology*, **76**, 882–891.

Sinclair, A.R.E. & Norton-Griffiths, M. (1982). Does competition or facilitation regulate populations in the Serengeti? A test of hypothesis. *Oecologia*, **53**, 364–369.

Sinclair, A.R.E. & Wells, M.P. (1989). Population growth and the poverty cycle in Africa: colliding ecological and economic processes? In *Food and Natural Resources* (Ed. by D. Pimentel & C. Hall), pp. 439–484. Academic Press, New York.

Sinclair, A.R.E., Dublin, H. & Borner, M. (1985). Population regulation of Serengeti wildebeest: a test of the food hypothesis. *Oecologia*, **65**, 266–268.

Sinclair, A.R.E., Hik, D., Schmitz, O.J., Scudder, G.C.E., Turpin, D. & Larter, N.C. (1995). Biodiversity and the need for habitat renewal. *Ecological Applications*, **5**, 579–587.

Soulé, M.E. (1991). Conservation: tactics for a constant crisis. *Science*, **253**, 744–750.

Talbot, L.M. & Swift, L.M. (1966). Production of wildlife in support of human populations in Africa. *Proceedings of the International Grasslands Congress*, 9, 1355–1359.

Tolba, M.K., El-Kholy, O.A., El-Hinnawi, E., Holdgate, M.W., McMichael, D.F. & Munn, R.E. (1992). *The World Environment 1972–1992: Two Decades of Challenge.* Chapman & Hall, London.

Turner, M. (1987). *My Serengeti Years* (Ed. by B. Jackson). Elm Tree Books/Hamish Hamilton, London.

Walker, B.H. (1976). An assessment of the ecological basis of game ranching in southern African grasslands. *Proceedings of the Grasslands Society of South Africa*, 11, 125–130.

Walker, B.H. (1979). Game ranching in Africa. In *Management of Semi-arid Ecosystems* (Ed. by B.H. Walker), pp. 55–81. Elsevier, Amsterdam.

Walker, B.H., Emslie, R.H., Owen-Smith, R.N. & Scholes, R.J. (1987). To cull; or not to cull: lessons from a southern African drought. *Journal of Applied Ecology*, 24, 381–401.

Watson, R.M. (1967). *The population ecology of wildebeest* (Connochaetes taurinus albojubatus, *Thomas*) *in the Serengeti*. PhD thesis, University of Cambridge.

Wells, M., Brandon, K. & Hannah, L. (1992). *People and Parks: Linking Protected Area Management with Local Communities.* The World Bank, Washington DC.

11. PHENOLOGY AND DYNAMICS OF AN AFRICAN RAINFOREST AT KORUP, CAMEROON

D. M. NEWBERY*, N. C. SONGWE[†] and G. B. CHUYONG[†]
* Department of Biological and Molecular Sciences, University of Stirling, Stirling FK9 4LA, Scotland, UK and Geobotanisches Institut, Universität Bern, Alternbergrain 21, CH-2013 Bern, Switzerland; [†] Institute de la Recherche Agronomique du Cameroun, Kumba Forestry Research Station, PMB 29 Kumba, SW Province, Cameroon

SUMMARY

1   In groves of ectomycorrhizal caesalpinaceous legumes in Korup, Cameroon, the most abundant of three co-dominant tree species, *Microberlinia bisulcata*, has had, and continues to show, very low recruitment. Replacement is likely to favour the two other guild members, *Tetraberlinia bifoliolata* and *T. moreliana*. All three species produce large sufficient seedling banks.

2   *Microberlinia bisulcata* attained the greatest tree sizes and had the most pronounced mast fruiting pattern (a 3-year cycle). Masting was associated with peaks in the previous dry-season radiation suggesting a requirement to accumulate carbon reserves before fruiting. More circumstantial evidence points to phosphorus supply as a co-controlling factor.

3   The dominance of *M. bisulcata* is most likely due to a recently unique, natural historical event, an epoch of unusually dry years, which allowed its shade-intolerant ectomycorrhizal seedlings to outcompete other species. The putative fungal connections to parents appear to be of no advantage (possibly a disadvantage) to the seedlings under such mature adults.

4   The composition of Korup is neither constant nor cycling but a fragment of change which might be reset by climatic fluctuations coincident with potential regeneration of a species also adapted to the low-phosphorus soils and strongly seasonal conditions of the site. Understanding the dynamics of such an African forest requires a stochastic view over several centuries.

INTRODUCTION

The present structure and floristic composition of a forest stand is partly determined by current conditions but largely by past processes and events.

History plays an important role in our understanding of forest dynamics and it is most unlikely that tropical forests are, or have been, in any form of equilibrium and constant composition (Maley 1996). The most useful model is one that admits substantial long-term change, and the key questions lie around what factors determine the changes, not so much what might hold composition in any fixed state. Studying what has controlled change in the past and determining the patterns seen today may enable predictions of forest composition in the future, especially in the context of global climate change. The detection of pattern and elucidating its underlying causes remains a strong inferential approach (Watt 1947, Greig-Smith 1983); and this is a considerable challenge in species-rich, heterogeneous and productive systems such as tropical rainforests. Thus, sampling limitations severely restrict what can be achieved; in many cases it has to be admitted that sampling has been inadequate in many studies. However, some tropical forest communities are less complex and detailed studies on them may give a better insight into general processes.

Dominance or codominance of a community by a few species is an attractive, simplifying feature. These species have evidently come to be selected at sites where the main ultimate causes are also likely to be few. Dominant species largely determine the whole community's dynamics, growth and recruitment, its carbon and nutrient cycles and water economy, and eventually its structure and floristic composition (Whittaker 1975, O'Neill, DeAngelis & Waide *et al.* 1986). From a sampling view-point they are more homogeneous, at least in space, and statistical inferences can be made more confidently than in a more species-rich forest. Models and predictions are likely to be simpler, firmer and more readily tested (Botkin 1993). This is not to say that they are temporarily homogeneous, since large forest trees are generally old and their population turnover times are very long.

Monodominant forests hold a particular fascination in that one species has been selected over all others locally available to survive best in what are invariably special and extreme environmental conditions (Richards 1952, Connell & Lowman 1989). These stands fall at the ends of coenoclines. Morphological, ecophysiological and life-history characteristics are often clear. At the other end of the diversity scale, lack of dominance provides little guidance to explanation, not just because there are so many species to account for simultaneously but because the forest is locally often very variable in time and space, far from any supposed equilibrium (Huston 1994). Therefore, relating current conditions to any one species can be misleading, and the subtleties of past events are unknown.

In between these extremes in diversity are forests which show codominance by a small number of species and, where these can be identified to have a guild structure, the system becomes more tractable in terms of definition, sampling, mechanisms and processes, and in modelling the dynamics. Not only are there sufficient individuals to examine the guild functioning but also there is the possibility of examining within-guild interactions in space and time. These codominants together will determine site and forest ecosystem processes in a broadly similar manner but each may differ in ways which can accommodate their co-occurrence and possibly also their coexistence. Small differences in characteristics leading to loss of fitness should be amplified in these more extreme and selective conditions. Codominance in a guild makes a system within a system, partitioning variation. Perhaps the best chance of measuring change reliably and establishing its causes in tropical forests lies in concentrating on these relatively species-poor communities.

The need to understand change in tropical forests is made more pressing by the increasing evidence for accelerating climatic influences at the global and regional levels (Woodward 1987, Maley 1991). Climate has changed and markedly altered tropical vegetation in the past and forms a key aspect of historical interpretation (Maley 1987, 1989). To know that species richness *per se* is changing is of limited scientific value: more interesting is which species are decreasing or increasing, and for what reasons. Present forest composition may be largely due to natural selection operating in the past centuries and millennia but we can expect the largest changes in recent decades to occur (through current selection) in the edaphically extreme sites, those already with limiting resources. A few dominant species in those sites make detection and understanding easier. The changes will also be more interpretable in evolutionary terms if the forest is a primary one.

For most forest sites in the tropics substantial studies of dynamics, even with records of 10–30 years, are short compared with the life-span of most canopy trees of the order 200–300 years, and often they cannot separate the short time-scale stochasticity (Hubbell & Foster 1983, 1986) from major trends. This forces a consideration of present-day size class distributions in order to infer what might have happened in the past and what is likely to happen in the future. It is necessarily only semi-quantitative as the exact ages of the trees will be unknown. However, where changes in composition are evident contemporaneous studies on phenology, growth and population dynamics of these species might highlight key differences between species which can then be applied to events in the past, or form a basis for a model of the future.

The forest in Korup National Park (9°E, 5°N) presents an interesting case in point. Within a mosaic of diverse primary forest can be found stands of large trees dominated by three ectomycorrhizal caesalpinaceous species (Gartlan, Newbery, Thomas & Waterman 1986, Newbery & Gartlan 1996).

They form distinct patches and one in particular, c. 1.5 km × 1 km in area, has been studied in some detail by us (Newbery, Alexander, Thomas & Gartlan 1988, Newbery, Alexander & Rother 1997). Other smaller patches exist and they are probably all remnants of the once much more widespread forest type, *la forêt biafriéene* (or Atlantic coastal forest) of Letouzey (1968, 1985). Korup lies at the western end of an arc of this forest that runs from the Niger Delta to Gabon in western Central Africa. (Details of the site can be found in Gartlan *et al.* 1986.) The three ectomycorrhizal species, *Microberlinia bisulcata* A. Chev., *Tetraberlinia bifoliolata* (Harms) Hauman and *T. moreliana* Aubrév., form a well-defined guild of ectomycorrhizal trees, with a characteristic phosphorus cycle (Newbery *et al.* 1997). Earlier vegetation analysis of an extensive set of plots within the Park (arranged along four transects labelled P to S) showed an association between the groves of these species and sandy, low phosphorus, low pH soils, of the more southern, lower elevation sites. The ectomycorrhizal status was confirmed in Newbery *et al.* (1988). Alexander (1989a) has discussed the taxonomy of this family in Africa with respect to mycorrhizal status. Outside the patches are other caesalpiniaceous species, also ectomycorrhizal, but these are at much lower densities throughout most of southern Korup. Our attention has focused since 1984 on the one main large patch on the so-called 'transect P', studying its structure, composition, dynamics, phenology and recruitment alongside a study of phosphorus cycling and the role played by ectomycorrhizal fungi (Newbery *et al.* 1988, 1997). It has become apparent in recent years that the most abundant species among the three, *M. bisulcata*, has been failing to replace itself on the site and this presents a paradox.

Forests that have large individuals of some species which also lack recruits are an interesting and distinctive feature of the African tropics (Richards 1952). Newbery and Gartlan (1996) showed for Korup and Douala-Edea (also in Cameroon; Newbery, Gartlan, McKey & Waterman 1986) that *M. bisulcata* (although the most clumped) was not alone in this predicament: 42 and 48% of species respectively which formed canopy and emergent populations apparently lack small stems sufficient to replace the adults, assuming a normal exponential model for mortality through the size classes. These so-called 'group 5' species have been modelled in Newbery and Gartlan (1996) and the deficit in small stems was shown to be substantial. From the seminal discussions by Aubréville (1938), reiterated in Aubréville (1971), it is clear from his examples that whilst *some* large-treed species lacked recruits close to them, small trees could be found elsewhere within the forest. To characterize the 'Aubréville phenomenon' further it is necessary to distinguish between species with no immediately local recruits and species with no recruits at all, even far away from adults. The former suggests perhaps a cyclic mosaic process (Richards 1952), but the latter means loss of a species from the site. Confidence in distinguishing between these two cases comes down to sampling. Is *M. bisulcata* elsewhere in Korup but our sampling has missed it? This seems unlikely because: (i) the sampling in Korup, and Douala-Edea, was extensive (135 and 104 plots respectively, each of 0.64 ha); (ii) travels by us and the guides and game guards within the southern part of Korup have not found other (patches of) small trees; and (iii) the fruits of *M. bisulcata* appear to be very poorly dispersed and if there were no saplings below the adults now they would not be elsewhere either. However, most of the other species highlighted by Newbery and Gartlan (1996) are more randomly or evenly dispersed as large trees. One key consideration is that possibly *M. bisulcata* was not so clumped in the past.

Lack of recruits can be due to the effects of many interacting factors on the serial stages of growth to maturity. To begin with, the length of the juvenile period determines when sexual reproduction is first possible. Then, seeds must be present to successfully form new seedlings, and these latter have to survive to grow on to pole-sized trees. Whilst competition and growth in natural tropical forests at the pole stage and beyond is almost entirely unknown, the most likely stages where natural selection might operate most decisively could be on seedling establishment and sapling growth and survival, or on seed production. Phenological processes determine the timing and production of the seeds; when seeds fall and where they germinate is crucial. For a tree species failing to regenerate the key considerations are: (i) whether enough viable seeds are produced; (ii) how often they are produced; (iii) whether they are shed at times to enable optimum germination and establishment; and (iv) whether the seedlings can grow on to saplings and poles. To establish whether or not a bank of seedlings is produced and that survivorship is sufficient would usefully rule out many other factors.

In this chapter we bring together new results on the size distributions of these three caesalps in Korup, their phenology, and a seedling demographic study, interpreting them with climatic data and integrating the findings into an improved synthesis of our understanding of caesalp-dominated forests in western Central Africa.

DYNAMIC NATURE OF THE FOREST

On transect P (1990–91) the large stand of ectomycorrhizal legumes was sampled within a 82.5-ha block (500 m N–S, 1650 m E–W) subdivided into 33 N–S 500 m × 50 m plots each of 10 subplots 50 m × 50 m. All trees ≥50 cm dbh in the whole block were mapped, tagged and measured (gbh) and identified. Within one randomly selected subplot in each plot all trees down to 10 cm dbh were tagged, measured and identified giving 33 subplots sampled in the class 10–50 cm gbh. Then, within each of these 33 subplots one N–S 50 m × 5 m strip was selected at random and in every odd-(plot-)numbered subplot ($n = 17$) all trees (saplings and small poles) down to 1 cm dbh were tagged and measured (dbh or gbh) and the ectomycorrhizal species only identified. In the even-numbered subplots the trees were simply counted in the 1–5 and 5–10 cm dbh classes and just the ectomycorrhizal species measured, tagged and identified.

The aim was not a complete enumeration of all trees ≥1 cm dbh but a tiered and stratified sample focused on understanding the regeneration of these three large ectomycorrhizal trees in relation to one another and all trees in their stand. Trees were identified in the field and confirmed with extensive botanical collecting, voucher collections now residing in the Herbier National de Yaoundé. Later, in 1992, many trees were double-checked and some recollected. After compilation of the data and preliminary analysis some trees just outside of the exact 10- or 50-cm cut-off dbhs were removed from the data set or relocated to the 10–<50 or ≥50 cm dbh classes. Attention is focused on the population structure of the ectomycorrhizal species in this chapter.

Taken together, the three ectomycorrhizal species accounted for 48.4% of all large trees (≥50 cm dbh) and 86.5% of very large trees ≥100 cm dbh, but only 3.6% of stems 10–<50 cm dbh and 5.0% of stems 1–<10 cm dbh (Table 11.1). Among the trees ≥50 cm dbh, *M. bisulcata* and *T. moreliana* had similar densities (*c.* 3.5 ha^{-1}) but *T. bifoliolata* less at 2.4 ha^{-1}, though among the very large trees *M. bisulcata* was dominant at *c.* 2 ha^{-1} with the two *Tetraberlinia* species at *c.* 0.5 ha^{-1}. This was reflected in the three dbh distributions of different shapes in Fig. 11.1. Among the intermediate-sized trees (10–<50 cm), *M. bisulcata* was clearly much less abundant than *T. moreliana* or *T. bifoliolata* (ratio 1:6) and this representation fell even lower in the small-tree size class (1–<10 cm) (ratio 1:27). The decline in densities of intermediate-sized trees of these three species is not exponential; between 25 and 50 cm dbh there was an approximately constant representation (Fig. 11.2). When the three large ectomycorrhizal species are compared (Fig. 11.3), *M. bisulcata* steadily decreased its contribution down the size classes, *T. bifoliolata* was predominant in the smallest classes and *T. moreliana* was mostly represented in the intermediate size classes. Given that all of these species can attain dbh >100 cm, future replacement, *ceteris paribus*, is likely to be lowest for *M. bisulcata*, with more *T. moreliana* and especially more *T. bifoliolata*.

Considering the distribution of densities of the three species in the 33 subplots for the intermediate size class (10–<50 cm) neither of the

TABLE 11.1. Densities of the three co-dominant ectomycorrhizal caesalpinaceous species of tree in increasing size classes in the 82.5-ha enumeration in southern Korup National Park.

Species	dbh class (cm)											
	1–<10 (Area 0.825 ha)			10–<50 (Area 8.25 ha)			≥50 (Area 82.5 ha)			≥100 (Area 82.5 ha)		
	n	ha^{-1}	%	n	ha^{-1}	%	n	ha^{-1}	%	n	ha^{-1}	%
Microberlinia bisulcata	4	5	0.1	9	1.1	0.3	295	3.58	18.3	170	2.06	58.8
Tetraberlinia bifoliolata	146	177	3.2	39	4.7	1.2	198	2.40	12.3	37	0.45	12.8
Tetraberlinia moreliana	76	92	1.7	66	8.0	2.1	287	3.48	17.8	43	0.52	14.8
Other species	4332	5251	95.0	3015	365.5	96.4	830	10.06	51.6	39	0.47	13.5
Total	4558	5525		3129	379.3		1610	19.52		289	3.50	

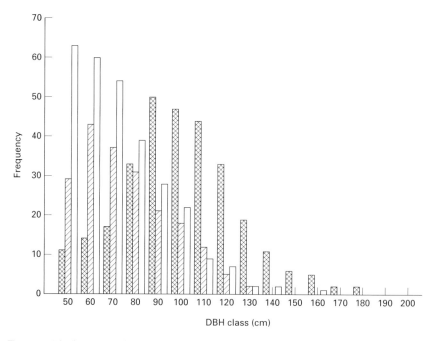

FIG. 11.1. The frequency distribution of large trees (≥50 cm dbh) in the 82.5-ha block enumerated in southern Korup National Park, in increasing dbh (diameter) classes (50 = 50–<60, etc.) for the three ectomycorrhizal caesalp species: *Microberlinia bisulcata*, ▨; *Tetraberlinia bifoliolata*, ▨; and *Tetraberlinia moreliana*, □.

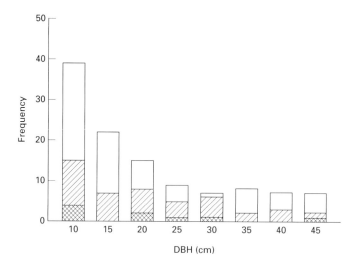

FIG. 11.2. The frequency distribution of intermediate trees (10–<50 cm dbh) in a 8.25-ha subsample of the main block enumerated in southern Korup National Park, in increasing dbh (diameter) classes (10 = 10–<15, etc.) for the three ectomycorrhizal caesalp species: *Microberlinia bisulcata*, ▨; *Tetraberlinia bifoliolata*, ▨; and *Tetraberlinia moreliana*, □.

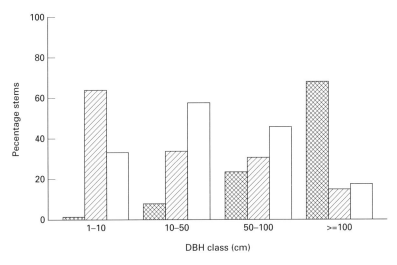

FIG. 11.3. Relative proportions of the three large caesalp species to one another within increasing major size classes: *Microberlinia bisulcata*, ▨; *Tetraberlinia bifoliolata*, ▨; and *Tetraberlinia moreliana*, □.

Tetraberlinia species differed from expectation of a Poisson distribution (i.e. randomness) ($\chi^2 = 0.26$, df = 3; $\chi^2 = 5.46$, df = 4; respectively, both $P > 0.05$), whilst there were too few trees to test the distribution of *M. bisulcata*. Similarly, among the small trees (1–<10 cm), there are again too few for *M. bisulcata*, but for the *Tetraberlinia* species there was no significant departure from randomness (*T. bifoliolata*, $\chi^2 = 8.95$, df = 5; *T. moreliana*, $\chi^2 = 1.97$, df = 4; $P > 0.05$). This suggests no clumping of the small and intermediate trees, and that the means based on the 33 subplots did not have overdispersed variances. The mean number of intermediate trees, of all species, was 94.8 per subplot with a SE of only 2.9, further suggesting a structurally homogeneous forest. Selecting five random and exclusive, stratified subsets of 80 1-ha squares within the main block (excluding the far west-most plot), none of the three ectomycorrhizal legumes showed significant departure from randomness ($\chi^2 = 0.035$ to 6.61, df = 2 or 3, $P > 0.05$ with Bonferroni adjustment) indicating further the homogeneity of this stand.

Combining the numbers of large and intermediate trees (×10) in 10-cm dbh classes above 10 cm dbh (to 180 cm) the dfd (diameter frequency distribution) double log vs. mid-class dbh plot regression was $Y = 2.51 - 0.00828$ dbh ($R^2 = 96.8\%$, $t_{slope} = -21.9$, df = 15, $P < 0.001$). This graph shows a change in slope between 70 and 110 cm dbh. This double-log model is indicative of the strong 'group 5' species component (Newbery & Gartlan 1996).

PHENOLOGY

Between June 1990 and March 1993, 10 marked randomly-selected trees (≥20 cm dbh) of each of 16 species were scored monthly for phenology along transect P. The 16 species included six ectomycorrhizal species plus 10 others common in the canopy. The features recorded were presence/absence of leaf flush, very high leaf fall, flowering and fruit fall. The essential species were *M. bisulcata*, *T. bifoliolata* and *T. moreliana* plus 13 others. Prior to this period general observations were made on the phenology of all the main caesalps (*M. bisulcata*, *T. moreliana* and *T. bifoliolata* especially) from August 1988 to April 1991 at *c.* 6-weekly intervals by J.A. Rother, and then again by our group from March 1994 to the end of 1995. Whilst the central period allowed a quantitative assessment of phenology, the other two periods were accurate enough to determine major flowering, fruiting and fruitfall events. Many visits to the forest site by J.J. Green and G.B. Chuyong between April 1993 and March 1995 were involved although phenology was not recorded specifically. However, we were always on the alert for heavy flowerings that might precede heavy seed falls. Knowledge of the Korup caesalps therefore spans the 7-year period from August 1988 to December 1995.

The main period of flowering for all species was January–July, and of fruiting June–October (Fig. 11.4). Most species flowered and set fruits to varying degrees most years. In many cases immature fruits were dropped with effectively no seed fall. In the 1992 peak month of flowering of April 19/160 (12%) of trees flowered (Fig. 11.4a), and the corresponding peak fruiting month of August involved 30/160 (19%) of trees (Fig. 11.4b). Fruiting covered a similar time-span to flowering.

In 1989, 1992 and 1995 there were mast fruitings throughout the forest, and especially for the three large ectomycorrhizal caesalps. The common interval of 3 years between main fruit falls is of note. Trees were able to flower most years, which suggests no limitation of pollinators. Pollination of the large canopy trees is thought to be achieved by small bees (D.M. Newbery, personal observation). In 1991 flowering was especially heavy for *M. bisulcata* but it resulted in very little seed. In 1994 some flowering for *M. bisulcata* was recorded but the two *Tetraberlinia* species and nine other canopy species which were monitored did not flower.

In the 1990–93 detailed study only *Klaineathus gaboniae*, *Coula edulis* and *Diospyros gabonensis* flowered and then fruited in 1992 masting year but did not do both in 1990, 1991 or 1993. *Microberlinia bisulcata*, *Berlinia bracteosa*, *Cola rostrata*, *Cola verticillata*, *Didelotia africana*, *Anthonotha fragrans* all flowered and fruited to some similar degree in each of the 3 years. *Oubanguia alata*, a very common lower-canopy species, did not flower in 1991. *Strephonema pseudocola*, *Strombosia glaucescens*, *Dichostemma*

Phenology and dynamics of rainforest 277

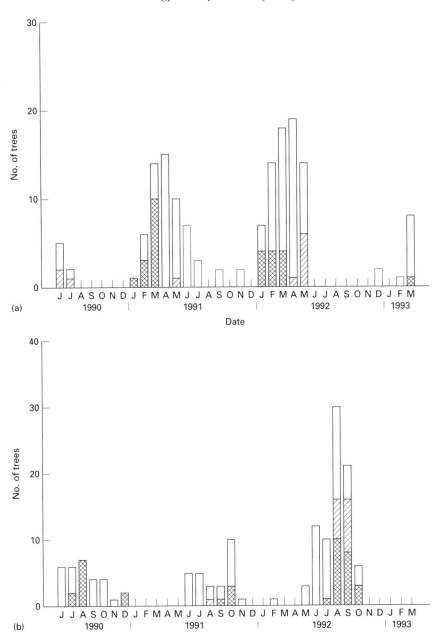

FIG. 11.4. Phenology of monitored trees in Korup, transect P, for (a) flowering and (b) fruiting of *Microberlinia bisulcata*, ▨; *Tetraberlinia bifoliolata*, ▧; and other species, □.

glaucescens and *Hymenostegia afzelii* barely flowered at all, with one or two individuals only in 1991; and *T. moreliana* flowered a very little in 1992 and 1993. The most pronounced peaks were in 1992, especially for *M. bisulcata* and *T. bifoliolata* flowering and then fruiting.

How well do the other records support this pattern? From late 1988 to early 1991 flowering and fruiting occurred almost entirely in 1989 with scattered reproduction for *M. bisulcata, Anthonotha fragrans, Berlinia bracteosa* and *Didelotia africana* in the other years. In 1989 the mast fruiting in the caesalps was made up of *M. bisulcata* and *T. moreliana* but not *T. bifoliolata*. In contrast to the other caesalps, *Anthonotha fragrans* has fruits which mature over two seasons, falling in the next early wet season. The flowering (but not fruiting) of *M. bisulcata* in 1991 was independently confirmed from litterfall results (Chuyong 1994).

In the period 1994 to early 1995 there was no mast fruiting and only limited flowering was seen in 1993 and 1994 for many species (especially *T. moreliana* and *T. bifoliolata*) though *M. bisulcata* flowered a little, producing some seed. This supported further the idea of *M. bisulcata* being able to flower annually but its seed crop varying substantially from year to year. In 1995 all three large caesalps, *M. bisulcata, T. bifoliolata* and *T. moreliana*, flowered, fruited and produced much seed. Many other species also produced flowers and fruits that year.

In summary, the pattern of fruiting observed was:

	1989	1992	1995
M. bisulcata	×	×	×
T. bifoliolata	–	×	×
T. moreliana	×	–	×

How could *M. bisulcata* achieve mast fruiting three times, every 3 years, but the other species only 2 out of 3 years in the period? There were strong coincidences between the three species in the mast years.

Microberlinia bisulcata began to flower in the dry season, or at the start of the wet season (January–March) each year (Fig. 11.4a), clearly shown for 1991–92. This phenological pattern was also noted for 1989. The species drops its seeds in August to October (7 months later) in all three mast fruiting events. *Tetraberlinia moreliana* followed a similar pattern (1989 and 1995 mastings). *Tetraberlinia bifoliolata* consistently flowered later than *M. bisulcata* in April–June, fruiting in August–September, but mostly completing its fruitfall earlier than *M. bisculata*. *Microberlinia. bisulcata* and *T. moreliana* are both deciduous and have microphyllous, pinnate leaves: *T. bifoliolata* is mesophyllous and less clearly seasonal in its leaf phenology (Chuyong 1994). After peak leaf-fall in the dry season the trees of *M. bisulcata* releafed, even within the dry season and therefore flowers were born on freshly leaved branches. *T. moreliana* was similar but the pattern was less pronounced, whilst *T. bifoliolata* showed the opposite trend, with marked flushing at the end of the wet season prior to the loss of old leaves in the dry season.

In 1989 *T. moreliana* and *M. bisulcata* both seeded in July. In 1992 the order was *T. bifoliolata* (early August) and *M. bisulcata* (late August), whilst in 1995 the order of seedfall was: *T. moreliana* (early July), *T. bifoliolata* (early August) and *M. bisulcata* (end August). The general order, compiling the 3 years' data, was: *T. moreliana* > *T. bifoliolata* > *M. bisulcata* spread over c. 4–6 weeks.

SEEDLING BANK

In March 1995 the composition of the seedling bank of the three large ectomycorrhizal species was sampled more extensively. In the eastern third of the main block (11 plots, 110 subplots), eight 2 m × 2 m quadrats were placed at random within each subplot (total $n = 880$) and the number of individuals of each species recorded in two classes (i) ≤30 cm height; and (ii) >30 cm and <1 cm dbh. These would have been very largely seedlings that survived from the 1992 fruiting, with some earlier masting. In November 1995 after that year's mast fruiting 91 4 m × 4 m permanent quadrats were marked, offset to each of the 50-m grid intersections within the eastern 10 plots (81) plus 10 others in a stratified and focused manner to incorporate cases of two or three caesalp species occurring together.

At the end of the 1992–95 masting interval, the three species had very similar densities and proportions in the ≤30 cm height class (Table 11.2). In

TABLE 11.2. Densities of the three co-dominant ectomycorrhizal caesalpinaceous species of tree in the seedling and sapling classes in southern Korup National Park.

	New cohort (November 1995) (Area 0.1456 ha)		Seedling bank (March 1995) (Area 0.3520 ha)			
			≤30 cm ht		>30 cm–<1 cm dbh	
Species	n	ha^{-1}	n	ha^{-1}	n	ha^{-1}
Microberlinia bisulcata	960	6593	297	844	214	608
Tetraberlinia bifoliolata	543	3729	344	977	606	1722
Tetraberlina moreliana	135	927	338	960	393	1116

the next class, those between 30cm height and 1cm dbh, *T. bifoliolata* increased its representation, *M. bisulcata* declined and *T. moreliana* did not change much. After the 1995 masting, among the new recruits, *M. bisulcata* and *T. bifoliolata* were at high abundances, seven- and four-fold greater than *T. moreliana* respectively (Table 11.2). Athough this was not a completely random sample it does indicate the approximate relative proportions of the species.

WEATHER

Monthly rainfall and radiation records (the latter derived from Gunn–Bellani evaporimeter readings) were available at Bulu, Ndian, for 1973–95. This site is 12 km SE from the centre of transect P in Korup (the main ectomycorrhizal stand) and at a similar elevation (40m asl). Bulu showed a pronounced seasonal pattern with one single-peak wet season (June–October) in which 71% of the annual total fell (Fig. 11.5) and a distinct dry season in the months of December–February with rainfall close to or less than 100mm month^{-1} and giving only 4.5% of the annual total. The rainfall peaked in August and was slightly negatively skewed annually. Radiation was more similar across the year but greatest in the early wet

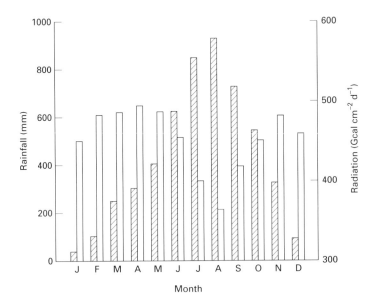

FIG. 11.5. The monthly pattern of rainfall (▨) and radiation (□) at Bulu, Ndian, averaged over the years 1973–95. (1 Gcal cm^{-2} (or langley) day^{-1} = 4.187 J day^{-1}.)

season (March–May), decreased by 34% in August (during the peak rains) and rose again in the late wet season (October–December). Radiation fell very slightly in the dry season due to the Harmattan dust.

Over the 23-year period 1973–95 total rainfall varied from a minimum of 4027 to a maximum of 6368 mm (mean = 5190, SE = 122). Three relatively dry years were 1984, 1985 and 1987. Radiation varied from 410 to 512 langleys (Gcal cm^{-2}) day^{-1} (mean = 452, SE = 5). During this period there were two 10-year phases (possibly relating to the sun-spot cycle: A. Hamilton personal communication) of increasing rainfall, from 1973 to 1983 and from 1984 to 1994/95, with the opposite trend in radiation in the second phase (overall, $r = -0.585$, df = 21, $P < 0.01$), a pattern clearly supported when monthly rainfall and radiation were plotted against time. Taking the dry season approximately as being January and February of a particular year plus the December of the previous year, and the peak part of the wet season as July–September (Fig. 11.5) dry and wet season rainfall and radiation were calculated for the years 1974 to 1995. Dry-season rainfall declined significantly over this period ($r = -0.447$, df = 20, $P < 0.05$) whilst dry season radiation increased more significantly ($r = 0.695$, df = 20, $P < 0.001$; Fig. 11.6).

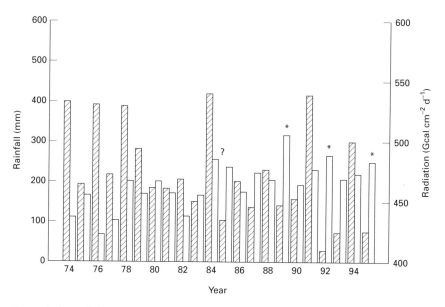

FIG. 11.6. Annual dry season rainfall (▨) and radiation (□) at Bulu, Ndian, for 1974–95. * indicates mast fruiting years. The season is defined here as December–February inclusively. Radiation units are as in Fig. 11.5. (Note that these values will not correspond exactly to the more precisely defined dry-season rainfall values in Table 11.3.) A retrospective suggestion for pre-1989 masting is indicated by the ? mark.

Neither wet season rainfall nor radiation showed any trend ($r = -0.189$ and 0.139, df = 20, $P > 0.05$). Across the dry seasons rainfall and radiation were not correlated ($r = -0.341, P > 0.05$) but for the wet seasons they were significantly and strongly correlated ($r = -0.741, P < 0.001$).

The start, duration and intensity of the dry season varied from year to year. To include the 3 mast years, daily rainfall and radiation records for 1 January 1987 to 31 December 1995 were examined for Bulu (Table 11.3). Dry season was more accurately defined by an end-run of 5–8 days of no rain (or no more than 1 day with <10 mm) and rarely a day of >20 mm 10–12 days in from either of these end-periods. Generally, this was readily applicable since the start and end of the dry season at Bulu (and Korup) were marked. For these 8 years the mean mid-dry-season date was 15 January, exactly mid-way through the period December–February (as defined above) and the mean duration was 86 days. This analysis supports the notion that on average the dry season consisted of those months December to February inclusively.

Longer rainfall records exist for Douala as complete years for 1888–1900, 1905–12 and 1922–88 (Chadwick-Healey Ltd 1992). Over these 88 years the mean rainfall was 3974 mm (range 2594 to 5327, SE = 57). From 1956–88 there was a decline from an estimated 4555–3360 mm (regression: $t_{slope} = -4.10$, df = 31, $P < 0.001$). The dry season during this period, with a mean rainfall of 150 mm, range 16 to 371 mm (SE = 14), also became drier, falling from an estimated 219 to 81 mm ($t_{slope} = -3.50$, df = 31, $P < 0.01$). In the long term there appears to have been a cyclic trend in rainfall but the data are not complete for harmonic analysis. Examining the three epochs for 1888–1900, 1928–40 and 1966–88 showed that mean rainfall

TABLE 11.3. Length and intensity of the dry season in years 1987–95 defined by daily rainfall records, and the number of days at 19°C or less (in December–February). M indicates a mast fruiting year.

Year	M	Start/end dates	Length (days)	Rainfall (mm)	Longest span no rain (days)	No. days ≤19°C	≤16°C
1987–88		9/12–5/2	59	46	28	3	2
1988–89	*	13/12–13/3	89	59	64	6	0
1989–90		22/11–29/3	128	158	39	0	0
1990–91		22/12–5/2	46	19	19	0	0
1991–92	*	3/12–7/3	96	30	58	18	2
1992–93		24/11–16/2	85	41	30	4	0
1993–94		7/12–18/2	74	80	24	2	0
1994–95	*	20/11–6/3	107	74	33	9	2

changed from 3968 to 3902 to 3764 mm in that order, but the dry season did not shift in position, or intensity at that scale, though it was possibly lower in the most recent epoch (130 for 1966–88 vs. 184 for 1888–1900 and 205 mm for 1928–40). The peak of the wet season rainfall moved $c.$ 1 month later over the century. Annual rainfall at Bulu and Douala were weakly but positively correlated for the period 1973–88 ($r = 0.434$, df = 14, $P < 0.10$).

The annual mean amount of water evaporated daily (from the Gunn–Bellani instrument) was 8.822 ml. Using the calibration of Pereira (1959) this value was converted to 452 Gcal cm^{-2} day^{-1} or langleys day^{-1} (18.9 mJ m^{-2} day^{-1} or 219 Wm^{-2}). This conversion will have over-estimated radiation when the volume evaporated was ≤ 2.0 ml. In the years 1987–95 for which daily data are available this occurred on 1.0% of days. Walsh (1996a) indicates that most tropical locations range from 350 to 500 langleys day^{-1} and Reading, Thompson and Millington (1995) suggest a mean of 20.3 mJ m^{-2} day^{-1} or 485 langleys day^{-1}. Therefore, Korup's annual radiation lies well within the expected range and, despite the heavy wet-season rainfall, radiation is not particularly low excepting some wet seasons. Monthly mean maximum temperatures, averaged over 1973–95, were highest in the dry season at 30.8°C, falling to 25.5°C in the wet season (annual mean 30.4°C). The corresponding minimum temperatures varied very little annually (mean 22.9°C) but were slightly lower in the dry season (21.3°C). The means of the monthly means of maximum and minimum varied very little (ranges and SE: 29.4–31.4, 0.09; 22.0–25.0, 0.15; $n = 23$) over the period 1973–95.

MAST FRUITING AND CLIMATE

Peak fruiting years were compared with non-mast ones for several climatic variables: dry-season rainfall and radiation (3-monthly defined); the start and duration of the dry season (precise) and the rainfall intensity and number of days without rain in that particular season; wet-season rainfall and radiation; and annual rainfall and radiation. None of these except dry-season rainfall and radiation showed any correspondence with masting for the current or previous years. Dry-season radiation peaked exactly in those 3 mast years of 1989, 1992 and 1995 (i.e. the dry season immediately prior to wet-season fruiting) and dry-season rainfall peaked the year before masting (i.e. in 1988, 1991 and 1994). After the masting year there was a steep fall in radiation and then a 2-year build-up to the next peak. Dry-season radiation was significantly positively correlated with the longest period of rainless days ($r = 0.756$, df = 6, $P < 0.05$). This variable was high in 1989 and 1992 but lower in 1995: the latter would have reached a value of 55 days comparable to the other 2 years if two adjacent days of 3–4 mm were discounted in the run. Those long runs of rainless days came early on in the dry season of masting years. They imply cloudless days of maximum radiation.

Among the temperature variables (monthly means of daily temperature maxima and minima in dry and wet seasons), minimum dry season temperature showed a striking correspondence with masting (Table 11.3). In 1992 and 1995, the monthly minima were the lowest at 16–19°C, with 1989 18–20°C. Returning to the daily minimum values for January 1987 to December 1995, the numbers of days in each dry season (December–February) which had particular temperatures were tallied. As can be seen in Table 11.3, 1992 and 1995 had many more days with temperatures of 19°C or less than the other years, with 1989 showing a smaller peak.

OIL-PALM YIELDS

From the neighbouring Ndian oil-palm plantations (PAMOL), fruit production for 1976–95 (20 years) was considered. These data came from 8 years of planting (1966–73 fields). The last year of palm production for the 1966 and 1967 fields was 1991, although the other fields were still yielding in 1995. The yields of the first 1, 2 and 3 years, respectively of the 1971, 1972 and 1973 planted fields were discounted since the palms were 5 years old or younger. Mean production was estimated for palms according to their ages (range 6–27 years) irrespective of planting date. Yield of fruits (fresh bunches,

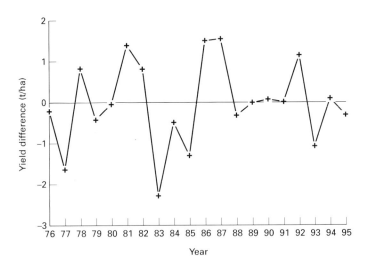

FIG. 11.7. Deviations from age-predicted yield of oil-palm bunches at Ndian for 1976–95, based on the mean of plantations established in each year of 1966–73. The mean yield was 8.37 t ha^{-1}.

t ha^{-1}) declined with palm age: it was described best by the equation; yield = 10.3 + 0.160 age − 0.0158 age^2 (R^2 = 96.1%, $F_{2,19}$ = 257, P < 0.001). Each year's yield in each field was then compared with the expected yield for its age from the general equation, and the mean deviation in yield per harvest year measured the relative over- or under-yielding. The palms yielded relatively better in 1981, 1986–87 and 1992 (Fig. 11.7) but not in 1989 and 1995 — the years in which the caesalps masted. The same climate conditions apply here as in Korup. Higher production showed a 5–6-year cycle.

SEEDLING DEMOGRAPHY

Two 2 × 2 m quadrats were randomly located in each of 33 subplots to either side of the central E–W line of the main 82.5-ha block. The quadrats were set up on 3 May 1990 (n = 132 quadrats) and all individuals (seedlings and small saplings) <50 cm in height were mapped and colour-tagged and inspected monthly (except November 1991 to February 1992) until December 1993 (40 occasions), recording mortality of the original population. From 5 November 1990, on every occasion, all recruits (new seedlings emerging) were mapped, number-tagged and their height (from soil surface to main apex) and number of leaves measured. On subsequent occasions survival was recorded and the height and leaf number remeasured. Observations were continued less frequently in May, August and December 1994. This provided a total period of 4 years 6 months of observation and 4 years 2 months in which new recruits were followed. Seedlings could rarely be identified, many dying too early, though some individuals surviving from October 1992 to the end could be named. (In two quadrats, only 8/c. 190 and 3/>c. 25 of the very abundant new *Oubanguia alata* were tagged.)

Of the 2977 seedlings present in the 132 quadrats at the start (May 1990) —mean and 95% confidence limits of 20.2, 17.9–22.7 individuals per quadrat, 343 (11.5%) had died 40 months later, 272 (9.1%) within the first 12 months. Mortality fell steeply and exponentially in those 12 months, and then a further peak in the wet season of the second year.

Over the 4 years monitored, 330 new seedlings established, of which 146 died within the period and 184 remained alive. The 1992 mast fruiting led to a large recruitment (243 individuals) whilst in the other 3 years it was poor. New seedlings most frequently died within 4 months (Fig. 11.8), in the first dry season after their emergence, with further smaller peaks in mortality in that following wet season and again in the wet season a year later. This general pattern was mirrored in the 1992 cohort (Fig. 11.8): there was a lack of deaths in the second dry season after emergence. Of the 1992 recruits, 138 (57%) were still alive after 23 months. A plot of survivorship vs. time (Fig.

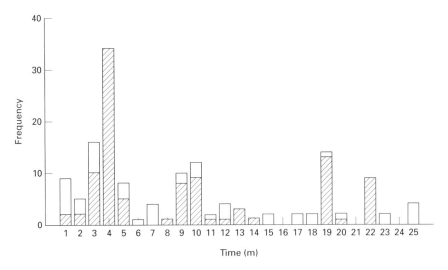

FIG. 11.8. The frequency of surviving ages to death of all seedling recruits, of which those of the 1992 mast-fruiting cohort are shown as ▨, monitored from May 1990–December 1993 in Korup on transect P.

11.9) showed the relatively large drop in February 1993. In the final extra interval (not shown on Figs 11.8 & 11.9) between August 1994 to January 1995 the overall number of new seedlings fell a further 17 to 167 (i.e. to 50.6%), and from that a tentative $t_{1/2}$ (based on the logarithmic model) of c. 4 years 3 months was calculated.

Height growth was extremely slow, and seedling recruits which died within the study period (47 months) grew at slightly, but not significantly, slower rates than those that were still alive at the end (at month 47): 1.09 ± 0.29 cm year^{-1} ($n = 146$) vs. 1.47 ± 0.17 cm year^{-1} ($n = 179$; five were new recruits in the last month recorded); $t = -1.12$, df$_{adj} = 239$, $P = 0.26$. Averaging the growth rates across seedlings within quadrats in which they occurred the means remained insignificantly different: 1.15 ± 0.27 ($n = 62$) and 1.60 ± 0.23 cm year^{-1} ($n = 72$), ($t = -1.28$, df$_{adj} = 126$, $P = 0.20$). The starting heights (plants with cotyledons or first leaf) differed little from the final heights, either the month prior to death or when still alive (recorded in the census), means $\pm$ SE ($n = 330$ quadrats): 18.3 ± 0.30 cm (range 6–41) and 20.2 ± 0.36 cm (range 7–60); $n = 330$.

Among the survivors at the end, *M. bisulcata* seedlings did not grow faster than those of *T. bifoliolata*: 1.14 ± 0.39 cm year^{-1} ($n = 30$) and 1.32 ± 0.38 cm year^{-1} ($n = 20$); $t = -0.34$, df = 46, $P = 0.74$ (27 and 19 respectively of these were from the 1992 mast fruiting cohort). There were only five *T. moreliana* identified among the survivors, but these had much higher and more variable growth rates (3.50 ± 1.34 cm year^{-1}). From characterizations of the

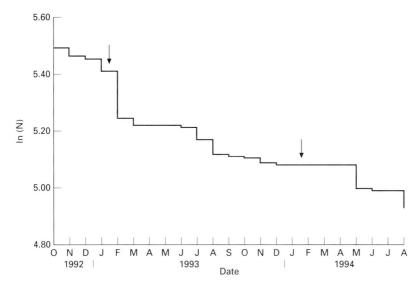

FIG. 11.9. Survivorship curve for the 1992 cohort of recruits monitored in Korup, transect P. The arrows indicate the middle of the dry seasons.

132 quadrat locations, only one seedling was in a better-lit gap: its single small recruit was 16 cm high, grew 1 cm and died after 19 months. The most growth (9.6 cm year^{-1}) was made by a larger 25-cm-tall recruit in a seasonal stream bed. For *M. bisulcata*, an approximate retrospective assessment of survivorship is possible. Taking those quadrats with identified *M. bisulcata* surviving to the end and no contrary indications that the other (dead) seedlings were not of that species, it is believed that these others, when recruited in October 1992 and of similar starting size, were highly likely to have been *M. bisulcata* too. On that basis in five quadrats, seven out of 25 *M. bisulcata* seedlings died, giving a 72% survivorship, a value above the general population average.

DISCUSSION

Phenology

The mast-fruiting pattern at Korup by the large caesalps is unusual for African tropical rainforest. Such a phenomenon has not been generally recognized for the extensive tracts of caesalp-dominated forests of central Africa (Lebrun 1936, Germain & Evrard 1956, Devred 1958, White 1983, Hart, Hart & Murphy 1989), nor for Cameroon (Aubréville 1968, 1970, Letouzey 1968, 1985). In West Africa, and Africa in general, seasonal and annual flowering and fruiting is the consistent finding (Lieberman 1982, Gautier-Hion, Duplantier, Emmons *et al.* 1985, Swaine & Hall 1986, Tutin & Fernandez 1993, Chapman, Wrangham & Chapman 1994, White 1994) though Gérard (1960) did note for *Gilbertiodendron dewevrei* (Caesalpiniaceae) in Zaïre that fruiting of this monodominant species was not always annual. On closer inspection Hart (1995) found mass fruiting of *G. dewevrei* at Ituri (Zaïre) occurring in different neighbouring patches in different years. Fruitfall in Gabon may vary from year to year but it is not supra-annual (e.g. Tutin & Fernandez 1993). In Nigeria *Triplochiton scleroxylon* was reported to flower irregularly but only occasionally to produce fruits, the loss of buds and immature fruits being ascribed to insect predation and fungal attack (Jones 1975). From a countrywide monitoring programme we noted that *T. scleroxylon* did not mast in 13 years (1962–75). Savill and Fox (1967) indicated that *Didelotia idae* in Liberia did not flower annually. Mast fruiting is such a pronounced phenomenon that it is most unlikely to have been otherwise overlooked by long-term observers.

In Central and South America the many studies of tree phenology in tropical rainforest reveal again that most species from these regions flower and fruit seasonally and annually, and many do so intermittently, continuously or subannually (Foster 1982, Milton, Windsor & Morrison 1982, Sabatier 1985, Mori & Prance 1987, Ter Steege & Persaud 1991, Lugo & Frangi 1993, Newstrom, Frankie & Baker 1994a, Newstrom, Frankie, Baker & Colwell 1994b, Peres 1994). (Ter Steege & Persaud 1991 evaluated a core 11-year period (1943–53) and Newstrom *et al.*, 1994a, followed phenology for 12 years (1969–81).) Interesting rare exceptions of mast fruiting have been reported for *Eperua grandifolia* (Caesalpiniaceae) in French Guiana by Forget (1992), at one site in 1983 and 1989, and at another in 1985 and 1989; and for nine species in the genera *Eschweilera* and *Lecythis* (Lecythidaceae) fruiting was highly gregarious in early 1982 in French Guiana (Sabatier 1985). Perhaps the clearest example was given by Wheelwright (1986) working in Costa Rica who recorded, within a 7-year period, heavy mast fruiting in the Lauraceae in 1979, 1981 and 1984, with poor crops between these years and in 1985. This latter is an example of 'alternate bearing', frequently documented for plantation fruit trees (Monselise & Goldschmidt 1982).

In SE Asia, where the climate is much more aseasonal than in Africa, Central or South America, mast flowering and fruiting has been well studied and documented (Wood 1956, Wycherley 1973, Janzen 1974, Cockburn 1975, Ng 1976, Janzen 1978, Appanah 1985, Van Schaik 1986, Ashton 1987, Ashton, Givnish & Appanah 1988), and between these events an aseasonal phenology recorded (Putz 1979). Although largely attributable to the dominant

Dipterocarpaceae of the lowland forests, the gregarious flowering every 3–10 years also included other families and the understorey (Appanah 1985). A remarkable feature of the dipterocarps is the local, staggered flowering of congenerics (Ashton et al. 1988, Yap & Chan 1990) suggesting competition for pollinators. Mass flowering is often followed by fruiting, but not always so (Wycherley 1973). The caesalps in Korup differ from the dipterocarps because the former's flowering is more regular but the fruiting is in mast years. The sequence and importance of controlling factors are likely to be different in a strongly seasonal site.

Outside of the tropics mast fruiting occurs in many large dominant tree species, most notably *Quercus* and *Fagus* spp. in the Fagaceae (Harper 1977, Silvertown 1980). In an 8-year study of three oak species in North America, Sork, Bramble and Sexton (1993) found that individuals within any one species tended to produce acorns in abundance (mast) together but the species had different intervals: *Q. velutina* 2 years, *Q. alba* 3 years and *Q. robur* 4 years; and this meant that in each year there were generally some acorns of one or more species at a site. Sork (1993) has speculated that mast fruiting might have evolved in tropical oaks in Central America but there is no actual evidence that these species exhibit supra-annual flowering and/or fruiting. A quite distinct occurrence of masting was recorded by Herrera, Jordano, Lopez-Soria & Amat (1994) for the mediterranean tree *Phillyrea latifolia* (Olacaceae) which produced large seed crops just twice in 15 years, in 1981 and 1989. Norton and Kelly (1988) reported on 33 years of phenological records for *Dacrydium cupressinum* (Podocarpaceae) in New Zealand finding that in six of the 15 mast years these followed a previous masting (i.e. several pairs of high-yielding years).

Evidence for masting, both tropical and temperate, has shown mixed patterns and characteristics in terms of synchrony and timing. This suggests that several causes operate, perhaps in varying combinations from site to site. The supra-annual fruiting pattern at Korup does not match any of the other examples mentioned as it involves three codominant species, which formed a guild through their ectomycorrhizal habit, and shared a 3-year cycle. Masting was not a completely forest-wide phenomenon, the three species if they did mast at all did so in the same year and it was not necessarily a consequence of mass flowering first. The mass effect appears to have been limited to the fruiting stage and thus, presumably related to the energy available for pod and seed maturation.

That mastings were precisely 3 years apart may have been a coincidence for those years observed (1988–95) but at least for those three events there was a clear association with dry-season peaks in radiation and minimum temperatures. These two parameters are not unrelated as clearer skies in the dry season mean greater night-time long-wave radiation. Extrapolation back over time from Fig. 11.6 to predict possible previous mast fruitings is less certain, as both 1984 and 1985 had relatively high dry-season radiation and, prior to these years, the peaks and troughs in radiation were lower overall and less clearly patterned. A prediction of masting in 1985 would match the findings of Tutin and Fernandez (1993) for high fruit crops in Gabon in 1985–86, and in 1989. Our prediction is that 1998 will be the next mast year for Korup.

Radiation may be directly and simply linked to fruit production. The considerable dry weight of the heavy fibrous pods and their seeds in these species (studies on its quantification are in progress) would be expected to create a considerable drain on tree resources and it may be that these large legumes need to accumulate reserves to a threshold level before fruits are produced. Several authors have commented on this question of carbon balance (Harper 1977, Longman & Jenik 1987, Norton & Kelly 1988, Sork *et al.* 1993, Van Schaik, Terborgh & Wright 1993). Wright and Van Schaik (1994) suggested that tropical rainforests in general are radiation-limited, especially under cloudy conditions, and in most seasonal habitats 1.4–3-fold more new leaves and flowers are produced in the dry than wet season; and White (1994) showed a positive correlation between number of sunshine hours and fruit maturation in the wetter months in Gabon. Wycherley (1973), and then Ng (1976), proposed that recent accumulation of sunshine hours was the underlying cause of mass flowering and fruiting in SE Asian dipterocarps (see also Van Schaik's (1986) re-analysis of Ng's data). Other support for the role of radiation comes from Baker (1965) noting that *Ceiba pentandra* in West Africa had annual flowering in dry habitats but supra-annual in wetter ones; and Yap and Chan (1990) commented that in general supra-annual flowering tended to increase from drier to wetter forests in Peninsular Malaysia.

Minimum temperatures may well cue flowering as suggested by Ashton *et al.* (1988) for dipterocarps and this was also used to account for variation in seasonal fruit crops in the dry forest in Gabon by Tutin and Fernandez (1993). However, as shown for Korup in 1991, heavy flowering was not always followed by mast fruiting in *M. bisulcata*. Presumably the resources were not sufficient, either immediate or stored, in those years of non-fruiting. The coincidence with minimum temperature and flowering is not convincing in our case since in the 1991 dry season there were no days with temperatures $\leq 19°C$ and the trees of *M. bisulcata* still flowered. Conversely, in 1988 there were two very low temperature (16°C) records (not made within the Korup stand of trees), combined with an otherwise warm dry season and there was no mass flowering. The low temperatures associated with fruiting years seem to be inversely related to the high radiation peaks, and the evidence is not convincing enough to suggest that these low temperatures alone cue flowering. From an evolutionary view-point low temperature would be an unreliable cue if subsequent fruiting were then energy- and resource-constrained.

An important consideration is that these masting trees are, in the main, very large, especially *M. bisulcata* which attains almost the largest dbhs of forest trees in SW Cameroon (Richards 1963, Letouzey 1985). Very large trees will suffer the greatest respiratory loads (Kozlowski, Kramer & Pallardy 1991) and have to support a large biomass of ectomycorrhizal fungi (Newbery *et al.* 1988). Furthermore, *M. bisulcata* and *T. moreliana* are deciduous and microphyllous, and there will be some loss in photosynthetic capacity during refoliation, though it is clear that they do refoliate in the dry season which is a time when the young leaves can make maximum use of the higher radiation. The timing of the new leaves (early or late dry season) could determine the energy gain in that period, and this probably depends initially on the pattern of rainfall in the dry season and its effect on tissue water potentials controlling growth (Borchert 1992, 1994). Our data are not detailed enough to test this aspect. The one pronounced strong annual dry season at Korup drives the cycle of litterfall for the whole forest, for these large canopy and emergent trees in particular, and it regulates the annual nutrient cycles that are dependent upon decomposition of the litter (Newbery *et al.* 1997).

With several slightly varying climatic parameters, every year will differ slightly from the next in the timing and intensity of the dry season. Given also that these trees vary in their size and precise topographic position, and will therefore differ in susceptibility to the dry period, it is surprising that the individuals of each species are synchronous and that the three species fruit together in mast years. However, not every mast year included all three species: two missed or skipped the event. If *T. bifoliolata* did not fruit heavily in 1989 why was the next fruiting not until 1992 and none in 1990 (with one more year of energy intake)? Likewise *T. moreliana* missed 1992 and could perhaps have fruited in 1993, not 3 years later in 1995.

Build-up of pollinators to large densities could be essential in mast years and might be started by the flowering of *M. bisulcata*. The sequence of flowering was *M. bisulcata* followed by *T. bifoliolata*, and rather tentatively *T. moreliana* could be placed between them. The suggestion is that *M. bisulcata* flowers each mast year, and the others may follow if ready. This separation may indicate competition for pollinators in the past leading to the tree species' temporal niche partitioning with the advantage to *M. bisulcata* in flowering first. Pollinators are believed to be small bees, which are very evident in the dry season (D.M. Newbery, personal observation) and, for such outcrossed caesalp flowers (Polhill, Raven & Stirton 1981), generalist bees are the most likely agents (Roubik 1989). Furthermore, the sequence of flowering (for *M. bisulcata* and *T. bifoliolata* at least) is the *reverse* of fruit maturation and seed fall: *M. bisulcata* needs the longest time to produce its pods, and *T. bifoliolata* the least. This indicates that *M. bisulcata* may be limited in carbohydrates from its tree reserves, even within a mast year. *T. bifoliolata* appears more 'efficient' in reproduction. Shorter pod maturation times would be an advantage in a seasonal site where an earlier germination leads to a better establishment before the dry season.

Considerable attention has been given to the predator-satiation hypothesis of Janzen (1974, 1978), developed largely to explain the mast flowering, and then fruiting, events for dipterocarps in SE Asia. Predators may similarly have shaped the phenological traits of many large and dominant temperate tree species (Silvertown 1980). It is highly likely that once a supra-annual pattern (i.e. one with events more than 1 year apart) has been established a mass of seeds could saturate predator requirements and lead to more survivors than a trait in which the energy required for fruiting had been allotted annually, over the mast and intermast years. However, the mechanism, in terms of tree physiology, remains completely unknown. Large trees, it can be postulated, need to build-up carbohydrate reserves for fruiting, over 3 or possibly 4–5 years, and this could be achieved for different individuals and species over different time-periods. Therefore, the cue or mechanism that allows energy build-up must be more fundamental, and it must be common to all species to enable synchrony. Since in Korup the year of masting is the one of peak radiation, it appears that this year has more weight than the previous ones because either (i) it is a direct cue to flowering; or (ii) the effect of energy intake in the previous (2) years is diminished over time because the energy is shunted elsewhere within the tree.

Reversing the perspective, it may be more the case that a previous masting leads to strong depletion and the intermast years are required to replenish the reserves (Harper 1977, Rathcke & Lacey 1985), and only when that adjustment in carbon balance has been made, can the tree respond to an immediate peak in radiation, perhaps combined with a temperature cue for flowering in a non-depleted year. Weather (i.e. radiation) and timing of the dry season are likely to entrain all large trees into a similar temporal pattern and pollination and predator satiation act as 'governors' on the system, further stabilizing and synchronizing it (Van Schaik *et al.* 1993, Waller 1993). The conclusion for Korup accords with Harper (1977) that weather acts as a coarse timer and predation 'fine-tunes' the system. The model of Lalonde and Roitberg (1992) bears out this notion. The important suggestion from the Korup data is that large trees have too great a demand on their maximum energy input to allow effective fruiting every year.

Radiation at Korup is not particularly limiting by tropical norms but the resource which is in short supply to forest growth is phosphorus (Newbery *et al.* 1997). After the 1989 mast fruiting there was a loss of the normal, annual, high peak in phosphorus concentration in the litter of ectomycorrhizal trees suggesting that they had depleted their phosphorus content by the reproduction. On that basis we can surmise that it is probably not just carbon that regulates masting but phosphorus within the tree. And possibly, further, it is phosphorus or the C:P balance that is the control. This idea is linked to the 'phenological and climatic ectomycorrhizal response' (PACER) hypothesis put forward by Newbery *et al.* (1997) to explain the short-term decline in labile soil phosphorus after mast fruiting of the large ectomycorrhizal caesalp trees. The relatively wetter dry seasons in the years prior to those in which masting occurred might also have meant an alleviation of water stress in those years, which facilitated nutrient uptake.

The oil-palm plantation 10 km away at Ndian forms an important comparison with the forest. That age-adjusted oil-nut yields did not match the masting of the caesalps suggests that radiation alone was not the controlling factor for the caesalps. Until 1982 these plantations were regularly fertilized with superphosphate, and then with smaller applications until it was stopped in 1986 (I.N. Timti personal communication). Whilst this might in part explain the curve of decreasing yield with age it is unlikely that phosphorus reached the very low, and putatively limiting, levels of the caesalp groves in the forest. This supposition tentatively points to phosphorus, as well as carbon, controlling mast fruiting in the forest. In Fig. 11.7 the very low yield in 1983 corresponded to a drought when many palms died (I.N. Timti personal communication).

It must be asked why mast fruiting has been so little noted elsewhere in Africa yet it is so pronounced at Korup. Is it a temporary recent phenomenon? Would stands of younger trees exhibit it? More especially, is the particular nature of the site at Korup, with its sandy, highly leached, low pH soil with low phosphorus content, combined with the ectomycorrhizal habit, deciduousness and the one strong dry season per annum, able to give an insight into mast fruiting generally? Van Schaik *et al.* (1993) have drawn attention to the possible role of low soil nutrients in masting. The dry season places emphasis on the role of carbon, water and phosphorus storage by the ectomycorrhizas (Alexander 1989b, Newbery *et al.* 1997) and if dry seasons have become more pronounced in recent decades, combined with the evident maturity of the stand of *M. bisulcata*, this may have inflicted further stress on trees in an already edaphically extreme site. In less phosphorus-limited forests mast fruiting is not evident and apparently follows the typical annual pattern. In SE Asia the mast fruiting is largely in the Dipterocarpaceae and this family is also ectomycorrhizal. In the temperate forests the Fagaceae (oaks and beech) are also strongly ectomycorrhizal (Harley & Smith 1983). This association between masting trait and ectomycorrhizal is possibly of wide ecological significance.

Other evidence of phosphorus being a controlling factor in tropical tree phenology is as yet very scarce. Zagt (1997) has recently suggested that limitation of phosphorus in poor soils at Mabura Hill, Guyana, may be the cause of biennial flowering (but not fruiting) in the locally abundant canopy tree *Dicymbe altsonii* (Caesalpiniaceae) as its flowers involve a high investment of that element. This species appears, however, not to be ectomycorrhizal (R. Zagt & H. Ter Steege, personal communications): indeed not all tribes of the Caesalpiniaceae are ectomycorrhizal (Alexander 1989a). Further comparisons between South America and Central Africa would be very valuable with respect to the phenology of caesalp-dominated stands and phosphorus-poor soils.

The Korup findings corroborate the hypothesis that weather variables act primarily on tree physiology and resource allocation and are the driving factors to mast fruiting, whilst predators and pollinators are moderators. The seeds of the caesalps are very poorly dispersed, the pods twist and crack open on hot afternoons in the wet season, and mostly fall within the area of the canopy. They germinate almost immediately and the pods fall later, making a layer of fibrous material. Seeds are not strongly mechanically protected or animal dispersed, which links with the very strong clumping of the three species. As yet we have no data on seed removal by predators (an exclosure study on the 1995 crop is in progress), but to test Janzen's (1978) predictions would require following the fate of seeds from isolated and asynchronously fruiting trees compared with masting ones. Primates are known to feed on immature caesalp pods and these later are found on the forest floor in some years (J.A. Rother & J.J. Green personal observations). This is better documented by Gautier-Hion, Gautier and Maisels (1993) and McKee *et al.* (1981) in Gabon and Cameroon pointing to the very likely selective role of predispersal predation, especially in former decades when animal populations in Korup were higher (Gartlan 1986, 1992). It is not uncommon to find seeds attacked by bruchid beetles (D.M. Newbery, personal observation). Similarly, post-dispersal predation by duikers and other animals may play, and may have played, a role; this and predispersal predation remain uninvestigated in Korup.

Harper (1977), Janzen (1974, 1978), Norton and Kelly (1988) and Waller (1979) have discussed the idea of 'economy of scale' which here means that it is more efficient (i.e. more seeds will survive to seedlings) to produce a large crop every few years rather than fewer seeds annually. (Predator satiation is one factor in this economy.) But this cannot explain the synchrony of mast fruiting although once cued it may contribute to entrainment. In different sites different economies may predominate, for instance wind pollination may be more efficient in mast years for temperate tree species (Norton & Kelly 1988, Sork *et al.* 1993; modelled by Smith, Hamrick & Kramer 1990).

Korup is considered as having been a remnant refugium of tropical rainforest in Africa in the late Quaternary (Maley 1987, 1989) and it lies at the centre of the Guinea–Congolean block. Thus, there is reason to believe that the caesalps evolved *in situ*. The deciduous nature of *M. bisulcata* and *T. moreliana* might indicate tolerance of drier marginal block conditions but it is unlikely that even during the last African dry period the refugium had an aseasonal climate (Maley 1987). Migration from outside or back into the Korup area seems unlikely. This historical aspect contrasts strongly with the case for the Dipterocarpaceae, which Ashton (1987) suggested originated in the Indian subcontinent and migrated into SE Asia, moving from a strongly seasonal (and drier) climate into an aseasonal and wetter one. With them the species retained (argue Ashton *et al.* 1988) the requirement for a temperature cue for flowering. This contrasting evolutionary situation begs a different hypothesis for the caesalps in Korup, unless all tropical seasonal habitats call for a temperature cue and this is over-ridden in Korup by the limitations of resources at the site.

Regeneration

The size distributions of trees ≤50 cm dbh (Fig. 11.2) suggest that *M. bisulcata* will not retain its dominance after the large trees have died; there will be very few replacements in the canopy as there are very few stems in the 1–10 cm dbh class (see Table 11.1). Even with 100% survivorship of the 5 ha^{-1} stems in that smallest class this would be insufficient. Combined with the absence of other *M. bisulcata* stems outside the groves the prediction is that this local population will decline to a few scattered individuals. The 10–50 cm dbh class is particularly poorly represented. The mode of the almost symmetrical size distribution of *M. bisulcata* in Fig. 11.1 is *c.* 90–100 cm dbh and without dating these trees cannot be aged. Part of the spread of this peak could be due to individuals of similar ages which vary in size as a result of differences in the resources they could access (especially nutrients) and their topographic positions. *T. bifoliolata* and *T. moreliana* show a different distribution being better represented in the 50–80 cm dbh class (Fig. 11.1) and in the 10–50 cm one (Table 11.1; Fig. 11.2) but with evidence of a peak for

T. moreliana in the intermediate sizes, and *T. bifoliolata* showing continuous replacement. Taken together these three species (as a guild) show a sequence of likely replacement: *M. bisulcata*, then *T. moreliana* and then *T. bifoliolata* (Fig. 11.3). Caution is required since survivorship between size (age) classes is unknown, and *T. moreliana* and *T. bifoliolata* may not be able to achieve the largest sizes of *M. bisulcata*. (*Large M. bisulcata* trees are always supported by very extensive buttresses.) The conclusion is that the ectomycorrhizal patch will pass from *M. bisulcata* dominated forest to one of *T. moreliana* and *T. bifoliolata* in perhaps 100–200 years time.

After the 1992 mast fruiting and just before the 1995 one (2.5 years later) all three species had appreciable densities of small saplings >30 cm in height to 1 cm dbh (Table 11.2) with *T. bifoliolata* the greatest, *T. moreliana* intermediate and *M. bisulcata* lowest. This order matches that found in the main 1990–91 enumeration of 1–10 cm stems (Table 11.1). These saplings would have been several years old and spanned at least the 1989 cohort and probably earlier from 1985/86. Similarly, the ≤30 cm height class in March 1995 would probably have been composed of 1989 and 1992 individuals. Given that 43% of the 1–10 cm dbh class in Table 11.1 were 1–2 cm dbh and assuming some correspondence across the non-sequential dates of measurement for successively larger (older) size classes, the decline in densities of all three species is appreciable, especially for *M. bisulcata*.

In 1995 all three species masted and the new recruits 1–2 months later gave higher densities for *M. bisulcata* and *T. bifoliolata*, but much lower for *T. moreliana*. This might simply have been variation in seed production that year, since densities of adult trees (taking those ≥50 cm dbh) are similar for *M. bisulcata* and *T. moreliana* but 32% lower for *T. bifoliolata* (Table 11.1). Nevertheless, in present times *M. bisulcata* can produce as many potential recruits as *T. bifoliolata* and *T. moreliana* but its survivorship is markedly poorer if the three classes in Table 11.2 are placed in an approximate time sequence before the 1–10-cm class in Table 11.1. Possibly post-dispersal predation of *M. bisculata* is higher than the other species. Whilst *T. moreliana* is lowest in density of new seedlings its survivorship is higher. *T. bifoliolata* achieved a similar density (as *T. moreliana*) at the pole size (mid 1–10 cm and 10–30 cm dbh classes) but by having a greater initial density and lower survivorship. The crucial stage in the differential selection of these three species appears to be between 30 cm height and the 1–10 cm dbh class, possibly in the 2–10 years age range, as saplings. The higher density of >30 cm height to 1 cm dbh saplings than ≤30 cm height seedlings probably reflects a successful cohort before 1989. In the absence of precise demographic data over 20 years this is as much as can be safely inferred.

The seedling demographic study was set up between mast fruitings,

6 months after that in 1989, and the survivorship of that bank of seedlings/saplings to 50cm height was a surprisingly high 88% after 40 months. Similarly, of the new cohort in 1992, 57% were alive after nearly 2 years. The largest initial hurdle for seedlings was the first dry season they encountered (Figs. 11.8 & 11.9), presumably many succumbing to water stress, and then in the first wet season more were lost probably from either low light levels in the shade or possibly, but less likely, flooding. The second dry season showed almost no losses, from which it is interpreted that the survivors had either a deep enough root system or, if mycorrhizal, were by then linked to the parents via hyphae. Further losses occurred in the second wet season. However, the 1993 dry season was not particularly dry nor was the 1993 wet season particularly wet. Unfortunately, the sample was neither large enough nor were the seedlings identified early enough to compare the three ectomycorrhizal species throughout. (Species-specific studies following the 1995 cohort are now in progress.) The demographic results support the enumeration and seedling census data showing that once established they mostly survive well but grow little in the first 3–4 years. The better growth rate of the few *T. moreliana* identified supports the idea that although of lower density this species survives the best.

The forest in the area of high ectomycorrhizal tree density is very closed. In the period 1990–93 there was just one new gap from a large treefall in the main plot which was being monitored. Therefore, the vast majority of the understorey is in shade (c. 91% of forest floor <2% of above-canopy photosynthetically active radiation; J.J. Green personal communication). The densities and closeness of large stems in the 82.5-ha plot also confirm this structure. However, the three large caesalps under discussion do respond to light. In a nursery experiment at Mundemba (close to Korup), and complementing a transplant experiment into the field (I.J. Alexander, J.A. Rother & D.M. Newbery unpublished data) 10 individual wildings from the forest, of the 1989 masting 1 month before, were grown from 4 October 1989 under palm-frond shading giving c. 20–30% sunlight. After c. 18 months (23 April 1991) the mean (and SE) above-ground dry weight yields were: *M. bisulcata* 28.0 ± 3.0 g, *T. moreliana* 22.6 ± 3.3 g and *T. bifoliolata* 17.8 ± 3.1 g. (The corresponding mean weights on 4 October 1989 were: 1.1, 1.0 and 0.8 g respectively.) Putting this together with the forest seedling data suggests that *M. bisulcata* is the most light-responsive and least shade-tolerant, and *T. bifoliolata* is the least light-responsive and relatively more shade-tolerant with *T. moreliana* in between and possibly more similar to *T. bifoliolata*. In Southern Bakundu Forest Reserve (south of Kumba and a much drier site with 1930 mm annually; Songwe, Fasehun & Okali (1988)) there is a well-established plantation stand of *M. bisulcata* originating from 1982 and growing in the open. *M. bisulcata* also occurs as scattered trees within this reserve.

The size distributions of the three large caesalps in the new 82.5-ha plot support the results of the earlier more widespread small-plot sampling of Gartlan *et al.* (1986), analysed further in Newbery and Gartlan (1996). The crucial issue is why *M. bisulcata* is presently not replacing itself and/or why there is an unusual peak in the frequencies of its largest trees today. This pattern can be contrasted with that for the two *Tetraberlinia* spp. Two hypotheses may be advanced: (i) that in the past there was a period of favourable conditions for *M. bisulcata* recruitment; and (ii) in recent decades, perhaps the last century, the conditions allowing *M. bisulcata* to be recruited have gradually diminished. Whilst size cannot be equated exactly with age, on this site of probably slow growth (on account of its limiting low soil fertility) it is unlikely that some trees in an age class will have had substantially improved increments over the others. Nevertheless, the peak distribution for *M. bisulcata* may be 'sharper' for ages than sizes, and it may be less positively skewed. The peak probably reflects a range in tree ages.

The first hypothesis can be supported if there is evidence of a major disturbance which allowed more light to the ground, not for just 1–3 years, but perhaps repeatedly over 10–20 years so that seedlings could establish into large saplings capitalizing on the increased light levels. A chance storm would create much damage yet like a fire lead to the rapid growth of a pioneer then secondary vegetation giving less light again to the forest floor. Such windthrows are not recorded for this part of Africa. In only one of 18 pits dug (Newbery *et al.* 1997) was there some charcoal but that could have been due to a lightning strike on a single tree (such deposits can be found on the forest floor; D.M. Newbery personal observation). *M. bisulcata* itself with its buttressing would be highly resistant to wind. As far as we are aware man has not interfered with this forest (Gartlan 1986) and its very poor soils are unlikely to have supported crops. The wet season and the rivers make it uninhabitable, especially when there are better soils just east of Mundemba, although the nearest village still within Korup is c. 8 km NW of the centre of transect P, on the Nigerian border. The large area of the patch on transect P and the existence of at least one other we have found suggest this was not an old village sites. No artefacts have been found. There are also very few trees of *Lophira alata* in southern Korup (Newbery & Gartlan 1996); this species was shown by Letouzey (1968) to be a strong indicator of past human disturbance in central-west Africa dating past cultures in *la forêt biafriéene*.

A much more likely cause is drought, or series of droughts, which substantially defoliated the forest for longer than the normal dry season. The importance of drought in determining the structure and floristic composition of lowland forests in NW Borneo has been put forward by Walsh (1996b) and Newbery, Campbell, Lee et al. (1992) and Newbery, Campbell, Proctor and Still (1996). The deciduousness of the caesalps at Korup, especially the microphyllous *M. bisulcata* and *T. moreliana*, and the ectomycorrhizal habit (suggested as partly an adaptation to dry periods; Newbery et al. 1997) mean that as adults they would have survived better. Quinn (1992) has analysed water-level records for the River Nile back to 1500 and related these to El Niño Southern Oscillations (ENSOs) (see also Diaz & Pulwarty 1992). There were two intense epochs of ENSOs (i.e. dryness) in Africa: 1765–99 with nine ENSOs (two very severe, five moderate/severe, two moderate) and 1692–1701 with three ENSOs (one very severe, one moderate/severe and one moderate) (Quinn 1992). We suggest that Korup, along perhaps with the coastal Atlantic forest in general, was affected by these events and the southern part of the National Park was particularly prone because of the very sandy, well-drained soil. The 1765–99 epoch seems the most pronounced.

The following scenario is suggested. In the 18th century the coastal Atlantic forests were more mixed and less gregarious than those recorded by Letouzey (1968, 1985). *M. bisulcata* formed occasional large trees. Given increased radiation leading to copious seed production (perhaps annually) there would have been many seedlings on the forest floor. These large trees can survive the dry periods, where many other species cannot, and their seedlings may well be attached to the parent hyphal system so that they can receive water and nutrients and have good survivorship. In this way the ectomycorrhizal species would be selected over non-ectomycorrhizal species in drier periods. Högberg (1986) has suggested a similar role for ectomycorrhizas in East African miombo woodlands. Is it therefore possible that the spread in dbhs from c. 80–130 cm dbh represents the recruitment of trees over that 35-year epoch. (On that basis the present trees should be clumped around the past parent locations.) Once the forests returned to a wetter regime, with shorter, less-intense, dry seasons, *M. bisulcata* would have lost its advantage and early establishment would have been prevented by heavier shade, at least until recently. The same arguments may apply to *T. moreliana* and *T. bifoliolata* because they too have relatively fewer trees in the 10–50-cm class (compare Table 11.2 densities × 10 with Table 11.1 and Fig. 11.1), but their dependence on light and subsequent susceptibility to shade may have been less than *M. bisulcata*. *T. moreliana*, like *M. bisulcata*, is microphyllous and deciduous and would be expected to respond similarly to dry periods. Being a closely related guild we would expect intense competition, especially at the pole stage for nutrients, and this might explain the smaller tree sizes of the *Tetraberlinia* spp. if they expanded under the same stimulus in the late 1790s. Nothing is known about the relative efficiencies of the ectomycorrhizas on the three host species.

The second hypothesis would require a factor which disadvantaged *M. bisulcata* recruitment compared with *T. moreliana* and *T. bifoliolata* gradually over time. It would imply that if the adult populations (trees ≥50 cm dbh) had the same density in previous times as today they also must have had a very considerable density of smaller trees to provide continuous replacement (Newbery & Gartlan 1996). This requirement could have been moderated by either very high survivorship (i.e. one not so steeply exponentially declining with age along with the forest in general) or the species had very high growth rates, neither of which are tenable. From the Douala rainfall data there is no evidence of a major change in dry season in the long term but over the past 30–40 years rainfall in the dry season decreased and this was paralleled at Korup by an increase in dry-season radiation. However, during this period and the last 100–150 years the stand has become progressively more closed and, whilst drier periods might select for *M. bisulcata*, this has probably been outweighed for the most part by the effect of the increasing canopy shade on recruits. In the past 10 years these lighter dry-season conditions might explain the recent high sapling bank as the balance moves again, temporarily perhaps, in favour of *M. bisulcata*.

Synthesis

Three patterns have been shown for the large caesalps in Korup: (i) they grow gregariously in large patches; (ii) they show a discontinuous size distribution of the most abundant species; and (iii) they exhibit mast fruiting. Statistically, the data do not yet have the confidence of replication across the landscape; similar studies in other patches are needed (one is in progress 5 km south of the main study site). Observations on masting have only been made for the past 10 years and the patch and size distributions are singular. However, these patterns are strong enough in Korup to form the basis for wider hypotheses. How may the three findings be linked together? Returning to the points in the Introduction, this forest type shows the relative simplicity of a codominated stand (spatial delimitation, floristic composition) and clear differences in size distribution (and arguably ages) between the three species. The guild of three codominants exhibits major characteristics interpretable as adaptations to the edaphic and climatic conditions of the site (*viz*. ectomycorrhizal habit – phosphorus stress, deciduousness/ seasonal litterfall – water), and the forest is in a state of non-equilibrium.

The structure and inferred dynamics of the forests in Korup and Douala-Edea within the Atlantic coastal forest, *la forêt biafriéene* of Letouzey (1968,

1985), in Cameroon were discussed recently by Newbery and Gartlan (1996) in more detail. Lack of recruitment of smaller trees (10–50 cm dbh) is much more widespread than previously acknowledged in the tropical literature and is certainly not restricted to *M. bisulcata*. For this species we do have the detailed data in the seedling/sapling and pole size classes (<10 cm dbh) which are lacking for the other species in the original surveys analysed in Gartlan *et al.* (1986) and Newbery *et al.* (1986) and therefore statements about their potential regeneration are made with much less confidence. For many of those more evenly and sparsely distributed canopy species, low sampling is a problem. However, in the Caesalpiniaceae, Letouzey (1968) has remarked on the gregariousness of many species within the caesalp-rich *forêt biafriéene*, with different species dominating or codominating patches to varying degrees and their regeneration in some places very good and in others absent. Korup does have a mixture of canopy codominance and different degrees of replacement of the three species and can serve as a model for the gregariousness in the forest type.

It is not known (and very unlikely to be known since conditions for pollen deposits are so scarce) whether these patches were more common in earlier millennia and whether those seen today are remnants, the last surviving patches. Alternatively, the forest may be continuing in an overall constant state and these patches have been cycling (*sensu* Aubréville 1938, Watt 1947, Richards 1952), species succeeding to dominance and then decaying through inhibition of their own seedlings and being replaced in some patches by the same or other species within the guild. Could *M. bisulcata* replace *T. bifoliolata* again in the future? Korup may well exhibit a combination of these two long-term dynamic processes.

The gregariousness is, to a large extent, undoubtably due to the poor dispersal of the caesalps, but then a large bank of seedlings is not a guarantee of replacement, as *M. bisulcata* illustrates. How then did new patches form on the landscape to enable a species which decayed in a patch to survive elsewhere? That may have involved agents of dispersal which perhaps no longer exist: could the defaunation of these African forests in the last two centuries have resulted in species with other, also poor, means of seed dispersal staying patch-bound?

Broader evidence and discussion for the 'Aubréville phenomenon' and a review of the atlantic coastal forests of Cameroon in this context can be found in Newbery and Gartlan (1996). Aubréville's ideas are central to the notion of equilibrium/non-equilibrium in tropical forests. A recent additional study by Poorter, Bongers, van Rompaey & de Klerck (1996) lends support to this phenomenon in Africa. In the Taï National Park, Côte d'Ivoire (also a primary forest) at three sites, size distributions for eight common canopy species showed two that lacked regeneration. However, Taï is a much drier site than Korup (2100 mm annual rainfall), is less seasonal, with better soils and consequently has fewer Caesalpiniaceae.

It has been supposed that an ectomycorrhizal network of mycelia may connect seedlings to adults and in this way transfer carbon to seedlings in the shade and prevent their death (Read 1989). However, if this mechanism actually exists it is presently of no evident benefit to *M. bisulcata* since its seedlings are not surviving to saplings and small trees. If the large emergent trees of this species can be viewed as 'over-mature' and their carbon costs are high compared with their fixation, then in the wet season the source–sink relationship might be reversed and the carbon would be moved from the seedlings to the larger trees. Mast fruiting involves a large drain on adult resources every 3–5 years and it is pertinent to note that seedlings survived until the next mast year and then appeared to decline. Being connected to the network of an ageing closed stand would be disadvantageous for the seedlings and could be viewed as an evolutionary cost for those benefits conferred in other stages of the tree-growth cycle. This forms a testable hypothesis for the Aubréville phenomenon, trenching experiments might be a first approach.

More likely direct benefits from being ectomycorrhizal in such a phosphorus-poor site and seasonal climate might be enhanced phosphorus acquisition, possible cycling of phosphorus in organic form and direct phosphorus access (yet to be definitely demonstrated), and the storage of water and nutrients to assist trees, especially saplings and poles, through the dry season (Newbery *et al.* 1997). The strongest positive connections may not be at the seedling stage but for the small trees when response to light and competition for onward growth are essential. This hypothesis would explain how light-responsive but shade-intolerant ectomycorrhizal species can form a dominant stand and then decline. A similar process might operate for *T. moreliana* and *T. bifoliolata* but they are not so large in size (as yet) and therefore do not exert the same average drain on resources per tree.

Since dispersal is poor most seedlings of *M. bisulcata* will be clustered around that species' adults (at the scale of 30–50 m radius) and so will the others species' seedlings similarly be close to the adults. Thus, two tests or predictions would be: (i) seedling mortality will decline with increasing size of adult *M. bisulcata*; and (ii) in places where neighbouring *T. bifoliolata* or *T. m. liana* trees happen to leave seedlings close to very big *M. bisulcata* trees their survivorship should be reduced too. The large ectomycorrhizal caesalps apparently share several species of fungal symbiont (I.J. Alexander personal communication) and their fungal hyphal connections probably interconnect; this poses a fascinating dimension to competition between the tree species and how natural selection operates within such a guild.

ACKNOWLEDGEMENTS

This research was financially supported by contract SDT2*0246 UK (SMA) from the European Commission (DGXII/G4). We are grateful for the assistance of P.J. Fraser, S. Lorenz, P. Taylor, E. Abeto and P. Ekondo in the field, to P. Mezili and the Herbier National de Yaoundé for taxonomic support, I.N. Timti of Plantations Pamol du Cameroun Ltd for making available the climatic data for Bulu and the Ndian oil-palm yields, to WWF–Korup Project in logistic matters, to J.A. Rother for valuable discussions and access to unpublished data, R.P.D. Walsh for making the Douala climatic data available to us, and to I.J. Alexander, N.D. Brown and J.J. Green for suggestions on the manuscript. The reviewers A. Hamilton, J. Hall, K.A. Longman and J.-P. Pascal offered valuable and constructive comments. We thank A.J. Ayuk-Takem, Director of the Institute de la Recherche Agronomique, for facilitation of our programme and the Ministries of Parks and of Education, Science and Technology for permission to work in Korup.

REFERENCES

Alexander, I.J. (1989a). Systematics and ecology of ecto-mycorrhizal legumes. In *Advances in Legume Biology* (Ed. by C.H. Stirton & J.L. Zarucchi), pp. 607–624. Monographs in Systematic Botany, Missouri Botanical Garden, MO.

Alexander, I.J. (1989b). Mycorrhizas in tropical forests. In *Mineral Nutrients in Tropical Forests and Savanna Ecosystems* (Ed. by J. Proctor), pp. 169–188. Blackwell Scientific Publications, Oxford.

Appanah, S. (1985). General flowering in the climax rain forests of South-East Asia. *Journal of Tropical Ecology*, **1**, 225–240.

Ashton, P.S. (1987). Dipterocarp reproductive biology. In *Tropical Rain Forests Ecosystems: Biogeographical and Ecological Studies* (Ed. by H. Leith & M.J.A. Werger), pp. 219–240. Elsevier, Amsterdam.

Ashton, P.S., Givnish, T.J. & Appanah, S. (1988). Staggered flowering in the Dipterocarpaceae: new insights into floral induction and the evolution of mast fruiting in the aseasonal tropics. *American Naturalist*, **132**, 44–66.

Aubréville, A. (1938). *La Forêt Coloniale: Les Forêts de l'Afrique Occidentale Française.* Société d'Editions Géographiques, Maritimes et Coloniales, Paris.

Aubréville, A. (1968). Les Caesalpinioidées de la flora Camerouno-Cangolaise. *Adansonia* (série 2), **8**, 147–175.

Aubréville, A. (1970). *Légumineuses–Césalpinioidées. Flore du Cameroun*, **9**. Muséum National d'Histoire Naturelle de Paris.

Aubréville, A. (1971). Regeneration patterns in the closed forest of Ivory Coast. In *World Vegetation Types* (Ed. by S.R. Eyre), pp. 41–55. Columbia University Press, Ithaca, NY.

Baker, H.G. (1965). The evolution of the cultivated kapok tree: a probable West African product. In *Ecology and Economic Development in Tropical Africa* (Ed. by D. Brokensha), pp. 185–216. Institute of International Studies, University of California, Berkeley.

Borchert, R. (1992). Computer simulation of tree growth periodicity and climatic hydroperiodicity in tropical forests. *Biotropica*, **24**, 385–395.

Borchert, R. (1994). Stem and soil water storage determine phenology and distribution of tropical dry forest trees. *Ecology*, **75**, 1437–1449.
Botkin, D.B. (1993). *Forest Dynamics: An Ecological Model.* Oxford University Press, Oxford.
Chadwick-Healey Ltd (1992). *World Climate Disc: Global Climate Change Data.* Disc, User Manual & Data Reference Guide. Chadwick-Healey Ltd, Cambridge.
Chapman, C.A., Wrangham, R. & Chapman, L.J. (1994). Indices of habitat-wide fruit abundance in tropical forests. *Biotropica*, **26**, 160–171.
Chuyong, G.B. (1994). *Nutrient cycling in ectomycorrhizal Legume-dominated forest in Korup National Park, Cameroon.* PhD thesis. Stirling University, UK.
Cockburn, P.F. (1975). Phenology of dipterocarps in Sabah. *Malaysian Forester*, **38**, 160–170.
Connell, J.H. & Lowman, M.D. (1989). Low-diversity tropical rain forests: some possible mechanisms for their existence. *American Naturalist*, **134**, 88–119.
Devred, R. (1958). La végétation forestière du Congo belge et du Ruanda-Urundi. *Bulletin de la Société Royale Forestière de Belgique*, **65**, 409–468.
Diaz, H.F. & Pulwarty, R.S. (1992). A comparison of Southern Oscillation and El Nino signals in the tropics. In *El Nino: Historical and Paleoclimatic Aspects of the Southern Oscillation* (Ed. by H.F. Diaz & V. Markgraf), pp. 175–192. Cambridge University Press, Cambridge.
Forget, P.-M. (1992). Regeneration ecology of *Eperua grandifolia* (Caesalpiniaceae), a large-seeded tree in French Guiana. *Biotropica*, **24**, 146–156.
Foster, R.B. (1982). The seasonal rhythm of fruitfall on Barro Colorado Island. In *The Ecology of a Tropical Forest: Seasonal Rhythms and Long-term Changes* (Ed. by E.G. Leigh, A.S. Rand & D.M. Winsor), pp. 151–172. Smithsonian Institution Press, Washington DC.
Gartlan, J.S. (1986). The biological and historical importance of the Korup forest. In *Workshop on Korup National Park* (Ed. by J.S. Gartlan & H. Macleod), pp. 28–35. WWF/IUCN Project 3206, Mundemba, Cameroon.
Gartlan, J.S. (1992). Cameroon. In *The Conservation Atlas of Tropical Forests: Africa* (Ed. by J.A. Sayer, C.S. Harcourt & N.M. Collins), pp. 110–118. IUCN.
Gartlan, J.S., Newbery, D.M., Thomas, D.W. & Waterman, P.G. (1986). The influence of topography and soil phosphorus on the vegetation of Korup Forest Reserve, Cameroun. *Vegetatio*, **65**, 131–148.
Gautier-Hion, A., Duplantier, J.-M., Emmons, L., Feer, F., Heckestweiler, P., Moungazi, A., Quris, R. & Sourd, C. (1985). Coadaption entre rhythmes de fructification et frugivorie en forêt tropicale humide du Gabon: mythe ou realité. *Revue d'Ecologie (Terre et la Vie)*, **40**, 405–434.
Gautier-Hion, A., Gautier, J.-P. & Maisels, F. (1993). Seed dispersal versus seed predation: an inter-site comparison of two related African monkeys. *Vegetatio*, **107/108**, 237–244.
Gérard, P. (1960). Etude écologique de la forêt dense à *Gilbertiodendron dewevrei* dans la région de l'Uele. *Publications de l'Institut National pour l'Etude Agronomique du Congo; Série Scientifique*, **87**, 1–159.
Germain, R. & Evrard, C. (1956). Etude écologique et phytosociologique de la forêt à *Brachystegia laurentii. Publications de l'Institut National pour l'Etude Agronomique du Congo Belge*, **87**, 1–105.
Greig-Smith, P. (1983). *Quantitative Plant Ecology.* Blackwell Scientific Publications, Oxford.
Harley, J.L. & Smith, S.E. (1983). *Mycorrhizal Symbiosis.* Academic Press, London.
Harper, J.L. (1977). *Population Biology of Plants.* Academic Press, London.
Hart, T.B. (1995). Seed, seedling and sub-canopy survival in monodominant and mixed forests of the Ituri Forest, Africa. *Journal of Tropical Ecology*, **11**, 443–459.
Hart, T.B., Hart, J.A. & Murphy, P.G. (1989). Monodominant and species-rich forests of the humid tropics: causes for their co-occurrence. *American Naturalist*, **133**, 613–633.

Herrera, C.M., Jordano, P., Lopez-Soria, L. & Amat, J.A. (1994). Recruitment of a mast-fruiting, bird-dispersed tree: bridging frugivore activity and seedling establishment. *Ecological Monographs*, **64**, 315–344.

Högberg, P. (1986). Soil nutrient availability, root symbioses and tree species composition in tropical Africa: a review. *Journal of Tropical Ecology*, **2**, 359–372.

Hubbell, S.P. & Foster, R.B. (1983). Diversity of canopy trees in a neotropical forest and implications for conservation. In *Tropical Rain Forest: Ecology and Management* (Ed. by S.L. Sutton, T.C. Whitmore & A.C. Chadwick), pp. 25–41. Blackwell Scientific Publications, Oxford.

Hubbell, S.P. & Foster, R.B. (1986). Biology, chance, and history and the structure of tropical rain forest communities. In *Community Ecology* (Ed. by J. Diamond & T.J. Case), pp. 314–329. Harper & Row, New York.

Huston, M.A. (1994). *Biological Diversity: The Coexistence of Species on Changing Landscapes.* Cambridge University Press, Cambridge.

Janzen, D.H. (1974). Tropical blackwater rivers, animals and mast fruiting by the Dipterocarpaceae. *Biotropica*, **6**, 69–103.

Janzen, D.H. (1978). Seeding patterns of tropical trees. In *Tropical Trees as Living Systems* (Ed. by P.B. Tomlinson & M.H. Zimmermann), pp. 83–128. Cambridge University Press, Cambridge.

Jones, N. (1975). Observations on *Triplochiton scleroxylon* K. Schum. flower and fruit development. In *Variation and Breeding Systems of* Triplochiton scleroxylon *K. Schum.* pp. 28–37. Symposium Proceedings, Federal Department of Forest Research, Ibadan, Nigeria, April 1975.

Kozlowski, T.T., Kramer, P.J. & Pallardy, S.G. (1991). *The Physiological Ecology of Woody Plants.* Academic Press, New York.

Lalonde, R.G. & Roitberg, B.D. (1992). On the evolution of masting behavior in trees: predation or weather? *American Naturalist*, **139**, 1293–1304.

Lebrun, J. (1936). La forêt équatoriale congolaise. *Bulletin Agricole du Congo Belge*, **27**, 163–192.

Letouzey, R. (1968). *Etude Phytogéographique du Cameroun.* P. LeChevalier, Paris.

Letouzey, R. (1985). *Notice de la Carte Phytogéographique du Cameroun Au 1:500 000.* Institut de la Carte Internationale de la Végétation, Toulouse.

Lieberman, D. (1982). Seasonality and phenology in a dry tropical forest in Ghana. *Journal of Ecology*, **70**, 791–806.

Longman, K.A. & Jenik, J. (1987). *Tropical Forest and its Environment.* Longman, Harlow, UK.

Lugo, A.E. & Frangi, J.L. (1993). Fruit fall in the Luquillo Experimental Forest, Puerto Rico. *Biotropica*, **25**, 73–84.

Maley, J. (1987). Fragmentation de la forêt dense humide africaine et extensions des biotopes montagnards au Quaternaire recent: Nouvelles données polliniques et chronologique; implications paleoclimatiques et biogéographiques. In *Paleoecology of Africa* (Ed. by J.A. Coetzee), pp. 307–334. Balkema, Rotterdam.

Maley, J. (1989). Late quarternary climatic changes in the African rain forest: forest refugia and the major role of sea surface temperature variations. In *Paleoclimatology and Paleometeorology: Modern and Past Global Atmospheric Transport* (Ed. by M. Leinen & M. Sarnthein), pp. 585–616. Kluwer, Dordrecht, The Netherlands.

Maley, J. (1991). The African rain forest vegetation and paleoenvironments during late quaternary. *Climatic Change*, **19**, 79–98.

Maley, J. (1996). The African rain forest — main characteristics of changes in vegetation and climate from the Upper Cretaceous to Quaternary. *Proceedings of the Royal Society of Edinburgh*, **104B**, 31–73.

McKey, D.B., Gartlan, J.S., Waterman, P.G. & Choo, G.M. (1981). Food selection by black colobus monkeys (*Colobus satanas*) in relation to plant chemistry. *Biological Journal of the Linnean Society*, **16**, 115–146.

Milton, K., Windsor, D.M. & Morrison, D.W. (1982). Fruiting phenologies of two Neotropical *Ficus* species. *Ecology*, **63**, 752–762.

Monselise, S.P. & Goldschmidt, E.E. (1982). Alternate bearing in fruit trees. *Horticultural Research*, **4**, 128–173.

Mori, S.A. & Prance, G.T. (1987). Phenology. In *The Lecythidaceae of a Lowland Neotropical Forest: La Fumeé Mountain, French Guiana* (Ed. by S.A. Mori *et al.*), pp. 124–136. *Memoirs of the New York Botanical Garden*, **44**.

Newbery, D.M. & Gartlan, J.S. (1996). A structural analysis of rain forest at Korup and Douala-Edea, Cameroon, *Proceedings of the Royal Society of Edinburgh*, **104B**, 177–224.

Newbery, D.M., Gartlan, J.S., McKey, D.B. & Waterman, P.G. (1986). The influence of drainage and soil phosphorus on the vegetation of Douala-Edea Forest Reserve, Cameroun. *Vegetatio*, **65**, 149–162.

Newbery, D.M., Alexander, I.J., Thomas, D.W. & Gartlan, J.S. (1988). Ectomycorrhizal rain-forest legumes and soil phosphorus in Korup National Park, Cameroon. *New Phytologist*, **109**, 433–450.

Newbery, D.M., Campbell, E.J.F., Lee, Y.F., Ridsdale, C.E. & Still, M.J. (1992). Primary lowland dipterocarp forest at Danum Valley, Sabah, Malaysia: structure, relative abundance and family composition. *Philosophical Transactions of the Royal Society, Series B*, **335**, 341–356.

Newbery, D.M., Campbell, E.J.F., Proctor, J. & Still, M.J. (1996). Primary lowland dipterocarp forest at Danum Valley, Sabah, Malaysia: species composition and patterns in the understorey. *Vegetatio*, **122**, 187–215.

Newbery, D.M., Alexander, I.J. & Rother, J.A. (1997). Phosphorus dynamics in a lowland African rain forest: the influence of ectomycorrhizal trees. *Ecological Monographs*, **67**, 367–409.

Newstrom, L.E., Frankie, G.W. & Baker, H.G. (1994a). A new classification for plant phenology based on flowering patterns in lowland tropical rain forest trees at La Selva, Costa Rica. *Biotropica*, **26**, 141–159.

Newstrom, L.E., Frankie, G.W., Baker, H.G. & Colwell, R.K. (1994b). Diversity of long-term flowering patterns. In *La Selva: Ecology and Natural History of a Lowland Tropical Rain Forest* (Ed. by L.A. McDade, K.S. Bawa, G.S. Hartshorn & H.A. Hespenheide), pp. 142–160. University of Chicago Press, Chicago.

Ng, F.S.P. (1976). Gregarious flowering of dipterocarps in Kepong, 1976. *Malaysian Forester*, **40**, 126–137.

Norton, D.A. & Kelly, D. (1988). Mast seedling over 33 years by *Dacrydium cupressinum* Lamb. (rimu) (Podocarpaceae) in New Zealand: the importance of economies of scale. *Functional Ecology*, **2**, 399–408.

O'Neill, R.V., DeAngelis, D.L., Waide, J.B. & Allen, T.F.H. (1986). *A Hierarchical Concept of Ecosystems*. Princeton University Press, Princeton, NJ.

Pereira, H.C. (1959). Practical field instruments for estimation of radiation and of evaporation. *Quarterly Review of the Royal Meteorological Society*, **85**, 253–261.

Peres, C.A. (1994). Primate responses to phenological changes in an Amazonian terra firme forest. *Biotropica*, **26**, 98–112.

Polhill, R.M., Raven, P.H. & Stirton, C.H. (1981). Evolution and systematics of the Leguminosae. In *Advances in Legume Systematics* (Ed. by R.M. Polhill & P.H. Raven), pp. 1–34. Royal Botanic Garden, Kew.

Poorter, L., Bongers, F., van Rompaey, R.S.A.R. & de Klerck, M. (1996). Regeneration of canopy tree species at 5 sites in West-African moist forest. *Forest Ecology & Management*, **84**, 61–69.

Putz, F.E. (**1979**). Aseasonality in Malaysian tree phenology. *Malaysian Forester*, **42**, 1–24.

Quinn, W.H. (**1992**). A study of Southern Oscillation-related climatic activity for AD 622–1900 incorporating Nile River flood data. In *El Nino: Historical and Paleoclimatic Aspects of the Southern Oscillation* (Ed. by H.F. Diaz & V. Markgraf), pp. 119–149. Cambridge University Press, Cambridge.

Rathcke, B. & Lacey, E.P. (**1985**). Phenological patterns of terrestrial plants. *Annual Review of Ecology and Systematics*, **16**, 179–214.

Read, D.J. (**1989**). Ecological integration by ectomycorrhizal fungi. In *Endocytobiology*, Vol. IV (Ed. by P. Nardon, V. Gianinazzi-Pearson, A.M. Granier, L. Margues & D.C. Smith), pp. 158–161. INRA, Paris.

Reading, A.J., Thompson, R.D. & Millington, A.C. (**1995**). *Humid Tropical Environments*. Blackwell Science, Oxford.

Richards, P.W. (**1952**). *The Tropical Rain Forest*. Cambridge University Press, Cambridge.

Richards, P.W. (**1963**). Ecological notes on West African Vegetation II. Lowland forest of the Southern Bakundu Forest Reserve. *Journal of Ecology*, **51**, 123–149.

Roubik, D.W. (**1989**). *Ecology and Natural History of Tropical Bees*. Cambridge University Press, Cambridge.

Sabatier, D. (**1985**). Saisonnalité et determinisme du pic de fructification en forêt Guyanaise. *Revue d'Ecologie (Terre et la Vie)*, **40**, 289–320.

Savill, P.S. & Fox, J.E.D. (**1967**). *Trees of Sierra Leone*. ODA, London.

Silvertown, J.W. (**1980**). The evolutionary ecology of mast seeding in trees. *Biological Journal of the Linnean Society*, **14**, 235–250.

Smith, C.C., Hamrick, J.L. & Kramer, C.L. (**1990**). The advantage of mast years for wind pollination. *American Naturalist*, **136**, 154–166.

Songwe, N.C., Fasehun, F.E. & Okali, D.U.U. (**1988**). Litterfall and productivity in a tropical rain forest, Southern Bakundu Forest Reserve, Cameroon. *Journal of Tropical Ecology*, **4**, 25–37.

Sork, V.L. (**1993**). Evolutionary ecology of mast-seeding in temperate and tropical oaks (*Quercus* spp.). *Vegetatio*, **107/108**, 133–147.

Sork, V.L., Bramble, J. & Sexton, O. (**1993**). Ecology of mast-fruiting in three species of North American deciduous oaks. *Ecology*, **74**, 528–541.

Swaine, M.D. & Hall, J.B. (**1986**). Forest structure and dynamics. In *Plant Ecology in West Africa* (Ed. by G.W. Lawson), pp. 47–93. J. Wiley & Sons, Chichester.

Ter Steege, H. & Persaud, C.A. (**1991**). The phenology of Guyanese timber species: a compilation of a century of observations. *Vegetatio*, **95**, 177–198.

Tutin, C.E.G. & Fernandez, M. (**1993**). Relationships between minimum temperature and fruit production in some tropical forest trees in Gabon. *Journal of Tropical Ecology*, **9**, 241–248.

Van Schaik, C.P. (**1986**). Phenological changes in a Sumatran rain forest. *Journal of Tropical Ecology*, **2**, 327–347.

Van Schaik, C.P., Terborgh, J.W. & Wright, S.J. (**1993**). The phenology of tropical forests: adaptive significance and consequences for primary consumers. *Annual Review of Ecology & Systematics*, **24**, 353–377.

Waller, D.M. (**1979**). Models of mast fruiting in trees. *Journal of Theoretical Biology*, **80**, 223–232.

Waller, D.M. (**1993**). How does mast-fruiting get started? *Trends in Ecology and Evolution*, **8**, 122–123.

Walsh, R.P.D. (**1996a**). Climate. In *The Tropical Rain Forest: An Ecological Study*, 2nd edn (Ed. by P.W. Richards), pp. 159–236. Cambridge University Press, Cambridge.

Walsh, R.P.D. (1996b). Drought frequency changes in Sabah and adjacent parts of northern Borneo since the late nineteenth century and possible implications for tropical rain forest dynamics. *Journal of Tropical Ecology*, **12**, 385–407.

Watt, A.S. (1947). Pattern and process in the plant community. *Journal of Ecology*, **35**, 1–11.

Wheelwright, N.T. (1986). A seven-year study of individual variation in fruit production in tropical bird-dispersed tree species in the family Lauraceae. In *Frugivores and Seed Dispersal* (Ed. by A. Estrada & T.H. Fleming), pp. 19–35. Junk, Dordrecht.

White, F. (1983). *The Vegetation of Africa*. UNESCO, Paris.

White, L.T.J. (1994). Patterns of fruit-fall phenology in the Lopé Reserve, Gabon, *Journal of Tropical Ecology*, **10**, 289–312.

Whittaker, R.H. (1975). *Communities and Ecosystems*. Macmillan, New York.

Wood, G.H.S. (1956). The dipterocarp flowering season in North Borneo, 1955. *Malayan Forester*, **19**, 193–201.

Woodward, F.I. (1987). *Climate and Plant Distribution*. Cambridge University Press, Cambridge.

Wright, S.J. & Van Schaik, C.P. (1994). Light and the phenology of tropical trees. *American Naturalist*, **143**, 192–199.

Wycherley, P.R. (1973). The phenology of plants in the humid tropics. *Micronesia*, **9**, 75–96.

Yap, S.K. & Chan, H.T. (1990). Phenological behaviour of some *Shorea* species in Peninsular Malaysia. In *Reproductive Ecology of Tropical Forest Plants* (Ed. by K.S. Bawa & M. Hadley), pp. 21–35. UNESCO, Paris.

Zagt, R.J. (1997). Pre-dispersal and early post-dispersal demography, and reproductive litter production, in a tropical tree *Dicymbe altsonii* in Guyana. *Journal of Tropical Ecology*, **13**, 511–526.

12. PRIMATES, PHENOLOGY AND FRUGIVORY: PRESENT, PAST AND FUTURE PATTERNS IN THE LOPÉ RESERVE, GABON

C. E. G. TUTIN* and L. J. T. WHITE[†]

*Centre International de Recherches Médicales de Franceville, Gabon, and Department of Biological and Molecular Sciences, University of Stirling, Stirling FK9 4LA, Scotland, UK; † The Wildlife Conservation Society, New York and Institute of Cell, Animal and Population Biology, University of Edinburgh, King's Buildings, Edinburgh EH9 3JT, UK

SUMMARY

1 Climate change in central Africa over the past 75 000 years has had a profound impact on vegetation of the Lopé Reserve in central Gabon. Present-day vegetation is dominated by closed-canopy tropical rainforest but a zone of savanna persists where forest recolonization has been blocked by regular burning. Eight species of diurnal primate share the forest habitats of the 50-km^2 study area where research has been in progress for 12 years.

2 The diets of the primates are dominated by fruit flesh and seeds, which account for between 55 and 77% of dietary items. Since 1986, the phenology of 63 species of tree which produce fruit eaten by primates has been monitored monthly. Analysis of these data show that the year can be divided into three periods with respect to fruit availability. During the 3-month dry season few ripe fruit are present; for 4 months after the onset of rains fruit availability is consistently high; while for the remaining 5 months of the year there are large interannual variations in the amount of fruit produced.

3 Flowering in some species with irregular production patterns is triggered by low temperatures which occur normally only in the dry season. Climate influences phenological patterns in other ways, for example one species of tree appears to require a combination of two climatic features (dry weather and sunshine) for flowering; and the absence of sunshine during the dry season inhibits ripening of fruit.

4 As fruit production is influenced by climate, both colder weather of the past and the likely warmer weather of the future will have a large and immediate impact on populations of frugivorous primates.

5 Climate change has in the past reduced, and is likely again in the future to reduce, the extent of closed-canopy forest at Lopé, but chances of recovery of primates from future population reductions are slim as both they and their tropical forest habitats are already under intense pressure from human activities such as selective logging, hunting and deforestation for agriculture.

INTRODUCTION

Changes in temperature and annual rainfall during the last glacial (c. 75 000–12 000 BP) had profound implications for the distribution and nature of tropical rainforests in Africa. Average temperatures were 3–6°C lower than today, annual rainfall was significantly less and the dry season more pronounced; tropical rainforests were reduced to a few isolated patches or refuges (Hamilton 1982, Livingstone 1982, Maley 1987, 1989). Similar changes have also been reported for the Amazon (Sanford, Saldarraga, Clark et al. 1985, Irion 1989, Van der Hammen 1992), SE Asia (e.g. Heaney 1992) and Australia (Kershaw 1992).

Today climate change may be occurring faster than at any point in the past 10 000 years (Houghton, Jenkins & Ephraums 1990). Phillips and Gentry (1994) surveyed tropical forests worldwide and found evidence of increased turnover rates (mortality and recruitment) in all tropical rainforest regions in the past decade. One possible explanation is that increased levels of atmospheric carbon dioxide have accelerated plant production. If this is true it will have wide-ranging implications for conservation of biodiversity, since species composition in tropical forests worldwide will gradually shift in favour of fast-growing trees and lianes. Furthermore, increased concentrations of the so-called 'greenhouse' gases are likely to result in average global warming of 1°C by the year 2025 and 2–3°C by 2100, which will cause alteration of regional rainfall patterns (Myers 1992, Houghton 1995).

The implications of future climate change for tropical rainforest ecosystems is a matter of great concern, both because these habitats contain the majority of the world's biodiversity and because of the role they play in regulating global climate. While the pace and amplitude of climate change cannot be predicted accurately, it is crucial to assess some of the likely effects on tropical flora and fauna.

In this chapter we concentrate on the patterns of food availability for rainforest primates. Apes and monkeys have been studied intensively, in part because of their phylogenetic proximity to man. At a number of sites, long-term research has monitored feeding ecology and behaviour over one, or several, decades. Primates are the most important arboreal consumers in tropical rainforests of Africa (Struhsaker & Leland 1979, Emmons, Gautier-

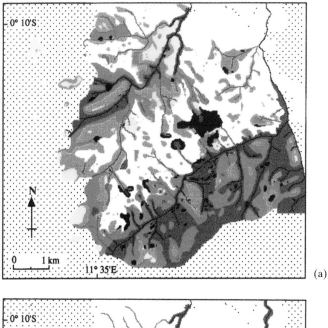

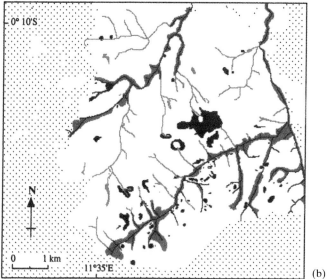

PLATE 12.1. This shows (a) present-day distribution of vegetation types in the Lopé study area and (b) the likely distribution of vegetation types in the Lopé study area c. 1400 years ago. The key for (a) shows Savanna (O); colonizing forest ( ); monodominant forest (●); Marantaceae forest (●); mixed Marantaceae (●); rocky forest (●); rivers with Caesalp. galleries (——); rivers with young galleries (——); unmapped areas (⋰⋱). The key for (b) shows Savanna (O); gallery forests dominated by Caesalps. (●); rock outcrops (●); water courses with forest cover (——); water courses without forest cover (——); unmapped areas (⋰⋱).

[facing page 310]

Hion & Dubost 1983, Galat & Galat-Luong 1985, White 1994a), South America (Terborgh 1983, Chapman 1987) and SE Asia (MacKinnon & MacKinnon 1980, Leighton & Leighton 1983). Almost 80% of primate species inhabit tropical forests where communities of 6–15 sympatric species are found. Diet is dominated by plants. Fruit and immature seeds are preferred foods, but other plant parts, particularly young leaves, are eaten regularly. The importance of fruit and seeds to primates means that changes in diet over the annual cycle are marked, as the majority of plants reproduce seasonally. Thus, frugivores are faced with a constantly changing array, and with variable quantities, of potential foods. Frugivorous primates have been shown to play an important role in the dynamics of tropical rainforest through seed dispersal (Howe 1980, Gautier-Hion, Duplantier, Quris *et al.* 1985, Estrada & Coates-Estrada 1991, Tutin, Williamson, Rogers & Fernandez 1991a, Julliot 1994) and the seeds of the majority of plant species in these habitats are dispersed by animals (White 1994b).

Climatic cues that act as environmental triggers to flowering and/or the production of new leaves by tropical plants include increased rainfall; increased temperature (also hours of sunlight, a measure likely to be correlated with temperature); and decreased temperature (Lieberman 1982, Corlett 1990, Fleming & Williams 1990, Tutin & Fernandez 1993a, Van Schaik, Terborgh & Wright 1993, Borchert 1996), but much remains to be discovered about tropical plant phenology. Even small changes in either temperature or patterns of rainfall may interrupt plant reproduction and change the amount of food available to primates. Climate can influence the amount of ripe fruit that is produced in other ways. Flowering does not inevitably lead to the production of ripe fruit and weather can act directly (e.g. by destroying flowers or immature fruit during storms), or indirectly (e.g. by affecting the population size of insect pollinators or predators), to influence the size of fruit crops. Hence, climate change is likely to disrupt phenological patterns and may contribute to changes in vegetation that occur during such periods.

In the Lopé Reserve, in central Gabon, eight species of diurnal primate share a habitat that is now dominated by tropical rainforest, although islands of savanna, dating from at least 10 000 years ago, remain where forest expansion has been blocked by regular burning (see below, Vegetation history and dynamics). Data have been collected on primate diet, tree phenology, forest dynamics and vegetation history (e.g. Williamson, Tutin, Rogers & Fernandez 1990, Tutin & Fernandez 1993b, Tutin, White, Williamson *et al.* 1994, White 1994b, White, Rogers, Tutin *et al.* 1995, White 1997), and archaeological research in the same area has shown a long history of human presence dating back at least 350 000 years (Oslisly 1993).

In this chapter we present data from Lopé related to primate food availability, climate and vegetation history in order to address the following points:
1 To what extent does the amount of food available to primates vary within and between years?
2 Can all, or any of this variation be related to climatic variables?
3 What was the major impact of past climate change on vegetation at Lopé?
4 What will be the likely short-term effects of warmer temperatures due to global warming on vegetation and thus on primate food availability?

STUDY SITE AND METHODS

Study area

The Lopé Reserve (5000 km^2) is shown in Fig. 12.1 as is the location of the main study area, which covers about 50 km^2 of forest and forest–savanna mosaic, centred at 0°12′S, 11°36′E. Mean annual rainfall is 1548 mm (1984–95) and the climate is characterized by a dry season of about 3 months from mid-June to mid-September. Figure 12.2 shows that rainfall tends also to be low in January and February, but the timing and duration of this short dry season vary from year to year. Temperatures vary little, but are lowest during the dry season when constant cloud cover during the daylight hours results in low evaporation rates, such that high relative humidity is maintained. Data on hours of sunshine from Makokou, in northeast Gabon, are included in Fig. 12.2 and illustrate the low levels of sunshine during the dry season.

Vegetation history and dynamics

The study area contains a complex mosaic of vegetation types which has resulted primarily from the process of savanna colonization during the course of the past 1400 years (White, Rogers, Tutin et al. 1995, White 1997, White, Oslisly, Abernethy & Maley 1997). Savannas in Lopé have been dated to more than 10 000 years BP using $\partial^{13}C$ measures (see Schwartz, Lanfranchi & Mariotti 1990) in combination with ^{14}C dating of archaeological remains (Oslisly, Peyrot, Abdessadok & White 1996). Savannas are currently maintained by regular fires lit by humans each dry season, but when protected from fire they are colonized by forest plant species. Between 1400 BP and 700 BP there was a human population crash in the region; during this 700-year period there is no evidence of human occupation and fires would have been

Primates, phenology and frugivory 313

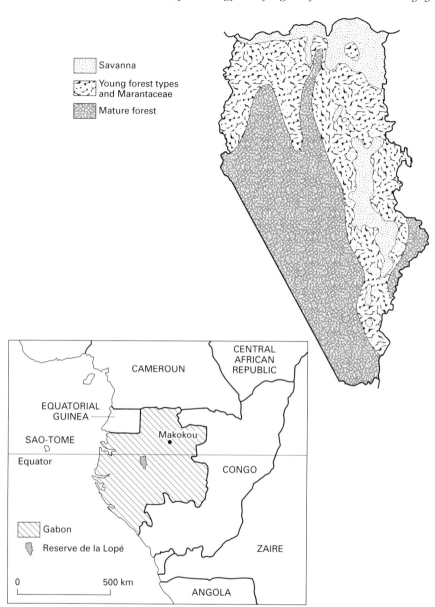

FIG. 12.1. The Lopé Reserve, Gabon: present-day distribution of major vegetation types and location of the study area.

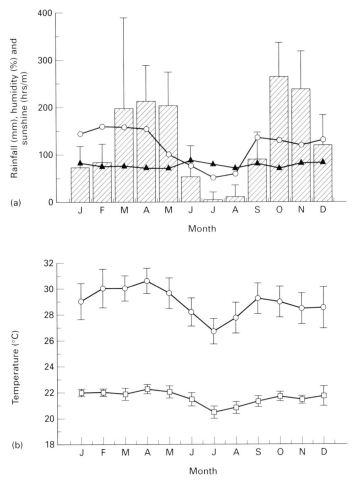

FIG. 12.2. (a) Mean monthly rainfall (hatched bars) with standard deviations and humidity (—▲—) at Lopé (1984–95) and mean monthly hours of sunlight (—○—) at Makokou, NE Gabon (1961–80). (b) Mean monthly maximum (—○—) and minimum (—□—) temperatures with standard deviations, Lopé (1984–95).

less frequent, allowing forest to advance into the savanna (Oslisly 1995, White 1997, White *et al.* 1997). Three tree species, *Aucoumea klaineana* (Burseraceae), *Lophira alata* (Ochnaceae) and *Sacoglottis gabonensis* (Humiriaceae) are amongst the early savanna colonizers. These species grow up to form dense, even-aged, stands of 'monodominant forest'. When these stands mature and the colonizers begin to die a dense thicket of herbaceous vegetation forms, dominated by species of Marantaceae and Zingiberaceae. The resulting forest type, 'Marantaceae forest', has a characteristic open structure, due to low stocking densities of medium-sized trees (Koechlin 1964, Letouzey 1968, White *et al.* 1995, White 1997). The colonizing trees do not regenerate within the forest, so they, therefore, gradually disappear and species composition becomes more diverse, resulting in the formation of 'mixed Marantaceae forest'.

The landscape in the study area prior to the phase of colonization which followed the human population crash has been recreated by mapping the distribution of tree species in the family Caesalpiniaceae, which disperse their seeds ballistically, by means of dehiscent pods. These species are useful indicators of forest refugia in Gabon, because of their relatively inefficient seed dispersal, limited to a number of tens of metres (Rietkerk, Ketner & de Wilde 1995). Plate 12.1(a) (facing p. 310) shows the present-day vegetation of the study area and Plate 12.1(b) (facing p. 310) shows it as it was thought to have been prior to the human population crash 1400 years ago. Marantaceae forest currently covers about half of the Lopé Reserve (Fig. 12.1), suggesting that the mosaic of savanna and gallery forests currently restricted to northern and eastern areas was once much more widespread. During cool, dry climatic phases of the past it is likely that this vegetation mosaic occurred over much of central Gabon, although there may have been patches of montane vegetation on some of the higher hills (>650 m) (Maley, Caballé & Sita 1990, Sosef 1991, White 1992).

Fruit availability

To date, 676 plant species have been identified in the study area and the total number probably approaches 1500 (Tutin *et al.* 1994, White & Abernethy 1996). Some data on plant phenology have been collected since January 1984, but systematic data on fruit production by 63 species eaten by gorillas and chimpanzees began on a monthly basis from October 1986. Data from 112 consecutive months (October 1986–December 1995) were analysed. Data were collected on the quantity of leaves (new, mature and senescent), flowers and fruit (immature and ripe) of individual trees. Normally 10 trees of each species were monitored, but this was not possible for some rare species ($N = 6$). Over 4–7 days at the beginning of each month, the 578 labelled trees were observed through binoculars and the quantity of leaves, flowers and fruit was recorded on a 10-point scale from 0 (none) to 10 (maximum crop). Ripe and immature fruit were recorded seperately. A mean fruit score was established *post hoc* each month for each species based on the individual scores for ripe fruit:

$$\text{Mean fruit score} = \frac{\Sigma FS \times 10}{N\,\text{Tr}}$$

where *FS* is the score for ripe fruit for each tree and *N* Tr the number of trees of that species monitored. Thus the maximum fruit score is 100 and the computation of *FS*, compensates for the minority of species with fewer than 10 monitored individuals. A total fruit score for each month was derived by summing the *FS* of all 63 species to facilitate comparisons of the quantity of ripe fruit available to primates.

'Poor' fruit months (i.e. periods of fruit scarcity) were defined as months when six, or fewer, species in the phenology sample had $FS \leq 5$; all other months were considered to be 'good' fruit months (Tutin, Fernandez, Rogers *et al.* 1991b).

The 63 species in the phenology sample include all major fruit foods of gorillas and chimpanzees (Tutin & Fernandez 1993b). There is a strong bias in this data set to species with succulent fruit (54 species, the other nine produce arillate fruit) as these are strongly preferred by primates at Lopé (Tutin, Ham, White & Harrison 1997a). However, a year-long study of all fruit found on five 5-km fruitfall transects (White 1994b) gave data on fruit production by the whole plant community at Lopé.

Primate diets

The eight diurnal primate species in the Lopé study area are: western lowland gorillas (*Gorilla g. gorilla*), chimpanzees (*Pan t. troglodytes*), mandrills (*Mandrillus sphinx*), grey-cheeked mangabeys (*Cercocebus albigena*), black colobus (*Colobus satanas*) and three species of guenon (*Cercopithecus nictitans, C. pogonias* and *C. cephus*). The diets of gorillas and chimpanzees have been documented by observation and faecal analysis over a 12-year period (Williamson *et al.* 1990, Tutin & Fernandez 1993b). Apes at Lopé have proven difficult to habituate to human presence, but indirect methods of monitoring diets have produced useful data (Tutin & Fernandez 1994). Harrison and Hladik (1986) quantified the diet of a group of black colobus using the feeding frequency method (Struhsaker 1975) during five consecutive days each month over 9 months from August 1983 to April 1984 and Ham (1994) used instantaneous scan samples (Altmann 1974) at 15-min intervals to study the diet of a group of grey-cheeked mangabeys during 5 days each month for 15 months between April 1991 and August 1992. Opportunistic data have been collected on foods consumed by all primate species over the 12-year study period. For each species, each food observed to be eaten was recorded. The plant food categories used in this study are:

fruit (flesh surrounding seed: mesocarp, exocarp or aril); seed; leaf; flower; pith; animal; and other (including bark, roots, galls and soil).

RESULTS

Seasonal and interannual variation in fruit availability

At Lopé most species of tree flower annually, but a minority flowers twice each year and some others have a supra-annual flowering periodicity. In some normally annual species, individual trees flower at best every other year. On a community scale there are two bursts of flowering: in March–April and September–November. The species that flower during the first period have immature fruit during the dry season which ripen from September to December while those that flower after the dry season bear ripe fruit between January and May.

The phenology sample is small compared to total floristic diversity at Lopé and biased towards species producing succulent fruit, but data on fruiting phenology of the entire community between June 1990 and May 1991 are available from White's (1994b) fruitfall study. Figure 12.3 compares the results obtained by the two phenology studies during this period and shows that the seasonal patterns of fruit diversity are similar. A significant correlation of seasonal variation exists between the two data sets (two-tailed Spearman's rank correlation: $r_s = 0.782; 0.002 < P < 0.005$).

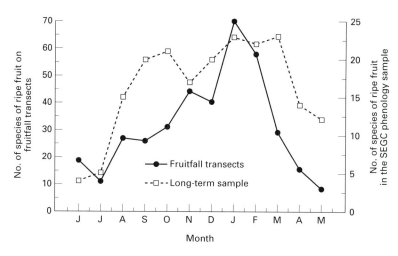

FIG. 12.3. Number of species bearing ripe fruit each month (06/90–05/91) as measured by five 5-km fruitfall transects and the long-term phenology sample of 63 species.

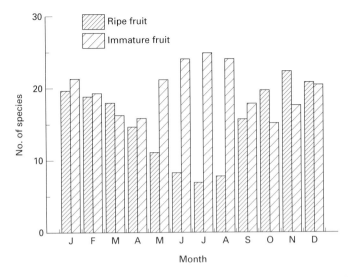

FIG. 12.4. Mean number of species in the long-term phenology sample with ripe and immature fruit (1986–95). A species with both ripe and immature fruit was scored only for ripe fruit.

The diversity of fruit available to primates at Lopé is reflected by the mean monthly number of species (of the 63 in the phenology sample) bearing ripe or immature fruit (see Fig. 12.4). A diverse array of ripe fruit (11–22 species) is available for most of the year. The dry season (June, July, August) is the exception, when few species bear ripe fruit, although immature fruit diversity peaks at this time of year. The quantity of ripe fruit available varies in a similar way over the year, see Fig. 12.5. The means of total ripe fruit scores (see Methods) for all of the species in the phenology sample show that only small quantities of ripe fruit are available in the dry season compared to other times of year. Some of the species that do have ripe fruit during the dry season are ones that flower and fruit asynchronously (e.g. *Ficus* spp. and *Duboscia macrocarpa*), while the arillate fruit of a group of seasonally synchronized species (*Xylopia* spp., *Pycnanthus angolensis*, *Staudtia gabonensis*) begin to ripen in August, the last month of the dry season.

Dividing months into good and poor fruit months (see Methods) is another way of looking at seasonal variation in fruit availability (see Fig. 12.5). The dry season months, June, July and August, were classed as poor fruit months in either 8 or 9 of the 9 years between 1986 and 1995, while no poor fruit months occurred in October, November or December. The months between February and May were usually good fruit months, but

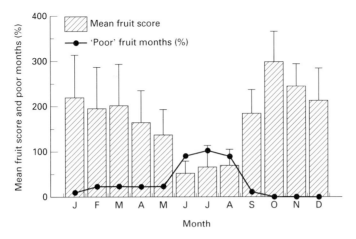

FIG. 12.5. Mean monthly total fruit score (summed *FS* for all species in the phenology sample bearing ripe fruit) with standard deviations, and percentage of 'poor' fruit months out of the 9 years: 1986–95.

poor months were recorded in 2 of the 9 years while only one poor month was recorded for both September and January. The year at Lopé can thus be divided into three periods with respect to the availability of ripe fruit: June–August when it is consistently scarce, October–December when it is always abundant, and February–May when it is usually abundant but occasional crop failures lead to fruit scarcity. This is illustrated in Fig. 12.6 with a comparison of the number of species bearing ripe fruit in the 'best' (1990) and the 'worst' (1991) years for fruit during the 9-year period. During the first 5 months of 1990 both the diversity (no. of species fruiting) and quantity (index derived from total fruit score divided by 10) of ripe fruit available to the Lopé primates were consistently greater than in the corresponding months of 1991. During the dry season, July–August, few ripe fruit were available in either year while in September–December, a high diversity (and large, though variable, quantity) of ripe fruit was present in both years. In terms of poor fruit months (those in which only six or fewer of the 63 species in the phenology sample had fruit scores ≤ 5), 1990 had only one (July) while 1991 had eight consecutive poor months (January–August).

Relationships between climatic variables and phenology

The large interannual variations in fruit availability that occur in January–May are the result of little or no flowering by some species in some years. A previous analysis showed that for at least eight tree species at Lopé,

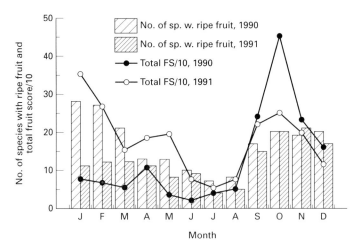

FIG. 12.6. Comparison of the number of species bearing ripe fruit each month during the best (1990) and worst (1991) years for fruit crops at Lopé.

flowering after the dry season occurs only in years when a critical low temperature has been reached (Tutin & Fernandez 1993a). The dry season is the coldest time of year (Fig. 12.2) and flowering in these species is triggered by a temperature of 19°C or less. A climatic cue for flowering stimulates and synchronizes flowering within a species, but it must be reliable; that is, occur only at a time of year that will result in pollination of flowers and maturation (and dispersal) of seeds. At Lopé, temperatures of 19°C or less have been recorded during the dry season in 10 of 12 years. However, twice (in 1985 and 1992), temperatures of 19°C also occurred in January and in both years some individuals of these eight species (and at least 15 additional species) showed aseasonal flowering in March–May and produced small crops of fruit.

In Fig. 12.7 there are examples of the pattern of fruiting of two of the species that appear to require temperatures of 19°C or less to flower: *Ganophyllum giganteum* (Sapindaceae) and *Diospyros dendo* (Ebenaceae). These two species were selected to illustrate the temperature-dependent pattern of reproduction as the former has the shortest interval between flowering and fruit maturation (6–8 weeks) and the latter, the longest (20–24 weeks). Following the warm dry season of 1987 (the other warm year was 1984 before systematic phenology data were collected), no fruit were produced but after the atypical cold spell in January 1992, both species produced small aseasonal fruit crops and flowered again at the normal time after the cool 1992 dry season.

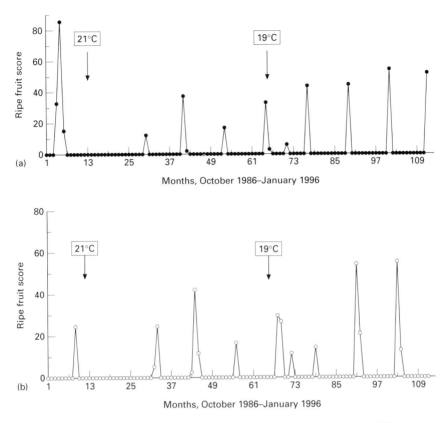

FIG. 12.7. Fruit scores for (a) *Ganophyllum giganteum*, and (b) *Diospyros dendo*. FS is the sum of individual tree's fruit scores, multiplied by 10 and divided by the number of trees monitored. Maximum value for each tree is 10, and maximum FS thus 100, see Methods (p. 312).

Fruiting patterns are shown in Fig. 12.8 for the same time period for two species that appear not to be influenced by low temperature. The first, *Pentadesma butyracea* (Clusiaceae) produces fruit on a regular seasonal basis but neither the warm dry season of 1987 nor the cool January of 1992 had any visible impact on fruit production. Fruit production by *Duboscia macrocarpa* (Tiliaceae) is aseasonal as individual trees flower asynchronously at 17-month intervals. Some ripe fruit was recorded in the phenology sample for 109 of the 112 months of data collection and no impact of patterns of minimum temperatures was apparent.

Figure 12.9 shows the phenology data for flowering and ripe fruit production by *Uapaca guineensis* (Euphorbiaceae). For this species, rhythms are

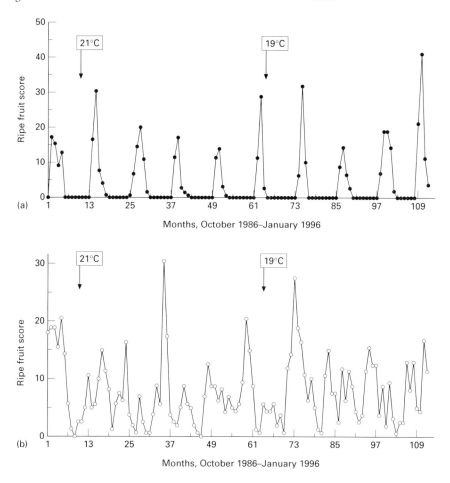

FIG. 12.8. Fruit scores for (a) *Pentadesma butyracea*, and (b) *Duboscia macrocarpa*.

less regular. In all years there was a peak of flowering between March and April and a corresponding peak in ripe fruit availability from October to December was recorded each year except 1993; however, in only 2 of the 9 years was reproductive activity limited to a single episode. In the other years, a second flowering episode in August–September resulted in a small crop of ripe fruit in March. Rhythms became blurred from 1993 with reduced synchrony of fruit ripening. From January 1993 (month 76) to December 1995 (month 111), ripe *Uapaca* fruit were recorded in 27 of 36 months compared to 15 months from January 1987 (month 16) to December 1989 (month 39).

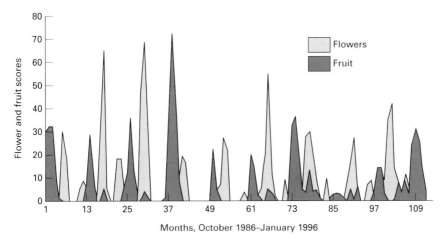

FIG. 12.9. Flower and fruit scores for *Uapaca guineensis*.

In an attempt to determine if flowering in *U. guineensis* was triggered by an environmental cue, we sought correlations between the timing of flowering and major climatic variables (minimum and maximum temperatures, rainfall patterns and humidity), but no clear relationships emerged. Absence of rain was implicated in a general way as all flowering episodes followed an extended period without rain, occurring either in July (the long dry season) or January (a relatively dry month, see Fig. 12.2). The relationship, however, is not direct as, were this the case, flowering would be concentrated in August–September, not in February–March. Although rainfall is generally low in January–February, no clear short dry season can be defined. The longest period without rainfall (drizzle of less than 1 mm day^{-1} was ignored) in the months of January–February, averaged 14 consecutive days (range 10–28). By comparison, the long dry season (mid-June to mid-September) saw an average of 69 consecutive dry days (range 28–97). Interannual variations in neither the number of consecutive days without rain, the number of heavy rains (≥10 mm), nor total rainfall during the drier weather, either by month, or blocks of months, correlated in any way with the timing or abundance of *U. guineensis* flowering.

A striking contrast exists, however, between dry weather in July–August and the dry spells of January–February as the former period is the coldest time of year (Fig. 12.2) due to persistent low cloud cover during the day and consequent absence of sunshine, whereas the latter period is sunny with mean temperatures 2–3°C higher. No data have been collected at Lopé on the number of hours of sunshine each month but such data exist for

Makokou, an area with similar annual rainfall, 160 km to the northeast (Figs 12.1 & 12.2). At Makokou, there was almost three times as much sunshine in January (143 h) as in July (52 h) and August was the second least sunny month with an average of 60 h of sunshine. The data suggest that flowering in *U. guineensis* may be triggered by a combination of dry and warm, sunny, weather; the absence of any correlation between temperature and flowering in *U. guineensis* suggests that it is the amount of sunshine that is important. Figure 12.9 shows that peak flowering occurred each year in March–April and in 7 years, a second burst of flowering occurred in August–September. In the other 2 years (1989 and 1990) no second flowering/fruiting episode occurred.

If our hypothesis of a combined dry–sunny cue is correct, we would expect a difference to emerge between these 2 years and the other 6 years. In terms of rainfall, 1989 and 1990 were 2 of only 3 years during which no rain fell in August (in 1992 a second burst of flowering did occur after a dry August). Rainfall in the dry season is usually associated with a break in the consistent cloud cover and some sunshine follows precipitation. Unfortunately without the corresponding data on sunshine from Lopé, which might differentiate between 1989 and 1990 compared to 1992, all we can do is suggest that dry spells of at least 10 consecutive days associated with sunshine might trigger flowering in *U. guineensis*. This hypothesis would also explain the blurring of synchrony of ripe fruit availability from 1993, as weather patterns were atypical: dry spells occurred in four consecutive sunny months (December–March) in late 1992–early 1993 and in three (December 93–February 94) the following year; compared to the usual pattern of such dry spells being limited to the month of January (in 4 of the 9 years), or to January–February (in 3 years). For tight synchrony within the species, the cue for flowering should occur only once. It appears that in the years when the proposed cue occurred more than once, *Uapaca* trees produced consecutive bursts of flowers over an extended period (flowers were recorded in 18 of 36 months in 1993–95 compared to 12 months in 1987–89). The small, staggered, crops of ripe fruit observed from 1993 to 1995 resulted from the loosening of synchrony in flowering.

Although our explanation of *U. guineensis* reproduction must remain a hypothesis for the present, the analysis illustrates the complexity of potential interactions between weather patterns and plant phenology. This has important repercussions for understanding and predicting the effects of changes in climate, but more data are needed. Long-term monitoring of individual trees is essential, as is collection of detailed meteorological data at the same site. Deviations from climatic norms present natural experiments that can confirm, or strengthen, hypotheses about causal relationships developed from correlations.

Taking the whole phenology sample of 63 species, the mean number of species bearing ripe fruit each month showed a significant positive correlation ($r_s = 0.881$, df = 10, $P < 0.001$) with the monthly average hours of sunshine recorded at Makokou. This relationship also held at the level of the whole plant community as the amount and diversity of ripe fruit counted over 12 months on fruitfall transects at Lopé showed significant positive correlations with the mean monthly hours of sunlight (White 1994b), suggesting that the long hours of sunshine experienced between rainfalls during the wetter months are important for fruit ripening. In contrast, it seems that the cloudy weather of the dry season does not permit ripening of the large amount of immature fruit present at this time of year (see Fig. 12.4). The interval between peaks of *Uapaca* flowering and fruit ripening varied depending on whether or not the dry season intervened. Comparing the length of the interval from flowering to fruit maturation (data from 1987–92, before rhythms became blurred) gave an average delay of 8.2 months ($N = 6$ years, range 8–9) for major crops which ripen after the dry season, and an average delay of 6.8 months ($N = 4$ years, range 6–7) for minor crops following end-of-dry-season flowering.

Primate diets

Diets of seven of the eight diurnal primates at Lopé are dominated by fruit in terms of number of different species eaten while the eighth species, the black colobus, has a diet dominated by seeds (Fig. 12.10). Quantitative data on the percentage of feeding time spent on each food class give similar results (Tutin *et al.* 1997a). The total number of foods recorded for the primate community was 397, including 362 plant foods from 202 different species. These 362 plant foods are divided amongst the food class categories as follows: 145 fruit, 74 seeds, 74 leaves, 28 flowers, 26 piths and 15 barks. All of the primate species have diverse diets including 46–220 different foods and none has a totally specialized diet, as each fed on three to five food classes for at least 5% of feeding time (Tutin *et al.* 1997a). Seasonal variation in diet has been documented for all of the eight primate species (Harrison & Hladik 1986, Rogers, Williamson, Tutin & Fernandez 1988, Ham 1994, Tutin *et al.* 1997a), with reduced frugivory during the annual dry season. Obviously, the primates survive this regular 3-month period of fruit scarcity but some weight loss occurs and none of the seasonally breeding species gives birth at this time of year (C.E.G. Tutin unpublished data). It is probable that the amount

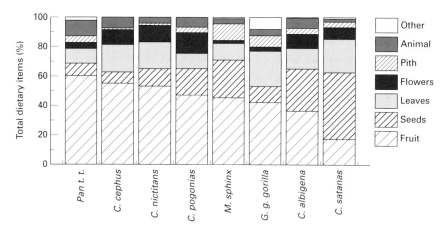

FIG. 12.10. Diets of eight sympatric primates at Lopé. Species are listed from left to right in decreasing order of frugivory. *Pan t. t.*, chimpanzee; *C. cephus*, moustached guenon; *C. nictitans*, spot-nosed guenon; *M. sphinx*, mandrill; *G. g. gorilla*, lowland gorilla; *C. albigena*, grey-cheeked mangabey; *C. satanas*, black colobus.

of food available during the dry season determines population densities, and the strong seasonality in fruit availability may, in part, explain the low primate biomass at Lopé compared to other sites in tropical Africa (White 1994a; Tutin *et al.* 1997a). Species of primate vary in their choices of alternate foods during periods of fruit scarcity. All show dietary flexibility but some species are able to switch to an almost completely folivorous diet (e.g. gorillas) while others (e.g. chimpanzees) appear to have a greater dependency on ripe fruit (Tutin *et al.* 1991b).

When succulent fruit is scarce, certain foods are heavily exploited, many of which are ignored at other times of year. These 'fallback' foods must, by definition, be available consistently either year round (as is the case for mature leaves, bark and fruit/seeds of common species that reproduce asynchronously) or during regular, seasonal, periods of succulent fruit scarcity (i.e. at Lopé, the 3-month dry season). The fallback foods identified for the primate community at Lopé include a few seasonally produced fruit and seeds as well as leaves, bark, flowers and insects (Tutin *et al.* 1997a). The seeds of six tree species are eaten; all but one are consumed when immature. Of the seven species of seasonal fruits eaten, only one is succulent — that of *Pseudospondias longifolia* (Anacardiaceae), a tree that fruits twice a year with fruit ripening in January and in August. The other fruit are either arillate, or fibrous, and, apart from the asynchronously fruiting *Duboscia* and *Elaeis guineensis* (Palmae), they ripen from August, the last month of the dry season. Figs are eaten when present, but none of the 12 species is common so, despite asynchronous fruiting, figs are not a dependable fallback food at Lopé (see also Gautier-Hion & Michaloud 1989). The terrestrial apes and mandrills eat large amounts of pith and young leaves of 11 species of herb (White *et al.* 1995). Gorillas eat large amounts of the bast (inner bark) of a common species of tree, *Milicia* (*Chlorophora*) *excelsa*, (Rogers, Tutin, Parnell *et al.* 1994), while chimpanzees rely on oil palm (*Elaeis*) and *Duboscia* fruit. Monkeys concentrate on small arillate fruit, immature seeds, the large flowers of *Pentadesma* (which has an extended flowering period, and produces copious nectar) and on insects (Tutin *et al.* 1997a).

DISCUSSION

The present: impact of weather on plant phenology and the availability of ripe fruit.

Data on fruit production, collected over 9 consecutive years, show a pattern of seasonality in the amount and in the diversity of ripe succulent fruit available to primates at Lopé. Two patterns recur each year: (i) a scarcity of ripe fruit during the dry season; and (ii) an abundance of ripe fruit from October to December; but fruit availability from February to June is subject to large interannual variation (Figs 12.5 & 12.6). We have identified some climatic variables that relate to these patterns (Figs 12.7–12.9), but full understanding must await analysis of the complete phenology data set.

Comparing Lopé phenology data with those from plant communities and individual species in other tropical areas may prove illuminating, but, unfortunately, few long-term data have been collected on individual species. Van Schaik *et al.* (1993) reviewed tree phenology at a community level using data from tropical sites around the world. They concluded that peaks of sunshine were the key factor influencing both flowering and leaf renewal except where water stress made this impossible. Taking the novel approach of obtaining phenological data from herbarium collections of certain Neotropical dry forest tree species, Borchert (1996) found that water balance was a key factor in the timing of leaf-loss and flowering for some, but not all, of the species for which sufficient collections existed.

At Lopé, the large amount of immature succulent fruit present during the dry season (Fig. 12.4) begins to ripen when rain and sunshine return in September. This suggests that the essential resources for producing sugar- and water-rich ripe fruit are absent during the cold, dry, cloudy months of the dry season. For a subset of species at Lopé, a critical low temperature triggers flowering and for *Uapaca guineensis* the cue appears to come from the coincidence of two climatic parameters: absence of rain for at least 10 days and sunny weather. We have not identified an environmental cue for the seasonal flowering of *Pentadesma* and the regularity of reproduction in this species over a decade, despite climatic variation, suggests control via a strong endogenous rhythm (Fig. 12.8). This is also likely to be the case for the minority of species, such as *Duboscia macrocarpa*, that fruit in an asynchronous manner.

Generalizing from these examples suggests that plant species at Lopé that have shown regular seasonal or asynchronous fruiting with no large interannual variations in crop size over the 9-year study period, reproduce at intervals determined by endogenous rhythms. For all other species, reproduction appears to be cued to one, or to a combination of, exogenous variables. If this reasoning is correct, the 40 species in the phenology sample that have shown irregular interannual variations in the size and/or timing of ripe fruit crops, depend on environmental cues in some way. This may be an overestimate as analysis of phenology data is complex: interactions between endogenous factors and external cues certainly occur (Borchert 1983) and these can mask causality. In at least some species, consistent individual differences exist in fruit crop size such that trees can be classed as 'good' or 'poor' fruiters (Tutin *et al.* 1991a), but it is unclear whether this is a result of genetic or environmental differences. Finally, some climatic variables covary (e.g. at Lopé: maximum and minimum temperatures; temperature and rain; and rainfall and sunshine (Tutin & Fernandez 1993a), making it difficult to isolate precise cues. Despite these provisos, the demonstrated influence of climatic variables on reproductive phenology has obvious implications for primates at Lopé as their diets are dominated by fruit or seeds.

Intuitively, it seems that tightly synchronized behaviour within a plant species will be linked to climatic cues. The number of reliable cues is limited in the tropical environment but several (including resumption of rainfall, increase in solar radiation and low temperature) have been identified as cues that trigger flowering in some species (Lieberman 1982, Borchert 1983, Van Schaik 1986, Tutin & Fernandez 1993a, Van Schaik *et al.* 1993). Plant species, and individuals within species, vary in their sensitivity to climatic cues and climatic patterns show interannual variation at most tropical sites. Supra-annual mast fruiting has been attributed to strong cues (such as a critical level of sunshine) that do not occur each year (Van Schaik 1986). Absence of a normally regular environmental cue can cause reproductive failure and, in some habitats, the resulting period of fruit scarcity can be so great as to cause mortality amongst frugivores (Foster 1982a). Foster (1982a,b) showed that on Barro Colorado Island, the absence of a clear dry season led to periodic years of mild or severe fruit scarcity and that rainfall during the dry season correlated with five severe famines that had led to mortality of frugivores in 1931, 1956, 1958, 1960 and 1970.

It is important to distinguish between the specific and general ways in which climate can influence plant phenology and fruit availability. Specific environmental cues that trigger flowering may be present or absent at a particular season or during a particular year giving clear-cut correlations between climatic events and plant reproductive behaviour. More general climatic variables (such as wet vs. dry weather or sunshine vs. cloud) will influence the physiological ability of individual plants to respond to an environmental cue and subsequently determine whether or not immature fruit are aborted or reach maturity. Little is known about: (i) indirect impact of climate on populations of insect pollinators; (ii) conditions that favour wind- and mechanical-dispersal of seeds; or (iii) climatic influences on germination and growth; but, undoubtedly, the weather intervenes at every stage and has exerted selective forces not only on the plant and animal components of tropical ecosystems but also on their varied interactions.

The past: reconstruction of the impact of climate change on Lopé vegetation

The Lopé ecosystem has emerged from a long history of dynamic changes as a result of past climate change and interactions with humans. Changes in vegetation at Lopé since the last glaciation have been reconstructed by a combination of dating transitions of $\partial^{13}C$ values in soil organic matter (which allow distinction between savanna and forest vegetation), pollen profiles and mapping of modern vegetation types that reflect varying ages of the recolonization of savannas by forest (Oslisly *et al.* 1996). Savannas have existed within the rainforest at Lopé throughout this period, but their extent has varied as a result of changes in both climate and human demography. The earliest known human occupation of the area dates from about 350 000 years ago (Oslisly 1993) and through their use of fire (Oslisly & Deschamps 1994), humans may have had a profound effect on vegetation dynamics, particularly in times of climatic stress.

Details of climate change in west-central Africa are well documented for the past 25 000 years with two major dry episodes occurring: first, 18 000–12 000 years ago the climate was drier and temperatures were at least 3°C lower than today; and second, 3000–2000 years ago, the climate was more seasonal (Maley 1987, 1992, 1993). During the first period cooler sea temperatures probably led to prolonged periods of cool, dry, cloudy weather extending the dry season beyond the present-day one of 3 months. As most succulent fruit (or at least those monitored in the present-day species pool at Lopé) cannot ripen in the dry season, the annual period of fruit scarcity would have lasted longer. In addition, the cooler temperatures, and perhaps also the drier weather, might have led to cues for flowering in some species becoming unreliable or irregular, thus loosening synchronous behaviour within these species or inhibiting reproduction altogether.

During the second period of stress 3000–2000 years ago, temperatures rose and higher sea temperatures are likely to have led to warm, sunny, weather during the dry season creating conditions similar to the present-day climate of 'dry' tropical forests in West Africa and in the Neotropics. The dry season might no longer have been the coldest time of year and the critical low temperature of 19°C required by some present-day species for flowering might never have been reached. Rhythms of fruit ripening are likely to have changed given the sunshine during the dry season. The probable disruption of patterns of ripe fruit availability during past periods of climate change would have had an immediate impact on primates and other frugivores.

It is known that forest area at Lopé was much reduced after the long period of drier, cooler, weather experienced 18 000–12 000 years ago (Oslisly *et al.* 1996, White 1997, White *et al.* 1997). Forest cover would have shrunk as trees died from physiological stress caused by reduced rainfall or through increased susceptibility to disease or to fire. Fires of human origin, lit in savannas at Lopé, do not spread into forest under present climatic conditions (Tutin, White & Mackanga-Missandzou 1996a), but in the drier climate of West Africa, fires do penetrate into forests especially those that have been selectively logged (Swaine 1992). Major forest expansion at Lopé began 1400 years ago when the area was void of humans (White *et al.* 1997). Before this, the forests of the Lopé study area were probably restricted to strips of riverine, or gallery, forest along watercourses (see Plate 12.1b) with patches of forest also persisting at altitudes above 650 m. As all of the primate species at Lopé depend on forest plants for food, the amount of suitable habitat would have been much less and their populations certainly reduced. The mosaic of present-day vegetation types at Lopé (Plate 12.1a) is exploited by primates on a seasonal basis as differences in plant-species composition result in differential fruit availability (Tutin, Parnell, White & Fernandez 1995, Tutin *et al.* 1997b). The decreased diversity of vegetation types in the savanna-dominated landscape of the past would have increased the vulnerability of frugivores to periodic crop failures.

Vegetation in ancient riverine forests at Lopé is dominated by trees of the family Caesalpiniaceae, which account for up to 55% of basal area (L.J.T. White unpublished data). In general, these trees do not produce succulent fruit but their dry seeds are major foods for primates at Salonga in Zaire in forests where few tree species produce fleshy fruit (Maisels, Gautier-Hion &

Gautier 1994). Seeds of several Caesalpiniaceae are heavily exploited by primates at Lopé in years when succulent fruit is scarce between February and May but few Caesalpiniaceae seeds are present during the annual dry season.

Primate populations at Lopé have had a turbulent history although details of the impact of past changes in vegetation on each species cannot yet be fully understood. Present-day patterns of primate use of the natural forest fragments in the remaining savanna vegetation at Lopé give some clues as to likely reactions of different species in the past and are, in the absence of fossils, the only way to reconstruct the likely history of the different species (Tutin *et al.* 1997b). Two of the *Cercopithecus* monkeys (*C. nictitans* and *C. cephus*) are able to live entirely within small forest fragments but other primates make only sporadic visits to the forest patches from the nearby continuous forest. The wide geographical distribution and wide range of habitats occupied by chimpanzees suggest that they could persist in a fragmented habitat. Mandrills, too, probably persisted and, indeed, the nomadic ranging patterns, large home ranges and formation of very large groups are adaptations that may have evolved in response to a fragmented habitat. However, our data indicate that the other four species (gorillas, black colobus, grey-cheeked mangabeys and crowned guenons) would have been restricted to refuges of continuous forest (Tutin *et al.* 1997b).

The future: impact of climate change on phenology and primates

Climate change through global warming in the future is likely to happen gradually and the initial impact will not involve dramatic increases in mortality of either plants or animals (but see Phillips & Gentry 1994, Condit, Hubbell & Foster 1996). Primates will, as in the past, be affected indirectly through changes in the availability of plant foods long before climate change induces physiological stress. Our data on plant phenology indicate that even small increases in temperature would disrupt flowering in certain tree species. Changes in the seasonal distribution of rainfall and/or sunshine will affect flowering, pollination, fruit set or fruit ripening in a range of other species. Undoubtedly, other unpredictable changes will occur as complex interactions and relationships will be upset.

If the amount of fruit being produced decreases dramatically, primates will seek alternative foods. Feeding on plant parts other than fruit is rarely advantageous to the plant and increased harvesting or leaves and bark may lead to death, or reduced fitness, in plant species with few chemical defences. Immature seeds are heavily exploited by most of the primates at Lopé whenever succulent fruit is scarce and the same is true of some mature wind-dispersed seeds. Levels of seed predation can reach very high levels depleting entire crops prior to ripening, and the strength and intelligence of primates allow them to breach normally effective seed defences such as hard shells or irritant hairs (Tutin, Parnell & White 1996b). Compared to fruit, many alternate foods are less nutritious, harder to digest, less easy to process and more time consuming to harvest. So even if sufficient calories could be obtained for survival, fitness of frugivorous species will be reduced and their reproduction adversely affected.

Primates are mobile, so if faced with a habitat that no longer produces sufficient food they can move in the hope of finding better areas. In a scenario of climate change through global warming, the probability of finding an unoccupied space would be small. The other option would be to adapt to the changed habitat which may be possible for four of the eight primate species (see above). The four species dependent on closed-canopy forest would be restricted to forest refuges, leading to geographical fragmentation of surviving populations.

CONCLUSIONS

The Lopé ecosystem has seen dramatic changes in climate and vegetation over the past 12 000 years — why then should we be concerned about the threat of future changes? There are several reasons for very deep concern and all are linked to the growth of populations of our own species. For the first time, climate change is occurring as a direct result of human technologies. At the same time biological diversity is under ever-increasing threat from actions directly related to human behaviour. At best, 7% of tropical habitats are being protected from logging, hunting and deforestation for agriculture. Pressure on tropical forests is growing daily and predictions of future trends are pessimistic; for example, the relationship between human population density and deforestation rates in tropical Africa suggests that in less than 100 years, forest habitat will persist in only three of the nine countries that now harbour wild gorillas (Harcourt 1995). Added to this, the short- and longer-term effects of climate change will cause many irreversible changes in tropical ecosystems. Recent studies using molecular genetic techniques suggest that some large mammal populations have in the past experienced extreme bottleneck periods where populations were reduced to a handful of survivors (e.g. Caro & Laurenson 1994). As conditions changed these survivors were founders of growing populations that spread into the empty habitat. What, though, is the probability of the handful of surviving gorillas or black colobus finding themselves within a protected area?

Conservation managers should perhaps take urgent action to identify likely forest refuges in the context of predicted global warming and create new protected areas.

Predicting the impact of climate change and global warming on natural ecosystems is not a new exercise, but long-term studies are now providing more information on how these changes might occur. There is a pressing need for more such studies and a multidisciplinary approach in order to better understand, predict and counter at least some of the negative impacts of man-made changes in climate. A discussion of the potential effects of global warming published in 1992 starts by stating 'It is unlikely that higher temperatures *per se* will be directly deleterious to tropical forest communities' (Hartshorn 1992). The analysis of our phenology data suggests otherwise as a group of important tree species that produce fruit foods for apes and monkeys will stop reproducing if the average temperature rises by only 1°C (Tutin & Fernandez 1993a). Similarly, recent analyses that suggest we can afford to wait for 30 years before tackling the root problems of global warming assume no serious impact of an increase in average temperature of 1.5–4°C (Ausubel 1991, Wigley, Richels & Edmonds 1996). These conclusions should be tempered: they are based wholy on anthropocentric arguments that consider human beings in splendid technocratic isolation.

The data now available on the role of climate in controlling the phenology of tropical plants indicate that if we do sit and wait for 30 years, perhaps the trees will wait too, but three decades of fruit scarcity will be very bad news indeed for the primates and other frugivores in the forests of central Africa.

ACKNOWLEDGEMENTS

First and foremost, we thank the Centre International de Recherches Médicales de Franceville, Gabon, for providing core funding for research at Lopé since 1983. Lee White is funded by the Wildlife Conservation Society and the ECOFAC (EU, DG VIII) programme provided invaluable technical and logistical support. We thank the following colleagues for contributing to data collection: Kate Abernethey, Rebecca Ham, Boo Maisels, Richard Parnell, Liz Rogers, Ben Voysey and Liz Williamson. For much stimulating discussion on topics covered in this chapter, we thank Michel Fernandez, Richard Oslisly, Jean Maley and Kate Abernethy. Constructive comments on the manuscript came from Herbert Prins and Robin Dunbar.

REFERENCES

Altmann, J. (1974). Observational study of behavior: sampling methods. *Behaviour*, **49**, 227–267.
Ausubel, J.H. (1991). Does climate still matter? *Nature*, **350**, 649–652.
Borchert, R. (1983). Phenology and control of flowering in tropical trees. *Biotropica*, **15**, 81–89.
Borchert, R. (1996). Phenology and flowering periodicity of Neotropical dry forest species: evidence from herbarium collections. *Journal of Tropical Ecology*, **12**, 65–80.
Caro, T.M. & Laurenson, M.K. (1994). Ecological and genetic factors in conservation: A cautionary tale. *Science*, **263**, 485–486.
Chapman, C. (1987). Flexibility in diets of three species of Costa Rican primates. *Folia Primatologia*, **49**, 90–105.
Condit, R., Hubbell, S.P. & Foster, R.B. (1996). Changes in tree species abundance in a Neotropical forest: Impact of climate change. *Journal of Tropical Ecology*, **12**, 231–256.
Corlett, R.T. (1990). Flora and reproductive phenology of the rainforest of Bukit Timah, Singapore. *Journal of Tropical Ecology*, **6**, 55–63.
Emmons, L.H., Gautier-Hion, A. & Dubost, G. (1983). Community structure of the frugivorous–folivorous forest mammals of Gabon. *Journal of Zoology (London)*, **199**, 209–222.
Estrada, A. & Coates-Estrada, R. (1991). Howler monkeys (*Alouatta palliata*), dung beetles (Scarabaeidae) and seed dispersal: Ecological interactions in the tropical rain forest of Los Tuxtlas, Mexico. *Journal of Tropical Ecology*, **7**, 459–474.
Fleming, T.H. & Williams, C.F. (1990). Phenology, seed dispersal, and recruitment in *Cecropia peltata* (Moraceae) in Costa Rican tropical dry forest. *Journal of Tropical Ecology*, **6**, 163–178.
Foster, R.B. (1982a). Famine on Barro Colorado Island. In *The Ecology of a Tropical Forest: Seasonal Rhythms and Long-term Changes* (Ed. by E.G. Leigh, A.S. Rand & D.M. Windsor), pp. 201–212. Smithsonian Institution Press, Washington DC.
Foster, R.B. (1982b). The seasonal rhythm of fruitfall on Barro Colorado Island. In *The Ecology of a Tropical Forest: Seasonal Rhythms and Long-term Changes* (Ed. by E.G. Leigh, A.S. Rand & D.M. Windsor), pp. 151–172. Smithsonian Institution Press, Washington DC.
Galat, G. & Galat-Luong, A. (1985). La communauté de primates diurnes de la forêt de Taï, Côte d'Ivoire. *Revue d'Ecologie Terre et Vie*, **40**, 3–32.
Gautier-Hion, A. & Michaloud, G. (1989). Are figs always keystone resources for tropical frugivorous vertebrates? A test in Gabon. *Ecology*, **70**, 1826–1833.
Gautier-Hion, A., Duplantier, J.M., Quris, R. Feer F., Sourd, C., Decoux, J.P. et al. (1985). Fruit characters as a basis of fruit choice and seed dispersal in a tropical forest vertebrate community. *Oecologia*, **65**, 324–337.
Ham, R.M. (1994). *Behavioural ecology of grey-cheeked mangabeys* (Cercocebus albigena) *in the Lopé Reserve, Gabon*. PhD thesis, University of Stirling.
Hamilton, A.C. (1982). *Environmental History of East Africa*. Academic Press, London.
Harcourt, A.H. (1995). Population viability estimates: Theory and practice for a wild gorilla population. *Conservation Biology*, **9**, 134–142.
Harrison, M.J.S. & Hladik, C.M. (1986). Un primate granivore: le colobe noir dans la forêt du Gabon; potentialité d'évolution du comportement alimentaire. *Revue Ecologique*, **41**, 281–29.
Hartshorn, G.S. (1992). Possible effects of global warming on the biological diversity in tropical forests. In *Global Warming and Biological Diversity* (Ed. by R.L. Peters & T.E. Lovejoy), pp. 137–146. Yale University Press, New Haven.
Heaney, L.R. (1992). A synopsis of climatic and vegetational change in Southeast Asia. In *Tropical Forests and Climate* (Ed. by N. Myers), pp. 53–61. Kluwer, Dordrecht.

Houghton, J.T., Jenkins, G.J. & Ephraums, J.J. (1990). *Climate Change: The IPPC Scientific Assessment*. Cambridge University Press, Cambridge.

Houghton, R.A. (1995). Tropical forests and climate. In *Ecology, Conservation and Management of Southeast Asian Rainforests* (Ed. by R.B. Primack & T.E. Lovejoy), pp. 263–290. Yale University Press, New Haven.

Howe, H.F. (1980). Monkey dispersal and waste of a Neotropical fruit. *Ecology*, **61**, 944–959.

Irion, G. (1989). Quaternary geological history of the Amazon lowlands. In *Tropical Forests: Botanical Dynamics, Speciation and Diversity* (Ed. by L.B. Holm-Nielsen, I.C. Nielsen & H. Balslev), pp. 23–34. Academic Press, London.

Julliot, C. (1994). Diet diversity and habitat of howler monkeys. In *Current Primatology*, Vol. I. *Ecology and Evolution* (Ed. by B. Thierry, J.R. Anderson, J.J. Roeder & N. Herrenschmidt), pp. 67–71. Université Louis Pasteur, Strasbourg.

Kershaw, A.P. (1992). The development of rainforest–savanna boundaries in tropical Australia. In *Nature and Dynamics of Forest–Savanna Boundaries* (Ed. by P.A. Furley, J. Proctor & J.A. Ratter), pp. 255–271. Chapman & Hall, London.

Koechlin, J. (1964). Marantacées, Zingiberacées. In *Flore du Gabon*, Vol. 9 (Ed. by A. Aubréville), pp. 15–158. Muséum National d'Histoire Naturelle, Paris.

Leighton, M. & Leighton, D.R. (1983). Vertebrate response to fruiting seasonality within a Bornean rain forest. In *Tropical Rain Forest Ecology and Management* (Ed. by S.L. Sutton, T.C. Whitmore & A.C. Chadwick), pp. 181–196. Blackwell Scientific Publications, Oxford, London.

Letouzey, R. (1968). *Etude Phytogéographique du Cameroun*. Editions Paul Lechevalier, Paris.

Lieberman, D. (1982). Seasonality and phenology in a dry tropical forest in Ghana. *Journal of Ecology*, **70**, 791–806.

Livingstone, D.A. (1982). Quaternary geography of Africa and the refuge theory. In *Biological Diversification in the Tropics* (Ed. by G.T. Prance), pp. 523–536. Columbia University Press, New York.

MacKinnon, J.R. & MacKinnon, K.S. (1980). Niche differentiation in a primate community. In *Malayan Forest Primates: Ten Years' Study in Tropical Rain Forest* (Ed. by D.J. Chivers), pp. 167–190. Plenum, New York.

Maisels, F., Gautier-Hion, A. & Gautier, J.P. (1994). Diets of two sympatric colobines in Zaire: more evidence on seed-eating in forests on poor soils. *International Journal of Primatology*, **15**, 681–701.

Maley, J. (1987). Fragmentation de la forêt dense humide africaine et extension des biotopes montagnards au Quaternaire récent: nouvelles données polliniques et chronologiques. Implications paléoclimatiques et biogéographiques. *Palaeoecology of Africa*, **18**, 307–334.

Maley, J. (1989). Late Quaternary climatic changes in African rain forest: Forest refugia and the major role of sea surface temperature variations. In *Paleoclimatology and Paleometeorology: Modern and Past Patterns of Global Atmospheric Transport* (Ed. by M. Leinen & M. Sarnthein), pp. 585–616. Kluwer, Dordrecht.

Maley, J. (1992). Mise en évidence d'une péjoration climatique entre ca. 2500 et 2000 ans BP en Afrique tropicale humide. *Bulletin de la Société Géologique de France*, **163**, 363–365.

Maley, J. (1993). The climatic and vegetational history of equatorial regions of Africa during the Upper Quaternary. In *The Archaeology of Africa: Foods, Metals and Towns* (Ed. by T. Shaw, P. Sinclair, B. Andah & A. Okpoko), pp. 43–52. Routledge, London.

Maley, J., Caballé, G. & Sita, P. (1990). Etude d'un peuplement résiduel à basse altitude de *Podocarpus latifolia* sur le flanc Congolais de Massif du Chaillu. Implications paléoclimatiques et biogéographiques. Etude de la pluie pollenique actuelle. In *Paysages Quaternaires de l'Afrique Centrale Atlantique* (Ed. R. Lanfranchi & D. Schwartz), pp. 336–352. ORSTOM, Paris.

Myers, N. (1992). *Tropical Forests and Climate*. Kluwer, Dordrecht.
Oslisly, R. (1993). *Préhistoire de la Moyenne Vallée de l'Ogooué (Gabon)*. ORSTOM, Paris.
Oslisly, R. (1995). The middle Ogooué valley, Gabon, from 3500 yrs BP: cultural changes and palaeoclimatic implications. *Azania*, **14/15**, 324–331.
Oslisly, R. & Deschamps, R. (1994). Découverte d'une zone d'incendie dans la forêt ombrophile du Gabon ca 1500 BP: Essai d'explication anthropique et implications paléoclimatiques. *Comptes Rendus de l'Académie des Sciences de Paris, série IIa*, **318**, 555–560.
Oslisly, R., Peyrot, B., Abdessadok, S. & White, L. (1996). Le site de Lopé 2: Un indicateur de transition écosystémique ca 10000 BP dans la moyenne vallée de l'Ogooué (Gabon). *Comptes Rendus de l'Académie des Sciences de Paris, série IIa*, **323**, 933–939.
Phillips, O.L. & Gentry, A. (1994). Increasing turnover through time in tropical forests. *Science*, **263**, 954–958.
Rietkerk, M., Ketner, P. & de Wilde, J.J.F.E. (1995). Caesalpinioideae and the study of forest refuges in Gabon: preliminary results. *Adansonia*, **1/2**, 95–105.
Rogers, M.E., Williamson, E.A., Tutin, C.E.G. & Fernandez, M. (1988). Effects of the dry season on gorilla diet in Gabon. *Primate Report*, **19**, 29–34.
Rogers, M.E., Tutin, C.E.G., Parnell, R.J., Voysey, B.C., Williamson, E.A. & Fernandez, M. (1994). Seasonal feeding on bark by gorillas: an unexpected keystone food? In *Current Primatology*, Vol. I. *Ecology and Evolution* (Ed. by B. Thierry, J.R. Anderson, J.J. Roeder & N. Herrenschmidt), pp. 37–44, Université Louis Pasteur, Strasbourg.
Sanford, R.L., Saldarraga, J., Clark, K.E., Uhl, C. & Herrera, R. (1985). Amazon rain-forest fires. *Science*, **227**, 53–55.
Schwartz, D., Lanfranchi, R. & Mariotti, A. (1990). Origine et évolution des savannes intra-mayombiennes (R.P. du Congo) I. Apports de la pédologie et de la biogéochimie isotopique (14C et 13C). In *Paysages Quaternaire de l'Afrique Centrale Atlantique* (Ed. by R. Lanfranchi & D. Schwartz), pp. 314–325. ORSTOM, Paris.
Sosef, M.S.M. (1991). New species of *Begonia* in Africa and their relevance to the study of glacial forest refuges. *Wageningen Agricultural University Papers, Studies in Begoniaceae*, **4**, 120–151.
Struhsaker, T.T. (1975). *The Red Colobus Monkey*. University of Chicago Press, Chicago.
Struhsaker, T.T. & Leland, L. (1979). Socioecology of five sympatric monkey species in the Kibale Forest, Uganda. *Advances in the Study of Behaviour*, **9**, 159–228.
Swaine, M.D. (1992). Characteristics of dry forest in West Africa and the influence of fire. *Journal of Vegetation Science*, **3**, 365–374.
Terborgh, J. (1983). *Five New World Primates*. Princeton University Press, Princeton, NJ.
Tutin, C.E.G. & Fernandez, M. (1993a). Relationships between minimum temperature and fruit production in some tropical forest trees in Gabon. *Journal of Tropical Ecology*, **9**, 241–248.
Tutin, C.E.G. & Fernandez, M. (1993b). Composition of the diet of chimpanzees and comparisons with that of sympatric lowland gorillas in the Lopé Reserve, Gabon. *American Journal of Primatology*, **30**, 195–211.
Tutin, C.E.G. & Fernandez, M. (1994). Faecal analysis as a method of describing diets of apes: Examples from sympatric gorillas and chimpanzees at Lopé, Gabon. *Tropics*, **2**, 189–198.
Tutin, C.E.G., Williamson, E.A., Rogers, M.E. & Fernandez, M. (1991a). A case study of a plant–animal relationship: *Cola lizae* and lowland gorillas in the Lopé Reserve, Gabon. *Journal of Tropical Ecology*, **7**, 181–199.
Tutin, C.E.G., Fernandez, M., Rogers, M.E., Williamson, E.A. & McGrew, W.C. (1991b). Foraging profiles of sympatric lowland gorillas and chimpanzees in the Lopé Reserve, Gabon. *Philosophical Transactions of the Royal Society, Series B*, **334**, 179–186.
Tutin, C.E.G., White, L.J.T., Williamson, E.A., Fernandez, M. & McPherson, G. (1994). List of plant species identified in the northern part of the Lopé Reserve, Gabon. *Tropics*, **3**, 249–276.

Tutin, C.E.G., Parnell, R.J., White, L.J.T. & Fernandez, M. (1995). Nest building by lowland gorillas in the Lopé Reserve, Gabon: Environmental influences and implications for censusing. *International Journal of Primatology*, **16**, 53–76.

Tutin, C.E.G., White, L.J.T. & Mackanga-Missandzou, A. (1996a). Lightning strike burns a large forest tree in the Lopé Reserve, Gabon. *Global Ecology and Biogeography Letters*, **5**, 36–41.

Tutin, C.E.G., Parnell, R.J. & White, F. (1996b). Protecting seeds from primates: Examples from the Lopé Reserve, Gabon. *Journal of Tropical Ecology*, **12**, 371–384.

Tutin, C.E.G., Ham, R.M., White, L.J.T. & Harrison, M.J.S. (1997a). The primate community of the Lopé Reserve, Gabon: Diets, responses to fruit scarcity and effects on biomass. *American Journal of Primatology*, **42**, 1–24.

Tutin, C.E.G., White, L.J.T. & Mackanga-Missandzou, A. (1997b). The use by rain forest mammals of natural forest fragments in an equatorial African savanna. *Conservation Biology*, in press.

Van der Hammen, T. (1992). Palaeoecological background: Neotropics. In *Tropical Forests and Climate* (Ed. by N. Myers), pp. 37–47. Kluwer, Dordrecht.

Van Schaik, C.P. (1986). Phenological changes in a Sumatran rain forest. *Journal of Tropical Ecology*, **2**, 327–347.

Van Schaik, C.P., Terborgh, J.W. & Wright, S.J. (1993). The phenology of tropical forests: Adaptive significance and consequences for primary consumers. *Annual Review of Ecology and Systematics*, **24**, 353–377.

White, L.J.T. (1992). *Vegetation history and logging disturbance: effects on rain forest mammals in the Lopé Reserve, Gabon*. PhD thesis, University of Edinburgh.

White, L.J.T. (1994a). Biomass of rain forest mammals in the Lopé Reserve, Gabon. *Journal of Animal Ecology*, **63**, 499–512.

White, L.J.T. (1994b). Patterns of fruit-fall phenology in the Lopé Reserve, Gabon. *Journal of Tropical Ecology*, **10**, 289–312.

White, L.J.T. (1997). Forest–savanna dynamics and the origins of 'Marantaceae Forest' in the Lopé Reserve, Gabon. In *African Rain Forest Ecology and Conservation* (Ed. by B. Weber, L.J.T. White, A. Vedder & H. Simons-Morland). Yale University Press, New Haven, in press.

White, L.J.T. & Abernethy, K.A. (1996). *Guide de la Végétation de la Réserve de la Lopé, Gabon*. ECOFAC, Libreville, Gabon.

White, L.J.T., Rogers, M.E., Tutin, C.E.G., Williamson, E.A. & Fernandez, M. (1995). Herbaceous vegetation in different forest types in the Lopé Reserve, Gabon: Implications for keystone food availability. *African Journal of Ecology*, **33**, 124–141.

White, L.J.T., Oslisly, R., Abernethy, K.A. & Maley, J. (1997). *Aucoumea klaineana*: A Holocene success story now in decline? In *Histoire de la Forêt Tropicale Africaine*, ORSTOM, Paris, in press.

Wigley, T.M.L., Richels, R. & Edmonds, J.A. (1996). Economic and environmental choices in the stabilization of atmospheric CO_2 concentrations. *Nature*, **379**, 240–243.

Williamson, E.A., Tutin, C.E.G., Rogers, M.E. & Fernandez, M. (1990). Composition of the diet of lowland gorillas at Lopé in Gabon. *American Journal of Primatology*, **21**, 265–277.

13. EFFECTS OF HABITAT FRAGMENTATION ON PLANT POLLINATOR INTERACTIONS IN THE TROPICS

S. S. RENNER
Department of Biology, University of Missouri—St Louis, St Louis, MO 63121-4499, USA

SUMMARY

1   This chapter reviews how tropical plant–pollinator systems respond to changes of the environment, especially habitat fragmentation due to deforestation; natural disturbances, such as hurricanes, also impact pollination interactions and are therefore also included. Of the 758 recent studies of habitat fragmentation, about two-thirds investigated how particular animal taxa respond to fragmentation while the remainder focused on landscape, habitat, or ecosystem characteristics from empirical or theoretical perspectives.

2   Of the 13 studies addressing the effects of large-scale habitat deterioration on plant–pollinator interactions and/or plant fitness, most provide qualitative, rather than quantitative, evidence. One study quantified effects by comparing fruit set, pollen deposition, pollen tube growth, or seed set in duplicated fragments of various sizes and in continuous habitat and found a negative effect of fragmentation in all but two of 16 species.

3   Five of the 13 studies report on pollinator extinction or presumed extinction; two report on pollinator absence due to cyclones or hurricanes, in one case followed by speedy recovery of the plant–pollinator interaction; and five report on pollination by other animals where the original pollinator was lost.

4   There are two ways to view the proposition that temperate and tropical latitudes differ in the vulnerability of their pollination mutualisms, and these should be explicitly tested in future studies. One perspective is that, because the proportions of various pollination syndromes differ with latitude (if altitude is held constant), a relative increase in more vulnerable kinds of pollination syndromes may make whole vegetation types particularly vulnerable. From this perspective it is possible to argue that since the relative frequency of pollination by birds and bats increases towards the equator this may make tropical vegetation more vulnerable than temperate vegetation, provided that bird or bat pollination is disproportionally sensitive. A second

perspective is that, because plants that are pollination specialists may be more vulnerable than plants that are pollination generalists, a relative increase in number of specialists with decreasing latitude may make vegetation at low latitudes more vulnerable.

5 The second perspective is more problematic than the first because studies comparing the relative proportions of pollination specialists and generalists in different vegetation and habitat types appear extremely difficult. Part of the problem stems from the asymmetry of pollination interactions, which precludes extrapolations, such as Tepedino's or Bond's, from floral-morphological specialization to specialization and dependence of the interaction. While the effects of habitat deterioration and fragmentation on plant–pollinator interactions will thus be hard to quantify in statistically manageable ways, it is feasible to monitor changes in seed set in carefully selected species in duplicated fragments of known age. Criteria for selection should be that species represent different mating systems, growth forms, and pollination syndromes, especially bee pollination, which is understudied relative to its quantitative importance.

INTRODUCTION

The objectives of this contribution were to review how tropical pollination systems respond to changes in their habitats. These changes might result from infrequent large natural disturbances or anthropogenic habitat fragmentation and degradation. It was clear that habitat fragmentation due to deforestation would be the most relevant change currently affecting plant–pollinator interactions in the tropics. Accordingly, I focus on documented effects of man-made habitat changes on pollination and seed set. Infrequent large natural disturbances, such as tropical hurricanes or climate deterioration, also produce habitat changes to which plant–pollinator interactions are responding in ways that are just beginning to be studied but that clearly are relevant to my topic.

Although it is often stated (Dirzo, Naeem, & Cushman 1995, Harrison 1995, Didham, Ghazoul, Stork & Davis 1996, Young, Boyle & Brown 1996) that there are many examples of plant–pollinator linkages breaking down due to habitat fragmentation, studies supporting this are scarce, as becomes clear in the present review (which also discusses problems with the one paper cited by all these authors as demonstrating such breakdown, namely Powell & Powell (1987)). Nevertheless, carefully designed studies are beginning to detect the negative effects of habitat deterioration on pollination predicted in the 1980s (e.g. Gilbert 1980, Feinsinger 1987, Janzen 1987), and their results permit a few generalizations.

Habitat fragmentation and pollinators   341

The primary data considered in this review are the direct effects of habitat fragmentation on pollination levels and pollinator abundances. Besides biotic changes, habitat fragmentation causes large changes in the physical environment (reviewed in Laurance & Bierregaard 1997), and it is clear that these affect pollination, but there are as yet no data on such indirect effects.

I begin with an assessment of current knowledge of the responses of different plant–pollinator interactions to habitat deterioration and fragmentation. Based on the results of these studies, I then discuss three possible generalizations about the vulnerability of plants differing in pollination syndrome, degree of pollinator dependence, and mating system. This is followed by a comparison of tropical and temperate latitudes in terms of their vulnerability to large-scale pollination disturbances.

THE DATA

Changes in biological interactions are difficult to measure (Connell 1983, Bronstein 1994), and it is therefore not surprising that the two surveys to date of the effects of habitat fragmentation on flower–pollinator interactions remark upon the scarcity of relevant data (Rathcke & Jules 1993, Olesen & Jain 1994). These reviews found only a handful of studies from both temperate and tropical latitudes. A recent edited volume on the ecology of tropical forest remnants (Laurance & Bierregaard 1997) also contains nothing on plant–pollinator interactions in relation to fragmentation. A January 1996 search of six electronic literature databases going back as far as the early 1970s retrieved 758 references for habitat fragmentation (Olesen personal communication). About 90% (689) of these papers appeared in the past 10 years. Seventy-five per cent (569) investigated how particular animal taxa, or species diversity, respond to fragmentation while the remainder (25%) focused on landscape, habitat or ecosystem characteristics from empirical or theoretical perspectives. Effects found most often were changes in animal movement patterns, animal abundance and community species-richness. How changes in animal abundance, or behaviour, affect plants in habitat remnants was less frequently studied; only 11% of the studies dealt with plants. Of 670 studies restricted to a particular region, 58% were from temperate latitudes, 17% from the subtropics (mostly Australia) and 26% from the tropics.

Thirteen studies that more or less directly addressed the effects of large-scale natural or anthropogenic habitat deterioration on plant–pollinator interactions and/or plant fitness (e.g. measured as seed set) are summarized in Table 13.1. Particularly important because of its careful sampling design is a study from Argentina by Aizen and Feinsinger (1994a); it is the first to investigate the effects of habitat fragmentation on pollination in a set of species occurring in replicated fragments, and its results will be discussed in detail below (cf. Bee Pollination).

We are thus far from a comprehensive picture of how plant–pollinator (and plant–disperser) interactions react to habitat fragmentation in the tropics or subtropics. In the six sections that follow, the available data are discussed systematically by pollinator group (in the sequence of Table 13.1).

POLLINATION SYSTEMS AFFECTED BY HABITAT FRAGMENTATION

Bird pollination

Fewer than 1% of all flowering plant species, or a few thousand species in perhaps 500 genera, are adapted for bird pollination (Porsch 1931, and my own estimate). However, in some tropical forests, such as that at La Selva in Costa Rica, as many as 15% of the species depend on hummingbirds for pollination (Kress & Beach 1994). Thus, a decline of hummingbird abundance or a loss of hummingbird species from forest fragments should adversely affect the reproductive success of a clearly defined guild of plants (identifiable because of the characteristic features of hummingbird-adapted flowers). This is borne out by the results of a study done in the Ecuadorean Andes, where A.B. Christensen, L.I. Eskildsen and J.M. Olesen (unpublished data) compared pollinator visits and fruit set in two populations of *Passiflora mixta* (Passifloraceae), one growing in a reserve, the other at a farmland site 40 km away. Flowers of *P. mixta* are very long-tubed and can be pollinated only by the sword-billed hummingbird, *Ensifera ensifera* (Snow & Snow 1980, A.B. Christensen, L.I. Eskildsen & J.M. Olesen, unpublished data). Individuals of *P. mixta* inside the reserve had significantly higher fruit set than the plants growing at the farmland site, where *E. ensifera* rarely visited them. The hummingbird, in turn, obtains nectar from a relatively widely distributed guild of long-tubed flowers, including *P. mixta*, *Brugmansia sanguinea* (Solanaceae) and some other species, all of which it pollinates.

A second study looking at hummingbird–flower interactions in fragments (Aizen & Feinsinger 1994a) demonstrated that other hummingbird species do cross inhospitable habitats to visit isolated fragments or plants and in fact prefer the light- and flower-rich fragments: of the two hummingbird-pollinated species included in the study, one exhibited no change in pollination level in fragments, the other had almost three times as many pollen tubes growing per style in fragments as in continuous forest. Also, the results of a 9-year sampling effort in Amazonian forest fragments (Stouffer & Bierregaard 1995) show that at least some forest understorey hummingbirds can persist in a matrix of fragments, secondary growth, active cattle pasture, and large (for a hummingbird) forest patches.

The remaining examples of plants suffering from the loss of their pollinating birds come from islands. *Stenogyne kanehoana* (Lamiaceae), an extremely endangered Hawaiian species with perhaps two or three individuals left in the wild (Weller 1994), has not set fruit in years. The species is endemic on the island of Oahu, and its flowers are almost certainly adapted for exclusive pollination by honeycreepers (Drepanidinae). The larger honeycreepers are all extinct on Oahu, and meliphagids are extinct throughout Hawaii (as the result of introduced avian malaria). While *Stenogyne* appears to be on the brink of extinction, *Clermontia arborescens* (Campanulaceae–Lobelioideae), a species thought originally to depend for pollination on a particular drepanid species, is now pollinated by the introduced Japanese white-eyes, *Zosterops japonica* (Lammers, Weller & Sakai 1987). Japanese white-eyes now also pollinate *Freycinetia arborea* (Cox 1983) and the native *Erythrina sandwicensis*, which is endemic to the dry forests in the lowland areas of all major Hawaiian islands and probably also was pollinated in the past by native drepanids (D. Neill personal communication).

Although coming from outside the tropics, the following case study from New Zealand provides a further example of plants suffering from damaged bird faunas that seems relevant to the tropics because the plant family involved is largely tropical. The New Zealand mistletoes in question have explosive flowers that need to be twisted open by nectarivorous birds (Ladley & Kelly 1995). Unvisited flowers sometimes self-pollinate when the petals undergo abscission, but in some species this is not possible. One such species, the New Zealand *Trilepidea adamsii*, is believed extinct (Ladley & Kelly 1995), and it has been suggested that reduced bird densities due to introduced mammalian predators may have contributed to its rapid decline. That the extinct species indeed depended on birds is suggested by the pollination mechanism of its extant close relatives, which are currently pollinated by meliphagids persisting locally at low abundances. Present-day bird populations in some areas are insufficient to open more than a minority of explosive mistletoe flowers (Ladley & Kelly 1995).

TABLE 13.1. Plant–pollinator interactions affected by large-scale habitat fragmentation.[1]

Plant	Animal	Study site	Plant response	Causes	References
Bird-pollinated plants					
Passiflora mixta (Passifloraceae)	*Ensifera ensifera*	Andes, Ecuador	Reduced fruit set	Low visitation rates in fragments	A.B. Christensen, L.I. Eskildsen and J.M. Olesen (unpublished data)
Stenogyne kanehoana (Lamiaceae)	Drepanidinae	Hawaii, Oahu	Absence of fruit set; plant almost extinct	Extinction of native bird pollinators	Weller (1994)
Clermontia arborescens (Campanulaceae)	*Vestiaria coccinea* (Drepanidinae)	Hawaii, Lanai	Pollinator shift (to *Zosterops*)	Extinction of native bird pollinators	Lammers *et al.* (1987)
Freycinetia arborea (Pandanaceae)	Drepanidinae and Corvidae	Hawaii	Pollinator shift (to *Zosterops*)	Extinction of native bird pollinators	Cox (1983)
Erythrina sandwicensis (Leguminosae)	Drepanidinae	Hawaii	Pollinator shift (to *Zosterops*)	Extinction of native bird pollinators	D. Neill (personal communication)
Isoplexis spp. (Scrophulariaceae)	Nectariniidae	Canary Islands	Pollinator shift	Extinction of native bird pollinators	Vogel *et al.* (1984)*, Olesen (1985)*
Lotus berthelotii (Leguminosae)	Nectariniidae	Canary Islands	Plant extinct in wild	Insufficient bird abundance or goats?	Vogel *et al.* (1984), Olesen (1985)
Canarina canariensis (Campanulaceae)	Nectariniidae	Canary Islands	Pollinator shift	Extinction of native bird pollinators	Vogel *et al.* (1984), Olesen (1985)
Peraxilla & *Alepis* spp. (Loranthaceae)	Meliphagidae	New Zealand	Reduced fruit set?	Insufficient bird abundance	Ladley & Kelly (1995)
Trilepidea adamsii (Loranthaceae)	Meliphagidae	New Zealand	Plant extinct	Extinction of native bird pollinators?	Ladley & Kelly (1995)
Ligaria cuneifolia (Loranthaceae)	*Sappho sparganura*	Argent. Chaco	Almost 3× as many pollen tubes per style	Opportunistic birds visit fragments	Aizen & Feinsinger (1994a)
Tillandsia ixioides (Bromeliaceae)	*Chlorostilbon aureoventris*	Argent. Chaco	None found	As for *Ligaria*	Aizen & Feinsinger (1994a)

Bat-pollinated plants					
e.g. *Ceiba pentandra* (Bombacaeae), *Syzygium* spp. (Myrtaceae)	Pteropodidae		Reduced fruit set?	Insufficient bat abundance due to hunting pressure by humans	Cox *et al.* (1991), Rainey *et al.* (1995)
Pteropus-pollinated/ dispersed plants	Pteropodidae	Guam	Reduced fruit set?	Cyclone: local extinction of bats	Pierson *et al.* (1996)*
Moth-pollinated plants					
Brighamia insignis (Campanulaceae)	Native sphingid moths	Western Samoa	Reduced fruit set; plant almost extinct	Insufficient moth abundance and other factors, such as goats	C. Gemmill *et al.* (unpublished data)
		Hawaii			
Butterfly-pollinated plants					
Dianthus deltoides (Caryophyllaceae)	Native butterflies	Sweden	Reduced seed set	Low visitation rates in fragments	Jennersten (1988)
Bee-pollinated plants					
Atamisquea emarginata (Capparidaceae)	Native and honey bees	Argent. Chaco	Reduced pollen growth	Increased deposition of self or genetically close pollen; all 3 spp. are self-incompatible	Aizen & Feinsinger (1994a)[2]
Prosopis nigra (Leguminosae)	Native and honey bees	Argent. Chaco	Reduced pollen growth		Aizen & Feinsinger (1994a)
Cercidium australe (Leguminosae)	Native and honey bees	Argent. Chaco	Reduced pollen growth		Aizen & Feinsinger (1994a)
Wasp-pollinated plants					
Ficus aurea (Moraceae)	Wasp: *Pegoscapus jimenezi*	Florida	None found	Hurricane: local extinction of wasps but quick recolonization	Bronstein & Hossaert-McKey (1995)*

[1] Studies marked with an * deal with large-scale natural disturbances or habitat deterioration, rather than habitat fragmentation due to man.
[2] The study comprised 14 insect-pollinated species, 13 of which suffered reduced seed output (defined as fruit set × seed set) in fragments compared to continuous forest. We include the three species with the strongest sampling designs (cf. Aizen & Feinsinger 1994a, p. 344).

Bat pollination

About 1000 species of angiosperms in 250 genera are adapted for pollination by megachiropterans or microchiropterans (Dobat & Peikert-Holle 1985); many additional species rely on bats for seed dispersal. At La Selva in Costa Rica, 10 of 276 species are pollinated by bats (Kress & Beach 1994), and bats disperse 16–18% of the tree species and up to 37% of the shrubs and treelets in a Peruvian rainforest (Foster, Arce & Wachter 1986). In SE Asia, flying foxes (Pteropodidae) have been identified as being important pollinators and seed dispersers, especially in oceanic island ecosystems with depauperate mammal faunas (Crome & Irvine 1986, Cox, Elmqvist, Pierson & Rainey 1991, Elmqvist, Cox, Rainey & Pierson 1992, Rainey, Pierson, Elmqvist & Cox 1995, Banack 1996). However, to my knowledge no studies have been published that directly demonstrate the effects of reduced bat abundance, or changes in movement patterns, on pollination or seed set in plants in habitat remnants. Perhaps because individual nectarivorous and frugivorous bats usually exploit a wide range of plants and show considerable habitat flexibility (e.g. Fenton, Acharya, Audet *et al.* 1992), bat–flower interactions are less readily affected than bird–flower interactions. Nevertheless, several scenarios relate human activites to the disintegration of bat–plant mutualisms (e.g. Cox *et al.* 1991, Rainey *et al.* 1995, Pierson, Elmqvist, Rainey & Cox 1996). Thus, in South Pacific islands, flying foxes are the principal pollinators of certain large-flowered tree species (Table 13.1), and anthropogenic extinctions of large vertebrates, including these bats and certain birds, on the islands have reduced the guild of pollinators and dispersers for these plants. As a result, bat-pollinated species are less diverse in areas that lack flying foxes.

Butterfly and moth pollination

Extratropical studies suggest that butterflies usually are comparatively unimportant as pollinators (e.g. Jennersten 1984). At La Selva, 12 of 276 flowering plants are pollinated by butterflies or skippers (Kress and Beach 1994). Butterflies are, however, among the best-studied organisms in terms of their reaction to habitat fragmentation, both in the temperate zone and in the tropics (e.g. Pollard 1982, Rodrigues, Brown & Ruszczyk 1993, Daily & Ehrlich 1995). A seminal study done in Sweden (Jennersten 1988) demonstrated that a decline in butterfly abundance directly affects pollination and seed set in some Caryophyllaceae.

No such direct data are available from the tropics. Research in forest fragments north of Manaus, Brazil (reviewed in Brown & Hutchings 1997), has shown, however, that the primary understorey butterflies, Ithomiinae, experience decreases in population sizes in fragments. It is also known that butterfly numbers may be limited by nectar sources for adults (Wood & Samways 1991, Rodrigues *et al.* 1993, and references therein), but it is unclear whether fragments in the tropics contain more, or fewer, nectar-yielding flowers than undisturbed vegetation. In the case of the Ithomiinae, females mainly feed on bird droppings, while males depend on weedy plants, such as *Heliotropium*, *Tournefourtia*, *Eupatorium* and *Senecio*, for pyrrolizidine alkaloids (DeVries 1987). Thus, the reduction in their numbers is unlikely to negatively affect the reproduction of the native trees and shrubs in the Manaus fragments.

More plant species in tropical forests depend on moths for pollination than on butterflies because their floral morphologies restrict them exclusively to insects with very long tongues (although it would be misleading to think of hawkmoths only as long-tongued specialists; Haber & Frankie 1989). Long-tongued large moths may have foraging ranges that include plant individuals as far as 10–15 km apart (W.A. Haber in Janzen 1983, p. 633, W.A. Haber in Chase, Moller, Kesseli & Bawa 1996). Unlike bees and hummingbirds, hawkmoths do not forage within narrow home ranges. Rather, they cruise large habitat patches, commonly visiting only a few flowers before moving to another patch, which may be hundreds of metres away (Haber & Frankie 1989). Both traits suggest that moth–flower interactions may be among the last to be affected, at least as long as moths cross secondary vegetation surrounding habitat remnants.

Even so, the low abundance or possible extinction of a pollinating sphingid moth species is implicated in the low fruit set of the endemic Hawaiian *Brighamia* (Campanulaceae). *Brighamia* has two morphologically similar species, *B. insignis* and *B. rockii*, both of which are on the US endangered-species list. Judging by floral morphology they depend(ed) for pollination on the green sphinx moth (*Tinostoma smaragditis* Meyrick), the Hawaiian sphinx (*Hyles calida* Butler) and/or the blackburn hawkmoth (*Manduca blackburni* Butler). Extant populations of *B. insignis* as restricted to a few disjunct sites, with a total of fewer than 200 individuals known to exist in the wild (C. Gemmill *et al.* unpublished data). Populations of both species are in immediate danger of extinction, mainly due to goats, but also because of insufficient natural pollination by these now rare moths. Natural pollination of *B. insignis* by wind or opportunistic flower-visiting insects produces only a few fruits per plant, each with 20–50 seeds, while hand pollinations produce significantly more fruits, each with *c.* 1000 seeds.

Bee pollination

Ninety per cent of all angiosperms (i.e. about 216 000 of 240 000 spp.) are adapted to pollination by insects, mainly beetles, butterflies, flies, wasps and bees, with the latter being the single most important group of pollinators in most habitats. Given this background, it is significant that only one study to date addresses how habitat changes affect bee-pollinated species (Table 13.1). Aizen and Feinsinger (1994a) compared pollination and seed output of 16 plant species residing in 5–30 year-old forest fragments and continuous expanses of forest in Argentinian Chaco vegetation (Fig. 13.1). Nine of the species was predominantly pollinated by bees, and significant declines in pollen grains per stigma or pollen tubes per style in habitat fragments

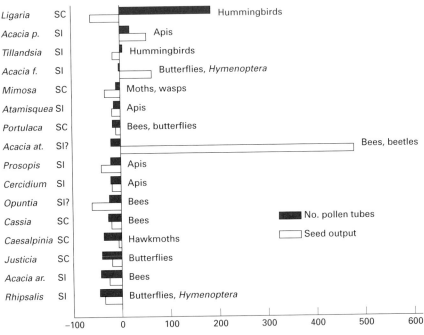

FIG. 13.1. Percentage change from values in continuous forest to values in fragment habitat units, ranked in descending order, for mean number of pollen tubes and seed output (fruit set × seed set) in 16 plant species studied in Argentine Chaco vegetation. For each species, compatibility system and principal flower visitor category have been added. Species visited by *Apis* received visits by other insects, too, but a large proportion of all visits to their flowers were by honey bees. For *Prosopis* and *Cercidium*, results from 1988 and 1989 were averaged. SC, self-compatible; SI, self-incompatible. (Modified by J.M. Olesen from Aizen & Feinsinger 1994a; see therein for plant species names.)

occurred in four of them. Seed output (seed set times fruit set) also declined significantly in 13 species (most of them pollinated by bees, but a few also pollinated by other insects), but percentage change in number of pollen tubes per style and percentage change in seed output were not significantly correlated. In three of the bee-pollinated species (see Table 13.1), both stigmatic pollen load and number of pollen tubes in styles were estimated. In all three, the percentages of grains that germinated were reduced in habitat fragments compared to continuous forest. The three species in question were self-incompatible, and therefore the reduced tube:grain ratios most likely reflect an increased proportion of self- or near-neighbour pollen on stigmas in fragments. As suggested by Aizen and Feinsinger (1994a) this decrease in pollen quality might be due to bees restricting most flights to plants within a fragment, which should increase the frequency of revisits to the same plant. Aizen and Feinsinger (1994b) also found that the frequency of visits to flowers by African honey bees (*Apis mellifera*) tended to increase from continuous forest to small fragments and that honey bees tended to remain foraging within single trees longer than did native bees. The total frequency of insect visits to bee-pollinated flowers varied little with fragment size.

The impact of forest fragmentation on a group of native bees was studied by Powell and Powell (1987) in forest fragments located north of Manaus in Brazil. Between 1982 and 1983, when previously continuous forest became isolated into a 100-ha, a 10-ha and a 1-ha fragment, numbers of male euglossines attracted during 39 sampling days to chemical baits declined significantly in four of the 12 morphospecies attracted. In three of these species, abundances diminished consistently with reserve size, and in the fourth species visitation rates to baits were significantly lower in the 1-ha fragments than in continuous forest. In the remaining species, visitation rates either did not differ among sites in continuous vs. fragmented forest, or bees were most numerous in clearings outside the fragments.

Unfortunately, this study was never followed up by monitoring pollination levels in male euglossine-pollinated plants in these fragments. Such monitoring would have been feasible for forest herbs known to be nectar-sources for male euglossines, such as Marantaceae. Becker, Moure, and Peralta (1991) working in the same fragments in 1988 and 1989 found no consistent relationship between forest size and euglossine abundance nor a decline in species richness between 1982 and 1989. As these authors and also Armbruster (1993) have pointed out, much more carefully designed sampling efforts will be needed in order to show that the male euglossine community in the Manaus fragments is negatively affected by forest fragmentation and that this is impacting plant reproduction in the fragments.

Wasp pollination

The obligate mutualism between figs and their wasp pollinators has long served as the paradigm of a vulnerable mutualism (McKey 1989, Bronstein, Gouyon, Gliddon *et al.* 1990). Recently, however, Bronstein and Hossaert-McKey (1995; cf. Table 13.1) showed that, contrary to expectations, catastrophic mortality of fig wasps during a hurricane did not lead to massive reproductive failure in the fig species dependent on these wasps. Rather, pre-hurricane ratios of pollinators to female-phase syconia in *Ficus aurea* were re-established within 5–16 months, suggesting that fig wasps migrated at least 60 km from intact trees not denuded by the hurricane. (Another possible explanation for the recovery of the interaction, involving plant phenology, is discussed below under general characteristics of especially vulnerable systems.) Isolated fig trees also sometimes set fruit, either due to other wasp species taking over as pollinators or due to long-distance attraction of wasps (Ware & Compton 1992, Bronstein & Hossaert-McKey 1995), which shows the resilience of fig reproduction.

Beetle pollination

Although no study has directly linked a decline in reproductive success to the absence of obligate beetle pollinators, work on nutmeg (*Myristica fragrans*) reproduction (Armstrong & Drummond 1986, Armstrong & Irvine 1989) is suggestive. Nutmegs rely on pollen-foraging beetles for pollination and seed production, and at some highly disturbed sites in SW India, a single beetle species now provides this service. Extensive loss of lowland rainforest may have reduced beetle diversity and may already have caused insufficient pollination levels in plantations (J.E. Armstrong personal communication). These ideas could be tested in Grenada where nutmeg cultivation is very important, and where the introduced *M. fragrans* must be borrowing pollinators among native beetle species.

GENERALIZATIONS AND CONCLUSIONS

Based on current knowledge, few generalizations are possible about how habitat deterioration and/or fragmentation in the tropics are affecting plant reproduction. To summarize: of the studies listed in Table 13.1 (including three from New Zealand, Florida and Sweden), most provide qualitative, rather than quantitative, evidence on the effects of fragmentation on plant pollination. Only one study (Aizen & Feinsinger 1994a) so far has quantified effects by comparing fruit set, pollen deposition, pollen-tube growth, or seed set in duplicated fragments of various sizes and in continuous habitat. Reductions in one or several of these parameters occurred in all but two of 16 species. The remaining studies listed in Table 13.1 provide qualitative evidence of plant responses to large-scale habitat deterioration (not necessarily fragmentation), and of these, five (Cox 1983, Lammers *et al.* 1987, Weller 1994, Ladley & Kelly 1995, C. Gemmill unpublished data) state that plant pollination changed because pollinators are extinct or presumed extinct. Two others report on effects of pollinator absence due to cyclones or hurricanes, in one case followed by the speedy recovery of the original state. Finally, there are five examples of pollination being taken over by other animals where the original pollinator is lost (Cox 1983, Vogel, Westerkamp, Thiel & Gessner 1984, Olesen 1985, Lammers *et al.* 1987, D. Neill personal communication; see also the sections on beetle and wasp pollination above). Bird-pollinated species especially seem to have been able to undergo such shifts. This may have happened during the Pliocene on the Canary Islands when nectariniids apparently were replaced by more opportunistic bird pollinators (Vogel *et al.* 1984, Olesen 1985). In historic times, such shifts occurred on Hawaii with the introduction of Japanese white-eyes in 1929, which for several species replace extinct Drepanidinae pollinators.

The following three generalizations need to be evaluated against these data. First, plants adapted to a few or a single animal species for pollination appear particularly sensitive to fragmentation. Second, plants pollinated by vertebrates, such as birds and bats, appear to have suffered more from the effects of anthropogenic disturbance, including habitat alteration and fragmentation, than any other pollination guild. Third, plants that are obligate outcrossers or self-incompatible species are particularly vulnerable to the effects of habitat fragmentation.

The first generalization, that dependence on a single animal species for pollination makes a plant vulnerable, forms the basis for a recent attempt by Bond (1994) to assess the impact of reproductive system on a plant species' likelihood of extinction. He ranks species as having a higher risk of extinction when they are pollination specialists than when they are generalists. Further entering the equation are the species' mating system (whether dioecious, self-incompatible, self-compatible, autogamous or apomictic; cf. the third generalization below), dispersal system (whether specialized or generalized) and life-span. Pollination specialists are defined as plants pollinated by one or a few ecologically similar animal species, while generalists are pollinated by several to many species, perhaps of widely diverse taxonomic origin. That specialist may be vulnerable receives limited support from the four species in Table 13.1 that are on the brink of extinction (*Stenogyne kanehoana, Trilepidea adamsii, Lotus berthelotii* and *Brighamia insignis*).

Three were (are?) bird-pollinated and one was sphingid-pollinated. That these plants were vulnerable because they depended entirely on one or a few animals for pollination may be concluded from the fact that their flowers attract no other animals, at least today. However, this may have been different in the past when these species were abundant because more abundant plants attract more foraging animals. It is at least possible that the pollinators of *Stenogyne*, *Trilepidea*, *Lotus* and *Brighamia* were forced to turn to other flowers for food at the same time as these species became rare for reasons unrelated to their pollination syndromes. In two of the four species, *L. berthelotii* and *B. insignis*, which survive only in botanical gardens, it seems that goats (introduced by humans to the islands where these plants are native) were extremely important in the demise of these defenseless plants.

Obligate interactions in which neither partner can reproduce without the other are undoubtedly rare (Bronstein 1994; figs and their wasps; *Yucca* and its moths). It is an open question whether one-to-one interactions are more vulnerable to habitat disturbance than asymmetric interactions in which only one partner is dependent on the other. Perhaps plants that, over the course of their evolution, became adapted to a single animal pollinator have evolved mechanisms allowing them to cope with natural variations in pollinator abundances. Thus, Bronstein and Hossaert-McKey (1995) suggest that the rapid recovery of the *Ficus aurea*–wasp interaction following a major hurricane may be explained not only by long-distance recolonization by wasps surviving elsewhere (as discussed above under Wasp Pollination) but may also have been due to phenological traits exhibited by *F. aurea*, namely within-tree flowering asynchrony. Such asynchrony may permit rapid recovery from climatically induced flowering failure and the associated loss of wasps. The average interval between hurricanes ranges from 6 to 21 years at different locations within the Caribbean region (Bronstein & Hossaert-McKey 1995), and this short cycle may well have selected for mechanisms on both the plant and the animal sides for recover from local extinctions of one's mutualist partner.

Support for the second generalization, that plants pollinated by large pollinators, such as birds and bats, are particularly vulnerable, is equally weak. Because anthropogenic habitat disturbance and/or fragmentation often entails hunting of large birds or (Old World) bats, the introduction of pets that hunt birds, or of parasites, species pollinated by large pollinators may be over-represented among plants whose reproductive fitness is suffering (Table 13.1). The absolute frequency of bird-pollinated species in the angiosperms probably barely reaches 1%, yet bird-pollinated species make up about half the case studies. It is possible that bird-pollinated plants are over-represented in the sample because researchers prefer to work on striking interactions likely to show effects, but it would still be legitimate to use such studies as models of how reproduction in other species may be effected by habitat fragmentation. Nevertheless, it is also possible that specialization on nectarivorous birds does heighten a plant's risk of extinction. According to one estimate (WCMC 1992), 115 species of birds have gone extinct since 1600 (I do not know how many of these were nectarivorous), while Milberg and Tyrberg (1993) list more than 200 species of extinct island birds only recorded as subfossils and which probably vanished due to prehistoric humans. Many extinct species belonged to the Hawaiian honeycreepers, and approximately half of the historically known nectarivorous passerines on Hawaii are now extinct or presumed extinct (Lammers *et al.* 1987). The effects of these historic pollinator losses may become visible with suitable research hypotheses and approaches to studying them.

The third generalization, that animal-pollinated obligate outcrossers (dioecious or self-incompatible species) are especially vulnerable to the effects of habitat fragmentation, cannot yet be evaluated because there are too few studies linking mating or compatibility system with vulnerability to fragmentation. Aizen and Feinsinger (1994a) found that the magnitudes of declines (in the forest fragments) in number of pollen tubes, fruit set and seed set were indistinguishable between the six self-compatible and the 10 self-incompatible species. Other, indirect evidence for the lack of a direct correlation between vulnerability and mating systems comes for an analysis by Weller (1994). Based on a sample of 84 rare plant species for which he had data, Weller found no relationship between mating system and rarity. Of course, this finding is only relevant here if there is a positive correlation between rarity and pollination-mediated vulnerability to habitat fragmentation.

Most tropical plants depend on animals for pollination (Ricklefs & Renner 1994), and many are self-incompatible or dioecious (e.g. Bawa & Opler 1975, Kress & Beach 1994, Renner & Ricklefs 1995). Effective populations in these species are expected to cover 100 or more square kilometres because densities of individual species are low. For example, in a 50-ha plot on Barro Colorado Island, the median density of 54 shrub species is 1.06 adults ha^{-1}; the median density of 73 small canopy tree species is 0.60 adults ha^{-1}; and the median density of 99 large tree species is 0.26 adults ha^{-1} (R. Condit, S.P. Hubbell and R.B. Foster personal communication; adults are defined as all plants ≥1 cm dbh for shrubs, 10 cm dbh for small canopy trees, and 30 cm dbh for large canopy trees). Genetic population structure of tropical tree species is largely unexplored, but K. Bawa, J. Hamrick and their collaborators have begun to use genetic markers to investigate gene flow in these species (e.g. Boshier, Chase & Bawa 1995, Chase *et al.* 1996, Stacy, Hamrick, Nason *et al.* 1996). Their results show that while small forest fragments, as well as isolated trees in pastures, may act as stepping stones for gene flow among larger forest remnants, preserves for many tree species will have to be larger than 60 ha to prevent significantly reduced breeding neighbourhoods.

Comparison of tropical and temperate latitudes

Based on the empirical knowledge presented in this review and the three, albeit problematic, generalizations on the vulnerability of different species (in terms of their pollination and mating systems) discussed in the previous section, there are two ways to view the proposition that temperate and tropical latitudes differ in the vulnerability of their pollination mutualisms. One perspective is that, because the proportions of various pollination syndromes differ with latitude (if altitude is held constant), a relative increase in more vulnerable kinds of pollination syndromes may make whole vegetation types particularly vulnerable. From this perspective it is possible to argue that since the relative frequency of pollination by birds and bats increases towards the equator this may make tropical vegetation more vulnerable than temperate vegetation. (A possible bias towards bird-pollination studies was discussed above.)

A second perspective is that, because plants that are pollination specialists appear more vulnerable than plants that are pollination generalists, a relative increase in number of specialists with decreasing latitude may make vegetation at low latitudes more vulnerable. This perspective is much more problematic than the first because we lack studies that compare the relative proportions of pollination specialists and generalists in different vegetation and habitat types, such as tropical forests, Mediterranean-type vegetation, temperate latitude prairies, etc. The different range of growth forms of these vegetation types would introduce a confounding effect into such comparisons, however, as would the hierarchical relationships of the plant lineages in them (even if the vegetation types were carefully matched for type of habitat) because of the effect of flower morphology on pollination specificity.

The lack of data on the relative frequencies of pollination specialists vs. pollination generalists in the tropics has encouraged contradictory assumptions. Thus, it has been stated that a large proportion of tropical plants are specialists in terms of their pollination (e.g. Bawa 1990, Renner & Feil 1993, Bond 1994), yet tropical trees are thought to be pollination generalists (Ashton 1969, Bawa & Opler 1975, Kress & Beach 1994, Stacy *et al.* 1996), and it has even been argued that '... the numerous species of stingless, social bees in the American Tropics, most of which are probably generalized in flower utilization, may indicate lower diversity of floral reproductive adaptations than we currently think' (Tepedino 1979) and that '... pollinator specificity may be crudely assessed from floral morphology' (Bond 1994). Part of the problem lies in a lack of field studies, but another part stems from a common misconception about plant–pollinator relationships. This misconception stems from a lack of appreciation of the asymmetry of pollination interactions (shown diagrammatically in Fig. 13.2). This asymmetry precludes extrapolations, such as Tepedino's or Bond's, from floral–morphological features to degree of specialization and dependence of the mutualistic partners. Thus, the animal can be dependent on a floral resource without the plant being dependent on that particular animal. An example are solitary bees that specialize (morphologically, ecologically and physiologically) on the pollen of a particular species, and are extremely efficient at

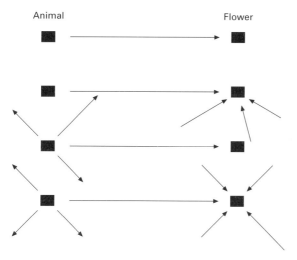

FIG. 13.2. Types of pollinator–flower mutualisms. Arrow heads indicate the direction of movement of an animal, and each flower visit is assumed to potentially result in pollination. From top to bottom: obligate mutualism with both partners exclusively dependent on each other; asymmetric mutualism with the animal unilaterally dependent on a particular species of flower; asymmetric mutualism with the flower unilaterally dependent on a particular species of animal; reciprocally facultative mutualism. Note that pollination specialists are often defined as plants pollinated by a few ecologically similar animal species, while this figure shows only a single species visiting the specialized flower. How narrow specialists are may be crucial for their survival. The pollination generalist is assumed to be effectively pollinated by several species of diverse taxonomic origin, and here again the ecological diversity of the potential pollinators may be the decisive factor in their survival under changing conditions.

harvesting it, but do not pollinate the flowers (J. Neff personal communication). Conversely, the plant can be dependent on an animal without that animal depending on the plant. For example, while the morphological specialization of *Passiflora mixta* to *Ensifera ensifera* may seem obvious, the sword-billed hummingbird does not depend on that species (cf. the section on Bird Pollination). Another example is provided by the kapok tree, *Ceiba pentandra*, which in Samoa seems entirely dependent on the flying fox *Pteropus tonganus* for pollination (Elmqvist *et al.* 1992) while elsewhere a diverse assemblage of vertebrates and invertebrates visits and pollinates the flowers of this pantropically distributed species. *Pteropus tonganus* in turn is a generalist.

If the generalization above holds that dependence on a single animal species for pollination makes a plant vulnerable, pollination generalists by implication should be less vulnerable to habitat fragmentation. A prediction from this would be that if canopy trees are indeed pollination generalists as often suggested, then they should be less vulnerable than understorey plants to pollinator loss through framentation. Future comparative studies of population genetic structure (using genetic markers and paternity analysis) in fragmented landscapes, on known pollination specialists and generalists, will have to show whether this is indeed the case.

Such studies will have to include field components on pollination and mating systems because extrapolation from floral morphological features to the plant's degree of dependence on the mutualist are unwarranted. Broad generalizations, such as the three discussed above, while providing guidance for the selection of species likely to be especially sensitive to habitat fragmentation, also need to be challenged. The effects of habitat deterioration and fragmentation on the plant–pollinator interactions as such are extremely difficult to quantify in statistically manageable ways, but it is feasible to monitor changes in seed set in carefully selected species in duplicated fragments of known age (e.g. Aizen & Feinsinger 1994a). Criteria for selection should be that species represent different mating systems, growth forms and pollination syndromes, especially bee pollination, which is understudied yet is the single most important pollination syndrome.

ACKNOWLEDGEMENTS

I am indebted to Jens Mogens Olesen, University of Aarhus, Denmark, for analysing references on fragmentation held in six electronic literature databases, for Fig. 13.1, and for many comments on the manuscript; and to the following for sending me published or unpublished papers and/or comments on the manuscript: S. Armbruster, University of Oslo, Norway; J. Armstrong,

Illinois State University; S. Banack, Brighham Young University; R. Bierregaard, University of North Carolina; T. Elmqvist, University of Umeå, Sweden; P. Feinsinger, Northern Arizona University, Flagstaff; C. Gemmill, University of Colorado, Boulder; D. Neill, the Missouri Botanical Garden, St Louis; R. Ricklefs, University of Missouri—St Louis; D. Roubik, Smithsonian Tropical Research Institute, Panama; M. Samways, University of Natal, South Africa; S. Weller, University of California, Irvine; C. Westerkamp, Bonn; and J. Ågren, University of Umeå, Sweden.

REFERENCES

Aizen, M.A. & Feinsinger, P. (1994a). Forest fragmentation, pollination, and plant reproduction in a Chaco dry forest, Argentina. *Ecology*, **75**, 330–351.

Aizen, M.A. & Feinsinger, P. (1994b). Habitat fragmentation, native insect pollinators, and feral honey bees in Argentine 'Chaco Serrano'. *Ecological Applications*, **4**, 378–392.

Armbruster, W.S. (1993). Within-habitat heterogeneity in baiting samples of male euglossine bees: possible causes and implications. *Biotropica*, **25**, 122–128.

Armstrong, J.E. & Drummond III, B.A. (1986). Floral biology of *Myristica fragrans* Houtt. (Myristicaceae), the nutmeg of commerce. *Biotropica*, **18**, 32–38.

Armstrong, J.E. & Irvine, A.K. (1989). Floral biology of *Myristica insipida* (Myristicaceae), a distinctive beetle pollination syndrome. *American Journal of Botany*, **76**, 86–94.

Ashton, P.S. (1969). Speciation among tropical forest trees: some deductions in the light of recent evidence. *Biological Journal of the Linnean Society*, **1**, 155–196.

Banack, S.A. (1996). *Flying foxes, genus* Pteropus, *in the Samoan Islands: interactions with forest communities*. PhD thesis, University of California, Berkeley.

Bawa, K.S. (1990). Plant–pollinator interactions in tropical rain forests. *Annual Review of Ecology and Systematics*, **21**, 399–422.

Bawa, K.S. & Opler, P.A. (1975). Dioecism in tropical forest trees. *Evolution*, **29**, 167–179.

Becker, P., Moure, J.S. & Peralta, F.J.A. (1991). More about euglossine bees in Amazonian forest fragments. *Biotropica*, **23**, 586–591.

Bond, W.J. (1994). Do mutualisms matter? Assessing the impact of pollinator and disperser disruption on plant extinction. *Philosophical Transactions of the Royal Society, Series B*, **344**, 83–90.

Boshier, D.H., Chase, M.R. & Bawa, K.S. (1995). Population genetics of *Cordia alliodora* (Boraginaceae), a neotropical tree. III. Gene flow, neighborhood, and population structure. *American Journal of Botany*, **82**, 484–490.

Bronstein, J.L. (1994). Our current understanding of mutualism. *Quarterly Review of Biology*, **69**, 31–51.

Bronstein, J.L. & Hossaert-McKey, M. (1995). Hurricane Andrew and a Florida fig pollination mutualism: resilience of an obligate interaction. *Biotropica*, **27**, 373–381.

Bronstein, J.L., Gouyon, P.-H., Gliddon, C., Kjellberg, F. & Michaloud, G. (1990). The ecological consequences of flowering asynchrony in monoecious figs: a simulation study. *Ecology*, **71**, 2145–2156.

Brown, K.S. & Hutchings, R.W. (1997). Disturbance, fragmentation, and the dynamics of diversity in Amazonian forest butterflies. *Tropical Forest Remnants* (Ed. by W.F. Laurance & R.O. Bierregaard, Jr), pp. 91–110. Chicago University Press, Chicago.

Chase, M.R., Moller, C., Kesseli, R. & Bawa, K.S. (1996). Distant gene flow in tropical trees. *Nature*, **383**, 398–399.

Connell, J.H. (1983). On the prevalence and relative importance of competition: evidence from field experiments. *American Naturalist*, **122**, 661–696.

Cox, P.A. (1983). Extinction of the Hawaiian avifauna resulted in a change of pollinators for the ieie, *Freycinetia arborea*. *Oikos*, **41**, 195–199.

Cox, P.A., Elmqvist, T., Pierson, E.D. & Rainey, W.E. (1991). Flying foxes as strong interactors in South Pacific island ecosystems: a conservation hypothesis. *Conservation Biology*, **5**, 448–454.

Crome, F.H.J. & Irvine, A.K. (1986). Two bob each way: the pollination and breeding system of the Australian rain forest tree *Syzygium cormiflorum* (Myrtaceae). *Biotropica*, **18**, 115–125.

Daily, G.C. & Ehrlich, P.R. (1995). Preservation of biodiversity in small rainforest patches: rapid evaluations using butterfly trapping. *Biodiversity and Conservation*, **4**, 35–55.

DeVries, P.J. (1987). *The Butterflies of Costa Rica and their Natural History*. Princeton University Press, Princeton.

Didham, R.K., Ghazoul, J., Stork, N.E. & Davis, A.J. (1996). Insects in fragmented forests: a functional approach. *Trends in Ecology and Evolution*, **11**, 255–260.

Dirzo, R., Naeem, S. & Cushman, J.H. (1995). Biotic linkages and ecosystem functioning. *Global Biodiversity Assessment* (Ed. by V.H. Heywood), pp. 427–433. Cambridge University Press, Cambridge.

Dobat, K. & Peikert-Holle, T. (1985). *Blüten und Fledermäuse*. Verlag von Waldemar Kramer, Frankfurt.

Elmqvist, T., Cox, P.A., Rainey, W.E. & Pierson, E.D. (1992). Restricted pollination on oceanic islands: pollination of *Ceiba pentandra* by flying foxes in Samoa. *Biotropica*, **24**, 15–23.

Feinsinger, P. (1987). Approaches to nectarivores–plant interactions in the New World. *Revista Chilena de Historia Natural*, **60**, 285–319.

Fenton, M.B., Acharya, L., Audet, D., Hickey, M.B.C., Merriman, C., Obrist, M.K. et al. (1992). Phyllostomid bats (Chiroptera: Phyllostomidae) as indicators of habitat disruption in the Neotropics. *Biotropica*, **24**, 440–446.

Foster, R.B., Arce, B.J. & Wachter, T.S. (1986). Dispersal and the sequential plant communities in an Amazonian Peruvian floodplain. In *Frugivores and Seed Dispersal* (Ed. by A. Estrada & T.H. Fleming), pp. 357–370. Junk, Dordrecht.

Gilbert, L.E. (1980). Food web organization and the conservation of neotropical diversity. In *Conservation Biology: An Evolutionary–Ecological Perspective* (Ed. by M.E. Soulé & B.A. Wilcox), pp. 11–34. Sinauer, Sunderland, MA.

Haber, W.A. & Frankie, G.W. (1989). A tropical hawkmoth community: Costa Rican dry forest Sphingidae. *Biotropica*, **21**, 155–172.

Harrison, S. (1995). Effects of spatial structure on ecosystem functioning. In *Global Biodiversity Assessment* (Ed. by V.H. Heywood), pp. 301–304. Cambridge University Press, Cambridge.

Janzen, D.H. (Ed.) (1983). *Costa Rican Natural History*. University of Chicago Press, Chicago.

Janzen, D.H. (1987). Insect diversity of a Costa Rican dry forest: why keep it and how? *Biological Journal of the Linnean Society*, **30**, 343–356.

Jennersten, O. (1984). Flower visitation and pollination efficiency of some North European butterflies. *Oecologia*, **63**, 80–89.

Jennersten, O. (1988). Pollination in *Dianthus deltoides* (Caryophyllaceae): effects of habitat fragmentation on visitation and seed set. *Conservation Biology*, **2**, 359–366.

Kress, W.J. & Beach, J.H. (1994). Flowering plant reproductive systems. In *La Selva: Ecology and Natural History of a Neotropical Rainforest* (Ed. by L.A. McDade, K.S. Bawa, H.A. Hespenheide & G.S. Hartshorn), pp. 161–182. University of Chicago Press, Chicago.

Ladley, J.L. & Kelly, D. (1995). Explosive New Zealand mistletoe. *Nature*, **378**, 766.

Lammers, T.G., Weller, S.G. & Sakai, A.K. (1987). Japanese white-eye, an introduced passerine, visits the flowers of *Clermontia arborescens*, an endemic Hawaiian lobelioid. *Pacific Science*, **41**, 74–77.

Laurance, W.F. & Bierregaard, Jr, R.O. (Eds) (1997). *Tropical Forest Remnants: Ecology, Management and Conservation of Fragmented Communities*. Chicago University Press, Chicago.

McKey, D. (1989). Population biology of figs: applications for conservation. *Experientia*, **45**, 661–673.

Milberg, P. & Tyrberg T. (1993). Native birds and noble savages—a review of man-caused prehistoric extinctions of island birds. *Ecography*, **16**, 229–250.

Olesen, J.M. (1985). The Macaronesian bird-flower element and its relation to bird and bee opportunists. *Botanical Journal of the Linnean Society*, **91**, 395–414.

Olesen, J.M. & Jain, S. (1994). Fragmented plant populations and their lost interactions. In *Conservation Genetics* (Ed. by V. Loeschcke, J. Tomiuk & S. Jain), pp. 417–426. Birkhäuser, Basel.

Pierson, E.D., Elmqvist, T., Rainey, W.E. & Cox, P.A. (1996). Effects of tropical cyclonic storms on flying fox populations on the South Pacific islands of Samoa. *Conservation Biology*, **10**, 438–451.

Pollard, E. (1982). Monitoring butterfly abundance in relation to the management of a nature reserve. *Biological Conservation*, **24**, 317–328.

Porsch, O. (1931). Grellrot als Vogelblumenfarbe. *Biologia Generalis*, **7**, 647–674.

Powell, A.H. & Powell, G.V.N. (1987). Population dynamics of male euglossine bees in Amazonian forest fragments. *Biotropica*, **19**, 176–179.

Rainey, W.E., Pierson, E.D., Elmqvist, T. & Cox, P.A. (1995). The role of flying foxes (Pteropodidae) in oceanic island ecosystems of the Pacific. In *Ecology, Evolution and Behaviour of Bats* (Ed. by P.A. Racey & S.H. Swift), pp. 47–62. *Symposia of the Zoological Society of London*, **67**.

Rathcke, B.J. & Jules, E.S. (1993). Habitat fragmentation and plant–pollinator interactions. *Current Science*, **65**, 273–277.

Renner, S.S. & Feil, J.P. (1993). Pollinators of tropical dioecious angiosperms. *American Journal of Botany*, **80**, 1100–1107.

Renner, S.S. & Ricklefs, R.E. (1995). Dioecy and its correlates in the flowering plants. *American Journal of Botany*, **82**, 596–606.

Ricklefs, R.E. & Renner, S.S. (1994). Species richness within families of flowering plants. *Evolution*, **48**, 1619–1636.

Rodrigues, J.J.S., Brown, K.S. & Ruszczyk, A. (1993). Resources and conservation of neotropical butterflies in urban forest fragments. *Biological Conservation*, **64**, 3–9.

Snow, D.W. & Snow, B.K. (1980). Relationships between hummingbirds and flowers in the Andes of Columbia. *Bulletin of the British Museum (Natural History) Zoology*, **38**, 105–139.

Stacy, E.A., Hamrick, J.L., Nason, J.D., Hubbell, S.P., Foster, R.B. & Condit, R. (1996). Pollen dispersal in low-density populations of three neotropical tree species. *American Naturalist*, **148**, 275–298.

Stouffer, P.C. & Bierregaard, R.O., Jr (1995). Effects of forest fragmentation on understory hummingbirds in Amazonian Brazil. *Conservation Biology*, **9**, 1085–1094.

Tepedino, V.J. (1979). The importance of bees and other insect pollinators in maintaining floral species composition. *Great Basin Naturalist Memoirs*, **3**, 139–150.

Vogel, S., Westerkamp, C., Thiel, B. & Gessner, K. (1984). Ornithophilie auf den Kanarischen Inseln. *Plant Systematics and Evolution*, **146**, 225–248.

Ware, A.B. & Compton, S.G. (1992). Breakdown of pollinator specificity in an African fig tree. *Biotropica*, **24**, 544–549.

WCMC (1992). *Global Biodiversity Status of the Earth's Living Resources.* World Conservation Monitoring Centre, Cambridge.

Weller, S.G. (1994). The relationship of rarity to plant reproductive biology. In *Restoration of Endangered Species* (Ed. by M.L. Bowles & C.J. Whelan), pp. 90–117. Cambridge University Press, Cambridge.

Wood, P.A. & Samways, M.J. (1991). Landscape element pattern and continuity of butterfly flight paths in an ecologically landscaped botanic garden, Natal, South Africa. *Biological Conservation*, **58**, 149–166.

Young, A., Boyle, T. & Brown, T. (1996). The population genetic consequences of habitat fragmentation for plants. *Trends in Ecology and Evolution*, **11**, 413–418.

14. A SPATIAL MODEL OF SAVANNA FUNCTION AND DYNAMICS: MODEL DESCRIPTION AND PRELIMINARY RESULTS

J. GIGNOUX*‡, J.-C. MENAUT†*, I. R. NOBLE‡
and I. D. DAVIES‡

*Laboratoire d'Écologie, Ecole Normale Supérieure, 46 Rue d'Ulm, 75230 Paris Cedex 05, France; †ORSTROM, 213 Rue la Fayette, 75480 Paris Cedex 10, France; ‡Ecosystem Dynamics, Research School of Biological Sciences, Australian National University, Canberra ACT 0200, Australia

SUMMARY

1 MUSE is a modular modelling environment which enables us to simulate linkages between structure, function and dynamics in savannas. MUSE is spatially explicit and flexible in time and space scales in order to deal with the various processes involved. It contains an extendable library of models allowing both to incorporate in MUSE new processes and compare results from models using different approaches to vegetation and soil simulations. It provides scenarios of ecosystem changes in response to intrinsic dynamic properties and to climatic or anthropogenic disturbances.

2 Present results focus on tree dynamics and spatial pattern, and on grass primary production in relation to water stress. Trees are treated as individuals (growth, reproduction, seed dispersal, recruitment) which interact with neighbours and grasses. Grass is treated as a homogeneous layer locally affected by trees. Disturbances are also spatially explicit, and affect (or are affected by) tree and grass coenoses. Impacts on (and feedback effects from) soil functioning (water, organic matter and nutrient dynamics) are incorporated.

3 MUSE is parameterized and validated on data collected during 30 years from a variety of sites at the Lamto Research Station (Côte d'Ivoire), and is being adapted to other savanna types (drier West African and Australian savannas).

MODELLING SAVANNAS: INTEGRATING STRUCTURE, FUNCTION AND DYNAMICS

The need of a spatially explicit model for savannas

Savannas constitute one of the largest biomes of the world (c. 1/6 of the emerged lands). They may significantly feed back to the global climate system since they display huge energy and water budgets, produce considerable biogenic and pyrogenic emissions of greenhouse gases and have the potential to sequester or release large amounts of carbon. The balance between trees and grasses and their relative spatial distributions are influenced by at least five factors (Frost, Medina, Menaut *et al.* 1986): competition for nutrients, competition for water, fire, grazing, and demographic strategies of grasses and trees. Small shifts in these driving processes can alter the balance between the tree and grass components and favour the development of a more dense woodland or an open grassland. The only tool enabling a coupling of the processes needed to predict the tree/grass balance across a wide range of climatic conditions is the simulation model because of the high complexity of even the simplest savanna model which has to incorporate at least two life forms (grass and trees) and the five driving processes.

Processes involved in the tree/grass balance in savannas are not only numerous but also very different (demography, competition for resources, disturbance), making the study of the tree/grass relationship in savannas a particularly difficult problem. Furthermore, savanna ecosystems present a marked structure, with two strata (trees and grasses), one of which (trees) is discontinuous, often displaying a very complex spatial pattern (with a mixture of tree clumps and scattered trees). Ecosystem function, dynamics and structure are linked. For example, the spatial pattern of trees (structure) in savannas locally affects the nutrient status (function), which in turn affects the recruitment of plants (dynamics). Any savanna simulation model should be able to represent these interactions between structure, dynamics and function of the ecosystem.

In simulation models, changes in available resources (i.e. ecological stresses) are usually considered to affect plant physiology at the stand level (i.e. ecosystem function: primary production, evapotranspiration, etc.) without mechanistic impact on stand dynamics and structure. In mixed life-form ecosystems, and especially in savannas, small changes in structure (spatial pattern) are likely to affect the balance between the various plant functional types. These may produce many more significant changes in state variables (biomass, water, energy status, etc.) than a direct effect of a change in photosynthetic rate, for example. Savannas are also characterized by recurrent, though irregular, disturbances directly affecting structural attributes (fire, herbivory, land-use practices) and consequently functional features. Extant ecosystem simulation models usually consider only ecosystem functioning or dynamics; the structure is fixed, incorporated in the model assumptions. Few attempts to model a changing structure, coupled either to function or dynamics, have been made (Menaut, Gignoux, Prado & Clobert 1990, Coughenour 1994, Hochberg, Menaut & Gignoux 1994, Jeltsch, Milton, Dean & Van Rooyen 1996). To actually couple structure, dynamics and function a spatially explicit model is needed, one which accounts for all plant positions and shapes because interactions between plants are local and cannot be averaged when the spatial structure of the ecosystem is heterogeneous in savannas trees within clumps and scattered trees do not grow in the same physical environment).

The aim of this chapter is to describe briefly a spatially explicit model specially designed to simulate savanna ecosystems and to present some preliminary results. We also show how the principles and concepts developed for this specific problem can apply to other systems and lead to a generic tool for simulation of vegetation dynamics. We intend to distribute the model publicly for scientific use as soon as a full reference manual is published. The complexity of the model does not allow us fully to document it here.

Modelling principles

The model derives from the very successful group of 'gap models' starting with JABOWA (Botkin, Janak & Wallis 1972) and developed further via the FORET and related models (Shugart & West 1977, Shugart, Hopkins, Burgess & Mortlock 1980, Shugart 1984) and more recently the FORSKA series (Prentice & Leemans 1990; Prentice, Sykes & Cramer 1993).

All these models typically simulate a small patch of forest (*c.* 0.1 ha) on the assumption that within a patch of this size all trees fully interact with each other; that is, each tree fully shades its neighbours and their roots occupy the same soil cylinder. JABOWA assumes that all foliage concentrates in a disc at the top of the stem; versions of FORET and FORSKA relax this assumption and distribute the foliage down the stem. Additional biological complexity has been added to further versions of FORET and FORSKA but the JABOWA heritage remains.

Although analytical models have been developed to explore the tree–grass balance in savannas (Walker & Noy-Meir 1982), there is no dynamic simulation model similar to the gap models efficiently used to simulate the dynamics of forested ecosystems. The main reason lies in the basic assumptions underlying the gap models: these are violated in savanna ecosystems. The most important one is the 'gap' itself. Gap models operate by simulating a small patch of ground corresponding to the area occupied by a single dominant individual tree, hence approximately the size of the gap created when that tree finally dies. In such an approach it is assumed that each individual in the patch interacts fully with every other individual, and thus no account needs to be taken of the detailed spatial structure of the patch of forest.

In savanna systems, tree spacing is much wider and much more variable. In some savannas, trees are spaced far apart in a roughly regular or random pattern; in others small clumps of trees occur. Most often both types of distribution coexist within the same species pool and each individual maintains specific interactions with its neighbours. The spatial coordinates of each tree must then be taken into account in order to predict the influence of one tree on another. The problem is all the more complex as the woody and grass strata strongly interact. Any savanna model must be able to capture the interactions within and between the two very different life-forms.

To simulate such couplings and reach an integrative understanding, one needs a spatial model which represents structure and is able to take into account matter and energy fluxes and their effects on plant demography. A generic, spatially-explicit, savanna vegetation model called MUSE (for *Mu*ltistrata *S*patially *E*xplicit model) has been developed to simulate linkages between structure, function and dynamics and the induced relative variations between them. The MUSE model is based on the view that many of the ideas used in gap models and the understanding arising from them can be applied to savannas by developing a spatially explicit model which can take into account the spatial patterning of the ecosystem, including plants, disturbances and environmental variables (Gignoux 1994).

DESCRIPTION OF THE MODEL

Such an approach has usually been seen as being prohibitively expensive in computing resources and input data. However, by adopting a careful set of simplifications, some very efficient computation algorithms and an object-oriented approach to describing the savanna communities, effective models of savanna dynamics can be developed. Rather straightforward simplifications allow its application to other vegetation types such as forest ecosystems, grasslands and old fields.

MUSE has been designed to represent savannas at the patch or small landscape scale. It is an individual-based spatial model with the interactions between individual plants based on their position in space and their shapes. MUSE is implemented in an object-oriented language (Borland Pascal

7.0™) which allows users to readily modify assumptions about the interactions between plants, change the descriptions of major biological processes and examine model output in comprehensive and flexible ways. Flexibility has been developed as far as possible on the grounds that it is easier to produce simple representations from the most complex spatial model than the reverse.

Model structure

The structure of MUSE derives from the functional analysis of a savanna ecosystem (Fig. 14.1). Potentially, all of the processes represented on Fig. 14.1 can be modelled, depending mainly on the available knowledge (i.e. when nothing is known about a process, there is no need to include it in the model). In a savanna, grasses and trees, as different life-forms, are usually described by completely different sets of variables and equations for growth, survival and reproduction (e.g. leaf area and biomass for grass; diameter at

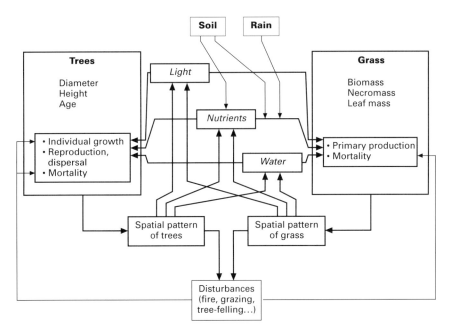

FIG. 14.1. Structure of a generic savanna model. Arrows represent the influence of each process on others. Plants (grasses and trees) have internal dynamics described by specific submodels (dynamics). They compete for resources (light, water and nutrients—function), the availability of which is affected by external factors such as soil types and climate, and the spatial structure of the plant populations (structure).

breast height (DBH), height and leaf area for trees). MUSE provides all the algorithms to deal with spatial interactions between plants and a general framework to deal with all the possible ecological processes present: the detail of the biological functions used is left to the user. MUSE is a modelling environment from which a series of models is derived. The savanna model described here uses the spatial formalism of MUSE with specific ecological functions for savannas.

The MUSE model allows users to apply a wide range of biological models (i.e. ways of calculating the availability of resources such as light and water; ways of converting this to net primary production; models of recruitment and mortality) to a range of model structures from the simplest gap model with limited vertical and no horizontal spatial detail to the most detailed spatially explicit version. This is useful for exploring the effect of taking spatial structure into account (or of not doing so) in the dynamics of the modelled ecosystem and in developing efficient modelling approaches for exploring management options or the response of ecosystems to global change. One aim of MUSE is to provide the possibility of exploring the effects of the level of spatial detail in a particular model without having to code the spatial interactions between plants (which is usually the most difficult task — and of little biological interest *per se*). MUSE enables the user to concentrate only on the biology, and provides the possibility of easily changing the spatial representation of the plants. As a consequence, scaling up and down is possible by adapting the degree of spatial detail in the model to the scale of observation of the studied site.

The sequence of actions executed at each simulation time step (Fig. 14.2) in MUSE directly reflects the above-mentioned features (Fig. 14.1). All of the main ecological processes are present: climatic inputs, resource dynamics, competition for resources as a result of the spatial arrangement of plants, and main biological processes (growth, reproduction, survival). They can be

FIG. 14.2. *(Opposite.)* MUSE main loop. The left series of boxes describes the sequence of events executed by the main code during a series of simulations. Arrows indicate the chronological sequence of function calls, loops are indicated by arrows coming back to the same box, with a letter for the list of objects on which the loop is performed (Sp, species loop; St, stages; N, neighbourhoods; P, plants; DT, disturbance types; C, cells). The competition loop is executed three times, once for each class of resource variables (global resources, above-ground and below-ground resources — see text). For all the 'ecological' computations, the program searches for user-defined functions, specific to the modelled ecosystem (if no user-defined function is found, it is skipped). The possible set of user-defined functions is represented by the series of boxes on the right (the names are those of the subroutines users can redefine for a particular model); each set defines a model, attached to the site. Each subroutine can be modified easily, so that they effectively behave as relatively independent modules.

Modelling savannas

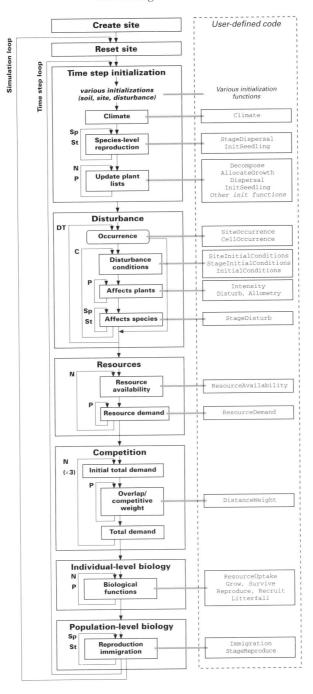

easily adapted to the available knowledge through a 'plug-in' programming interface facility. This enables, for example, use in the same simulation of a primary production- and physiology-based model for the grass cover and an individual-based model for trees in a savanna.

Spatial representation: plants and their neighbourhoods

Plant shape is represented by a pile of canopy discs and a pile of root system discs. This enables the generation of any canopy/root profile. It is possible to change the plant shape without affecting the other aspects of the model, due to a complete separation of geometry and biology in the program. In more detail, a plant is represented by two sets of variables: those describing its geometry (root system radii in each stratum, canopy radii in each stratum, total height, bole height, root depth — all measured in metres), and those describing its biology (e.g. biomass, leaf area per stratum of canopy, root density per soil stratum, dbh or any other relevant variable). This enables a change of one set of variables without affecting the other: the spatial representation of plants can be changed for the same physiological model.

The shape of a plant defines its neighbourhood; that is, the space in which it takes up its resources. Competition only takes place within this neighbourhood. Owing to the possible coexistence of plant types with completely different biological variables, resource variables must be common to all plant types. This mimics reality: the real plants compete for common resources such as water, nutrients and light, whatever their species and the way the plants are measured and described. Resources are stored in, and accessed from, neighbourhoods only (locally). Overlapping plants (and only them) are considered to be neighbours.

To simulate variation in the physical environment, the modelled site is divided into square cells where topography, soil type and disturbance conditions can vary. Plants are considered to grow in the cell in which their stems stand.

Interactions between plants

A plant has only access to resources within its immediate neighbourhood and resources are proportionally limited by the amount of overlap with other plants (stratum by stratum) within the neighbourhood.

The effect of a neighbour on a plant is proportional to the sum of the overlap of each of its strata on that plant's strata. Plants that do not overlap do not influence each other. This takes into account neighbour size as well as distance to the neighbour. Depending on the type of resource considered, overlaps can be computed between discs located in the same stratum only (e.g. nutrient uptake by roots), or may incorporate effects from strata located above (e.g. when considering light penetration in the canopy). This leads to a classification of resources into two groups, which we call simply 'stock' and 'flow' resources. As an example, light will always be treated as a flow resource, while nutrients will be treated as a stock. The distinction is relative to the time step used in simulations: water can be treated as a stock resource in a daily model but as a flow in a yearly model.

For all resources, competition is based on the following algorithm, used at each time step for each plant:

1 Computation of the available resource in the neighbourhood, from the external inputs (climate) and the possible stock from the previous time step.
2 Computation of the resource demand by the plant (and its neighbours), from the plant's current state and according to its species/stage.
3 Computation of the total resource demand in the plant's neighbourhood, taking into account the demand of each of the neighbours (weighted by their overlap sum).
4 Computation of the plant resource uptake from the three previous values. This last function can be user-defined so that no specific competition model is imposed.

This algorithm is fairly general and can be applied to any type of resource variable. It can even be used to compute less physiologically-based competition indices, like the weighted number of neighbours (Table 14.1).

The necessary generality of an ecosystem model for savannas

One of the key problems in modelling savannas lies in the interaction of different life-forms, with completely different individual sizes (a few cm^2 for grasses, a few m^2 for adult trees), different variables used to describe them (biomass, phytomass and leaf area index (LAI) for grass; diameter, height and LAI for trees), and even different functions used to model their growth, survival and reproduction.

In MUSE, grass is modelled as a 'green slime', a homogeneous layer of grass material within each cell (possibly with strata). Trees are modelled using the spatial representation described in the above sections. The 'grass object' is either associated with a tree which influences it (to model the grass growing within the shade or rooting zone of a tree) or can stand alone (to model grass in the open). The spatial representation of the 'grass objects' is derived from that of the 'tree objects' through a straightforward simplification, enabled by the separation between the geometric and biological description of plants. The combination of these two types of spatial represen-

TABLE 14.1. Examples of use of the resource competition algorithm of MUSE.

Model	Source	Resource	RA	RD	TD	RU
FORSKA 1	(Prentice & Leemans 1990)	Light	PAR $400\,\mu\text{mol}\,\text{m}^{-2}\text{s}^{-1}$	Sum of LAI above $\sum_{x>z} LAI_x$	Weighted sum of LAI $\sum_i RD_i(z) \cdot OV(i,j)$	Beer's law $RA \cdot e^{-k \cdot TD(z)}$
Lamto grass model	(Le Roux et al. 1996)	Water	Reserve + rainfall – runoff $R_{60} + R_{170} + Rf$-Ro	Potential evaporation* EP_c	Potential evapotranspiration* EP	Evapotranspiration* E
		Light	PAR $\varepsilon_c \cdot R_s$	LAI	Weighted sum of LAI $\sum_i RD_i(z) \cdot OV(i,j)$	Absorbed PAR $\varepsilon_a \cdot RA$, with $\varepsilon_a = 1 - e^{-0.479 \cdot TD}$
Lamto tree model	(Menaut et al. 1990)	Space	—	Height H	No. of taller neighbours $\sum_j 1_{(H_j > H)}$	$e^{-\alpha \cdot TD}$

RA, resource available in neighbourhood; RD, resource demand; TD, resource total demand in neighbourhood; RU, resource uptake. Variable names: PAR, photosynthetically active radiation; LAI, leaf area index; z, elevation of a given stratum; $OV(i,j)$, overlap between plant i and neighbour j; k, extinction coefficient; R_x, soil water reserve down to depth x; Rf, rainfall; Ro, runoff; ε_c, conversion efficiency; R_s, solar radiation; ε_a, absorption efficiency; α, competition coefficient.
* Details of computations not shown (too complex).

TABLE 14.2. *Current library of models of MUSE.*

Model class	Model	Modelled ecosystem	Reference
Spatial models	'FRENCH'	West African savanna (trees)	(Menaut *et al.* 1990)
		Mediterranean grassland	I. Noy-Meir unpublished
Gap models	JABOWA	North American boreal forest	(Botkin *et al.* 1972)
	FORET	North American boreal forest	(Shugart & West 1977)
	KIAMBRAM	Australian rainforest	(Shugart *et al.* 1980)
	BRIND	Australian dry-deciduous forest	(Shugart & Noble 1981)
	FORSKA 1	Swedish boreal forest	(Prentice & Leemans 1990)
	FORSKA 2	Swedish boreal forests with response to climate	(Prentice *et al.* 1993)
Non-spatial models	Competitive lottery	Theoretical model	(Chesson & Warner 1981)
		Japanese temperate rainforest	(Kohyama 1993)
Savanna subsystems		North Australian savanna (annual grass)	O. Ronce unpublished
		West African savanna (perennial grass)	(Le Roux *et al.* 1996)
		West African savanna (soil organic matter)	L. Abbadie & C. Garcia unpublished

tations proved to be very rich as it enabled exploration of various spatial configurations of ecosystems.

Other ecosystem models can be derived from savanna models by simplification: for example, functional forest models are deduced from Fig. 14.1 by deleting the grass component; purely demographic models, by deleting the resource components; non-spatial models, by ignoring the effects of the spatial patterns on the local resource availabilities. As different interacting processes can have different temporal and spatial scales, the model is able to run different processes with different time-steps and spatial ranges.

As a result, virtually any biological system can be implemented into MUSE, via a programming interface that enables the user to 'plug in' any new set of biological functions to the main program. Beyond a simple model, MUSE is a modelling environment able to manage a library of models (Table 14.2). The current models of the library are published models which we used to design and test the various spatial configurations of MUSE. These models range from spatial models with one or two life-forms to gap models and non-spatial models.

INCORPORATING SPECIFIC MODULES: CASE EXAMPLES FOR SAVANNAS

Tree population dynamics: effects of tree resistance to fire

We incorporated the FRENCH model into MUSE (see Menaut *et al.* 1990) for a complete description of model functions and parameters) of tree population dynamics of a West African humid savanna (Lamto, Côte d'Ivoire) to relate fire and competition between trees. This model is characterized by:

1 an individual-based representation of trees as flat discs at the top of a stem, with empirical relationships between basal diameter, height and crown radius based on field measurements;

2 an empirical growth model taking into account the effect of the number of taller neighbours on the diameter increment of a tree, summarizing competition for light between trees;

3 a single 'synthetic' tree species;

4 an annual time-step;

5 an implicit grass layer for which mean height is a direct function of local tree density (this implicitly assumes that grass is a poorer competitor for light);

6 fire sensitivity of trees included in the grass layer (<2 m in height) and fire resistance of bigger trees;

7 a resprouting model giving a young tree, whose aerial parts have been killed, the ability to resprout from below-ground parts the following year; and
8 variability of fire in space and time, translated as a variations in flame height according to local tree density, the year, and a random probability of escaping fire by chance for each individual tree.

This model focuses on the demographic aspects of tree dynamics, especially at the recruitment stage. In the MUSE framework, it uses the fully spatial representation of plant shape, with only one canopy disc, and a simple index of competition (the number of taller neighbours of a tree) as a 'resource' variable.

Using recent field data (Gignoux 1994), we were able to have better estimates of model parameters (adult and juvenile mortality, fecundity) than in the original version, which produced tree densities and spatial patterns more comparable to field data (Fig. 14.3). In particular, it was possible to control tree density with the fire heterogeneity parameter (i.e. probability of a tree being unburnt by chance: Fig. 14.3). For very small values of this parameter (constant fire), trees were driven to extinction; for average values (0.2 probability of escaping fire), densities similar to observed densities were obtained; and for large values (>0.3), tree density increased to a level reached on one of our savanna plots, where density was high, and, as a result, grass and fire were locally excluded. The simulation results of Fig. 14.3 demonstrate that there is a kind of 'phase transition' driven by fire heterogeneity, between a fire-driven system when the parameter is low, and a competition-driven system when it is high enough: in the first case, trees are either driven to extinction or stabilize to relatively low densities (0–800 trees ha^{-1}), while in the second they reach much higher densities (between 2100 and 3100 trees ha^{-1}). The boundary between the two regimes is difficult to characterize in terms of density, since the intrinsic variability of the model (Fig. 14.3) produces large confidence intervals. However, there must be a threshold of fire heterogeneity (between 0.2 and 0.3) separating the two regimes. Lamto savannas would then be a patchwork of fire-driven and competition-driven tree populations.

This raises the important question of the stability of such savannas under the current fire regime. Since grass is excluded from a competition-driven tree population, there is no possibility (apart from other disturbances such as clear-cutting or catastrophic late fires) to revert to a fire-driven tree population, while the transition from a fire-driven to a competition-driven tree population is always possible (when fire heterogeneity is not strictly constant, it is a quite reasonable hypothesis). We should thus expect these savannas to be totally invaded by trees on a very long time-scale. Field evidence of

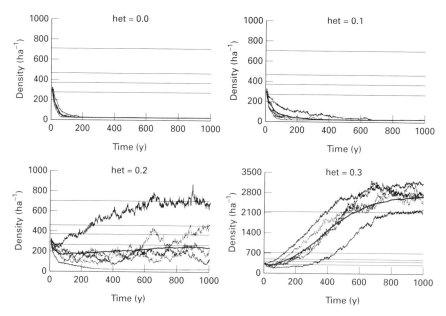

FIG. 14.3. Results of the FRENCH model (savanna tree population dynamics in relation to fire): simulated tree density (annual fire with an average flame height of 2 m and no interannual variation of flame height). Top and bottom curves, 95% Monte-Carlo confidence limits computed from 100 simulations; thick solid line, average of 100 simulations; horizontal lines show the observed densities of five permanent savanna plots in Lamto; other curves are three single simulations drawn from the 100 simulations, to show possible trajectories. The het parameter is fire spatial heterogeneity; that is, the probability that a tree escapes fire by pure chance.

a doubling in tree density over 20 years in Lamto (Gautier 1990), satellite data showing that at the regional scale, only 80% of the surface is burnt annually (Menaut, Abbadie, Lavenu *et al.* 1991), and other studies on similar systems (White, Abernethy, Oslisly & Maley 1996) suggest that this model result also holds for the real system, and that Guinean savannas are not stable in the long term without episodic fires that would burn more than 70–80% of the surface.

This stability issue is extremely difficult to assess with field data, because the range of observed tree densities is wide, quite comparable to the confidence limits of simulation results (Fig. 14.3). Given their variability, observed tree densities are of little help in analysing the stability of the savanna. Simulations with a stable tree population or an increasing population could produce the same densities and fluctuations of density in a single simulation encompassing the same range of densities. Fortunately there are different ways of tackling this uncertainty.

The first approach is to use a number of different types of simulation outputs to compare to field data, such as spatial patterns and stand-size distributions. This implies thorough analysis which is currently in progress with available demographic data (Gignoux 1994). As an example, for one of the cases shown on Fig. 14.3 (the case with a fire heterogeneity of 0.2), we saved the 100 simulated final spatial patterns and analysed them using a Monte-Carlo test based on Ripley's K-function (Ripley 1981, Moeur 1993). Of the 100 simulations, 22 went extinct before 1000 years, 76 had an aggregated spatial pattern according to Ripley's test (at $P < 0.05$), with a distance of maximal discrepancy d_{max} between observed and expected K-values <2 m for 70 of them, and two had a random spatial pattern. For comparison, spatial patterns of two of our permanent savanna plots were significantly aggregated using the standard test statistic for this method (Moeur 1993), with small distances d_{max} of 1.67 and 2.5 m.

$$L_{max} = 1.803, P = 0.004; \text{ and}$$

$$L_{max} = 1.360, P = 0.004, \text{ with}$$

$$L_{max} = \sup_d \left| \sqrt{K(d)/\pi} - d \right|$$

The simulated and observed spatial patterns appear similar (Fig. 14.4), although the simulated ones may be slightly more clumped, as revealed by the slightly larger distances obtained for the field data. The test presented here is a first attempt but certainly more complete studies using the spatial pattern of different tree species and size classes, and their patterns of spatial associations would constitute much stronger tests for the simulation model.

The second approach consists of studying in much more detail fire behaviour and tree response to fire, a key driving force in the model. Recent experimental work (Gignoux, Clobert & Menaut 1997) showed that trees could have very different strategies in resisting fire, implying trade-offs between fire resistance and growth rate, and with probable relationships to their spatial patterns (Gignoux 1994). Observed spatial patterns could then be the result of complex interactions between very fire-resistant and less fire-resistant species.

The main findings of our preliminary simulations (confirming previous results of Menaut *et al.* 1990) are that: (i) some variability of fire is needed to maintain trees in the savanna; (ii) fire causes or enhances tree clumping; and (iii) tree invasion is possible if fire does not burn more than 70–80% of the savanna each year, resulting in an unstable system. These results constitute an *a posteriori* justification for using a spatial formalism, since the arrange-

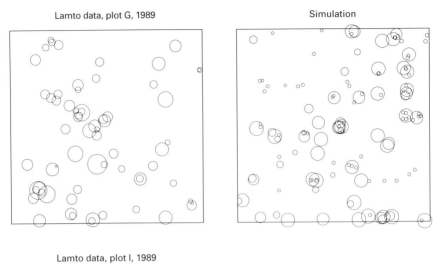
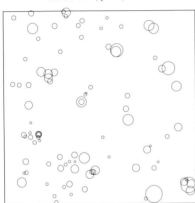

FIG. 14.4. Spatial pattern generated by the FRENCH model after a 1000-year simulation, compared to two observed savanna plots. The same model parameters as Fig. 14.3, with het = 0.2, were used. Circles represent tree crowns. Observed and simulated plots are 50 m × 50 m.

ment of trees in space, resulting from and imposing variations in fire intensity, is a key factor in explaining tree maintenance.

The results we present here are preliminary and imply further work on parameter estimation, a full sensitivity analysis of the major driving parameters, multi-species scenarios, and comparisons of simulated and observed size distributions and spatial patterns. We intend to use this model to explore the effects of tree spatial patterns, dispersal and spatial and temporal variations in intensity of fire on savanna stability.

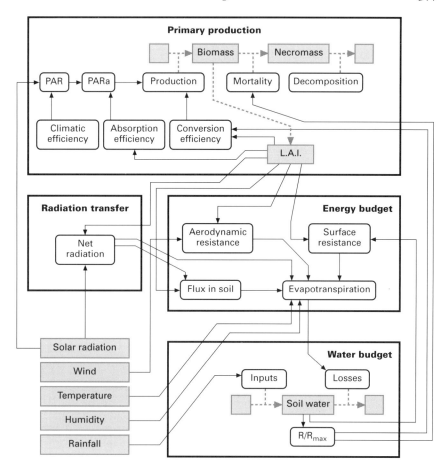

FIG. 14.5. Perennial grass primary production model (Le Roux *et al.* 1996). Functional relations between the submodels. Dashed arrows show matter flows; black arrows show the influence of a process on another one. Note the central position of the LAI in the coupling of the submodels. Variable names: PAR, photosynthetically active radiation; PAR_a, PAR absorbed by the vegetation; R, soil water reserve; R_{max}, soil water reserve at field capacity; LAI, leaf area index. Unlabelled shaded boxes represent matter inputs from compartments not modelled here (for example, the shaded box before the 'biomass' box represents the carbon and nitrogen input from some external source; the shaded box after the 'soil water' box represents the water drained from the soil). Typically, a whole submodel could replace any of these boxes.

Perennial grass primary production, energy and water budget

We have incorporated into MUSE a model of grass primary production for the same West African site (Le Roux 1995, Le Roux, Tuzet, Zurfluh *et al.* 1996, Le Roux, Gauthier, Bégué & Sinoquet 1997). The model (Fig. 14.5) is explicitly based on the functional relationships linking water and energy budgets, and to phenology and primary production. The model couples:

1   a water budget submodel (adapted from Tuzet, Perrier & Masaad 1992), simulating soil water reserve in two horizons (0–60 and 60–170 cm);
2   an evapotranspiration submodel (adapted from Tuzet *et al.* 1992);
3   a production submodel, (derived from Monteith 1977) simulating the

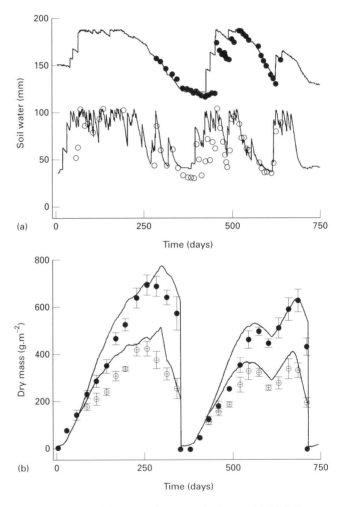

FIG. 14.6. Preliminary results of the grass primary production model. (a) Soil water content for the 0–60 cm deep soil layer (bottom curve) and the 60–170 cm soil layer (top curve). (b) Grass biomass (bottom curve) and phytomass (top curve). Solid lines: simulation outputs; dots: field data; error bars represent one standard deviation for dry mass data ($n = 8$).

seasonal variation of grass biomass and necromass, and the variation of dead and living leaf area.

The model represents grass as a single homogeneous layer, under the control of the simulated water-budget model. The water-budget model and the evapotranspiration model are under the control of the primary production model via phenology. The time-step is one day. Within MUSE we used the 'filler' shape objects to represent the grass layer.

One of the original aspects of this model is the effective coupling between the water budget and the primary production via a precise modelling of LAI dynamics (Fig. 14.5). The model correctly reproduces the observed biomass and soil water reserve patterns (Fig. 14.6). Although humid savannas do not seem to be water-limited (1300 mm year^{-1} rainfall at Lamto), grass phenology is driven by the onset of short drought periods, as exemplified in Fig. 14.6.

In a next stage, we shall couple this model, or a derived version with a greater time-step, to the tree-dynamics model to produce estimates of savanna primary production whilst controlling for heterogeneity of tree spatial pattern and dynamics.

Annual grass primary production

This model simulates the dynamics of the dominant annual grass *Sorghum intrans* in North Australian humid savannas. The model is derived from Watkinson, Lonsdale and Andrew (1989) and focuses on *Sorghum* recruitment as a function of the presence of tree litter, tree cover, and presence/absence of fire. Major phases of the life-cycle have been parameterized with data from field experiments (O. Ronce & G. Cook unpublished data).

To achieve our objective of an integrated model of savanna function and dynamics, we shall couple the types of models described here. As a preliminary exercise, we have coupled the models of tree and *Sorghum* dynamics in a mixed life-forms model to demonstrate the ability of MUSE to handle these situations.

FUTURE DEVELOPMENTS

Savanna models

Our savanna results, although preliminary, revealed that:
1 Tree densities in wet savannas are subject to important stochastic fluctuations which, however, does not enable us to state directly whether wet savannas are stable or unstable.

2 The interaction between spatial patterning of trees and fire variability in space and time has a strong influence on tree population density (Jeltsch et al. 1996). There is a 'phase transition' between a fire-driven tree population and a competition-driven tree population, caused by small changes in a single parameter, fire heterogeneity (the probability of a tree being unburnt by chance during a fire). Consequently, Guinean savannas should not be stable in the long term.

3 Generating spatial patterns comparable to field data is a key test for simulation models. Our approach was able to obtain scattered trees in addition to easily generated tree clumps, contrary to simpler cellular-automata based approaches (Hochberg et al. 1994, Jeltsch et al. 1996).

4 Although water does not seem limiting, grass phenology is driven by the occurrence of even short periods of drought.

Future development concerns the building of new models, mainly by the coupling of the existing savanna submodels within a global savanna model taking into account the main factors involved in tree/grass coexistence, and performing detailed sensitivity analyses of the models presented here. The main effort will concern the adaptation of the grass and tree submodels to each other—that is, we shall need in the tree model to produce outputs on leaf area and root biomass to couple it with the grass model; we shall probably also need to adapt the time-steps of the two models relative to one another. Other related work includes implementing a soil organic matter turnover model. All this will lead to a model coupling many processes, where transitions between driving processes are expected, just as we have shown in the tree dynamics model here.

A further step will be to generalize the concepts, parameterize and run the models developed for Lamto savannas with other savanna systems along the SALT transect in West Africa, a rainfall gradient from Guinean savannas to Sahel.

MUSE as a generic tool for ecosystem modelling

Starting with the development of a savanna model, we ended by developing a general spatial framework for vegetation models without a large extra effort in programming. We were able to adapt already published simulation models to MUSE with a minimal of recoding effort. The major improvements of the MUSE approach, compared to former works are: (i) the detail in the spatial representation of plants; (ii) the ability to mix completely different life-forms such as trees and grasses; and (iii) a great flexibility enabling us to change the spatial representation of plants, or to mix different spatial representations without changing the fundamental biological equations. Applications of such a model call for the choice of the best spatial representation for a specific problem, scaling up from patch to landscape, and comparing different models within the same conceptual framework. MUSE can handle almost any terrestrial vegetation type, from grassland to forest, including heterogeneous systems such as savannas and old fields. MUSE will be fully referenced and made publicly available soon.

The domain of application of MUSE is bounded by three major assumptions: (i) variations are described down to the individual level (to an individual's canopy or root system disc), but not further; (ii) the vertical dimension is always stratified — that is, overlaps of volumes are computed by discrete approximation; and (iii) spatial objects must have clear boundaries — that is, fuzzy boundaries given in the ecological field theory of (Wu, Sharpe, Walker & Penridge 1985) are not possible. We believe that these assumptions are acceptable for a large range of ecological applications. Modelling every individual is a necessary step to handling spatial heterogeneity but it is not the goal of ecosystem studies to describe in full detail every individual in a system. A compromise between the good representation of spatial heterogeneity and computing power must be achieved.

MUSE will keep developing. Some algorithms can still be optimized. We wish to implement automatic model simplification algorithms (Fulton 1991, Moore & Noble 1993, Stage, Crookston & Monserud 1993), more detailed spatial representations of plants, and the ability to handle lateral resource flows (e.g. runoff in a small catchment).

REFERENCES

Botkin, D.B., Janak, J.F. & Wallis, J.R. (1972). Some ecological consequences of a computer model of forest growth. *Journal of Ecology*, **60**, 849–871.

Chesson, P.L. & Warner, R.R. (1981). Environmental variability promotes coexistence in lottery competitive systems. *American Naturalist*, **117**, 923–943.

Coughenour, M.B. (1994). *Savanna — Landscape and Regional Ecosystem Model. Documentation.* Colorado State University, Fort Collins.

Frost, P.G.H., Medina, E., Menaut, J.C., Solbrig, O., Swift, M. & Walker, B.H. (1986). Responses of savannas to stress and disturbance. *Biology International, Special Issue No. 10*, 1–82.

Fulton, M.R. (1991). A computationally efficient forest succession model: design and initial tests. *Forest Ecology and Management*, **42**, 23–34.

Gautier, L. (1990). Contact forêt-savane en Côte d'Ivoire centrale: évolution du recouvrement ligneux des savanes de la réserve de Lamto (sud du V baoulé). *Candollea*, **45**, 627–641.

Gignoux, J. (1994). Modélisation de la coexistence herbes/arbres en savane. PhD thesis, Institut National Agronomique Paris-Grignon, Paris.

Gignoux, J., Clobert, C. & Menaut, J.C. (1997). Alternative fire resistance strategies in savanna trees. *Oecologia*, **110**, 576–583.

Hochberg, M.E., Menaut, J.C. & Gignoux, J. (1994). The influences of tree biology and fire in the spatial structure of the West African savanna. *Journal of Ecology*, **82**, 217–226.

Jeltsch, F., Milton, S.J., Dean, W.R.J. & Van Rooyen, N. (1996). Tree spacing and coexistence in semiarid savannas. *Journal of Ecology*, **84**, 583–595.

Kohyama, T. (1993). Size-structured tree populations in gap-dynamic forest — the forest architecture hypothesis for the stable coexistence of species. *Journal of Ecology*, **81**, 131–143.

Le Roux, X. (1995). Etude et modélisation des échanges d'eau et d'énergie sol-végétation-atmosphère dans une savane humide (*Lamto, Côte d'Ivoire*). PhD thesis, Université de Paris-6, Paris.

Le Roux, X., Tuzet, A., Zurfluh, O., Gignoux, J., Perrier, A. & Monteny, B. (1996). Modélisation des interactions surface/atmosphère en zone de savane humide. In *Interactions Surface Continentale/Atmosphère: l'Expérience HAPEX-Sahel* (Ed. by M. Hoepfner, B. Monteny & T. Lebel), pp. 303–318. ORSTROM, Paris.

Le Roux, X., Gauthier, H., Bégué, A. & Sinoquet, H. (1997). Radiation absorption and use by humid savanna grassland, assessment using remote sensing and modelling. *Agricultural and Forest Meteorology*, **85**, 117–132.

Menaut, J.C., Abbadie, L., Lavenu, F., Loudjani, P. & Podaire, A. (1991). Biomass burning in West Africa. In *Global Biomass Burning — Atmospheric, Climatic and Biospheric Implications* (Ed. by J.S. Levine), pp. 133–142. MIT Press, Cambridge, MA.

Menaut, J.C., Gignoux, J., Prado, C. & Clobert, J. (1990). Tree community dynamics in a humid savanna of the Côte d'Ivoire: modelling the effects of fire and competition with grass and neighbours. *Journal of Biogeography*, **17**, 471–481.

Moeur, M. (1993). Characterizing spatial patterns of trees using stem-mapped data. *Forest Science*, **39**, 756–775.

Monteith, J. (1977). Climate and the efficiency of crop production in Britain. *Philosophical Transactions of the Royal Society, Series B*, **281**, 277–294.

Moore, A.D. & Noble, I.R. (1993). Automatic model simplification: the generation of replacement sequences and their use in vegetation modelling. *Ecological Modelling*, **70**, 137–157.

Prentice, I.C. & Leemans, R. (1990). Pattern and process and the dynamics of forest structure: a simulation approach. *Journal of Ecology*, **78**, 340–355.

Prentice, I.C., Sykes, M.T. & Cramer, W. (1993). A simulation model for the transient effects of climate change on forest landscapes. *Ecological Modelling*, **65**, 51–70.

Ripley, B.D. (1981). *Spatial Statistics*. Wiley, New York.

Shugart, H.H. (1984). Gap models. In *A Theory of Forest Dynamics* (Ed. by H.H. Shugart), pp. 49–67. Springer, Berlin.

Shugart, H.H. & Noble, I.R. (1981). A computer model of succession and fire response of the high-altitude *Eucalyptus* forest of the Brindabella Range, Australian Capital Territory. *Australian Journal of Ecology*, **6**, 149–164.

Shugart, H.H. & West, D.C. (1977). Development of an Appalachian deciduous forest succession model and its application to assessment of the impact of the chestnut blight. *Journal of Environmental Management*, **5**, 161–179.

Shugart, H.H., Hopkins, M.S., Burgess, I.P. & Mortlock, A.T. (1980). The development of a succession model for subtropical rain forest and its application to assess the effects of timber harvest at Wiangaree State Forest, New South Wales. *Journal of Environmental Management*, **11**, 243–265.

Stage, A.R., Crookston, N.L. & Monserud, R.A. (1993). An aggregation algorithm for increasing the efficiency of population models. *Ecological Modelling*, **68**, 257–271.

Tuzet, A., Perrier, A. & Masaad, C. (1992). Crop water budget estimation of irrigation requirement. *ICID Bulletin*, **41**, 1–17.

Walker, B.H. & Noy-Meir, I. (1982). Aspects of the stability and resilience of savanna ecosystems. In *Ecology of Tropical Savannas* (Ed. by B.J. Huntley & B.H. Walker), pp. 556–590. Springer, Berlin.

Watkinson, A.R., Lonsdale, W.M. & Andrew, M.H. (1989). Modelling the population dynamics of an annual plant *Sorghum intrans* in the wet–dry tropics. *Journal of Ecology*, **77**, 162–181.

White, L., Abernethy, K., Oslisly, R. & Maley, J. (1996). L'okoumé (*Aucoumea klaineana*): expansion et déclin d'un arbre pionnier en Afrique centrale au cours de l'Holocène. In *Dynamique à long Terme des Écosystèmes Forestiers Intertropicaux*, pp. 195–198. CNRS-ORSTROM, Bondy.

Wu, H., Sharpe, P.J., Walker, J. & Penridge, L.K. (1985). Ecological field theory: a spatial analysis of resource interference among plants. *Ecological Modelling*, **29**, 215–243.

15. EVOLUTION AND DIVERSITY IN AMAZONIAN FLOODPLAIN COMMUNITIES

P. A. HENDERSON, W. D. HAMILTON and W. G. R. CRAMPTON

Animal Behaviour Research Group, Department of Zoology, University of Oxford, South Parks Road, Oxford OX1 3PS, UK

SUMMARY

1 The physical environment of the Rio Solimões and Rio Japurá floodplain confluence is shown to be highly variable both spatially and temporally at scales ranging from 0.01 to 10000 m and 1 h to 100 years.

2 In the case of fish the floodplain probably holds >300 species; however, almost all these species can be found in other local habitats such as main river channels, black-water lakes or forest streams.

3 For both plants and fish it is argued that factors positive for gene-flow and negative for patch persistence make speciation in the floodplain uncommon.

4 Speciation is much more common under the isolation afforded by side streams and peripheral areas; floodplain species and clades are probably usually derived from these areas.

5 Lack of speciation combines with the high rate of temporal change to demand plastic adaptability in most floodplain inhabitants.

6 Low speciation is not low macroevolution; indeed much remarkable floodplain adaptation is present. 'Genetically assimilable' plasticity often precedes radical novelty.

7 Examples are given of past important macroevolutionary developments that may have originated in tropical floodplains, including the likely origins of angiosperms (Taylor & Hickey 1995), and of tetrapods (Ahlberg & Milner 1994). Radical innovations of the present and recent past in the Amazon floodplain are illustrated.

INTRODUCTION

The landscape of the Upper Amazon floodplain is a palimpsest written by the recent erosions and depositions of the river. Land is cut and relaid to maintain an intricate mosaic of forest, scrub, marshes, lakes and channels. Owing to the annual flood regime, all habitats flood occasionally. Most habitats flood every year, some submerging by up to 11 m. Some floating communities do not submerge but are still subject to mixing and disruption by the floods. We discuss how this restless physical setting affects its ecological and evolutionary patterns, particularly those of fish and plant life.

All habitats are dominated by the physical activities of the river. Much of the 'land' begins as ridges alternated with swamps or lakes. The ridges are created by annual floods and sedimentation at the sides of major channels. As land distances from the main channels, fine deposits in-fill this relief and poor drainage in the low-water season forms backswamps. In such interior areas the vegetation cannot progress to the high forest of levees and sandbar ridges but forms lower swamp forest (*chavascal*). Subsequently, this land is recut by the return of the previous channel or another. Therefore, in sharp contrast to the adjacent unflooded forest no climax vegetation type is ever reached (Salo, Kalliola, Hakkinen *et al.* 1986). Likewise, no old waterbodies develop. Along straight transects, open-water, swamp and forest communities change abruptly, while along the more natural lines of the water channels changes are smoother but still show hierarchies of rather discrete patches at many scales.

Richly vegetated floodplains of tropical rivers were probably present in one continental site or another ever since macrophytes and large animals evolved in the early Palaeozoic. Many important episodes of fossilization are due to such riverine environments. Freshwater flood-prone forest habitats contributing to coal measures, for example, have been well studied. In spite of their importance for the geological record and for evolution, we are not aware that either the general evolutionary characteristics of the biota of such riverplains, or how these are likely to differ from those of contemporary habitats of other kinds, have been discussed. A thread of very relevant research on floodplain ecology and evolutionary outcome, however, has been extended for the single beetle family Carabidae (Erwin & Adis 1982) and we return to this later. General questions that arise include, how typical are such habitats of the biodiversity of their times? Are they rich or poor in the incipient radical novelties that may later come to demarcate major groups?

Based on our experience over 6 years at a nature reserve in western Brazil (Estação Ecológica Mamirauá) (Fig. 15.1) we discuss ecological and evolutionary characteristics of the plants and fish of the particularly broad section of the alluvial plain of the Rio Solimões (Amazon) that surrounds its junction with a large ex-Andean tributary, the Rio Japurá. First, we describe the physical aspects of the floodplain, the template (Southwood 1977) upon which the diversity and life-histories of the biota are formed. Second, we illustrate the biodiversity supported using data on plants and fish. Finally, we

Evolution and diversity 387

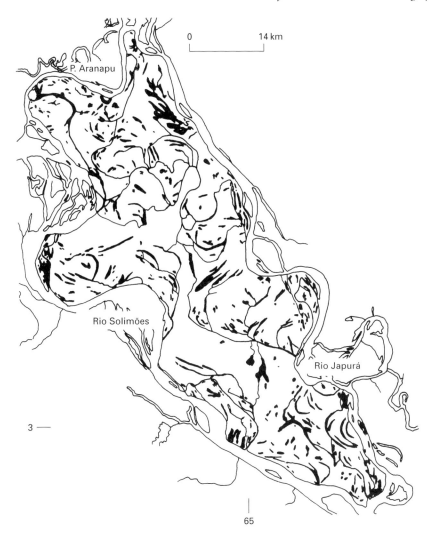

FIG. 15.1. Map of the floodplain of the Rios Solimões and Japurá showing the main rivers referred to in the text. The map shows that part of the floodplain designated as the focal area of Reserva Mamirauá. The networks of lakes and channels within the reserve are filled in black.

consider the scope and evidence for species formation and macroevolution within the *várzea* (white-water floodplain).

HABITAT TYPES

The Andean rocks supply both the sediments from which the floodplain is formed and the dissolved nutrients which allow high biomass productivity. Lower productivity black-water habitats fed by nutrient-poor forest catchments exist nearby. These two types of seasonally flooded habitat occurring inside and outside of the outermost levees of the river and known respectively as *várzea* and *igapó* are often sharply contrasted even when spatially close (Prance 1979). *Igapó* is not confined to the neighbourhood of the floodplain; it is also found around the edges of the sometimes very large impeded ('black-water') lakes and tributaries (the whole lower stretch of the Rio Negro is one example) and is scattered as poorly drained and temporarily flooding patches within the matrix of non-flooding *terra firme* forest. In contrast, *várzea* habitats and vegetation are strictly confined to silty 'white-water' floodplains.

Within the general category of *várzea* it is useful to distinguish several habitat subtypes. These are defined in Table 15.1 (for Upper Amazon comparable habitats see Kalliola, Puhakka, Salo *et al.* 1991a). It is inevitable that all the named landscape features intergrade, and contrasting the simplicity of Table 15.1 with the complexity of Fig. 15.1 (with even the last obviously far from exhausting the habitats that could be shown by a closer mapping)

TABLE 15.1. Floodplain habitats with their equivalent local names in italics.

Mainly terrestrial	Wholly or mainly aquatic
Island and point sand bars—*praias*	Main river channels—*rios*
Levees—*restingas* (silt-heightened riverbanks and bars, on which grows tall forest)	Lesser river channels—*paranas* (usually flowing but sometimes static during low water)
Low swampy woodland—*chavascal* (usually behind levees)	Lake channels—*canos* (as above but connecting to lakes rather than to other channels)
White-water floodplain—*várzea*	
Black-water floodplain—*igapó* (i.e. around lakes and streams outside the flanking levees of a *várzea*)	Lakes—*lagos* (waterbodies of variable shape and size but usually fluvial in origin holding static or slow moving water)
Never inundated land—*terra firme* (land flanking the floodplain)	

Evolution and diversity 389

reaffirms an unavoidable vagueness in the definitions. A rapid-flowing *parana* at flood season may transform to a series of lakes at low water; lakes are often elongated like *paranas* and may be not much wider than their own channels; and so on. At high water, the entire floodplain can be under water that is for the most part nearly static. At such a time only the pattern of the emergent vegetation allows the different types of waterbodies to be distinguished. Even when not all is covered, it is useful to remember that all land has been deposited by the river and therefore *várzea* includes nothing, excepting upper parts of substantial trees, that is never flooded.

Large-scale processes

All of the physical features mentioned range greatly in size, reflecting the variety of processes that form them. Those on the largest scale gain their characteristic forms over time scales of $>10^4$ years. The most important of these events for the Reserva Mamirauá area have been the eustatic ocean changes of the Holocene (Irion 1984). During periods of lower sea water level, rapid channel incision occurred. Over the past 4000 years a rise in sea water level has widened the floodplain and aggrading silt has slowly limited the blocked-valley ('black-water') lakes along the border of the reserve to their present shape. So-called 'black-water' of such blocked valleys and side-streams is characteristically red-tinted with forest-derived humic material. Margins of the black-water lakes support the *igapó* flooded forest and have little floating meadow.

Medium-scale processes

The patchy, often laminar, structure so conspicuous in radar and satellite photographs of the floodplain is created over time-scales of 10^2–10^4 years. Main channel migration over this time-scale results in the general pattern of lakes and channels we see. Channel migration can be rapid. Puhakka, Kalliola, Rajasilta and Salo (1992) give an average erosion rate of 115 m day^{-1} and a maximum of 300 m day^{-1} for the Amazon in Peru. Other estimates are 400 m year^{-1} near the Peruvian border with Brazil (Mertes 1985; Kalliola, Salo, Häme *et al.* 1992) and 200 m year^{-1} beach growth at the Solimões side of the Mamirauá reserve (Pedro Santos personal communication). Within the *várzea*, new waterbodies are formed in various ways. Meander migration results in a scroll-swale topography within the meander loops. The low-lying region between the meander bar and the inside bank initially forms a crescent lake which later becomes a back swamp because of overbank sedimentation. Many lakes within the reserve are crescent lakes of this kind and they offer a series of successional habitats (Kalliola, Salo, Puhakka & Rajasilta 1991b). Larger oxbow lakes are formed by the cutting of whole meanders. These vary considerably in age and size. The largest result from the abandonment (avulsion) of a main river channel and may include a number of meanders. Within the reserve it is likely that its eponymous Lago Mamirauá is an abandoned main river channel, a notion supported by its maximum depth of about 40m, which is similar to that of the adjacent main channel of the Rio Solimões.

The continual reworking of the floodplain in the manner described produces a characteristic scaling of lake size. Hamilton, Melack, Goodchild and Lewis (1992) showed that for Amazonian floodplain lakes size distribution is self-similar over a size range from 0.74 to 1000 km² with an estimated fractal dimension of 1.8. Using 1:50000 maps for the Mamirauá reserve focal area, which is the triangle of floodplain bounded by the Rio Solimões, Rio Japurá and the Parana Aranupu, the Pareto plot in Fig. 15.2 shows that this self-similar structure extends to waterbodies as small as 10 ha. Below this size the Pareto plot is non-linear, probably reflecting census bias against smaller waterbodies; over the approximately linear part of the curve and slope is −1.27, exactly the same as Hamilton et al. (1992) calculated for Orinoco channel lakes and confirming the fractal patchiness of water distribution at this scale.

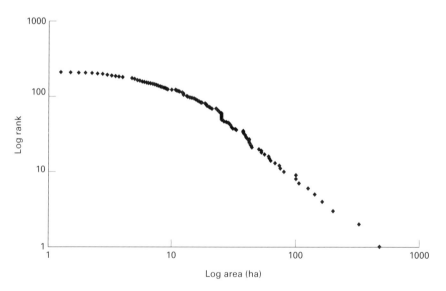

FIG. 15.2. The Pareto plot of lake area distribution within the triangle of floodplain within the Rios Solimões and Japurá and the Parana Aranapu.

Evolution and diversity

Short-term processes

Over time-scales of 1–10^2 years, shear stress and sediment deposition channel structures such as undercut banks, sand bars and levees. Erosion and sedimentation create, change and destroy both aquatic and 'land' habitats. For example, bank erosion fells and sinks trees to create submerged wood habitats which are characteristic for many animal and plant species. Sedimentation changes flow patterns so that well-oxygenated flowing channels transform to languid, oxygen-poor water.

At time scales of <1 year by far the most important process is the seasonal change in river flow which creates the annual flood. Seasonal variation in water depth is shown in Fig. 15.3. The 0-m datum in the Mamirauá water-level series was originally selected as a level to which the water was likely to fall during the low-water season.

As for other rivers (Mandelbrot 1982), the course of the annual flood cycle is highly variable. Variation within the annual cycle alters both the amount and duration of habitat availability. Marked on Fig. 15.3 is the depth at which the forest surrounding Lago Mamirauá becomes completely inundated. Forest-dwelling fish must adapt to an ever-changing habitat availability. A much-longer time series 1902–86 is available for the Solimões–Negro confluence at Manaus collected by the Brazilian Port Authority PORTOBRAS. A simple conversion was possible by comparing the flood waveforms from the Mamirauá reserve in 1983–84 in Ayres (1993) with the same period in the PORTOBRAS series. Using these data, Table 15.2 shows the

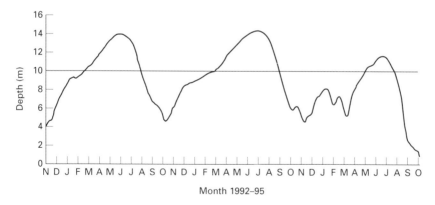

FIG. 15.3. Seasonal variation in water depth for the hydrograph station in Systema Mamirauá. This is one of the main lake systems within the focal area of Reserva Mamirauá in the Rios Solimões–Japurá floodplain. The horizontal line at about 10 m marks the level at which all the surrounded forest is under water. (Modified after Henderson & Crampton 1997.)

TABLE 15.2. The frequency at which the water level declines below selected water levels during the annual low water season between 1902 and 1984. The values are given for an arbitrary 0-m datum and the frequencies were calculated using PORTOBRAS data.

Water depth (m)	Frequency
−2	Once every 25 years
−1	Once every 10 years
0	Every 4–5 years
1	Every 2–3 years
2	Every 2 years
3	8 out of 10 years
4	9 out of 10 years

frequency at which different water depths were experienced in the Mamirauá reserve over the 20th century.

These data suggest that the drying up of shallow lakes which have a bed at +3 m is an almost consistently annual phenomenon, and that channels which largely have a bed at 0 m dry out on average about every 4–5 years (about 22% of low-water seasons). Throughout the *várzea* most lakes have beds at −1 to −2 m and would be expected to be dry once every 10–50 years. However, some channels in the Mamirauá reserve, such as the upper Cano do Lago Mamirauá, have stretches with a bed depth of about 15 m. These areas, of limited extent, act as refugia during the most extreme conditions.

SPATIAL AND TEMPORAL VARIATION IN WATER QUALITY

Water quality has been relatively easy to measure and we report on this as perhaps exemplifying the situation for other physical factors. As pointed out, 'terrestrial' habitats of the study area are never completely non-aquatic over time and at least comparable niche complexity will doubtless eventually be found to apply to them.

Much limnological research in Amazonia has shown the importance of dissolved oxygen, temperature, turbidity and dissolved nutrient levels to animal presence and abundance (e.g. Schmidt, 1973, Fittkau, Irmler, Junk *et al.* 1975, Junk, Soares & Carvalho 1983, Robertson & Hardy 1984). The overall nutrient content is reasonably summarized by water conductivity. As these four main variables are influenced by the physical structure of the waterbodies and by the temporal variation in flow and flooding mentioned in the previous section we can anticipate that these too must show spatial

Evolution and diversity

and temporal variation. However, the variables now in focus reflect new factors: they are influenced by the extracts, excretions and decay products of organisms. These themselves are intrinsically patchily distributed, so creating even more fragmented spatial and dynamic temporal patterns than would follow from physical processes alone. Below we limit the discussion to the temporal-spatial variation of these variables at scales which may influence species presence and community diversity.

Solar heat and oxygen enter water via the free surface. Thus, still water typically shows vertical gradients for both. Oxygen is also generated by photosynthesis and is consumed by decomposition. This enhances the vertical stratification as light is normally only available near the surface and most decomposition occurs within the sediments. The depth, shape, orientation and flow of lakes and channels influence temperature and dissolved oxygen resulting in large-scale spatial variation in these variables. Small-scale variation is also evident. Typical examples of vertical profiles for oxygen and temperature are shown in Fig. 15.4. Another example of the dramatic small-scale spatial variation in oxygen appears in Fig. 15.5 where oxygen levels are shown for localities under forest, floating meadow and in open water only 20 m apart. Temporal variation can be equally extreme. Because of the role of vertical mixing in determining oxygen concentration and temperature there is long-term variation seasonally and between years in line with the flood cycle. For example, flowing water in *paranas* is vertically mixed and deeper water here holds more dissolved oxygen than in lakes. The deeper *paranas* become anoxic during exceptional low water periods when they cease to flow with main river water. Oxygen availability always varies diurnally and may change dramatically even by the hour. Figure 15.6 shows the variation in dissolved oxygen and temperature in surface waters within Cano do Mamirauá during October 1995. Following an afternoon rainstorm, the water went from supersaturation during the day to almost total anoxia at night, resulting in a mass fish mortality. Similar mortalities have been seen several times during our limited visits. On a slightly longer time-scale these events emphasize the high productivity generating the huge numbers of fish that are always present to be killed.

The meeting of waters of different temperature, such as when a flowing *parana* enters a lake, creates a sharp spatial discontinuity because of incomplete mixing. Such frontal systems can range in scale from a few metres to many kilometres wide at the meeting of major rivers such as the Japurá and Solimões. They offer unique sets of conditions and are often zones of enhanced productivity.

Turbidity and conductivity also vary spatially and temporally at many scales. The Rios Japurá and Solimões, because they collect waters from

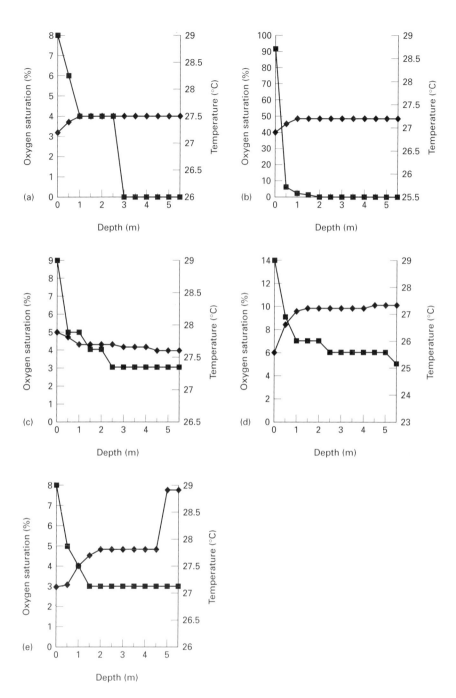

FIG. 15.4. Examples of typical vertical profiles for oxygen and temperature within the Rio Japurá–Solimões floodplain waters. (a) The Apara Parana along which Rio Japurá water flows; (b) the lake Lago Mamirauá; (c), (d) and (e) different localities within the mixing zone between the Apara Parana and Lago Mamirauá Key: ♦ temperature, ■ oxygen.

Evolution and diversity 395

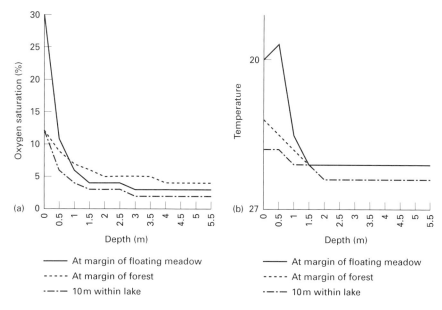

FIG. 15.5. Small-scale variation in oxygen and temperature profiles with in Systema Mamirauá: (a) dissolved oxygen; (b) temperature. Data were collected between 11.35 am and 12.35 pm on 20 July 1994.

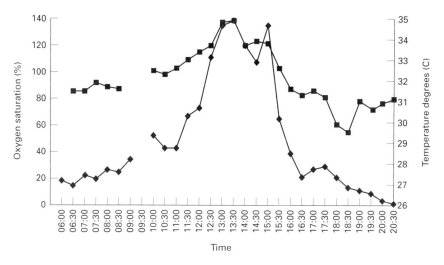

FIG. 15.6. Diurnal variation in surface dissolved oxygen and temperature in Cano do Mamirauá over 24 October 1995. A light rain shower occurred at 14.00 h and heavier rain fell between 15.30 and 16.30 h. The first distressed fish were seen at the surface at 20.15 h. Key: ■ temperature, ♦ oxygen.

different geographical regions, differ in conductivity and suspended sediments. Above the Auati-Parana (the first channel via which the Japurá receives Solimões water though occasionally the flow is reversed), the Japurá and Solimões have conductivities of $8\,mS\,cm^{-1}$ and $80\,mS\,cm^{-1}$ respectively, a 10-fold difference. Thus, the floodplain has the interesting large-scale feature that it is bounded by, and receives, waters of different character. The main river inputs, however, are not the only factor determining conductivity, which varies considerably between lakes and within a lake between seasons. Some lakes during the low-water period have conductivities greater than $180\,mS\,cm^{-1}$ at $20°C$. High conductivities are probably linked to both evaporative concentration and decomposition of floating meadow. Turbidity is linked to flow, wave action, bed sediments, plankton and macrophytes, all of which are continually changing.

As will by now be evident, we have sketched a set of habitats with intermittent potential for high biological productivity but which is extremely variable and unstable on all scales. For sufficiently agile species, it could provide a wealth of niches, just on its physical characteristics alone. We now turn to the life-forms present, especially plants and fish.

SOME EXAMPLES OF DIVERSITY IN THE FLOODPLAIN

Plants

Several lists of plants, emphasizing the trees, are available for our region but a full list is not, herbs especially being neglected. However, it can be inferred with fair confidence from comparisons of *várzea*, *igapó* and *terra firme* forests at other Amazonian sites (Kalliola et al. 1991a, A.J.G. Ferreira personal communication) that Mamirauá *várzea* will have a reduced tree list compared to its nearby *terra firme*. This list is probably about equal in size to that of the local *igapó* which occupies, in this region, a much smaller area. Given the much greater fertility and productivity of *várzea* compared to *igapó*, the result of this comparison may seem surprising but the first point to note is that it conforms with the so-called 'paradox of enrichment'—the frequent observation that nutrient enrichment can lead to reduced biodiversity (Rosenzweig 1971, Riebesell 1974, Verhoeven, Koerselman & Meulman 1996).

When one descends to the lower storeys of the forest or moves to habitats outside, the flora remains poor by absolute Amazonian standards but becomes more special. The ground flora of forests of levees and *chavascal* seems similar in floristic richness to that of *igapó* and *terra firme*. As leaf-litter and soil emerge after the flood, a forest-floor semblance of the spring activity of herbs of north temperate forest arises in the *várzea* with the same tendency to local dominance by a few species. In *restinga* forest, a small *Calathea* for example, leafing from bulbs that a few weeks previously were underwater and had remained so by a metre or more for several months, creates an effect of the dense stands of *Endymion* or *Anemone* of a European deciduous wood, while elsewhere clumps of Cyperaceae simulate the temperate woodland grasses like *Deschampsia*, *Brachypodium* and *Melica* as well as sedges like *Carex pendula*.

Occasional true grasses are present in these shady sites; however, grasses are vastly more dominant in the early stages of succession outside of forest, especially in the very characteristic 'floating meadow' which grows over open shallow water and across slow-flowing channels. The largest of all grasses, of course (potentially common on river banks and even in mature *chavascal*) are two *Guadua* bamboos, while second and third in size are the giant reed *Gynerium sagittatum* and *Laziacis procerrima*. The last two are colonists respectively of high sand banks (Kalliola *et al.* 1991b) and of forest edges and gaps. After these in size come the main floating grasses. 'Floating meadows' are often tens to hundreds of hectares in extent but they contain only very few species (Junk 1970, 1973). *Paspalum repens* and *Echinochloa polystachya* are equally dominant, but *Hymenachne amplexicaule* and *Oryza glumipatula* are also present locally. A less robust grass, *Leersia hexandra*, forms smaller floating mats in less eutrophic (but not 'black') waterbodies.

Significant grass populations have existed along the Amazon for at least 9000 years (Absy 1985) but it is possible that even among the few species listed above are recent additions, as the following points suggest:

1   Absy showed pollen in recent deposits currently on a rising trend and at the maximum she has found.
2   The *Oryza glumipatula* in our area is very close to *O. rufipogon*, which is regarded as a wild (or feral) species of Africa. Local opinion claims this '*arroz bravo*' to have increased greatly during living memory.
3   Pires and Prance (1985) treat *Echinochloa polystachya* as absent or at least uncommon above Manaus and abundant below it until the Atlantic Ocean.

This is far from the case now: in the Tefé region *E. polystachya* now floats in dense largely monospecific and parasite-free stands over larger areas than its main rival, *Paspalum repens*, which by contrast is often severely damaged by insects and fungi. The situation with related *Echinochloa* suggests that *E. polystachya* comes from Africa ultimately but perhaps indirectly to the Amazon via more northern neotropical areas and with hybridization and/or polyploidy involved. All the grasses so far mentioned are perennial but at low water a half dozen or so annual species plus a similar set of annual Cyperaceae (including several *Cyperus*) flourish on open mudbanks with a modest variety of annual dicot herbs intermixed. All the species sometimes make dense uniform stands. Conspicuous among the herbs are *Sphenoclea zeylanica*, some five or so species of *Ludwigia* (Onagraceae), and three or four of *Aeschynomene* (Papillionaceae).

Floating and truly aquatic macrophytes make up the most remarkable communities and show the most unexpected diversity in *várzea* plant-life. Three types are represented. First, there are plants typically rooted on the bottom but with floating parts that for some or most of the flood-cycle photosynthesize at or above the water surface. Second, there are plants always submerged but floating unrooted. Third, there are plants floating at the surface and photosynthesizing above it.

The first class includes the large grasses already mentioned. When water is present these grow extremely fast, easily keeping pace with the rising flood and spreading rapidly laterally across the water surface at the same time. When dropped back on the mud at the end of the flood season they are greatly reduced by disease, decay and consumption by fish, but do not necessarily die or cease growing completely (Junk 1973). Only *Oryza* (later, when floods return, to vie with *Hymenachne* as the least buoyant floater) grows substantially during the 'terrestrial' phase. Less common but similarly semi-aquatic are *Polygonum acuminatum* and *Caperonia castanaeifolia*. Fully aquatic and requiring a pool that persists even at low water, is the characteristic and magnificent *várzea* water lily, *Victoria amazonica*. Dubiously also in this category fall a species of *Callitriche* starwort and a *Riccia* liverwort. It is noteworthy that each of these very different rooted plant types seems to have only one species and this is true of most other genera. Besides the exceptions already mentioned in *Cyperus*, *Ludwigia* and *Aeschynomene*, however, small annuals of the genera *Bacopa* and *Lindernia* also show several species each.

The second class, that of free-floating but submerged plants, has surprisingly few representatives and again genera are mostly monospecific in our area, for example *Ceratophyllum*, *Najas* and *Wolffiella*, while *Utricularia* has two species. Perhaps the paucity of submerged aquatics arises because such plants are unusually at the mercy of: (i) the currents that affect most habitats at one season or another—they are less likely to be held amongst attached vegetation than full floaters; and (ii) the abundance, diversity, specialism and hunger of the low-water fish population. Herbivorous fish such as anostomids are to be observed eating leaves and shoots of *Paspalum*. Fish may act against any fully exposed floating plants like piranhas may against ducks, which are another at first sight surprisingly under-represented group.

The third class, the surface floaters, make up the most immediately striking community. It is remarkable in the following respects:

1 While the community comprises at least 23 species, about a half dozen are so abundant that at the margins of any open area of still water they can usually be found in a few minutes' search.

2 In a 'floating lawn' as we prefer to call an assemblage of these free-floating small species, plants often exist interspersed as a true community in which the smaller fill gaps between the larger. Thus, in descending size, may be observed: *Eichhornia* > *Pistia* > *Limnobium* > *Salvinia* > *Azolla* > *Ricciocarpus* > *Spirodela* > *Lemna*. The 'floater' genus *Wolffia* (Lemnaceae), containing the smallest of all angiosperms, is recorded from Amazonia but we have not encountered it. Lawns of single species occur but are less common than mixtures.

3 On all floating lawns phytophagous insects are abundant and cause obvious damage. Fungal and microbial diseases are also evident.

4 So far as can be judged by presence of sexual structures, all plants of these floating lawns reproduce sexually.

5 All the species also reproduce prolifically by vegetative means.

6 While some species may have long been pantropical, several from the community have been spread by human agency from seemingly just these kinds of communities in tropical South America where in natural waters they are virtually never pests, to other continents (Cook 1985) where they quickly become pests due to their uncontrolled multiplication and the formation of dense monospecific mats (Cook 1987, Pieterse & Murphy 1990).

7 Soon after so migrating they are found to have partially or wholly lost sexuality (Cook 1987, Barrett & Husband 1989), either through having become inbreeders and/or almost exclusively reliant on vegetative propagation. Among the water pests to have originated in this habitat or others closely similar elsewhere in S. America are: *Gymnocoronis*, *Eichhornia*, *Pistia*, *Utricularia*, *Salvinia* and *Azolla*.

Similar habitats occurring elsewhere in South America need co-emphasis in the above claim. In a trend which applies to water plants generally and especially to small ones (probably because propagules or pieces of these are most easily carried by migrating birds), ranges of *várzea* water plants are often very wide. In the case of Amazon water plants a particularly common range extension is into the *pantanal* and *chaco* areas of the Parana river basin to the south.

Returning more generally to the plants in our region, no local endemic has yet been found and the most local species seem not to be from other habitats than *várzea*. For example, a collection so far unidentified to species is a tree in the genus *Bucida* (Combretaceae) whose species are only otherwise from coastal forests of the western Caribbean. This particular small tree is from the *igapó* of a small black-water lake adjacent to the reserve. While such an unexpected genus is easily noticed, our survey cannot generally resolve what might be new species from mere local varieties: our knowledge of the plants cannot match, for example, the discrimination achieved by Ayres (1985) and Yonenaga-Yassuda and Chu (1985) for the new squirrel monkey, *Saimiri vanzolinii*, a truly endemic species of the *várzea* of the Mamirauá area. As noted in a subsequent paper (Ayres & Clutton-Brock 1992) major river channels in Amazonia quite commonly serve as barriers to gene flow for terrestrial large mammals which are poor swimmers; the rivers thus delimit species ranges and control speciation. Migration of islands and channel avulsions in the unstable Amazon floodplains, however, generally work against isolation by water. In the Tefé area this factor seemingly has not been sufficient to prevent speciation in the case of *Saimiri*, and has also been insufficient in the case of a social stingless bee, *Melipona seminigra* (Camargo 1994). So far, we lack any parallel case for a plant.

Fish

At present, 286 species of fish have been identified within the floodplain 'reserve' (Fig. 15.1) and its immediate vicinity, an area of approximately 2×10^3 km². This is about 10% of the total species known from the Amazon basin, which has an area of 7×10^6 km². Probable upper limits for species diversity within the Solimões-Japurá floodplain and the Amazon are 300 and 3000 species respectively (Val & Almeida-Val 1995). High as these numbers may seem, a more impressive diversity is that of form and life-style shown by floodplain fish and this is discussed further below. Floodplain diversity seems not to be based on the rapid speciation of a few ancestral forms such as is found, for example, in the remarkably species-rich cichlid communities of the African great lakes (see Ribbink (1994) for a recent review of cichlid speciation).

So far, as with the plants, no species endemic to the Japurá-Solimões floodplain has been identified. Floating meadow and floating lawn, especially abundant and extensive in the Tefé area and almost confined to the floodplain, might seem to provide promising habitats for endemism. The former especially has been the subject of intensive studies (Henderson & Hamilton 1995, Crampton, 1996, Henderson & Crampton 1997) and yet no example has been found. The importance of electric fish, a group characteristic of these habitats, has been discussed by Crampton (1996). A complete survey of the gymnotids of floating meadows, for example, revealed 14 species with seven Hypopomidae and five Gymnotidae. All of the species with confirmed identifications are known from other regions of the Amazon basin and similar forms to the unidentified species have been seen in museum collections from other areas. Thus, there is as yet no compelling evidence that any gymnotids are endemic to the immediate vicinity of Tefé.

Diversity at deeper phyletic levels is also high. Table 15.3 lists the 42 families and number of species of fish in each family within the Mamirauá reserve. The dominant groups in terms of species number are the characin families Curimatidae, Serrasalmidae, Characidae, the electric fish groups Eigenmaniidae, Hypopomidae, Apteronotidae, the catfish groups Doradidae, Auchenipteridae, Pimelodidae and Loricaridae, and the Cichlidae. This list is typical of Amazonian floodplain habitats (Lowe-McConnell 1987). In addition, the floodplain also supports a number of species-poor families of predatory fish, the Osteoglossidae, Arapaimidae, Synbranchidae, Erthyrinidae and Electrophoridae. The following single representive of each family, *Osteoglossum bicirrhosum*, *Arapaima gigas*, *Synbranchus marmoratus*, *Hoplias malabaricus* and *Electrophorus electricus*, are important members of the fauna both in number and biomass. All five occur along the full length of the lowland Amazon floodplain.

In general, a spatial and temporal patchiness fit to accommodate a large fauna is not difficult to imagine. Looking only to the most major divisions within the river corridor we recognize four principal aquatic domains: forest streams (S), river channels (R), blocked valley lakes (B) and floodplain waters (V). How fish species distribute among these major domains can be conveniently represented in a Venn diagram (Fig. 15.7). Several features are at once striking. The set R + V with 239 members represents the fish of the white waters of all kinds. The great majority are found in flowing channels, V $\cap$ R = 236 and V − R = 8. Of the set V − R only two species, an undetermined *Gymnotus* knife fish and the lungfish *Lepidosiren paradoxa*, have been found to reside solely in static 'white' water and even here it is possible that the gymnotid may move outside of the *várzea* as it has not been captured at low water. Also notable is that the set of fish only caught in static water (B + V) − (C + S) = 9 is small; evidently most species prefer, or at least tolerate, flowing water at some time. In comparison, the much more poorly sampled forest stream habitat holds 13 species not found in C, V or B. Finally, the set of generalist species R $\cap$ V $\cap$ C $\cap$ S holds eight species and includes three of the most dominant in terms of biomass *S. marmoratus*, *H. malabaricus* and *E. electricus*. Within this particular Amazon habitat it seems that these fish manage to be both jacks and masters of all trades. Two important features that they share is a predatory life-style and having adaptations to cope with low dissolved oxygen.

Each of the four aquatic domains can be divided into a large number of habitats each offering patches widely differing in size. A good example is the

TABLE 15.3. The number of fish species subdivided into families recorded from the Rios Solimões-Japurá floodplain. Data were collected between 1990 and 1996 using a wide range of fishing techniques.

Family	Number of species
Potamotrygonidae	2
Clupeidae	1
Engraulidae	2
Arapaimidae	1
Osteoglossidae	1
Erythrinidae	3
Ctenolucidae	1
Crenuchidae	1
Characidiidae	6
Anostomidae	9
Hemiodidae	3
Lebiasinidae	9
Curimatidae	16
Gasteropelecidae	2
Serrasalmidae	12
Characidae	48
Sternopygidae	1
Eigenmanniidae	11
Rhamphichthyidae	6
Hypopomidae	10
Apteronotidae	12
Gymnotidae	6
Electrophoridae	1
Doradidae	16
Auchenipteridae	10
Aspredinidae	2
Pimelodidae	25
Ageneiosidae	4
Cetopsidae	1
Hypophthalmidae	3
Trichomycteridae	3
Callichthyidae	4
Loricariidae	17
Belonidae	1
Cyprinodontidae	3
Synbranchidae	3
Sciaenidae	2
Cichlidae	21
Gobiidae	1
Soleidae	1
Tetraodontidae	1
Lepidosirenidae	1
Total species number	286

Evolution and diversity

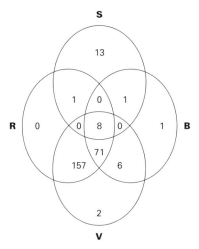

FIG. 15.7. Venn diagram showing the distribution of fish species between the four principal aquatic domains in the vicinity of the Solimões–Japurá floodplain; S, forest streams; R, flowing channels; B, black-water lakes; V, floodplain (*várzea*) lakes.

lake size distribution already discussed. Open waters hold specialized fish communities changing with the size of the habitat. Another important example is floating meadow (Henderson & Hamilton 1995, Henderson & Crampton 1997). Both shelter and food amongst the roots and rhizomes are favourable for small fish including the juveniles of important commercial species (Saint-Paul 1979, Goulding & Carvalho 1982). During quantitative seine net sampling of floating meadow 79 species have been captured. The species accumulation curve shows an almost linear increase in species number up to a sampled area of 400 m^2 (Henderson & Crampton 1997). Given the lack of asymptotic behaviour and the recorded presence of further small species using hand nets, there can be no doubt that the total species number for this habitat will exceed 100. Thus, this one habitat holds more than one-third of the total species complement for the region (presently 286 species with those of forest steams and blocked valley lakes included) and about half of all the species found within the *várzea* lakes and channels. However, no single site ever holds more than a small part of such species diversity. The floating meadows with the most species occur along the edge of flowing channels such as the Apara-Parana which carries water from the Rio Japurá into Lago Mamirauá. The maximum species number collected at one station in this channel was 23 in 10 m^2. Probably the most species-rich localities have complements of about 2.5 species m^{-2}.

Rank-abundance graphs and dominance indices can be found in Henderson and Crampton (1997). Very few species contribute more than 90% of the biomass and these dominants, *Hoplias malabaricus*, *Aequidens* cf. *tetramerus*, *Hypopomus* spp. and *Parauchenipterus* sp., can be classified as floating-meadow generalists since they were found at almost all the stations sampled. The majority of species of floating meadow are infrequently caught and rarely abundant. It is probable that they use the habitat for only one period within the annual cycle or only when other more favoured areas such as flooded forest are unavailable.

The flood-cycle greatly affects fish density and standing crop within the meadows. At low water the fish are concentrated in the lakes. For example, a floating meadow site within the reserve sampled at low water had an average fish standing crop of 195.4 g wet weight m^{-2}, whereas at high water the value was only 17.3 g m^{-2}. Order of magnitude changes in fish density are probably typical.

From the preceding it can be seen that even where floodplain habitat may appear uniform it is exceptionally rich in aquatic vertebrate species. For comparison, 10 m^2 of a simple floating meadow habitat dominated by at most two plants (*Paspalum repens* and *Echinochloa polystachya*) had a species total of fish almost equal to the total for NW Europe. Clearly, the fish data are telling us that *várzea* niches are abundant. But this is not to say that there are as many species as there could be. Within a fluvial system the presence of large numbers of species within one area does not imply that they have been generated *in situ* to fill niche gaps nor that they have filled all of them. Indeed, we argue below that probably for all groups the floodplain is like a battery which is charged by a constant stream of forms originating in forest streams and headwaters. This battery accumulates endlessly (though presumably with some dissipation/extinction), but very occasionally generates dramatic exterior pulses – that is, macroevolutionary changes appear capable of making important contributions elsewhere.

Details of the fish fauna offer some support for this view. It was noted above that the Japurá-Solimões floodplain held no known endemic fish species, suggesting little *in situ* speciation. The argument that temporally variable habitats do not favour fish speciation has been made before (Fryer & Iles 1972, Ribbink 1994) but has been little discussed. In our case three factors seem to conspire to prevent genetic separations. First, spatially all the habitats are physically interconnected, both by channels and seasonally by the mantle of flood water. Second, all habitats are ephemeral on the various longer time scales we have discussed. Third, at low water many populations are forced out of the floodplains into the adjacent channels. Finally, there is the mass transfer of fauna in drifting islands of meadow (Henderson

& Hamilton 1995). Jointly, these factors determine that: (i) it is almost inevitable that the lowland Solimões will show many fish species ranging over the full floodplain length; and (ii) as predicted by the general direction of flow, adaptation and speciation will be increasingly restricted further down the river. As shown by the greater species list in Henderson and Crampton (1997) for the Solimões-Japurá junction compared to that of Junk et al. (1983) for *várzea* in the vicinity of Manaus, 800 km downstream, our present indications seem to fit this. However, given the number of large tributaries converging in the Upper Amazon Basin, the excess may (also or instead) be a further indication of the effect already suggested, that the number of tributaries rather than Amazon channel size or even local extent of floodplain is the significant factor in adding to regional species lists.

It was noted above that the long-term 'charging' of the species diversity of the floodplains may derive from a slow trickle out of the surrounding hinterland of headwaters. Two conditions are needed to fortify this suspicion. The first is that the headwaters are capable of generating new species. The second is that these species are sometimes able, perhaps only very occasionally, to invade floodplain habitats. The reality of these conditions is considered in turn.

While each stream holds a comparatively low diversity fish fauna in comparison to the *várzea*, samples from two forest streams gave 13 fish species not found in any floodplain or river habitat. In comparison, the considerably larger *várzea* community was found to hold only two species not found in either river, lake or stream, and one of these is doubtful. This simple example gives some measure of the potential for isolation and speciation within forest headwaters compared with *várzea*.

Two cases serve to suggest how fish species may 'evolve' from headwaters to floodplains. Both are based on known 'pre-adaptations'.

The first concerns the gymnotiformes. A large *várzea* floodplain subtends on each side a comb of local forest streams within each of which infaunal, cryptic, gymnotiform fish are largely isolated. Such isolation offers ideal conditions for speciation within groups such as *Gymnotus*, *Steatogenys* and *Hypopomidae*. In our area only two species are shared between floodplains and streams. These are the electric eel *Electrophorus electricus* and *Sternopygus macrurus*; in other areas *Gymnotus* cf. *carapo* is also known to be shared. Interestingly, each of these three species belongs to a different, ancient, clade branching near the base of the phylogeny of the gymnotiformes. However, none of the other 58 or so gymnotiformes known from the area is shared between streams and floodplains and indeed six species are known from streams and nowhere else. Despite the restriction of many species to streams there are strong reasons to believe that they could be pre-adapted to floodplain habitats and could hence be candidates to become future invaders. Most of the gymnotiformes of the streams occur either in dense leaf-litter banks or among hanging roots. These habitats are structurally very similar to the hanging root mass of floating meadows. The stream-dwelling gymnotiformes, which live in leaf-litter banks or root masses, resemble their floodplain infaunal counterparts of the floating meadow both in being small and, by merit of their anguilliform shape and electric sense, in their admirable adaptation to manoeuvre and forage in the dense reticulate habitat of a floating meadow. Furthermore, although streams on the whole tend to be better oxygenated and lower in temperature than the floodplain waters, the leaf-litter banks are atypical, being often not only hypoxic but also relatively warm (Henderson & Walker 1986, 1990) Hence, conditions in the leaf-litter bank pose similar physiological challenges to the floating meadows of the *várzea*. Experimental investigations of three common leaf-litter bank specialists: *Microsternarchus bilineatus*, *Hypopomidae* sp. nov, and *Gymnotus* cf. *anguillaris* showed that all were tolerant to protracted periods of anoxia (W.R.G. Crampton unpublished observations). In summary, it would appear as if these three species have attributes pre-adapted to life in the floodplain.

An examination of species distribution within the family Curimatidae also shows the potential for speciation in isolated catchments and headwaters. This group makes a good example as its members are some of the commonest fish within the *várzea* (Araujo-Lima, Agostinho & Fabre 1995). They occur in large numbers, grazing upon algae, vegetation and detritus. As detritivores they are able to exploit a food source which is available in all forest streams and is at its most plentiful within the floodplain. The taxonomy has been recently revised by (Vari 1989a,b, 1991, 1992) and 95 species are now recognized mostly with limited distributions. About nine species are known from the Solimões-Japurá floodplain. Vari (1988) hypothesized that the present distribution could be explained by allopatric speciation within tributaries and headwaters with occasional cases of wide-scale dispersal. The present distribution of species shows a clear tendency for headwaters and tributaries to differ in their species while the main stem of the river holds a set of widely distributed abundant forms.

Beetles

Erwin and Adis (1982) claim that the beetle family Carabidae is more diverse in inundation habitats of Amazonia than in the *terra firme*. Their assessment includes, and is possibly mainly based on, *igapó* habitats: they record *igapó* as higher in carabid diversity than *várzea*, whereas, as noted above, the situation for plants and fish turns out more equal. When discussing the local generation of species, they conjecture that great variation in extent of *igapó* is likely to have accompanied world climatic (and in particular sea level) changes during the Cenozoic, and that this might, through a kind of 'mini-refuge' scenario, account for a high rates of speciation in *igapó*. While patches of *igapó* around mouths of side streams may indeed expand, fuse and contract during climatic changes, it is harder to see this happening for *várzea*. Continuance of the Amazon river plus a regime of floods almost guarantees, differently from *igapó*, a minimal corridor of floodplain existing at all times along the main valley. Thus again *várzea* seems unlikely to provide sufficient barriers for allopatry, strengthening the impression given by Erwin and Adis (1982) that it is *igapó* rather than *várzea* which contributes most to floodplain carabid diversity.

MACROEVOLUTION

What next? Does nothing move out of *várzea*? Of course 'nothing' is perhaps unlikely for any well-filled habitat but, given the above-argued lack of speciation, it is striking to find hints that those groups that very rarely do emerge into other habitats may be very important. Strong claims on these lines for the last-mentioned group, Carabidae, have been made and qualitatively they concur remarkably with our own data for plants and fish.

Erwin and Adis (1982) indicate that carabids of inundation forest are characterized by generalism and include species occupying broad and sometimes even several different niches. For example, some carabids of *igapó* alternate during the seasonal flood-cycle between tree tops and forest ground litter. Although little is said on mechanism (and that somewhat indefinitely) the authors imply that this generalism (or 'plesiotypy') creates an important evolutionary potential, with the result that on a geological time-scale floodplains generate 'taxon pulses' (Erwin 1979) that spread out to carry new wetland-originating carabid forms, speciating and specializing as they go, right to the ultimate limits of life of the insect biosphere in deserts, mountains, the far north, and so forth.

We argue similarly that absence of common speciation in the floodplain actually abets rather than denies the possibility that the habitat has been important for macroevolution. Being 'stretched' over several niches is conducive to polymorphism (Moran 1988; 1991; 1992) and polymorphism in turn is conducive to radical changes (West-Eberhard 1989). Extremely rarely, developmental morphs of plastic species may transform into new and more efficient adaptations enabling descendants to invade new habitats. Examples where wing polymorphism may have been antecedent to origin of major groups (e.g. ants, anthocorid and cimicid bugs) are shown by Hamilton (1979). (A wingless very ant-like bethylid female, seemingly somewhat communal in its social life, is common under bark of recently floating or submerged dead logs in levee forest of Mamirauá; immature stages have not been seen.) Henderson and Bamber (1987) argued that the adaptations for plasticity required of fish in estuarine conditions (which are not unlike those of *várzea*) allow rapid invasion of fresh water when niches became available. Judging by the presence of typically marine fish families such as the Clupeidae, Soleidae, Sciaenidae, Belonidae or Tetraodontidae, in the channels of the study area, 2600 km from the sea, this transition has frequently been accomplished. Many similar examples of ex-estuarine invasions of rivers by fish could be given for other regions — for example, atherinids in Mexico (Bamber & Henderson 1989). As regards specific *várzea* potentials for novelty, it is notable that the one species of fish of the study area which seems to be a complete floodplain specialist is the lungfish *Lepidosiren paradoxa*. The unique set of adaptations of this animal is particularly directed at survival in low oxygen, seasonal environments. Its air breathing is unlikely to open new evolutionary avenues to it now but this is because terrestrial habitats are so well occupied by what are, in effect, its remote relatives. We are not of course suggesting *Lepidosiren* here as a perfect 'frozen' ancestor of the land vertebrates: other lungfish exist in seasonally flooded habitats in Africa and Australia and are not very closely related. Combined, the three merely suggest likely characters of the common ancestor. Instead, what is significant for the present theme is that all three combine to implicate seasonally flooded freshwater habitat.

Prior occupancy of the land did not apply in the early Palaeozoic, either for animals or, as we will discuss later, for plants. While the invasion of land by vertebrates is still poorly understood, there is a general consensus that tetrapods originated from sarcopterygian fish during the late Devonian (Ahlberg & Milner 1994), probably from shallow freshwater forms although Bray (1985) suggests that coastal marine swamp habitats could have been important. By the Visean–Namurian epochs of the early Carboniferous 'most tetrapod assemblages [were in] lowland swamps or lakes on coastal plains, and are all situated in the continent of Euramerica close to the palaeoequator of the time' (Ahlberg & Milner, 1994). Tetrapod fossils from the Lower Carboniferous in Scotland are found in freshwater limestone, black shales and tuffs in association with scorpions, eurypterids, millipedes, harvestmen, ostracods and plant material. The presence of ostracods as the only aquatic group is indicative of temporary flooding. Thus, the general picture we presently have of the originating habitat of tetrapods bears similarities to the recent *várzea*.

Evolution and diversity 409

Put more theoretically, a habitat's demands for plasticity could well accelerate macroevolution via genetic assimilation (provided—the usual difficulty for this concept — plasticity can be sufficiently maintained as the process goes on). A lungfish that has to live as an adult in an expanse of drying mud as well as living at other times in a shallow lake may find itself, via genetic assimilation, within reach of macroevolutionary slopes that are unattainable to a pure shallow-lake specialist. Its advance may then occur in ways not necessarily connected with aestivation in mud (see Dennett (1995) for a good outline explanation of these ideas although the difficulty concerning the maintenance of plasticity has not yet been treated). Likewise for tambaqui, *Colossoma macropomum*, which varies from being a predator when young to a seed eater in sparsely populated flooded woodland as an adult (Goulding & Carvalho 1982). As an adult this fish displays a remarkable further plasticity in switching seasonally from feeding on seeds crushed by strong teeth at high water to planktivory using gill-rakers at low water. Such an example may help in part to explain a related fish group whose origin may indeed be in the floodplain. The biting piranhas (family Serrasalmidae) are found currently in many aquatic habitats of tropical South America. They include at least one representative, the group-living, predatory *Serrasalmus nattereri*, that has to be regarded as essentially a floodplain species even though its young are also found in river margins. The piranha method of feeding has its special advantages in two circumstances: first, during the overcrowded conditions of a *várzea* low-water lake when the species attacks large prey in groups (as well as certainly continuing its normal and individualistic fin and tail biting habit); second, at high water within the forest where its dentition allows it to scavenge flood-stricken victims. Plausibly ancestors of Serrasalmidae at some point specialized from the plastic repertoire of a tambaqui to become unique neotenous chunk-biting predators, thus initiating the radiation of form and behaviour that we see. The observation that most neotropical fish have an African analogue does not hold for the piranhas, and the idea of the arrival of a fertile pair of piranhas in an African river leads to interesting evolutionary speculations. These cases are only a few out of a list that could be given of thwarted or promising adaptive transitions in *várzea* fish.

In plants, parallels exist for both the situations mentioned for fish, that is both for 'frozen' ancestors and new expansions. In the first class, and perhaps like *Lepidosiren* now thwarted by the very extent of their ancestral triumphs, we find *Gnetum leyboldi* and *Ceratophyllum demersum*, one a liane dangling above the water surface and the other commonly snagged on twigs just beneath it. Both occupy very basal positions in most suggested seed-plant phylogenies. *Ceratophyllum* is so thoroughly an aquatic it is a little hard to imagine either its ancestors or descendants being anything else. Nevertheless, in *várzea* it is regularly dumped on wet mud and so is being 'given a chance'. *Gnetum* is not quite so thoroughly a liane. Its wood is weak under bending and tension and it is not difficult to imagine a herbaceous ancestor for the Amazonian riverine species. This could have lived on channel-side mud and sprouted rather *Piper*- or *Saururus*-like jointed stems, their weakness here probably costing the plant little. Instead speed of growth was the essence and more significantly for their success we can imagine such herbs unfurling paired broad leaves with unprecedented rapidity. These leaves would be the best solution for leaf design to have evolved on earth to date. To summarize, the imagined growth style would be very similar to that of various of the recently focused 'palaeoherbs' of today, to which group some modern phylogenies actually count *Gnetum* to be the immediate sister. Four further points seem worth making about *Gnetum leyboldi*. First, its adaptations to seed dispersal by fish show thorough commitment to life in flooded forest (Kubitski 1985). Second, mature lianes in the *várzea* show capability to create long horizontal snake-like shoots at ground level suggestive (whether for past, future or both) of a rhizomatous potential (as in a *Piper* sp. of the same area): so far as we know both this feature and the fish seed dispersal are unique to the flooded-forest species. Third, as seen at a glance the floodplain *Gnetum* easily passes as a dicot liane; close attention is needed to distinguish if from various opposite-leafed vines such as *Salacia* with which the gymnosperm is often intermingled; whether by convergence or homology *Gnetum* has very angiospermous leaves. Fourth and finally, we must add a caution to all the above: *Gnetum* species occur pantropically and often in much less watery habitats; therefore without review of other members a too-glowing portrait of a single species calls for reserve. A similar caution needs to be applied to our discussion of *Ceratophyllum* which is far more cosmopolitan still and which so far as we know has no special features to distinguish *várzea* representatives.

Too much wind- and water-borne gene dispersal for local genetic differentiation plus competition with such angiospermous vines as *Salacia* could explain why *Gnetum* itself is held to a fairly rigid niche at present. However, a particular advanced angiosperm of the *várzea* might reveal something of the opportunities that might have been open to a *Gnetum* ancestor in the Jurassic. *Phyllanthus fluitans* is classified into a very large, almost cosmopolitan, genus (more than 600 species) that includes very numerous herbs, some bushes, and a few small trees. Yet *P. fluitans* is a tiny floating plant mingling with equally small *Azolla*, *Salvinia* and *Spirodela* in floating lawns. It extends only a few internodes of stem along the water surface and expands only a few

Evolution and diversity

of its rounded, crinkly leaves before breaking into pieces that become independent plants. When stranded it puts down shallow roots into the soil but is at best static in this situation and virtually never grows an upward stem. Its tiny white and separately male and female flowers presumably provide the characters for its generic assignment, but it is easy to imagine that the reduced flowers of such a reduced plant could lead eventually to a radical new arrangement so that if the plant ever evolved to be large again, and its flowers with it, its placement might be obscure. Some superficially radical characters are present in the genus *Phyllanthus*, at least for the untechnical observer; for example, the absence of typical euphorb latex and the curious arrangement of flowers on side shoots that makes them seem to spring from the axils of petiolules of a compound leaf. However, it does not seem that *P. fluitans* can be a basal species with respect to these developments since, while the wood of members of the subtribe including *P. fluitans* is indeed stated to be unusually uniform and distinct (Mennega 1987), which is what might be expected if secondary thickening had been reinvented after a 'fluitans' phase, *P. fluitans* is placed in a section of the subtribe seemingly not treated as primitive and containing many non-aquatic members (Punt 1987). Only one other member of this huge genus is aquatic. Obviously, this favours *P. fluitans* being a derived outlier from either a herbaceous or a woody ancestor, quite the contrary to itself showing the ancestral form. A similar situation holds for the ephemeral herb *Aciotis aequatorialis*, a member of the Melastomaceae which is otherwise almost entirely woody in Amazonia as elsewhere. Flowering sometimes on a monopodial stem at about 5 cm tall, *A. aequatorialis* grows on waterside mud and along forest edges and is not aquatic, while above it in such sites may arch a *Miconia* plant, member of a huge, woody (thus far more typical) genus of the same family. Examples of floral innovation in small (new?) taxa of small plants of open habitats, with such taxa likewise allied to large genera, occur in *Aphanes* and *Sibbaldia* of the Rosaceae. The corresponding larger sister taxa of which are *Alchemilla* and *Potentilla*. As an example of a structural transposition that could be important for future flower form, in the case of *Aphanes* (formerly *Alchemilla*) the stamens are outside rather than inside the flower's nectar-secreting disk.

A similar rationale to that suggested above for *P. fluitans* and *A. aequatorialis* (as also for *Aphanes* and *Sibbaldia*) could be applied to other small floaters mentioned, including the heterosporous ferns. An actual example of novelty in very small *várzea* plants is a 'spike-rush', *Eleocharis radicans*, which forms creeping mats over wet sandy mud at low water. Inflorescence-destined stems from the small caespitose plants arch down to touch the mud at a very early stage of the initiation of their flower spikelet. Roots, one or more new shoot meristems, and more leaves then develop simultaneously with further expansion (and upward reorientation) of the inflorescence spike. Such 'inflorescence plantlets' are often seen growing detached from their parents following early decay or breakage of the connecting stem: the plant then has the paradoxical appearance of having flowered before it produced its first leaves. Effectively, this form achieves rapid vegetative spread and seed production in a single and very economical process. Touching the mud is not essential in the above process. Inflorescence plantlets are often held suspended in the mat of hair-like leaves but still emit leaves and onward-colonizing flower stalks.

Such speculations of evolutionary importance for small, short-lived forms must be admitted to be quite opposite to a long-held botanical preference which traces the major developments of higher plant evolution through woody forms and even sometimes designates aquatics specifically as end states (Church 1919, Heywood 1978). Fossil findings, however, have recently reversed this belief for the ultimate origins of angiosperms and have led to designation of a set of 'palaeoherbs' that are likely never to have had woody angiospermous ancestors. The idea is currently much discussed (Taylor & Hickey 1995) referring mainly to an environmental background in Cretaceous North America that is strongly suggestive of *várzea* (e.g. see figure 9.2, Taylor & Hickey 1995). This encourages a further suggestion that plant macroevolution may have involved important herbaceous phases at other points, with woodiness intermittently evolving. Besides implicating all monocotyledons as never-arborescent descendants of palaeoherbs (and thus embracing a substantial set of the *várzea* water plants already mentioned), the family Nymphaceae is often specially focused upon in the discussions. Within Nymphaceae *Victoria amazonica* is even more *várzea*-adapted than *P. fluitans*, if less currently abundant. Like *P. fluitans*, *V. amazonica* is unusual for its group in several ways. Some of these suggest past evolutionary states and others suggest current plasticity. In spite of its giantism (and unlike most other water lilies) *V. amazonica* can sometimes (here in a kind of parallel to *Aciotis?*) behave as an annual. Its very large spine-protected flowers are pollinated by bulky dynastine beetles (Prance & Arias 1975). On other grounds beetle pollination is suggested to be the oldest entomophilous system (Crepet & Friis 1987).

Comparing *V. amazonica* to *Brasenia* or *Cabomba*, genera generally considered even more 'primitive' than any Nymphaceae, we find *Victoria* to be a pachycaul plant (Corner 1954), contrasted to small herbaceous leptocauls while at the same time we see a great increase in size. In this respect two other perhaps 'promising' *várzea* plants deserve notice as possible parallels.

First, the aroid *Montrichardia arborescens* is a pachycaulous and nearly monopodial giant herb growing up to about 5 m tall. It is a common early colonist of newly deposited sand especially along the Lower Amazon (in our area it is widespread but not abundant). Second, *Cecropia membranacea* and *C. latiloba* are pachycaul trees, the former attaining to perhaps 25 m. As is necessary for such a stature, *Cecropia* are woody, and in fact in our area they provide the first abundant tree colonists of new levees and islands. Closely related to *Cecropia* and sometimes in our area appearing on open riverbanks is a genus of mostly forest trees, *Pourouma*. Current botanical interpretation typically sees *Pourouma*, the tree, as the primitive form in a sequence: *Cecropia* by this interpretation is an intermediate while a large herb, like *Urera baccifera* again in our study area (Urticaceae, close and even possibly sister group to Cecropiaceae), could be taken to illustrate what might eventuate from the 'descent' of a *Cecropia* (Taylor & Hickey 1995). However, another view is possible. *Monotrichardia* is very obviously not a descending tree and it thus suggests a reversed view for *Cecropia*. The alternative sequence would have an *Urtica*-like herb at the start and *Pourouma* a 'neo-tree' at the end, in this case *entering* the forest. Strengthening this point, the characteristics of island evolution suggest woody *Urtica ferox* in New Zealand as far more likely ascended from a herb, similar to the many woody 'speedwells' (*Hebe*) and arborescent 'daisies' of the same region, rather than descended from a tree. The moderately pachycaulous *Urera baccifera* (and perhaps still better *U. carascara* of the same region) can be a model of an intermediate life-form which, in a parallel context, might have led from an *Urtica* (or a *Dorstenia*) towards the state of a *Cecropia* and then a *Pourouma*. In Africa the cecropiacean *Musanga*, showing short-lived but well-built trees, could represent a further development in this line. Remarkably in this context *Cecropia membranacea*, which of the two main *várzea* species favours the richer and therefore more 'nettly' habitats, actually has fragile, irritant hairs on its leaves and young stems. (Yet another parallel *Cecropia* has with many urticaceous 'nettles' is dioecy, but this is a character of all Cecropiaceae.)

Surprisingly, a theory that evolutionary enlargements and lignifications in herbs have been common could revive attention to various positive features of Corner's neglected 'Durian' theory of plant evolution (Corner 1949, 1954), the main difference being that the stem of angiosperm evolution would lie largely through short-lived fast-evolving herbs which, from time to time (probably usually following radical biochemical innovations, see Berenbaum & Seigler 1992) would send up lines to become arborescent. As already mentioned for New Zealand, there seem to be many trends in plants especially on islands (including mountains as 'islands' in land areas) where this can be seen happening today (Carlquist 1966). Herbs evolving to become woody do not have to proceed through a pachycaul habit, of course, as many of the island examples and other woody oddities occurring within herbaceous genera show (e.g. in *Bupleurum, Mimulus, Tabacum*) but they are perhaps especially likely to become pachycaul when there is a premium on either water conservation or on fast growth in rich soils. The last would apply to *várzea*. Here leaves would be expected to be large and these and the few meristems producing them would need to be well protected by latex, novel poisons, stinging hairs, physical defences (as with the spines in *Victoria*), or, as in *Cecropia*, by ants. *Macaranga*, the Asian life-form and ecological parallel to *Cecropia*, extends its parallel even to the possession of ants. As a woody *Urtica* might stand in a transition towards *Cecropia*, so in the euphorb case various co-tribal genera could be considered model transitions through herbs, treelets and even lianes, as exemplified in the genera *Ricinus, Manihot, Acalypha* and *Dalechampia*.

Since palms worldwide tend to be tolerant of flooding, palms as a group might be suspected to have floodplain origins. This suspicion might find supporting evidence in other continents but it seems to gain none from Amazonian *várzea* or *igapó*. There palms are less common and diverse in the flooded habitats than they are on adjacent *terra firme*. In SE Asia the genera *Nypa* and *Metroxylon* with their unusual rhizomatous tendencies show themselves long adapted to forest-water edges and swampy situations. This is especially true of *Nypa* which has a pollen record of over 100 million years to the Cenomanian, a record not excluding the New World. *Elaeis oleifera*, the western oil palm, may also approximate to life-styles of *Nypa* and *Metroxylon* in the places where it is native in Central America. In Amazonia it is local and perhaps only adventive in sites of pre-Colombian human introduction; however, it still always occupies 'low-lying wet areas' (Henderson 1995). Apart from sometimes having air roots or stilt roots, the palms of the *várzea* and of neighbouring *igapó* do not show obvious adaptation to floods. Spruce (1871) argued long ago that biogeographic evidence suggests Amazon palms came into their present flood-prone habitats from origins in the shield-rock areas to north and south of the basin or else from Andean foothills. Spruce's seemingly unchallenged opinion for this group obviously converges to ours for fish and to that which we hold, as yet more tentatively, for the plant life generally. One palm exception that has been found to Spruce's claim is monotypic *Barcella odora*, a small acaulous palm of semi-open campina scrub. This species is truly endemic to central Amazonia; but it is on *terra firme*. The only local endemic palm of the study area, *Bactris tefensis* (a single member of a large Amazonian genus) is likewise from *terra firme* (Henderson 1995). It seems, therefore, that the adaptations of floodplain palms were probably acquired by ancestors in wet sites near to or beyond the periphery of the Amazon basin rather than acquired near its centre. The very successful lepidocaryoid palms *Mauritia flexuosa* and *Mauritiella armata* currently grow on fringes of floodplains of the Amazon tributaries along most of their length but they are also dispersed all over the basin and also beyond it. Notably, they are common in extremely different communities along streams and in marshes of grassy savannas both to the north and south of the forest area.

As with the groups discussed earlier in this chapter, however, there seems to be at least one exception to offset what is for palms a repetition of our generally negative theme. In our area as in others along the main rivers, species of the genus *Desmoncus* are typical of forest-water edges and swampy liane forests. Phyletically *Desmoncus* occupies an undistinguished position very close to the upright bactridine palms (Henderson 1995). The life-form of *Desmoncus*, however, is entirely different from the typical bactridines since all its members are lianes. In this case, biogeography and ecology seem at present consistent with a floodplain origin. (However, the existence of huge suitable non-floodplain areas of *chavascal* and liane forest that are distant from navigable channels, and therefore botanically unexplored, should be borne in mind.) As suggested by its systematic placement and paucity of species (five in Amazonia, two others elsewhere, as stated by Henderson 1995), the evolution of *Desmoncus* is evidently recent. Nevertheless, if at least two strikingly parallel but much older unrelated developments in the palms in SE Asia are taken as guide, the emergence of *Desmoncus* in Amazonia could have importance for the future. In the eastern realm several hundred species of rattan of the palm genus *Calamus* provide forests and scrub habitats with dominating, smothering lianes of very diverse sizes and types, while yet another eastern genus of spiny liane palms, *Korthalsia*, has begun, uniquely for the whole order, an alliance with ants.

As already mentioned, the evolution of *Desmoncus* in *várzea* fits with the broken, semi-open quality of the vegetation which is very favourable to lianes and generated a long list. Among many other examples of incipient or recent scandent forms in *várzea* two other plants deserve mention, both like *Desmoncus* standing out from the growth habits typical of their family. One is *Byttnera ancistrodonta*, a species from the only 'scandent' genus within Sterculiaceae (as a juvenile the plant is a bush). The other is *Scleria secans* in Cyperaceae. In this second case the climbing habit is unique within the genus, all other *Scleria* having caespitose or reedy forms that are much more typical for their 'sedge' family. *S. secans* ('tiririca' in our region, well known for its backward-cutting sharp leaves) climbs high into trees where it forms dense festoons. It is especially common in disturbed *igapó* which is not very 'black'.

To conclude, the patchwork of ephemeral *várzea* habitats selects for colonizing attributes such as short life-cycles, rapid growth and life-style adaptability. In spite of its multiplicity of niches it is not favourable to high specialism or to *in situ* processes of speciation. In its inhabitants, simplification of body form and phenotypic plasticity are common outcomes. In both plants and fish this commonly manifests in weedy life-styles, short lives and size-change experiments both towards large and small. Many fish demonstrate striking plasticity in their ability to change their physiology, behaviour and body form in response to anoxia and factors of the flood-cycle. In plants, as just one example, diverse rooted life-forms swiftly develop arenchyma and become floaters when they are flooded. Many other examples have been shown. Simplified, responsive organisms are a clay from which evolution may be moulding not so much abundance of species as novel forms.

REFERENCES

Absy, M.L. (1985). Palynology of Amazonia: the history of the forests as revealed by the palynological record. In *Key Environments: Amazonia* (Ed. by G.T. Prance & T.E. Lovejoy), pp. 72–82. Pergamon Press, Oxford.

Ahlberg, P.E. & Milner, A.R. (1994). The origin of tetrapods. *Nature*, **372**, 507–514.

Araujo-Lima, C.A.R.M., Agostinho, A.A. & Fabre, N.N. (1995). Trophic aspects of fish communities in Brazilian rivers and reservoirs. In *Limnology in Brazil* (Ed. by J.G. Tundisi, C.E.M. Bicudo & T. Matsumura-Tundisi) 384 pp., ABC/SBL, Rio de Janeiro.

Ayres, J.M. (1985). On a new species of squirrel monkey, genus *Saimiri*, from Brazilian Amazonia (Primates, Cebidae). *Papeis Avulsos de Zoologia*, **36**, 147–164.

Ayres, J.M. (1993). *As Matas de Várzea do Mamirauá*. CNPq, Brasilia.

Ayres, J.M. & Clutton-Brock, T.H. (1992). River boundaries and species range size in Amazonian primates. *American Naturalist*, **140**, 531–537.

Bamber, R.N. & Henderson, P.A. (1989). Pre-adaptive plasticity in atherinids and the estuarine seat of teleost evolution. In *Fish in Estuaries* (Symposium proceedings: Ed. by P.A. Henderson & A.R. Margetts). *Journal of Fish Biology*, **23**(A), 17–23.

Barrett, S.C.H. & Husband, B.C. (1989). The genetics of plant migration and colonisation. In *Plant Population Genetics: Breeding and Genetic Resources* (Ed. by A.H.D. Brown, M.T. Clegg, A.L. Kahler & B.S. Weir), pp. 254–277. Sinauer, Sunderland, MA.

Berenbaum, M. & Seigler, D. (1992). Biochemicals: engineering problems for natural selection. In *Insect Chemical Ecology: An Evolutionary Approach* (Ed. by B.D. Roitberg & M.B. Isman), pp. 89–121. Chapman & Hall, New York.

Bray, A.A. (1985). The evolution of terrestrial vertebrates: environmental and physiological considerations. *Philosophical Transactions of the Royal Society, Series B*, **309**, 289–322.

Camargo, J.M.F. (1994). Biogeografia de Meliponini (Hymenoptera, Apidae, Apinae): a fauna Amazônica. *Anais do 1° Encontro sobre Abelhas, Ribeirão Preto, SP, Brasil*, pp. 46–59.

Carlquist, S. (1966). Wood anatomy of Compositae: a summary with comments on factors controlling wood evolution. *Aliso*, **6**, 25–44.

Church, A.H. (1919). *Thalassiophyta and Subaerial Transmigration.* Clarendon Press, Oxford.
Cook, C.D.K. (1985). Range extensions of aquatic vascular plant species. *Journal of Aquatic Plant Management*, **23**, 1–6.
Cook, C.D.K. (1987). Vegetative growth and genetic mobility in some aquatic weeds. In *Differentiation Patterns in Higher Plants* (Ed. by K.M. Urbanska), pp. 217–225. Academic Press, London.
Corner, E.J.H. (1949). The Durian theory or the origin of the modern tree. *Annals of Botany New Series*, **13**, 367–414.
Corner, E.J.H. (1954). The evolution of tropical forest. In *Evolution as a Process* (Ed. by J.S. Huxley, A.C. Hardy & E.B. Ford), pp. 34–46. Cambridge University Press, Cambridge.
Crampton, W.G.R. (1996). Gymnotiform fish: an important component of Amazonian flood plain fish communities. *Journal of Fish Biology*, **48**, 298–301.
Crepet, W.L. & Friis, E.M. (1987). The evolution of insect pollination in angiosperms. In *The Origins of the Angiosperms and their Biological Consequences* (Ed. by E.M. Friis, W.G. Chaloner & P.R. Crane), pp. 181–202. Cambridge University Press, Cambridge.
Dennett, D.C. (1995). *Darwin's Dangerous Idea.* Penguin, London.
Erwin, T.L. (1979). *Taxon Pulses, Vicariance and Dispersal: An Evolutionary Synthesis Illustrated by Carabid Beetles.* Columbia University Press, New York.
Erwin, T.L. & Adis, J. (1982). Amazonian inundation forests. Their role as short-term refuges and generators of species richness and taxon pulses. In *Biological Diversification in the Tropics* (Ed. by G.T. Prance), pp. 358–371. Columbia University Press, New York.
Fittkau, E.J., Irmler, U., Junk, W., Reiss, F. & Schmidt, G.W. (1975). Productivity, biomass and population dynamics in Amazonian water bodies. In *Tropical Ecological Systems* (Ed. by F.B. Golley & E. Medina), pp. 289–311. Springer, New York.
Fryer, G. & Iles, T.D. (1972). Alternative routes to evolutionary success as exhibited by African cichlid fishes of the genus *Tilapia* and the species flocks of African great lakes. *Evolution*, **23**, 359–369.
Goulding, M.C. & Carvalho, M.L. (1982). Life history and management of the tambaqui (*Colossoma macropomum*, Characidae): an important Amazonian food fish. *Revista Brasileira de Zoologia São Paulo*, **1**, 107–133.
Hamilton, W.D. (1979). Wingless and fighting males in fig wasps and other insects. In *Reproductive Competition, Mate Choice and Sexual Selection in Insects* (Ed. by M.S. Blum & N.A. Blum), pp. 167–220. Academic Press, London.
Hamilton, S.K., Melack, J.M., Goodchild, M.F. & Lewis, W.M. (1992). Estimation of the fractal dimension of terrain from lake size distributions. In *Lowland Floodplain Rivers: Geomorphological Perspectives* (Ed. by P.A. Carling & G.E. Petts), pp. 145–163. Wiley, New York.
Henderson, A. (1995). *The Palms of the Amazon.* Oxford University Press, New York.
Henderson, P.A. & Bamber R.N. (1987). On the reproductive strategy of the sand smelt *Atherina boyeri* Risso and its evolutionary potential. *Zoological Journal of the Linnear Society of London*, **32**, 395–415.
Henderson, P.A. & Crampton, W.G.R. (1997). A comparison of fish diversity and density between nutrient rich and poor lakes in the Upper Amazon. *Journal of Tropical Ecology*, **13**, 175–198.
Henderson, P.A. & Hamilton, H.F. (1995). Standing crop and distribution of fish in drifting and attached floating meadow within an Upper Amazonian várzea lake. *Journal of Fish Biology*, **47**, 266–276.
Henderson, P.A. & Walker, I. (1986). On the leaf litter community of the Amazonian blackwater stream Tarumã-Mirim. *Journal of Tropical Ecology*, **2**, 1–17.
Henderson, P.A. & Walker, I. (1990). Spatial organisation and population density of the fish community of the litter banks within a central Amazonian blackwater stream. *Journal of Fish Biology*, **37**, 401–411.

Heywood, V.H. (1978). *Flowering Plants of the World.* Mayflower, New York.
Irion, G. (1984). Sedimentation and sediments of Amazonian rivers and evolution of the Amazonian landscape since the Pliocene times. In *The Amazon. Limnology and Landscape Ecology of a Mighty Tropical River and its Basin* (Ed. by H. Sioli), pp. 201–243. Junk, Dordrecht.
Junk, W.J. (1970). Investigations on the ecology and production-biology of the floating meadows (*Paspalum—Echinochloetum*) on the Middle Amazon, *Amazoniana*, **2**, 449–495.
Junk, W.J. (1973). Investigations on the ecology and production-biology of the floating meadows (*Paspalum—Echinochloa*) on the Middle Amazon. *Amazoniana*, **4**, 9–102.
Junk, W.J. & Piedade, M.T.F. (1993). Biomass and primary-production of herbaceous plant communities in the Amazon floodplain. *Hydrobiology*, **263**, 155–162.
Junk, W.J., Soares, G.M. & Carvalho, F.M. (1983). Distribution of fish species in a lake of the Amazon river floodplain near Manaus (Lago Camaleão) with special reference to extreme oxygen conditions. *Amazoniana*, **7**, 397–431.
Kalliola, R., Puhakka, M., Salo, J., Tuomisto, H. & Ruokolainen, K. (1991a). The dynamics, distribution and classification of swamp vegetation in Peruvian Amazonia. *Annales Botanici Fennici*, **28**, 225–239.
Kalliola, R., Salo, J., Puhakka, M. & Rajasilta, M. (1991b). New site formation and colonising vegetation in primary succession on the Western Amazon floodplains. *Journal of Ecology*, **79**, 877–901.
Kalliola, R., Salo, J., Häme, T., Räsänen, M., Neller, R., Puhakka, M. et al. (1992). Upper Amazon channel migration: implications for vegetation perturbance and succession using bitemporal Landsat MSS images. *Naturwissenschaften*, **79**, 75–79.
Kubitski, K. (1985). Ichthyochory in *Gnetum venosum*. *Anais da Academia brasileira de Ciência*, **57**, 513–516.
Lowe-McConnell, R.H. (1987). *Ecological Studies in Tropical Fish Communities.* Cambridge University Press, Cambridge.
Mandelbrot, B.B. (1982). *The Fractal Geometry of Nature.* W.H. Freeman, San Francisco.
Mennega, A.M.W. (1987). Wood anatomy of the Euphorbiaceae. In *The Euphorbiaceae. Chemistry, Taxonomy and Economic Botany* (Ed. by S.L. Jury, T. Reynolds, D.F. Cutler & F.J. Evans), pp. 111–126. Academic Press, London.
Mertes, L.A.K. (1985). *Floodplain development and sediment transport in the Solimões–Amazon River, Brazil.* PhD thesis, University of Washington, Washington.
Moran, N.A. (1988). The evolution of host plant alternation in aphids: evidence that specialisation is a dead end. *American Naturalist*, **132**, 681–706.
Moran, N.A. (1991). Phenotype fixation and genotypic diversity in the life cycle of the aphid *Pemphigus betae*. *Evolution*, **45**, 957–970.
Moran, N.A. (1992). The evolutionary maintenance of alternative phenotypes. *American Naturalist*, **139**, 971–989.
Nevo, E. (1993). Adaptive speciation at the molecular and organismal levels and its bearing on Amazonian biodiversity. *Evolución Biolóica*, **7**, 207–249.
Pieterse, A.H. & Murphy, K.J. (1990). *Aquatic Weeds. The Ecology and Management of Nuisance Aquatic Vegetation.* Oxford University Press, Oxford.
Pires, J.M. & Prance, G.T. (1985). The vegetation types of the Brazilian Amazon. In *Key Environments: Amazonia* (Ed. by G.T. Prance & T.E. Lovejoy), pp. 109–145. Pergamon Press, Oxford.
Prance, G.T. (1979). Notes on the vegetation of Amazonia III. The terminology of Amazon forest types subject to inundation. *Brittonia*, **31**, 26–38.
Prance, G.T. & Arias, J.R. (1975). A study of the floral biology of *Victoria amazonica* (Poepp.) Sowerby (Nymphaceae). *Acta Amazonica*, **5**, 109–139.

Puhakka, M., Kalliola, R., Rajasilta, M. & Salo, J. (1992). River types, site evolution and successional vegetation patterns in Peruvian Amazonia. *Journal of Biogeography*, **19**, 651–665.

Punt, W. (1987). A survey of pollen morphology in the Euphorbiaceae with special reference to *Phyllanthus*. *Botanical Journal of the Linnean Society*, **94**, 127–142.

Ribbink, A.J. (1994). Biodiversity and speciation of freshwater fish with particular reference to African cichlids. In *Aquatic Ecology, Scale, Pattern and Process* (Ed. by P.S. Giller, A.G. Hilldrew & D.G. Raffaelli), pp. 261–288. Blackwell Scientific Publications, Oxford.

Riebesell, A. (1974). Paradox of enrichment in competitive systems. *Ecology*, **55**, 183–187.

Robertson, B.A. & Hardy, E.R. (1984). Zooplankton of Amazonian lakes and rivers. In *The Amazon, Limnology and Landscape Ecology of a Mighty River and its Basin* (Ed. by H. Sioli), pp. 57–102. Junk, Dordrecht.

Rosenzweig, A. (1971). Paradox of enrichment: destabilisation of exploited ecosystems in ecological time. *Science*, **171**, 385–387.

Saint-Paul, U.B.P. (1979). A situação da pesca na Amazonia Central. *Acta Amazonica*, **9**, 109–114.

Salo, J., Kalliola, R., Hakkinen, L., Makinen, Y., Niemela, P., Puhakka, M. et al. (1986). River dynamics and the diversity of Amazon lowland forest. *Nature*, **322**, 254–258.

Schmidt, G.W. (1973). Primary production of phytoplankton in the three types of Amazonian waters. The limnology of a tropical flood-plain lake in central Amazonia (Lago do Castanho). *Amazoniana*, **4**, 139–203.

Southwood, T.R.E. (1977). Habitat, the template for ecological strategies? *Journal of Animal Ecology*, **46**, 337–365.

Spruce, R. (1871). Palmae Amazonicae. *Journal of the Proceedings of the Linnean Society, Botanical*, **4**, 58–63.

Taylor, D.W. & Hickey, L.J. (1995). Evidence for and implications of an herbaceous origin for Angiosperma. In *Flowering Plant Origin, Evolution and Phylogeny* (Ed. by D.W. Taylor & L.J. Hickey), pp. 232–266. Chapman & Hall, New York.

Val, A.L. & Almeida-Val, V.M.F. (1995). *Fishes of the Amazon and their Environment. Physiological and Biochemical Aspects*. Springer, Berlin.

Vari, R.P. (1988). The Curimatidae, a lowland neotropical fish family (Pisces: Characiformes); distribution, endemism and phylogenetic biogeography. In *Neotropical Distribution Patterns: Proceedings of a Workshop* (Ed. by P.E. Vanzolini & W.R. Heyer), pp. 106–137. Academia Brasiliera de Ciências, Rio de Janeiro.

Vari, R.P. (1989a). A phylogenetic study of the neotropical Characiform family Curimatidae (Pisces: Ostariophysi). *Smithsonian Contributions to Zoology*, **471**, 1–71.

Vari, R.P. (1989b). Systematics of the neotropical genus *Psectrogaster* Eigenmann and Eigenmann (Pisces: Ostariophysi). *Smithsonian Contributions to Zoology*, **481**, 1–42.

Vari, R.P. (1991). Systematics of the neotropical Characiform genus *Steindachnerina* Fowler (Pisces: Ostariophysi). *Smithsonian Contributions to Zoology*, **507**, 1–118.

Vari, R.P. (1992). Systematics of the neotropical Characiform genus *Curimatella* Eigenmann and Eigenmann (Pisces: Ostariophysi), with summary comments on the Curimatidae. *Smithsonian Contributions to Zoology*, **533**, 1–48.

Verhoeven, J.T.A., Koerselman, W. & Meulman, A.F.M. (1996). Nitrogen- or phosphorus-limited growth in herbaceous, wet vegetation: relations with atmospheric imputs and management regimes. *Trends in Ecology and Evolution*, **11**, 494–497.

West-Eberhard, M.J. (1989). Phenotypic plasticity and the origins of diversity. *Annual Reviews of Ecology and Systematics*, **20**, 249–278.

Yonenaga-Yassuda, Y. & Chu, T.H. (1985). Chromosome banding patterns of *Saimiri vanzolinii* Ayres, 1985 (Primates, Cebidae). *Papeis Avulsos de Zoologia*, **36**, 165–168.

16. COMMUNITY DYNAMICS OF ARBOREAL INSECTIVOROUS BIRDS IN AFRICAN SAVANNAS IN RELATION TO SEASONAL RAINFALL PATTERNS AND HABITAT CHANGE

P. JONES

Institute of Cell, Animal and Population Biology, University of Edinburgh, King's Buildings, Edinburgh EH9 3JT, UK

SUMMARY

1 Palearctic migrants overwinter in highly seasonal African savannas. When present, they make up about 20% of the savanna avifauna. Many perform further, long-distance intra-African migrations during their stay; up to 40% of Afrotropical savanna species may also be intra-African migrants. This chapter discusses what is known of the community dynamics of this highly mobile avifauna, whose patterns of species' interactions vary greatly between seasons and between habitats.

2 Migration patterns of savanna birds are determined by the sequence of alternating wet and dry seasons caused by the annual north–south movements of the Inter-Tropical Convergence Zone (ITCZ). The ITCZ and most Afrotropical migrants move northwards in the first half of the calendar year and southwards in the latter half.

3 Most Palearctic migrants are insectivores, which occupy the semi-arid northern tropical savannas throughout the dry season, while the majority of Afrotropical insectivores withdraw southwards. Few Palearctic species favour broadleaved savannas during the northern tropical dry season, or the wet season conditions in the savannas of eastern and southern Africa. Nevertheless, some 'itinerant' Palearctic species make a further, intra-African migration southwards to wet season conditions. In the northern tropics there is thus a seasonal replacement of Afrotropical by Palearctic insectivores, whereas south of the equator Afrotropical rains migrants and Palearctic immigrants coincide in their occupancy of the wet season savannas.

4 The smaller Afrotropical arboreal insectivores do not migrate. North of the equator they are joined and outnumbered (in terms of both species and individuals) by immigrant Palearctic warblers. In southern Africa Palearctic

warblers occur at lower densities and are outnumbered by Afrotropical species. Palearctic migrants mostly occupy different feeding niches from Afrotropical species but the extent to which interspecific competition during the rains might limit the penetration of Palearctic migrants into southern Africa remains unknown.

5 Habitat degradation causes a shift in the relative proportions of Afrotropical and Palearctic species. More open habitats favour Palearctic birds but the overall effect is a reduction in bird densities and impoverishment of the avifauna; the changes in interspecific interactions and community dynamics by which this occurs remain unstudied.

INTRODUCTION

Two major intercontinental bird migration systems, the Nearctic–Neotropical and the Palearctic–Afrotropical, have long interested ecologists who have sought to understand how a wide range of taxa in several feeding guilds can be accommodated, often for more than 6 months of the year, within the bird communities of their tropical reception areas (Moreau 1972, Keast & Morton 1980, Rappole 1995). In many respects, the community dynamics of Nearctic migrants have become much better understood than those of Palearctic birds overwintering in Africa (Rappole 1995).

In the Americas, the taxa which have attracted particular attention are the insectivorous passerines, foremost amongst which are the small arboreal leaf-gleaning insectivores, the parulid wood-warblers and tyrannid flycatchers. They are of considerable ecological and evolutionary interest because they contain a large number of Nearctic migrants (46 warblers, 23 flycatchers) that are morphologically rather similar amongst themselves, and which overwinter in Neotropical woodlands and rainforests that are already occupied by numerous closely related and morphologically similar residents. The integration of these migrants into the Neotropical resident avifauna is now known to be achieved through complex patterns of geographical and spatial separation (e.g. Keast 1980a,b).

The reasons for this similarity are historical. Most Nearctic migrant passerines had a tropical forest origin and have remained in contact with their tropical and subtropical counterparts; almost half have conspecifics breeding in the Neotropics and a further 30% have congeners (Mönkkönen, Helle & Welsh 1992). The Palearctic and Afrotropical avifaunas are much more distinct; Palearctic migrants are originally birds that evolved in seasonal temperate environments (Snow 1978). Unlike their Neotropical counterparts, which occupy predominantly forested habitats, most Palearctic migrants in Africa occupy the extensive and highly seasonal semi-arid savannas; only a few enter broadleaved habitats and almost all avoid rainforest.

A further crucial difference from the Nearctic–Neotropical migration system is that, although many immigrants remain resident throughout their stay in Africa, others perform further long-distance migrations within Africa after arrival. In doing so, they join a large number of Afrotropical savanna birds that also perform similar seasonal intra-African migrations. The savanna avifauna, of which Palearctic migrants become a seasonally integral part, is therefore a highly dynamic assemblage of species. Their movements within Africa, and thus the extent to which they meet and interact, are closely tied to the alternating pattern of wet and dry seasons across the continent. As a result of this complexity, our understanding of resource partitioning and niche separation within the Palearctic–Afrotropical migration system remains less well advanced than in the Neotropics.

This chapter cannot redress this imbalance but sets out instead to examine some of the ecological complexities of the African environment that have made a better understanding more elusive. It describes the seasonally changing dynamics of bird communities of African savannas, focusing primarily on the arboreal insectivores. As in the Nearctic–Neotropical migration system, these comprise one of the most important feeding guilds among Palearctic migrants and Afrotropical species alike.

AFRICAN RAINFALL PATTERNS AND VEGETATION TYPES

The migration patterns of African savanna birds are ultimately determined by the annual sequence of alternating wet and dry seasons, acting via the effects that the relative lengths of these periods have on the character of the vegetation and the seasonal availability of food. Furthermore, because the timing of rainfall differs markedly between the northern and southern tropics, it greatly influences where different species of Palearctic migrants can best overwinter.

Rainfall patterns

Rainfall in Africa derives primarily from the Inter-Tropical Convergence Zone (ITCZ), the 'meteorological equator' where northeast and southeast tradewinds converge in an extensive low pressure zone of warm, humid air.

This produces rain as it rises and cools. The resulting high altitude dry air moves north and south away from the equator, eventually descending again at the subtropical high pressure zones to cause severely arid conditions at the earth's surface — the Sahara and Kalahari/Namib deserts. Dry air from the two high pressure zones then returns to the equator as the tradewinds, their easterly component given by the earth's rotation. The northeast tradewind remains dry as it passes mainly over land, the southeast is oceanic and humid. Rainfall beyond the deserts on the North African and Cape coasts derives from the Atlantic and sub-Antarctic low pressure areas that lie to the north and south of the subtropical high pressure zones.

The geographical positions of the ITCZ and its associated cells of recirculating air masses do not remain stationary, however, but move seasonally with the changing zenith position of the sun. The ITCZ moves north of the equator during the first half of the calendar year and returns to cross it southwards during the latter half (Fig. 16.1).

Within about 4° north and south of the equator the successive northwards and southwards movements of the ITCZ causes a bimodal peak in annual rainfall. Where total annual rainfall exceeds c. 1200 mm, as in the western half of equatorial Africa, the rainy seasons merge and there is only small seasonal variation. In eastern equatorial Africa, however, rainfall is generally less. As a result, there is a more marked double peak in rainfall, with two wet seasons separated by two dry seasons each year, but the region's topographical complexity produces considerable variability in rainfall from place to place. In drier parts, both of the rainy seasons are short and unreliable and the dry seasons are of varying severity. North or south of the equator one of the two rainy seasons is longer and more predictable (the 'long rains'), and the other less so (the 'short rains').

Away from the equator, both north and south, rain falls in a single wet season annually. The northern tropics receive rain between March and October as the ITCZ moves slowly northwards before retreating more rapidly southwards. Because of this pattern of movement, the wet season at higher latitudes becomes progressively shorter, begining later and ending sooner, while the dry season is correspondingly longer. Total rainfall is therefore also much reduced at higher latitudes and the dry season more severe (Fig. 16.1).

As the ITCZ crosses the equator southwards in October–November, the northern tropics enter the dry season. At the same time the rains begin over much of southern Africa and the wet season continues until April–May. As in the northern tropics, the rains in southern Africa generally begin later and end sooner with increasing latitude, though the rainfall pattern is rather more complex. This is because the southeast coast receives its first rain in

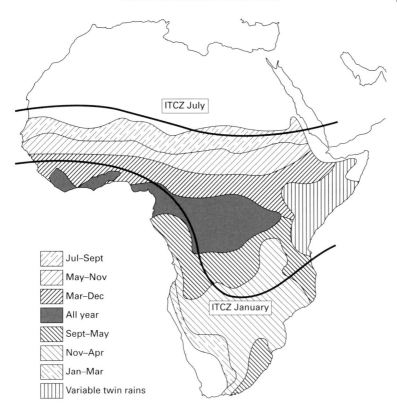

FIG. 16.1. Seasonal rainfall distribution in Africa, showing the January and July positions of the Inter-Tropical Convergence Zone (ITCZ). (From Jones 1995.)

September, rather earlier than the interior, when the southeast tradewinds re-establish at this latitude and bring in humid air from the Indian Ocean (Fig. 16.1). A further complication is that the southern tip of South Africa receives its rain during the austral winter from the southern hemisphere low pressure zone, which moves northwards as the ITCZ and its associated air masses move into the northern tropics. The Cape region is therefore wet while the rest of southern Africa is dry, and vice versa.

In broad terms, therefore, and despite some regional complexity, the rainfall distributions north and south of the equator are mirror-images of one another in both timing and amount. It is important to remember that the north-temperate winter is equivalent in timing to the dry season in the northern tropics and to the wet season south of the equator.

Vegetation types

Because the duration and amount of rainfall decrease and the severity of the dry season increases with increasing latitude, the latitudinal distribution of African vegetation zones also mirror each other, in structure if not in species composition, in the northern and southern tropics (Keay 1959, White 1983).

The region of almost year-round rainfall in equatorial west and central Africa supports rainforest. Where this forest has been cleared it has given rise to a 'derived savanna' of remnant secondary forest and grassland. The next zone comprises various types of deciduous broadleaved savanna, known in the northern tropics as Guinea savanna and dominated in its northern part by doka (*Isoberlinia*) woodland, though *Combretum* is also characteristic. The equivalent vegetation type to *Isoberlinia* woodland south of the equator is miombo (dominated by the closely related genus *Brachystegia*) but southern Africa also has extensive areas of other types of broadleaved woodland (*Combretum, Terminalia, Baikaea*, etc., and mopane *Colophospermum mopane*). Further from the equator occur mixed broadleaved/*Acacia* woodlands (known as the Sudan zone in the northern tropics). These give way to increasingly arid *Acacia*-dominated savannas (the Sahel in the northern tropics), which reach the margins of the northern (Sahara) and southern (Kalahari/Namib) deserts.

The significant exception to this broad vegetational pattern is equatorial East Africa, where its very varied topography and twin rains produce a complex mosaic of vegetation types from rainforest to semi-arid *Acacia* savanna. East Africa therefore differs from much of the rest of the continent in that this geographical complexity makes it difficult to demonstrate clear latitudinal trends in rainfall distribution, vegetation zones, or the resulting patterns of birds' movements.

The areas of these different habitat types across sub-Saharan Africa are difficult to estimate, partly because different classification systems have been used, in which vegetation formations were combined in different ways, and partly because large areas have been considerably modified by man. The figures given here are based on Brown, Urban and Newman (1982) who used Keay's (1959) map and MacKinnon and MacKinnon (1986) who used White's (1983) classification.

The seasonal savannas cover a vast area. Semi-arid *Acacia*-dominated savannas total $4-5 \times 10^6$ km^2, with those of the northern tropics being about twice as extensive as in southern Africa. The broadleaved savannas are more extensive still, covering $8-9 \times 10^6$ km^2, with areas of rainforest–savanna mosaic adding a further 1.5×10^6 km^2; these broadleaved habitats are roughly equally represented north and south of the equator. The rainforests themselves cover almost $3 \times 10^6 \text{km}^2$ and the complex mosaic of forest and savanna in equatorial East Africa covers $c. 2 \times 10^6 \text{km}^2$.

THE NUMBERS OF BIRD SPECIES INVOLVED

Estimates of the relative proportions of resident and migrant species in Africa vary depending on which regions/habitats or taxonomic/ecological groupings of species are used. For the present purpose it is convenient to use figures in Moreau (1966, 1972); they are illustrative rather than definitive.

Moreau (1966) estimated that some 1481 Afrotropical landbird species occur in sub-Saharan Africa. Of these, 409 are non-migrant residents of lowland and montane evergreen rainforests and 74 are confined to montane non-forest habitats. Such habitats are barely used by Palearctic migrants, so their bird species can be ignored in the following comparisons. The remaining 998 non-forest lowland birds are joined by 187 species of Palearctic immigrants for the northern winter (Moreau 1972), which therefore comprise $c. 16\%$ of the savanna avifauna of Africa as a whole. These visitors are not distributed evenly, however. The greatest number of Palearctic species occur in a belt across the northern tropics in the Sahel and Sudan zones, on average comprising $c. 26\%$ of savanna species (Moreau 1972). The richest area of all for Palearctic immigrants is in the east of this northern tropical belt, including Sudan, Ethiopia, Uganda and Kenya, whereas species numbers fall off markedly further south into south–central and southern Africa (Alerstam 1990, Newton 1995).

Of the 998 species of non-forest lowland Afrotropical birds, Moreau (1966) recognized 91 (9%) as intra-African migrants. Since then our knowledge of intra-African migration has greatly increased, especially through the work of Elgood and coworkers in Nigeria, who have provided the most comprehensive account of migration among a regional avifauna (Elgood, Fry & Dowsett 1973, Elgood, Heigham, Moore *et al.* 1994). They found no firm evidence of migration among Nigerian forest or montane species (apart from some seasonal altitudinal changes in the latter) but showed that approximately 20% ($n = 417$) of lowland savanna species were definite seasonal migrants and a further 10% were likely to be so. An equivalent analysis of Ghanaian birds showed that 23% ($n = 310$) of savanna landbird species are intra-African migrants (Grimes 1987). However, if additional known migrants, which were overlooked by Elgood *et al.* (1994), and the more fragmentary data available for many other species are included, it is likely that up to 40% of West African landbirds may perform regular intra-African migrations (Elgood *et al.* 1973, 1994, Jones & Ward 1977, Jones 1985, P. Jones unpublished observations).

Intra-African migration is widespread amongst savanna birds elsewhere in Africa too, but our knowledge of the numbers of species involved is still patchy across the continent as a whole. This is for two main reasons. First, migration becomes a more conspicuous feature of the avifauna at higher latitudes away from the equator, as savanna environments become increasingly seasonal (Fig. 16.1). Conversely, in more equable habitats nearer the equator, the same species may be only partially migratory, with varying numbers of individuals remaining present all year. Second, large-scale seasonal displacements of migrant populations from one vegetation zone to another are much more easily detected where a single annual rainfall peak and relatively uniform topography produce a simple zonation of habitat types over large geographical areas.

For both these reasons, intra-African migration has been generally well documented in western and southern Africa but much less so in eastern Africa. East Africa's much more varied topography and complex twin rainfall system, its resulting mosaic of different habitats in close juxtaposition, and the probably common occurrence of partial migration, all combine to obscure any relatively simple underlying patterns.

PATTERNS OF BIRD MIGRATION WITHIN AFRICA

Afrotropical species

The commonest migration pattern throughout Africa seems to be one in which birds in the northern tropics follow the ITCZ northward from March–April onwards, returning southward in October–November once the rains end. It is mirrored south of the equator by birds moving further southwards with the rains as they begin in October–November, returning northwards in April–May. The pattern is essentially the same whether birds remain wholly north or south of the equator, or perform a trans-equatorial migration (Alerstam 1990). Such migration occurs in virtually all feeding guilds and is common, for example, amongst raptors, waterbirds and granivores. This discussion focuses on the insectivorous taxa, however, where the phenomenon is particularly clearly seen. Almost all the insectivorous families have intra-African migrants, including cuckoos, nightjars, bee-eaters, rollers, kingfishers, swifts, swallows, cuckoo-shrikes, chats, flycatchers, sunbirds, shrikes, orioles and starlings.

The clear seasonal pattern of intra-African migration suggests that many insectivorous Afrotropical birds can exploit arid and semi-arid savannas only during the rains. Dry season conditions in these habitats are presumably in some way inhospitable, most likely because suitable food is lacking. The few quantitative data available show that arthropod biomass in *Acacia* savanna is generally some 2–10 times greater during the rains compared with the dry season (Morel 1968, Sinclair 1978, Lack 1986a) but the biomass of arthropods living in the foliage of woody vegetation may increase up to 60-fold in the rains (Morel 1968). Evidence that the wet season is particularly favourable for Afrotropical insectivores is that the majority of species show a peak of breeding activity during the rains, and that the majority of intra-African migrants follow the rains in order to breed.

A re-analysis of Nigerian data in Elgood *et al.* (1994) shows that 98 out of 134 species (73%) in the insectivorous taxa listed above are wet season breeders, if this is taken to include those that begin breeding at the end of the dry season with the flush of new leaves just before the rains begin (Tables 16.1). Of these 98, 41 (42%) move north with the rains to breed. Two others are trans-equatorial migrants that breed in the rains in southern Africa but spend the non-breeding period in the rains of the northern tropics. A similar analysis of the data in Maclean (1984) for southern Africa shows that 263 (96%) out of 274 insectivorous species breed just before or during the wet season, of which 44 (17%) are intra-African migrants that move south with the rains to breed (Table 16.2); another rains migrant is a non-breeding visitor from the northern tropics.

Only a minority of insectivores breed in the dry season, and apparently fewer do so in southern Africa than in West Africa (Tables 16.1 & 16.2). In West Africa, 36 of 134 species (27%) are dry season breeders and, of these, nine (26%) are intra-African migrants that move north with the rains after breeding, when they presumably moult, for example the grey-headed kingfisher *Halcyon leucocephala* (Jones 1980). In southern Africa, only 11 out of 274 insectivores (4%) breed in the dry season. No dry season breeder in southern Africa is an intra-African migrant.

In comparison to West Africa, more savanna insectivores in southern Africa breed in the wet season; the reasons for this are unknown. Relatively

TABLE 16.1. Numbers of migrant and non-migrant insectivorous savanna birds in Nigeria breeding in the wet and dry seasons; re-analysed from data in Elgood *et al.* (1994).

	Migrant	Non-migrant	Total
Wet-season breeders	41	57	98
Dry-season breeders	9	23	36
Total	51	84	134

TABLE 16.2. Numbers of migrant and non-migrant insectivorous savanna birds in southern Africa breeding in the wet and dry seasons, re-analysed from data in Maclean (1984).

	Migrant	Non-migrant	Total
Wet-season breeders	44	219	263
Dry-season breeders	0	11	11
Total	44	230	274

fewer of them are migrants: although more than 30 species in both western and southern Africa migrate with the rains to breed, there are a greater number of resident insectivores in southern Africa, so that the proportion of migrants is lower. The difference is accounted for mainly by a greater number of larks and chats in southern Africa, most of which are non-migratory, compared with the Nigerian data set used here.

It might seem especially advantageous among insectivores for wet season breeders to move with the rains, because they could extend their breeding opportunities by doing so. However, there is no significant difference in the proportion of wet and dry season breeders that are migrants (Nigeria: $\chi^2_1 = 2.51$, ns; southern Africa $\chi^2_1 = 1.26$, ns).

Among the arboreal insectivorous feeding guild, surprisingly few species are intra-African migrants. Some of the larger species do migrate, such as the cuckoos and the sulphur-breasted bush-shrike *Telophorus sulfureopectus*, but the other bush-shrikes do not. Furthermore, given that sylviid warblers constitute such an important component of the immigrant Palearctic avifauna, and that many of these undertake further, intra-African, migrations during the course of their stay, it is curious that so few Afrotropical sylviids migrate (Fig. 16.2). The common African warblers encountered almost anywhere in *Acacia* or broadleaved savannas — one or other of the species of *Apalis*, *Cisticola*, *Camaroptera*, *Eremomela*, *Phyllolais*, *Prinia* or *Sylvietta* — all seem to be year-round residents almost everywhere (Morel 1968, Maclean 1984, Elgood *et al.* 1994, P. Jones unpublished observations). The only Afrotropical sylviid for which I can find evidence of seasonal migration is the yellow-breasted apalis *Apalis flavida*, which appears to be a rains migrant in Togo (Cheke & Walsh 1996) and may be so in parts of Kenya (Lack 1985, 1987).

Other ecologically similar small insectivores, such as the tits and penduline tits *Parus* and *Anthoscopus*, are all residents. Only among the sunbirds are seasonal movements common, and then the migration may be only

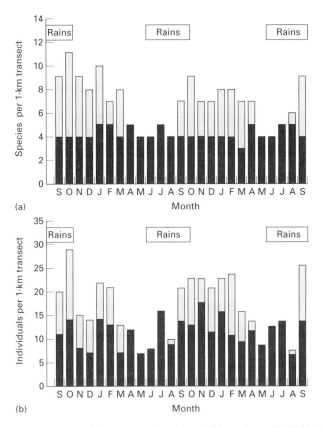

FIG. 16.2. Seasonal changes in (a) numbers of species and (b) numbers of individuals of Afrotropical (solid bars) and Palearctic (shaded bars) arboreal insectivores in Sudan/Sahel savanna near Maiduguri, northern Nigeria. Data derive from a standard 1-km transect surveyed weekly during 1973–75 and values indicate the minimum number of species/individuals recorded in each month (P. Jones unpublished observation). Afrotropical species: buff-bellied warbler *Phyllolais pulchella*, grey-backed camaroptera *Camaroptera brevicaudata*, northern crombec *Sylvietta brachyura*, red-pate cisticola *Cisticola ruficeps*, green-backed eremomela *Eremomela pusilla* and tawny-flanked prinia *Prinia subflava*. Palearctic species: olivaceous warbler *Hippolais pallida*, Bonelli's warbler *Phylloscopus bonelli*, willow warbler *P. trochilus*, wood warbler *P. sibilatrix*, common whitethroat *Sylvia communis*, lesser whitethroat *S. curruca*, garden warbler *S. borin* and subalpine warbler *S. cantillans*.

partial and not readily detectable everywhere in Africa; their movements may perhaps be related more to the availability of nectar sources rather than insect prey. In West Africa, five of the seven savanna species are migrants: the pygmy sunbird *Anthreptes platurus* and the variable, beautiful, scarlet-chested and copper sunbirds *Nectarinia venusta*, *N. pulchella*, *N. senegalensis* and *N. cuprea* (Elgood *et al.* 1973, 1994, P. Jones unpublished observations). However, the copper sunbird is the only species definitely known to be a rains visitor in southern Africa (Maclean 1984), though there are suggestions that five other *Nectarinia* species may also undertake seasonal movements (Benson 1982). The five common sunbird species in Lack's (1985) study in southeast Kenya were all resident.

For the most part, therefore, Afrotropical members of the arboreal insectivorous guild are year-round residents that occur together with Palearctic immigrants, whether in wet or dry season conditions. A few Afrotropical species breed in the dry season of the northern tropics while large numbers of Palearctic migrants are present; in eastern and southern Africa, birds breeding in the local wet season must all do so whilst sharing their habitats with Palearctic immigrants.

Palearctic species

Most Palearctic passerines migrating to Africa are small insectivores and their favoured habitats are the seasonal savannas. It should be expected, therefore, that their habitat choice and overwintering distribution will be constrained by similar ecological factors to those that dictate whether Afrotropical insectivores migrate or not.

Palearctic migrants arrive in Africa south of the Sahara between August and October, and depart between March and May. The 6-month difference in the timing of the wet seasons in the savannas north and south of the equator is of the utmost importance to the migrants using them. Birds crossing the Sahara in autumn reach the northern savannas shortly after the rains have ended and the region is still green and productive. However, the Sahel and Sudan zones quickly dry out and appear inhospitable within a few weeks, though the Guinea savanna woodlands remain green for longer into the dry season. At the same time, however, parts of East Africa and the savannas south of the equator are entering the rains, offering apparently much better feeding opportunities for the largely insectivorous immigrants.

Four wintering strategies are available to Palearctic migrants:
1 they could fly directly to their wintering areas in the northern tropical savannas and remain there throughout the winter in the local dry season, which is more arid in the Sahel and Sudan zones than in the Guinea savannas farther south;
2 they could pass rapidly southwards across the equator to wet season savannas, appearing only on passage in the northern tropics;

3   they could occupy the Sahel and Sudan zones for as long as these regions remain wet but then move the relatively short distance southwards to less arid habitats in Guinea savanna; or

4   they could initially occupy the northern tropics while these remain wet but, once they dry out, migrants could move to wet conditions in the local rainy seasons in eastern or southern Africa.

Strategy 1

Surprisingly, despite becoming increasingly arid during the ensuing dry season, the Sahel and Sudan zones at the southern margins of the Sahara constitute the most important area for wintering Palearctic migrants; about two-thirds of the individuals and 40% of the species remain north of the equator (Leisler 1992, Newton 1995). The dry season occupation of this superficially inhospitable area by so many migrants, which seem to avoid both the apparently richer broadleaved woodlands of the northern tropical dry season and the wet season south of the equator, became known as 'Moreau's Paradox'. The paradox is of course more apparent than real (see later and Morel 1973, Fry 1992). Palearctic visitors that do remain in the dry northern tropics include *Sylvia* warblers, which eat substantial amounts of fruit as well as gleaning abundant insects from trees that remain in leaf during the dry season, and also include the ground-feeding chats and wheatears, which feed on abundant small terrestrial insects (Pearson & Lack 1992). In contrast, the insectivores that predominate in Guinea savanna are those that glean insects from dense vegetation or which feed on aerial insects.

Strategy 2

Rather few migrants fly directly to winter quarters in the local wet seasons of eastern and southern Africa, presumably because the distances involved are too great to travel without refuelling. Furthermore, conditions there may not become fully suitable until rather later in November–December, when the local rains have properly begun. Aerial feeders such as European swallows *Hirundo rustica* may move steadily southwards, feeding as they go and appearing at intermediate points only as passage migrants. Icterine warblers *Hippolais icterina* and golden orioles *Oriolus oriolus* also seem to appear only as passage migrants as they move south; other species may spend longer periods at intermediate stopover points where conditions are favourable for refattening. When such species spend more substantial periods in favourable conditions *en route*, they may be classed as 'itinerants', as below.

Strategies 3 and 4

The strategy of utilizing more than one area during the course of the winter (other than for mere stopovers) was called 'itinerancy' by Moreau (1972) and it appears to be common. Many migrants remain in the northern tropics for the first couple of months after arrival, before performing a second, midwinter migration further south. This involves either a relatively short distance within the dry northern tropics to overwintering areas in Guinea savanna (strategy 3), or a longer journey to wet season conditions in the eastern and southern African savannas (strategy 4). Birds occupying the eastern end of the Sahel–Sudan zone can reach wet season savannas relatively quickly in equatorial East Africa but those moving out of West Africa must first overfly the large tract of unsuitable rainforest in the Congo basin before reaching the savannas of southern Africa.

In West Africa, consistent patterns from year to year in the timing of arrival, brief residency, premigratory fattening and departure suggest widespread itinerant behaviour among migrants (Elgood, Sharland & Ward 1966, Jones 1985, Morel & Morel 1992, P. Jones unpublished observations). Many species occupy the Sahel and Sudan zones in September–October before moving to the southern Guinea and derived savannas of West Africa (e.g. the nightingale *Luscinia megarhynchos*), or the rains of southern Africa (e.g. the spotted flycatcher *Muscicapa striata*; Jones 1985). Evidence of refattening in West Africa for the second stage of the migration is now becoming available for a few species, for example the great reed warbler *Acrocephalus arundinaceus* (Hedenström, Bensch, Hasselquist *et al.* 1993).

Similarly, on the eastern side of the continent, many species spend the first 2–3 months after arrival in the *Acacia* savannas between 11°N and 15°N in areas where it has recently been raining (Pearson, Nikolaus & Ash 1988). As in the western Sahel, many of these species move on as the dry season begins but their departure coincides with extensive areas of equatorial East Africa entering the rains and good feeding opportunities becoming available (Lack 1983). This second step of the southward migration peaks between November and January and is remarkable for the quite clear segregation between species in the routes they take through East Africa (Pearson & Lack 1992). Many species, mainly ground-feeders, then remain in East Africa for the rest of their stay, while others, especially vegetation-gleaning species, pass rapidly on, carrying amounts of fat consistent with a trans-equatorial migration to wintering grounds 2000–3000 km away in southern Africa (Pearson 1990, Pearson & Lack 1992). Yet others, such as garden warblers *Sylvia borin*, stop and establish territories in Kenya and Uganda for a month or so in November–December, during which they refatten before departing

(Pearson 1990). Brief though their stay may be, even these species become a temporarily important part of the local avifauna, in the same way as they and other itinerants had utilized the Sahelian savannas further north some 2–3 months earlier.

RELATIONSHIP BETWEEN THE MOVEMENTS OF PALEARCTIC AND AFROTROPICAL SPECIES

Because of the difference in timing of the rains north and south of the equator, there are differences in the extent to which the Palearctic and Afrotropical bird communities meet and interact. North of the equator the general pattern is that a large proportion of the native African species withdraws southwards at the end of the rains, to be replaced by immigrant Palearctic arrivals. Six months later, towards the end of the dry season, the reverse process occurs when Palearctic migrants depart northwards, being replaced as the rains begin with intra-African migrants from further south. Data from Nigeria show that the movements of the two sets of species are largely complementary (Fig. 16.3).

The withdrawal southwards out of the Sahel and Sudan zones by large numbers of migrant Afrotropical insectivores as the wet season comes to an end would seem to be easily understood in terms of a seasonal reduction in the availability of suitable food. Presumably the Guinea savannas offer better conditions to such Afrotropical species during the dry season.

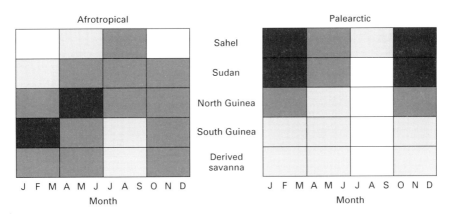

FIG. 16.3. Seasonal distribution of Afrotropical ($n = 52$ species) and Palearctic migrants ($n = 60$) in each vegetation zone in Nigeria. Shading indicates percentage of total landbird species present in each zone in each 3-month period: none, 1–20%; light, 21–40%; dark, 41–60%; black, 61–80%. (Data derived from Elgood et al. 1966, 1973.)

Nevertheless, abundant arthropod prey must remain available, since the exodus of Afrotropical species from the Sahel/Sudan zone is followed immediately by the massive influx of immigrant insectivores from the Palearctic, which do not penetrate the Guinea savanna to such a great extent. Why Guinea savanna is less important for Palearctic species than the habitats further north is not clear. It may simply be because Guinea savanna has already been reoccupied at this time by Afrotropical migrants withdrawing southwards.

As the rains begin again, the general tendency to move north suggests that there is a corresponding reduction in the total number of birds present in the Cuinea savannas of West Africa during the wet season (Elgood et al. 1973). These areas will have lost all their Palearctic migrants and large populations of wholly and partially migrant African birds as well, though this must be partly offset by reproduction among the resident wet season breeders. Nevertheless, it raises the question of whether the Guinea savannas become inhospitable during the rains, or whether the wet season opportunities further north, in habitats now vacated by Palearctic visitors, are simply so much better.

The situation is clearly rather different in eastern and southern Africa. Birds moving southwards out of the eastern Sahel as it dries out soon reach equatorial East Africa, where the rains begin as they arrive or shortly afterwards. In southeastern Kenya, the main peak in insectivore numbers occurs in December–January and is due to Palearctic immigrants (Lack 1987). However, some other larger Afrotropical insectivores are also commoner in the rains; for example, the African cuckoos move into the region in both wet seasons but are commoner in December–January, when the rains are more predictable, than in April (Lack 1987). In eastern Africa, therefore, immigrations of Palearctic and Afrotropical migrants coincide, though most of the leaf-gleaning Palearctic birds soon move on further south (Lack 1983).

Like equatorial East Africa, the southern African savannas are also entering their local wet season in November and begin to receive their first Palearctic immigrants at the same time as Afrotropical migrants begin to arrive from further north. It has been frequently noted, however, that in comparison with the equatorial savannas, and the northern tropics in particular, the southern African savannas are not utilized by many Palearctic species (see discussion in Newton 1995). Possible reasons would seem to be either: (i) most wintering populations are adequately accommodated further north without any need to extend southwards; (ii) the costs of migration over the extra distance may be too high; (ii) wet season conditions are somehow unsuitable; or (iv) the wet season presence of Afrotropical migrants excludes

Palearctic species. There seem to be no adequate data to test these suggestions at present.

RELATIVE DENSITIES OF AFROTROPICAL AND PALEARCTIC SPECIES

Moreau (1972) guessed that more than 4×10^9 Palearctic passerines might enter Africa each autumn but he had no adequate means of estimating the bird populations breeding south of the Sahara for comparison. Regrettably, there are still only very few quantitative studies of the relative densities of African and Palearctic passerines, particularly for those species that might be expected to have similar ecological requirements.

In mid dry season in the northern Nigerian Sahel, Jones, Vickery, Holt and Cresswell (1996) found Palearctic warblers to be about twice as common as African sylviids, though if Palearctic species were compared with all African birds in the same arboreal–insectivorous feeding guild, Palearctic migrants were only half as common. Over the course of the dry season in one area of the Senegalese Sahel, the combined density of 10 species of immigrant Palearctic warblers was 5–20 times greater than that of the three resident Afrotropical warbler species (Morel 1968).

Some data sets are also available from areas of Africa experiencing the wet season while Palearctic migrants are present. In southeastern Kenya, Palearctic migrant insectivores occurred at densities of 1.0–2.5 birds ha^{-1}, only about half the density of their Afrotropical counterparts (Lack 1987), while in the Kenyan Rift Valley, Rabøl (1987) found that Afrotropical warblers outnumbered the only common Palearctic immigrant, the willow warbler *Phylloscopus trochilus*.

Farther south in Zambia, Palearctic migrants were hardly represented at all in intact *Brachystegia* woodland. Willow warblers, the only species present at 0.05 birds ha^{-1} were greatly outnumbered by a diverse assemblage of Afrotropical arboreal insectivores at a density of 2.13 birds ha^{-1} (reanalysed from Ulfstrand & Alerstam 1977). In Zambian *Acacia* woodland willow and icterine warblers *Hippolais icterina* were present together at 0.42 birds ha^{-1} but were still outnumbered by Afrotropical species at 1.75 birds ha^{-1}.

Although data are scarce, they suggest a general pattern through Africa of changing relative densities between Palearctic and Afrotropical insectivores, with Palearctic birds occurring at relatively lower population densities farther south. This pattern reflects that shown by the numbers of species of Palearctic migrants overwintering in different parts of Africa (Newton 1995). As discussed above, it remains a matter of conjecture whether this is due to Palearctic species simply petering out because their populations can be adequately accommodated further north, or if adverse ecological factors associated with wet season conditions south of the equator and/or competition with Afrotropical species are the cause.

COMPETITION AND RESOURCE PARTITIONING BETWEEN AFROTROPICAL AND PALEARCTIC SPECIES

The huge numbers of Palearctic visitors arriving south of the Sahara might be expected to have a significant effect on the food resources and habitat distributions of local species. Moreau's (1972) initial analysis concluded that some 28% of Palearctic migrants might be in putative competition with related native species, after excluding those which differed significantly in size or which used different feeding stations. Palearctic species also find themselves in the company of other immigrants with which they could potentially compete; Lack (1971) suggested that 10% of Palearctic migrants could be potential competitors with each other. Both these factors might dictate which wintering areas can be shared and which can only be occupied alone.

Leisler (1992) suggested that non-overlapping winter ranges of Palearctic birds in Africa may be the result of past migrant–migrant competition. The allopatric winter ranges of warblers in the genus *Hippolais* and among *Ficedula* flycatchers have been cited as evidence of this, but in general the range overlaps between congeneric species are extensive (Newton 1995; see maps in Moreau 1972 and Curry-Lindahl 1981). Leisler further suggested that an inverse relationship between the abundances of Afrotropical and Palearctic species could indicate past migrant–resident competition. What evidence there is suggests the opposite, however. On a continent-wide scale, some of the richest areas for overwintering migrants are also areas of high Afrotropical bird diversity (compare maps in Crowe & Crowe 1982, Pomeroy 1993 with Newton 1995). At a local level at study sites in southeastern Kenya and the Nigerian Sahel, comparisons of densities of Palearctic and Afrotropical arboreal insectivores showed a strong positive correlation between them, the numbers of both increasing with the density of vegetation (Lack 1987, Jones *et al.* 1996).

Palearctic warblers, in whatever habitat they overwinter in Africa, find themselves among a resident suite of ecologically similar Afrotropical warblers or other insectivores, whose generic composition is much the same everywhere, though the local species may differ. For the most part, however, the African insectivores are appreciably smaller than the visitors (Moreau

1972) and/or their feeding stations rather different (Sinclair 1978, Lack 1985, Rabøl 1987, P. Jones personal observation): crombecs *Sylvietta* spp., walk and run along horizontal twigs taking insects from bark; *Camaroptera* spp. exploit the middle interior and bases of medium- and small-sized trees, taking most prey from twigs and stems; *Eremomela* spp. feed throughout the outer parts of the canopy interior; *Apalis* spp. use the tops and outside of trees of all sizes; *Prinia* spp. use the outsides of bushes and low trees and frequently leave cover and forage into grassland, as do *Cisticola* spp., while the buff-bellied warbler *Phyllolais pulchella* takes minute prey from just inside the tree canopy and does not use bushes; tits and penduline tits, as well as taking insects from twigs and leaves, also search in bark and seed pods. The sunbirds too feed in exposed positions at the top and outsides of trees, and especially where flowers provide a nectar source.

A significant part of the arboreal insectivore guild, however, comprises rather larger-bodied species such as the bush-shrikes and cuckoos. Like the smaller birds discussed above, most bush-shrikes are resident and therefore co-occur with Palearctic immigrants when they are present, but they take much of their food on the ground. Of these residents, only the northern brubru *Nilaus afer* gleans from foliage and branches high in the canopy. The other mainly arboreal bush-shrike, the sulphur-breasted, and the cuckoos are all rains migrants, so that Palearctic warblers would encounter them only in the wet season savannas of East and southern Africa. It is unknown to what extent they might compete; their larger body size suggests that the bush-shrikes take larger prey. Like the cuckoos, the sulphur-breasted bush-shrike takes hairy caterpillars (Maclean 1984) that are almost certainly avoided by Palearctic migrants.

Among the Palearctic migrants, *Sylvia* spp. remain within the canopy interior where they glean insects from leaves; they also eat substantial amounts of fruit if available, whereas none of the Afrotropical warblers does so. Competition between migrants and residents must therefore be greatly reduced. Moreau (1972) thought that there was more potential for competition within the genus *Sylvia* than between these and native species, although, for example, in West Africa subalpine warblers *S. cantillans* treat both pygmy and beautiful sunbirds as potential competitors by frequently giving aggressive chase to them (P. Jones personal observation). In northern Nigeria the only three Palearctic warblers that are at all common — subalpine warbler, lesser and common whitethroats *S. curruca* and *S. communis* — all glean insects from *Acacia* and especially *Balanites* trees. These three species differ in size, however, and common whitethroats do not enter this habitat in large numbers until late in the dry season after the other two have departed (Vickery *et al.*, unpublished data).

In southeast Kenya there seemed to be clear segregation between species pairs in both *Hippolais* and *Sylvia* (Lack 1985). Upcher's warblers *H. languida* arrived about a month later than olivaceous warblers *H. pallida* and had different habitat preferences, being able to tolerate rather drier conditions. Likewise, barred warblers *Sylvia nisoria* reached peak numbers a month after common whitethroats and occupied a wider range of habitats. Marsh warblers *Acrocephalus palustris* resembled the *Hippolais* species in their feeding stations but were present almost only as passage migrants. Willow warblers behaved differently from all these, being much more aerial in their feeding and taking more prey from the canopy edge. While the Palearctic species appeared to be well segregated amongst themselves, Lack (1985) suggested that together they might exclude the Afrotropical yellow-breasted apalis, whose feeding behaviour most resembled an 'average' Palearctic warbler and which was perhaps most similar in feeding station to a *Sylvia* species. The apalis was absent from Tsavo while the Palearctic migrants were present and was only seen there, though uncommonly, during the May–October dry season.

Further west in Kenya, however, the yellow-breasted apalis is an important component of a warbler community whose sole Palearctic member is the willow warbler (Rabøl 1987). In contrast to Lack's study area in southeast Kenya, the *Hippolais* and *Sylvia* species were absent. Willow warblers arrived in large numbers in late September and the apalis population was augmented by an influx in October–November. The willow warbler's arrival coincided with apparent niche shifts among the two African species ecologically closest to the immigrant, the apalis and the grey-backed camaroptera *C. brevicaudata*. Both moved to the canopy interior and foraged lower down when willow warblers were present, while the apalis used a less fluttering method of prey capture. The apalis also showed a switch in diet at about the time the willow warblers arrived, from lepidopteran larvae to beetles as caterpillars apparently became depleted. Later, the willow warblers also began taking more flies, again reflecting a depletion of caterpillar prey.

Rabøl's and Lack's studies constitute virtually the only quantified data sets on resource partitioning between resident and immigrant Palearctic birds in Africa. It is not certain from Rabøl's data, however, whether his observed changes in foraging behaviour were brought about by past or present competition, or represent seasonal changes that might have occurred anyway. Furthermore, his observations did not extend to what further niche shifts might have occurred when the Palearctic migrants departed.

Comparable data sets for the dry season savannas of the Sahel and Sudan zones are still entirely lacking, despite the interest in solving 'Moreau's Paradox', as they are also for the wet season in the *Acacia* savannas of southern Africa.

THE EFFECTS OF CLIMATE CHANGE AND HABITAT DEGRADATION ON BIRD DENSITIES

African habitats are not stable in the long-term. The semi-arid savannas in particular can experience marked annual and longer-term fluctuations in rainfall, but the effects of such varying conditions on savanna bird populations are still poorly known. Some data are available for Palearctic migrants dependent on Sahelian wetlands, including the herons *Ardea purpurea*, *Ardeola ralloides* and *Nycticorax nycticorax*, the white stork *Ciconia ciconia* and sedge warbler *Acrocephalus schoenobaenus*, whose overwinter survival rates are strongly correlated with rainfall in the Sahel the previous wet season (Den Held 1981, Cavé 1983, Kanyamibwa, Schierer, Pradel & Lebreton 1990, Peach, Baillie & Underhill 1991, Fasola, Hafner, Prosper *et al*. 1997). Such a relationship for these species and similar trends for other insectivorous migrants such as the common whitethroat, redstart *Phoenicurus phoenicurus* and sand martin *Riparia riparia*, strongly suggest that their dramatic population crashes since the late 1960s are the direct result of longer-term drought in the Sahel (Winstanley, Spencer & Williamson 1974, Cowley 1979, Marchant, Hudson, Carter & Whittington 1990).

The desertification of the semi-arid northern tropics has been brought about by a combination of climate change and increasing human pressure (Sinclair & Fryxell 1985). Much of the Sahel and Sudan zones is intensively farmed or overgrazed and heavily exploited for fuelwood, even within legally protected forest reserves (P. Jones personal observation). These changes have undoubtedly caused a contraction and impoverishment of habitats used by migrants. (Ledant, Roux, Jarry *et al*. 1986, Grimmett 1987, Morel & Betlem 1992). Overwinter survival of common whitethroats and sedge warblers has been shown to be density dependent (Baillie & Peach 1992), which suggests that their Sahelian habitats are occupied to the full. The consequence of habitat degradation on such an enormous scale must be that far fewer birds can now overwinter successfully.

The lack of detailed information on resource use and niche partitioning among Palearctic and Afrotropical birds means, however, that we are still far from understanding the immediate mechanisms by which habitat change on this scale may affect bird numbers. Palearctic migrants make greater use of more open and edge habitats than Afrotropical species. This is true on a regional scale, as for example between the different vegetation zones of West Africa (Elgood *et al.* 1966), and on a local scale between habitats (Ulfstrand & Alerstam 1977, Lack 1986b), as well as between microhabitats within a single vegetation type (Rabøl 1987). The extent to which these relationships are duplicated when habitats are fragmented by human disturbance, or degraded by drought and overgrazing, is less clear. The few available data for bird densities at different degrees of habitat degradation suggest that the effect might vary with habitat.

In a still relatively intact *Acacia–Balanites* forest reserve in northern Nigeria, mid dry season densities of Palearctic warblers were 7.0 birds ha^{-1}, whereas in semi-degraded and heavily degraded woodlands of the same type these densities dropped to 1.6 and 0.7 birds ha^{-1} repectively (Fig. 16.4; Jones *et al.* 1996). An even more dramatic decline occurred in sunbird densities in the same woodlands, whereas Afrotropical warblers, which were anyway less common, showed much less change. Other habitats surveyed all had similarly low population densities of Afrotropical warblers and sunbirds.

Surprisingly, a comparatively important mid dry season habitat for migrant *Sylvia* warblers in northern Nigeria was farmland. Their density was 1.0–2.1 birds ha^{-1}, provided native trees remained at field margins or as isolated mature trees within fields. In contrast, arboreal Afrotropical insectivores were all but absent from this habitat (Fig. 16.4; Jones *et al.* 1996, Stoate 1997).

The only comparable information of bird densities in relation to habitat

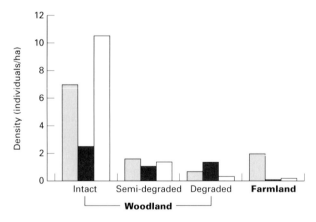

FIG. 16.4. Densities of Palearctic warblers (shaded bars), ecologically similar Afrotropical insectivores (solid), and sunbirds (open) in *Acacia–Balanites* woodland at different degrees of habitat degradation, and in farmland, in the Sahel of northern Nigeria. (From Jones *et al.* 1996.)

disturbance in the wet season savannas comes from Zambia (Ulfstrand & Alerstam 1977). Undisturbed *Brachystegia* woodland seemed to be hardly penetrated by Palearctic migrants, where willow warblers were the only species present (0.05 birds ha^{-1}). Migrants were much better represented, however, in degraded *Brachystegia* and *Acacia* woodlands, where willow and icterine warblers showed up to an eight-fold increase in density (0.22–0.42 birds ha^{-1}); Afrotropical species showed the opposite trend, decreasing from 2.13 to 1.19 birds ha^{-1} (Ulfstrand & Alerstam 1977). Elsewhere in Zambia, Kelsey (1992) recorded densities an order of magnitude greater in secondary woodland on abandoned cultivation, although Afrotropical species still outnumbered Palearctic immigrants at 13.4 to 9.3 birds ha^{-1}, respectively.

Loss of natural habitats in East Africa also seems to have taken its toll of some arboreal insectivores. Lack (1985) suggested that the warblers have been more subject to changes in status and occurrence than most other bird families, and he reported anecdotal evidence that the loss of *Acacia* and *Commiphora* woodland in the Tsavo National Park in Kenya has been accompanied by a great reduction in numbers of at least five species of Afrotropical warbler since the early 1960s. How East African habitats may have changed in their suitability for Palearctic migrants over the same period is unknown.

CONCLUSIONS

This chapter has attempted to bring together what is known of the broad pattern of intra-African migration by Palearctic and Afrotropical insectivores, with studies on the behavioural and ecological processes by which migrant and resident bird communities coexist and interact. The few detailed field studies attempted in Africa have revealed the complex and often transient relationships between immigrants and native species, resulting from the much more highly seasonal nature of the environments they occupy and the consequent mobility of the savanna avifauna. We are still far from understanding these interactions in the same depth as among the migrant insectivore assemblage of the Neotropics (Rappole 1995).

Questions originally arising from Moreau's (1972) synthesis, and restated more recently by Newton (1995), remain to be answered. Perhaps foremost amongst these, and relevant to the resolution of 'Moreau's Paradox', is why so few migrants extend their wintering ranges further south. It may be that they are adequately accommodated in the north, because their major potential competitors withdraw southwards themselves, leaving the northern semi-arid savannas 'vacant' for the dry season. The extent to which they might have competed with the absent Afrotropical migrants, and how they integrate with the remaining residents in an inhospitable environment, remains unassessed. The second possibility, but not mutually exclusive with the first, is that only a few species continue onwards to southern Africa because the wet season occupancy of the southern savannas by Afrotropical rains migrants might exclude Palearctic species. By analogy, few Nearctic migrants in the arboreal insectivorous guild penetrate far into South America, apparently because the South American forests are already occupied by a large and diverse assemblage of closely-related resident wood-warblers and leaf-gleaning tyrannids (Keast 1980a).

To answer these fundamental questions of community dynamics, we still need much more comprehensive data on population densities of Afrotropical and Palearctic species, and on how resource partitioning and niche separation are achieved between them in relation to the productivity of different habitats across Africa during the wet and dry seasons. Intra-African migrations are not well documented for much of Africa and the niche displacements and competitive interactions that may occur as populations arrive and depart remain almost unrecorded.

The disruptions to these interactions that may occur with habitat degradation through over-exploitation and climate change are completely unknown, though the long-term population consequences are already visible among Palearctic migrants. In the Americas, the better understanding of the interactive behaviours and ecological requirements of Nearctic migrants is beginning to yield insights into the processes that have resulted in species declines in response to increasing environmental disturbance. We are still far from understanding such complexities in the Afrotropics.

ACKNOWLEDGEMENTS

The unpublished data from Nigeria were collected with the late Peter Ward, who stimulated my interest in this subject. The chapter has benefited greatly from comments by Alan Kemp, Peter Lack, Catriona MacCallum, Ian Newton, Herbert Prins and Juliet Vickery. Catriona MacCallum produced the figures.

REFERENCES

Alerstam, T. (1990). *Bird Migration.* Cambridge University Press, Cambridge.
Baillie, S.R. & Peach, W.J. (1992). Population limitation in Palaearctic–African migrant passerines. *Ibis*, **134**, (Suppl. 1), 120–132.
Benson, C.W. (1982). Migrants in the Afrotropical region south of the equator. *Ostrich*, **53**, 31–49.

Brown, L.H., Urban, E.K. & Newman, K. (1982). *The Birds of Africa*, Vol. 1. Academic Press, London.
Cavé, A.J. (1983). Purple heron survival in tropical West Africa. *Ardea*, **71**, 217–224.
Cheke, R.A. & Walsh, J.F. (1996). *The Birds of Togo*. BOU Checklist No. 14. British Ornithologists' Union, Tring.
Cowley, E. (1979). Sand martin population trends in Britain, 1965–1978. *Bird Study*, **26**, 113–116.
Crowe, T.M. & Crowe, A.A. (1982). Patterns of distribution, diversity and endemism in Afrotropical birds. *Journal of Zoology, London*, **198**, 417–442.
Curry-Lindahl, K. (1981). *Bird Migration in Africa: Movements Between Six Continents*, 2 vols. Academic Press, London.
Den Held, J.J. (1981). Population changes of the purple heron in relation to drought in the wintering area. *Ardea*, **69**, 185–191.
Elgood, J.H., Sharland, R.E. & Ward, P. (1966). Palaearctic migrants in Nigeria. *Ibis*, **108**, 84–116.
Elgood, J.H., Fry, C.H. & Dowsett, R.J. (1973). African migrants in Nigeria. *Ibis*, **115**, 1–45, 375–411.
Elgood, J.H., Heigham, J.B., Moore, A.M., Nason, A.M., Sharland, R.E. & Skinner, N.J. (1994). *The Birds of Nigeria*, 2nd ed. BOU Checklist No. 4. British Ornithologists' Union, Tring.
Fasola, M., Hafner, H., Prosper, P., van der Kooij, H. & Schogolev, I.V. (1997). Population changes in European herons: relationships with African climate. *Proceedings of the 9th Pan-African Ornithological Congress*, in press.
Fry, C.H. (1992). The Moreau ecological overview. *Ibis*, **134**, (Suppl. 1), 3–6.
Grimes, L.G. (1987). *The Birds of Ghana*. BOU Checklist No. 9. British Ornithologists' Union, London.
Grimmett, R. (1987). *A Review of the Problems Affecting Palearctic Migratory Birds in Africa* ICBP Study Report, **22**. International Council for Bird Preservation, Cambridge.
Hedenström, A., Bensch, S., Hasselquist, D., Lockwood, W. & Ottosson, U. (1993). Migration, stopover and moult of the great reed warbler *Acrocephalus arundinaceus* in Ghana, West Africa. *Ibis*, **135**, 177–180.
Jones, P.J. (1980). The timing of wing moult in the greyhooded kingfisher in Nigeria. *Ostrich*, **51**, 99–106.
Jones, P.J. (1985). The migration strategies of Palearctic passerines in West Africa. In *Migratory Birds: Problems and Prospects in Africa. Report of the 14th Conference of the European Continental Section, ICBP, 1983* (Ed. by A. MacDonald & P. Goriup), pp. 9–21. International Council for Bird Preservation, Cambridge.
Jones, P.J. (1995). Migration strategies of palearctic passerines in Africa. *Israel Journal of Zoology*, **41**, 393–406.
Jones, P.J. & Ward, P. (1977). Evidence of pre-migratory fattening in three tropical granivorous birds. *Ibis*, **119**, 200–203.
Jones, P., Vickery, J., Holt, S. & Cresswell, W. (1996). A preliminary assessment of some factors influencing the density and distribution of palearctic passerine migrants wintering in the Sahel zone of West Africa. *Bird Study*, **43**, 73–84.
Kanyamibwa, S., Schierer, A., Pradel, R. & Lebreton, J.-D. (1990). Changes in adult annual survival rates in a western European population of the white stork *Ciconia ciconia*. *Ibis*, **132**, 27–35.
Keast, A. (1980a). Spatial relationships between migratory parulid warblers and their ecological counterparts in the Neotropics. In *Migrant Birds in the Neotropics: Ecology, Behavior, Distribution and Conservation* (Ed. by A. Keast & E.S. Morton.), pp. 109–130. Smithsonian Institution, Washington DC.
Keast, A. (1980b). Migratory Parulidae: what can species co-occurrence in the north reveal about ecological plasticity and wintering patterns? In *Migrant Birds in the Neotropics:*

Ecology, Behavior, Distribution and Conservation (Ed. by A. Keast & E.S. Morton), pp. 457–476. Smithsonian Institution, Washington DC.

Keast, A. & Morton, E.S. (1980). *Migrant Birds in the Neotropics: Ecology, Behavior, Distribution and Conservation.* Smithsonian Institution, Washington DC.

Keay, R.W.J. (1959). *Vegetation Map of Africa.* Oxford University Press, Oxford.

Kelsey, M.G. (1992). Conservation of migrants on their wintering grounds: an overview. *Ibis*, **134**, (Suppl. 1), 109–112.

Lack, D. (1971). *Ecological Isolation in Birds.* Blackwell Scientific Publications, Oxford.

Lack, P.C. (1983). The movements of palaearctic landbird species in Tsavo East National Park, Kenya. *Journal of Animal Ecology*, **52**, 513–524.

Lack, P.C. (1985). The ecology of the land-birds of Tsavo East National Park, Kenya. *Scopus*, **9**, 2–23, 57–96.

Lack, P.C. (1986a). Diurnal and seasonal variation in the biomass of arthropods in Tsavo East National Park. *African Journal of Ecology*, **24**, 47–51.

Lack, P.C. (1986b). Ecological correlates of migrants and residents in a tropical African savanna. *Ardea*, **74**, 111–119.

Lack, P.C. (1987). The structure and seasonal dynamics of the bird community in Tsavo East National Park, Kenya. *Ostrich*, **58**, 9–23.

Ledant, J.P., Roux, F., Jarry, G., Gammel, A., Smit, C., Bairlein, F. *et al.* (1986). *Aperçu des Zones de Grand Interêt pour la Conservation des Espèces d'Oiseaux Migrateurs de la Communauté en Afrique.* Rapport EUR 10878FR. Commission des Communautés Européenes: Environnement et Qualité de la Vie. Luxembourg.

Leisler, B. (1992). Habitat selection and coexistence of migrants and Afrotropical residents. *Ibis*, **134**, (Suppl. 1), 77–82.

MacKinnon, J. & MacKinnon, K. (1986). *Review of the Protected Areas System in the Afrotropical Realm.* International Union for the Conservation of Nature and Natural Resources/United Nations Environment Programme, Gland.

Maclean, G.L. (1984). *Roberts' Birds of Southern Africa.* John Voelcker Bird Book Fund, Cape Town.

Marchant, J.H., Hudson, R., Carter, S.P. & Whittington, P. (1990). *Population Trends in British Breeding Birds.* British Trust for Ornithology, Tring.

Mönkkönen, M., Helle, P. & Welsh, D. (1992). Perspectives on Palearctic and Nearctic bird migration; comparisons and overview of life-history and ecology of migrant passerines. *Ibis*, **134**, (Suppl. 1), 7–13.

Moreau, R.E. (1966). *The Bird Faunas of Africa and its Islands.* Academic Press, London.

Moreau, R.E. (1972). *The Palaearctic–African Bird Migration Systems.* Academic Press, London.

Morel, G.J. (1968). Contribution à la Synécologie des Oiseaux du Sahel Sénégalais. *Mémoires, ORSTOM* **29**, Office de la Recherche Scientifique et Technique Outre-Mer, Paris.

Morel, G.J. (1973). The Sahel Zone as an environment for Palaearctic migrants. *Ibis*, **115**, 413–417.

Morel, G.J. & Betlem, J. (1992). The use of *Acacia nilotica* riverine forests of Senegal by birds, particularly Palearctic migrants. *Proceedings of the VII Pan-African Ornithological Congress*, (ed. L. Bennun) pp. 495–502.

Morel, G.J. & Morel, M.-Y. (1992). Habitat use by palaearctic migrant passerine birds in West Africa. *Ibis*, **134**, (Suppl. 1), 83–88.

Newton, I. (1995). Relationship between breeding and wintering ranges in Palaearctic–African migrants. *Ibis*, **137**, 241–249.

Peach, W.J., Baillie, S.R. & Underhill, L. (1991). Survival of British sedge warblers *Acrocephalus schoenobaenus* in relation to west African rainfall. *Ibis*, **133**, 300–305.

Pearson, D.J. (1990). Palaearctic passerine migrants in Kenya and Uganda: temporal and spatial patterns of their movements. In *Bird Migration: Physiology and Ecophysiology* (Ed. by E. Gwinner), pp. 44–59. Springer, Berlin.

Pearson, D.J. & Lack, P.C. (1992). Migration patterns and habitat use by passerine and near-passerine migrant birds in eastern Africa. *Ibis*, **134**, (Suppl. 1), 89–98.

Pearson, D.J., Nikolaus, G. & Ash, J.S. (1988). The southward migration of Palaearctic passerines through northeast and east tropical Africa: a review. *Proceedings of the 6th Pan-African Ornithological Congress*, pp. 243–261.

Pomeroy, D. (1993). Centers of high biodiversity in Africa. *Conservation Biology*, **7**, 901–907.

Rabøl, J. (1987). Coexistence and competition between over-wintering willow warblers *Phylloscopus trochilus* and local warblers at Lake Naivasha, Kenya. *Ornis Scandinavica*, **18**, 101–121.

Rappole, J.H. (1995). *The Ecology of Migrant Birds: A Neotropical Perspective*. Smithsonian Institution Press, Washington DC.

Sinclair, A.R.E. (1978). Factors affecting the food supply and breeding season of resident birds and movements of palaearctic migrants in a tropical African savannah. *Ibis*, **129**, 480–497.

Sinclair, A.R.E. & Fryxell, J.M. (1985). The Sahel of Africa: ecology of a disaster. *Canadian Journal of Zoology*, **63**, 987–994.

Snow, D.W. (1978). Relationships between the European and African avifaunas. *Bird Study*, **25**, 134–148.

Stoate, C. (1997). Abundance of whitethroats *Sylvia communis* and potential invertebrate prey in two Sahelian sylvi-agricultural habitats. *Malimbus*, **19**, 7–11.

Ulfstrand, S. & Alerstam, T. (1977). Bird communities of *Brachystegia* and *Acacia* woodlands in Zambia. *Journal für Ornithologie*, **118**, 156–174.

White, F. (1983). *The Vegetation of Africa: A Descriptive Memoir to Accompany the UNESCO/AETFAT/UNSO Vegetation Map of Africa*. Unesco, Paris.

Winstanley, D., Spencer, R. & Williamson, K. (1974). Where have all the whitethroats gone? *Bird Study*, **21**, 1–14.

17. SPECIES-RICHNESS OF AFRICAN GRAZER ASSEMBLAGES: TOWARDS A FUNCTIONAL EXPLANATION

H. H. T. PRINS and H. OLFF

Department of Terrestrial Ecology and Nature Conservation, Wageningen Agricultural University, Bornesteeg 69, 6708 PD Wageningen, The Netherlands

SUMMARY

1 A theoretical analysis was made on the underlying causes of variation in species-richness of herbivore assemblages in Africa at different spatial scales. The study was restricted to grazers. We explored why many species of grazers are able to co-occur in the same area, given that they all utilize the same resource (grass). An attempt is made to explain the species composition and species-richness of grazer assemblages by resource competition and facilitation interactions.

2 Simple theoretical arguments are developed regarding the dependence of gross and net daily food intake on body mass, and the consequences of body mass for numerical response patterns on an annual basis. Experimental data are reviewed for the interdependence of primary production and quality of primary production, and for the functional response patterns of differently sized herbivores. We concluded that forage quality may decrease with increasing plant standing crop, imposing an important constraint on net nutrient and energy intake by grazers.

3 Both theoretical arguments and experimental evidence support the conclusion that large grazers are better suited in handling high biomass/low quality forage than smaller species. The presence of large herbivores therefore can decrease the standing crop and improve forage quality, leading to facilitation for smaller grazers. However, when the body mass difference is too large, the smaller species is not expected to benefit from the presence of the larger species. When two species in a herbivore assemblage are 'too similar' in body mass, then resource competition is expected to prevail over facilitation interactions. Therefore, some 'optimum' difference between body weights of subsequent species in assemblages is expected.

4 Experimental data on grazer assemblages on different spatial scales (whole of Africa, whole ecosystems, parts of ecosystems) show that body masses indeed show a certain regularity. Each species in an assemblage is, on

average, a constant proportion larger than the next smaller one (i.e. the species show a constant weight ratio). These data thus provide evidence that body masses of different species in herbivore assemblages do not represent a random selection from a uniform distribution, but rather have a certain regularity, which might be related to the balance between competition and facilitation interactions.

5 Observational evidence on the effect of the removal and addition of grazers in African national parks is reviewed. The scarce evidence available suggests that grazers more similar in body mass are more likely to compete, and that large grazers may facilitate for smaller ones.

6 A map of potential species-richness of African grazer assemblages is compiled through a GIS overlay of the distribution maps of 96 species. This map shows that the highest potential species-richness of grazers is found in East Africa and in the border area of Zimbabwe and Mozambique. Furthermore, we describe weight ratios and species-richness patterns for a large number of ecosystems. The causes of regional and local variation are discussed in terms of area size, primary productivity, habitat isolation, fragmentation, and competition and facilitation interactions.

7 We conclude that more insight into resource competition and facilitation is needed for a better understanding of the structure and species-richness of grazer assemblages.

INTRODUCTION

During the Late Pleistocene the Palaearctic, South America and Australia experienced a tremendous decline in mammalian species-richness, whilst Africa to a large extent retained the full diversity of its large mammal species. In North America 75% of all herbivore genera with body masses heavier than 5 kg went extinct; in South America this loss was 76% and in Europe 45%. In contrast, Africa lost only 13.5% of its larger herbivore genera. Generic extinction outside Africa was positively correlated with body mass. Thus, all herbivore species heavier than 1000 kg became extinct, 76% of the genera between 100 and 1000 kg, 41% of the genera between 5 and 100 kg, and less than 2% of smaller genera (Owen-Smith 1988, p. 281 *et seq.*). Within Africa, though, herbivore species-richness remains very high, with, for example, *c.* 76 species of bovids (Dorst & Dandelot 1972, Haltenorth & Diller 1979) and 96 species of vertebrates heavier than 2 kg relying, to a larger or smaller extent, on grasses (Table 17.2). This means that Africa offers at present an unparalleled opportunity for studying patterns of herbivore species-richness and mechanisms of coexistence. African ecosystems (especially those of savannas) are very rich in species as exemplified by the Tanzanian National Parks such as Serengeti (Sinclair & Arcese 1995, p. 639 *et seq.*) or Lake Manyara (Prins & Douglas-Hamilton 1990).

The study of the patterns and causes of species-richness has been the focal point of ecology for many a decade. Hence, we hark back to methods and principles established many years ago which we will put into a new perspective. In this chapter we explore the relationships between body size and instantaneous food intake, daily intake and seasonal intake, the relationships between grass biomass and quality, and between seasonal intake rates and grazer reproduction. These we link to ideas of resource competition and facilitation, and of species packing, to arrive at conclusions about the causes of variation in species-richness in grazer assemblages.

LINKING GRASS FEATURES WITH HERBIVORE POPULATION DYNAMICS

In our investigation of species-richness of African grass-eaters we have been heavily influenced by the desire for simplicity in order to be able to detect patterns, and by the works of G.E. Hutchinson in the 1950s, R.H. McArthur in the 1960s and 1970s and by that of R.M. May. We use a simple approach directed towards the falsification of models, because we want to 'represent current knowledge as faithfully as possible so that this knowledge can be tested for sufficiency in generating an overall behaviour of ecosystems, and thus reveal lacunae in our current knowledge' (Hogeweg & Richter 1982). Too often, studies of large herbivore community interactions 'merely describe patterns of resource use and partitioning among sympatric species' (Putman 1996, p. 7). First, concerning simplicity, we have limited our investigation to wild vertebrates heavier than 2 kg, native to Africa and feeding to a greater extent on monocotyledons. This we have simplified to 'grass', and we have assumed that grass forms a single resource dimension. We focus on species that have grass as an important component in their diet, and explore the theoretical consequences of a situation where grass would be the *only* forage available. Although some of the investigated species are mixed feeders, which is a potential source of niche differentiation, this differentiation is outside the scope of our study.

We thus focus on a group of species exploiting the same class of environmental resource in a similar way; such a group has been termed a guild (Root 1967, Begon, Harper & Townsend 1990, p. 718, Simberloff & Dayan 1991, De Kroon & Olff 1995). Examples of such classes of environmental resources for herbivores are fruits, seeds, tree leaves, herbs and grasses. For herbivores, it is expected that interspecific competition and facilitation occurs, especially within guilds. A resource class can often be split into several resource dimensions (Root 1967, Platt & Weiss 1977). Fruits and seeds come in different sizes, and each size class can be viewed as a different resource. One of the ways in which animals, utilizing these resources composed of discrete particles, are thought to coexist is by differences in their body sizes. Differently sized animals are expected to eat differently sized fruits or seeds until there is enough non-overlap to allow coexistence (Hutchinson 1959, MacArthur & Levins 1967, May 1973, Schoener 1974, Wilson 1975, Case 1981). For browsing animals this limiting similarity principle may also apply; species sufficiently different in size or climbing ability may coexist by utilizing differently sized trees or different parts of trees.

However, for the guild of herbivores utilizing the graminaceous resource the situation is different. It is difficult to think of grass as a resource class composed of more than one dimension, since it does not consist of discrete food items (at least for species with a body mass exceeding 2 kg), and plant parts are not positioned out of range for smaller species. Hence, the question arises as to how several species of grazers can coexist on this single resource. In order to answer this question we need further insight into the consumer–resource interactions of herbivores and grass. We believe that for grazers body mass differences are also a crucial determinant of species' interactions but in a somewhat different way than in frugivores, seed-eaters or browsers.

From a herbivore's point of view, the graminaceous resource has three important features, namely: (i) standing crop; (ii) crude protein content; and (iii) digestible energy content. Above-ground biomass of grasses has been reported to be negatively correlated with (crude) protein content (Mattson 1980) and with digestibility (Crampton & Harris 1969, p. 285 et seq.). For several reasons, it will be advantageous for consumers of plant tissues with low plant protein concentrations to evolve a large body mass, as outlined by Mattson (1980) or Van Soest (1982, p. 342 et seq.). Large body mass permits wide-scale foraging for high-volume harvesting (Du Toit & Owen-Smith 1989). Furthermore, a larger body mass offers mechanical advantages for removing and macerating plant tissues. Finally, larger body mass may allow for the development of highly elaborate gastrointestinal fermentation and N-recycling systems that are necessary for digesting low-quality food. Given the negative correlation between above-ground biomass and protein concentrations, high biomass is expected to be utilized especially by herbivores with a larger body weight. Digestibility (which to a large extent determines the grazer's energy intake; Van Soest 1982, p. 278 et seq.) is little affected by the species of grazer but it is mainly food specific. Indeed, dry matter digestibility figures are similar for different wild African herbivores (De Bie 1991, tables 7.24 and 7.25: mean 60%, SD = 9.4, n = 12). These fall within the range of values for other non-sheep/cattle ruminants of 58.1% (SD = 6.6, n = 31), and for sheep and cattle of 55.9% (SD = 5.6, n = 31), feeding on roughage (calculated from Van Soest 1982, p. 338). Hence, existing differences do not appear to warrant species-specific treatment from our perspective, and in this we treat digestibility as a vegetation-specific parameter. For this analysis we also treat protein content as a vegetation-specific parameter. Although we are fully aware that leaf table height, leaf stem ratio, secondary compounds or other parameters play an important role in understanding food selection (Prins 1996) we chose to ignore them in the present study.

For an understanding of the relationship between population numbers and interspecific competition, it is useful to distinguish three responses of herbivores to forage availability, namely scaling up from: (i) response of gross instantaneous intake rate; to (ii) response of net daily intake rate; to finally (iii) response of annual net changes in numbers or density.

Population dynamics is of course governed by annual net changes in numbers, but studies of the instantaneous and daily food intake rates can provide insights into the mechanisms of population dynamics. We consider body mass to be a critical parameter for understanding response patterns at either time-scale. Body mass (weight) determines on the one hand energy and protein requirements of the herbivore and on the other hand its capacity to ingest and digest the vegetation. Therefore, body mass has important consequences for competition between, and for, facilitation grazing species at either time-scale.

Gross instantaneous food intake: functional response patterns

The form of the functional response for instantaneous food intake is usually a continuously increasing function: if food density is higher then the intake will be higher too (Illius & Gordon 1987). A commonly used function (e.g. Fryxell 1991) for calculating the gross instantaneous consumption rate (C_i) per herbivore is:

$$C_i = m_i V / (e_i + V) \qquad (1)$$

where m_i is the maximum instantaneous rate of consumption per herbivore, V is the vegetation abundance per unit area and e_i is the vegetation density at which the consumption is one-half of the maximum (half-saturating constant). Both parameters m_i and e_i are expected to be related to body mass. Larger animals can take larger bites, and chew more per unit time than smaller ones, resulting in a higher maximum instantaneous consumption rate (Spalinger & Hobbs 1992). Smaller animals can handle smaller plant parts, and their intake is expected to saturate at a lower available biomass (lower e_i), for example, by grazing down a sward to a shorter height (Illius & Gordon 1987). Thus, both e_i and m_i are functions of body mass (W).

An indication of the form of the functions $m_i(W)$ and $e_i(W)$ (with W in kg) can be obtained from the data presented by Gross, Shipley, Hobbs et al. (1993) who performed a feeding trial with boreal herbivores of different body mass. The animals were offered trays with plants of different sizes, and the removal rates were quantified. Both maximum intake rate and half-saturation plant biomass were linear functions of body mass with a zero intercept (Fig. 17.1a, b). Replacing m_i by $m_i W$ and e_i by $e_i W$ in eqn (1) provides a simple function for the dependence of instantaneous intake rate on both body mass and available plant biomass:

$$C_i = m_i WV/(e_i W + V) \qquad (2a)$$

or expressed as intake per unit body mass,

$$C_i/W = m_i V/(e_i W + V) \qquad (2b)$$

Using the parameter values obtained from the study of Gross et al. (m_i = 0.1209 and e_i = 0.0013, with W in kg, V in g plant^{-1}, m_i in g min^{-1} and e_i in g plant^{-1}) we can draw this dependence of instantaneous gross intake rate on plant biomass for different body masses (Fig. 17.1c). This figure shows that in an absolute sense (i.e. per animal) intake rates of larger herbivores are always higher. However, per unit body mass (eqn 2b) intake rates at low plant biomass are higher in smaller herbivores (Fig. 17.1d). At high plant biomass, all species converge to the same value. This result implies that when populations of a small and a large herbivore species are competing for a food resource at the same herbivore biomass density (weight/unit area), the instantaneous intake of the smaller herbivore population will be higher (lower half-saturation constant), especially when the food is depleted, and the smaller species is thus expected to be the superior competitor (Fig. 17.1d). In the feeding trials of Gross et al. (1993) the small animals did not have a disadvantage when foraging for high plant biomass. However, in field situations they might show a lower instantaneous gross intake rate per unit body mass at high vegetation biomass because of a vigilant behaviour towards potential predators, because of physical problems of walking in dense vegetation or because of problems in feeding in a vegetation with a low bulk density (Stobbs 1973a,b, Van de Koppel, Huisman, Van de Wal & Olff 1996).

Gross daily intake rate: functional response patterns

The simplest way to calculate the daily net food intake rate would be to sum the gross instantaneous energy or protein intake rate (eqn 2) over the length

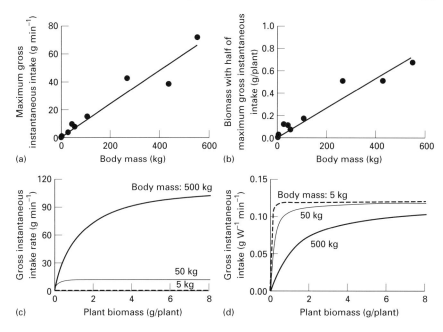

FIG. 17.1. Relationships between maximum gross instantaneous intake rate (g min⁻¹) and body weight (a), of half-saturating plant biomass (g plant⁻¹) with body mass (b), and the absolute (c) and specific (d) intake rates as a function of plant biomass at different body mass as calculated from the first two graphs, using a Michaelis–Menten function. Regression equation for (a): $y = 0.1209x$, $r^2 = 0.93$ and for (b): $y = 0.0013x$, $r^2 = 0.92$; resulting in the equation $y = 0.1209Wx/(0.0013W + x)$ used for (c) and (d), where W is body mass in kg. (Data from Gross et al. 1993.)

of the potential foraging period within 24 h, and to subtract daily losses and expenses. However, several complications arise when looking at a daily time scale. The actual foraging period can be shorter than the potential period because of capacity limitation: when the food has a poorer quality (energy and/or nutrient concentration), the herbivore will need more time to digest the food before it can eat again. This means that when the digestibility of the food decreases, the gross daily intake rate is reduced because of a digestive constraint (e.g. Sibbald, Maxwell & Eadie 1979, Van Soest 1982, Fryxell 1991). Experimental data show that the crude protein concentration of African grasslands decreases at higher biomass, with a greater decrease on nutrient-poor soils (Mali, West African Sahel) than on nutrient-rich soils (Tsavo, Kenya) (Fig. 17.2). In fact, the protein concentration in high biomass (high rainfall) regions in the Sahel even drops below the minimum requirements of ruminants which is c. 7–8% crude protein independent of body mass (Prins 1996, p. 261 et seq.). In addition, in species where the throughput rate does not decrease at lower forage quality (e.g. hindgut fermenters) the

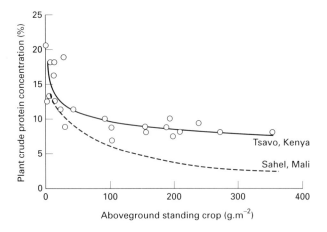

FIG. 17.2. Dependence of plant crude protein concentration (%) on above-ground standing crop (g m^{-2}) for Tsavo National Park, Kenya regression: $y = 20.879x^{-0.1697}$. (Data from Van Wijngaarden 1985 and the Sahel, Mali data from Breman & de Wit 1983.)

advantages of a higher intake at higher biomass might be exceeded by the disadvantages of poorer forage quality.

The gross intake of energy or nutrients is thus found by taking the minimum of the value dictated by constraints imposed by food availability and digestion and passage from the gut. Together with the aforementioned decrease in quality with increasing standing crop, this results in an optimum in the gross daily nutrient or energy intake rate at intermediate vegetation biomass (Fryxell 1991). Since a larger body mass allows for a more extensive fermentation and digestion system (Mattson 1980), we hypothesize that these digestive constraints are much stronger in smaller-sized herbivores, resulting in a greater decrease in gross intake at higher vegetation biomass. The only way small herbivores can utilize a low-quality food (as found at high plant biomass) is either by having a low gross intake (limited by body mass) with a long digestion time, or a high gross intake with a high passage rate (low retention efficiency). In both cases, the *net* daily intake rate (gain) per unit body mass will be low.

Some experimental evidence for the dependence of gross daily intake rate on vegetation biomass for species with different body mass is provided by Short (1985) who measured the daily food intake of rabbits ($W = 1.25$ kg), kangaroo ($W = 26$ kg) and sheep ($W = 58$ kg) in Australian paddocks. For all three species, Short fitted a continuously increasing function for the dependence of daily gross intake per unit body mass (C_d/W) on vegetation biomass:

$$C_d/W = m_d\left(1 - \exp(-V/e_d)\right) \qquad (2)$$

where m_d denotes the maximum daily intake rate per unit body mass, V is the vegetation biomass and e_d is the vegetation biomass at which the intake is at 65% of the maximum. This model has approximately the same form as a Michaelis–Menten equation. Short (1985) then concluded that the form of eqn 2 was not different for the three herbivore species. A closer look at his published data, however, reveals that although this model fitted well for sheep (Fig. 17.3c), for rabbit and kangaroo it was less appropriate since gross daily intake decreased again at higher vegetation biomass (Fig. 17.3a,b; see also Stobbs 1973a,b) which was in line with our theoretical prediction. Also the fact that some minimum vegetation biomass must exist which animals need in order to forage (Fig. 17.3a,b,c) is lacking from eqn 2, which calls for

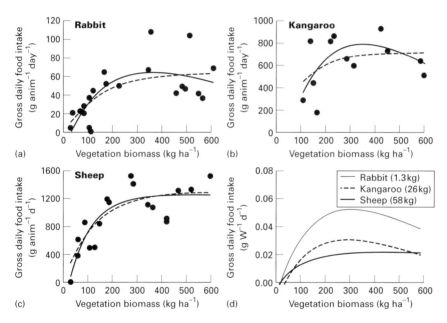

FIG. 17.3. Dependence of gross daily food intake per animal (a–c) and per unit body weight (d) on above-ground vegetation biomass of rabbits (a) kangaroo (b) and sheep (c). (Data are from Short 1985.) Regression models were recalculated using a continuously increasing model (dashed lines in a–c) and a model including an intake reduction at high biomass and a minimum required biomass (solid lines in a–c). The solid lines of (a–c) were graphed per unit body weight in (d). Regression equations: rabbit, dashed line: $y = 65(1 - \exp(-x/125))$, $r^2 = 0.46$; rabbit, solid line: $y = 45355(1 - \exp(-(x - 21.5)/67779))$, $r^2 = 0.50$; kangaroo, dashed line: $y = 723(1 - \exp(-x/91))$, $r^2 = 0.21$; kangaroo, solid line: $y = 214017(1 - \exp(-(x - 40.4)/20967))\exp(-0.0040x)$, $r^2 = 0.33$; sheep, dashed line: $y = -1293(1 - \exp(-x/128))$, $r^2 = 0.70$; sheep, solid line: $y = 1248(1 - \exp(-(x - 23.3)/91))$, $r^2 = 0.72$.

an adjustment of the equation. Adding a minimum required biomass (l_d) and a reduction term for intake at high biomass (h_d) to eqn 2 yields the following empirical model which is conceptually more appropriate:

$$C_d = m_d \left(1 - \exp\left(-(V - l_d)/e_d\right)\exp(-h_d V)\right) \quad (3a)$$

or per unit body mass,

$$C_d/W = \left(m_d \left(1 - \exp\left(-(V - l_d)/e_d\right)\exp(-h_d V)\right)\right)/W. \quad (3b)$$

Using non-linear regression, we found that eqn 3a fitted Short's data better than eqn 2, both for rabbit ($r^2 = 0.50$ vs. 0.46, Fig. 17.3a) and for kangaroo ($r^2 = 0.33$ vs. 0.21, Fig. 17.3b). For sheep, no reduction of intake occurred at higher biomass, so only l_d was added to the model ($r^2 = 0.72$ vs. 0.70, Fig. 17.3c). The resulting functions for rabbit, kangaroo and sheep are compared in Fig. 17.3d, with gross daily intake expressed per unit body weight. The smaller herbivores had a higher daily intake rate per unit body weight at low vegetation biomass. At higher vegetation biomass, daily intake decreased in the two smaller species, but not in sheep. That the specific intake rate of rabbits and kangaroo did not decrease below intake rates of sheep might be because the study of Short (1985) was performed in a very low biomass trajectory (<60 g m^{-2}). Little difference between species was found in the minimum level of biomass required for foraging (Fig. 17.3d).

Gross daily intake rate and body mass

The parameters m_d, l_d, h_d and e_d may all be a function of body mass. The parameters l_d and h_d are expected to be lower in smaller species (see section 'Gross instantaneous food intake'). For m_d we can speculate on the form of the function. A simple mathematical expression in use for the dependence of daily maximal (at sufficient food) gross intake rates (C_d) on body mass (W) is:

$$m_d = aW^b. \quad (4)$$

The question then is whether daily intake rates are a constant proportion of body mass (that is, $b = 1$) or proportional to metabolic body mass (that is, $b = 0.75$). It should be noted that maximal daily intake per unit body mass decreases with W when $b < 1$. The justification for $b = 0.75$ rests on the assumption that metabolic requirements are related to metabolic size. This is based on the general observation that the basal metabolic rate (BMR) is equal to $293\,\text{kJ} \times W_{kg}^{0.75}$ (see Moen 1973). In reality the fasting metabolic rate shows quite some variation between species (see Owen-Smith 1979 who provides data for wildebeest, eland, Coke's hartebeest, cattle and sheep). The expression of intake as a linear function (or constant proportion) of body mass ($b = 1$) is favoured by those who observe little advantage in relating intake behaviour to metabolic body mass. The use of $b < 1$ is inconsistent with the observations that gastrointestinal capacity and rumination are themselves related to the power 1.0 of body mass; the weight of the fermentation contents of both ruminants and non-ruminants in Africa is approximately constant at 13% (calculated from Van Soest 1982, fig. 20.1). However, metabolic requirements may be related to power 0.75 or less (Thonney et al. 1976 cited in Van Soest 1982, Garland 1983).

Published daily food intake figures in feeding trials (assumed to be maximal intake) for different adult African herbivores are described by the equation $m_d = 0.0396 W^{0.8343}$ (Fig. 17.4a). These data show that food intake per unit body mass decreases with body mass from c. 2.5% in Thomson's gazelle to c. 1% in African elephant (Fig. 17.4b). The calculated value of 0.83 of parameter b in eqn 4 might imply that intake is determined both by metabolic and capacity constraints. A general figure for the daily dry matter intake for free-ranging grazers is 2.5% of body mass (Prins 1996, p. 261 et seq.) which is higher than most values in Fig. 17.4b but which includes food intake of growing and lactating individuals. It should be noted that in ruminants the value of C_d/W may be reduced either by food availability or the digestible energy content of the vegetation related to biomass (eqn 3). This dependence on food quality was implicitly taken into account in eqn 3 by assuming that digestible energy content decreased at higher vegetation biomass. If quality is low, daily intake will drop to c. 1% of body mass only (if

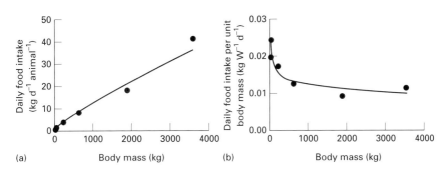

FIG. 17.4. The dependence of daily food intake per animal (a) and per unit body mass (b) on body mass for different free-ranging African grazers. (Data are from Delany & Happold 1979, table 11.12). Regression equation for (a): $y = 0.0396 x^{0.8343}$, $r^2 = 0.99$. Species are (from small to large body mass): Thomson's gazelle, wildebeest, buffalo, hippopotamus and savanna elephant.

the digestible energy is lower than about 7100 J kg^{-1}) but it can be higher than 3% when food quality is high (digestible energy content higher than c. 11 000 J kg^{-1}) (Crampton & Harris 1969, table 14.5). This drop in daily intake due to bad quality at high plant biomass is expected to be more important for smaller species. The higher sensitivity of smaller species to decreasing quality can be incorporated in eqn 3 by letting e_d be some increasing function of W. More data are needed to make the dependence of m_d, l_d, h_d and e_d on body mass explicit.

Net daily intake rate: functional response patterns

Finally, we discuss in this section the net daily intake, being the difference in gross intake minus costs and losses. The net daily intake of protein and energy is found by subtracting the daily losses from the gross daily intake. Protein is lost by defecation (with additional N-loss via urine), while energy is lost in defecation, urine production and respiration (energy costs of digestion, maintenance, motion) with an additional loss in ruminants through methane production. Multiplying the gross daily food intake (Fig. 17.3d) by its energy content and subtracting the defecation and heat losses provides a net daily intake of energy (which may be negative if food is in short supply or at too low quality (Fig. 17.3d)). Energy requirements are a function of metabolic body mass (see above). Since we argued that potential (maximum) energy intake is proportional to body mass (W^1), while expenditures are proportional to metabolic body mass ($W^{0.75}$), the net difference potential between intake and expenditure per unit body mass will decrease with increasing body mass for different species using the same food. This provides an extra argument why small animals have to restrict themselves to higher quality food and, thus, show a high selectivity. In the case of grazers, high food quality will often be found in low biomass areas (Fig. 17.2).

Based on the insights hitherto gained, a tentative form of the dependence of net daily food intake (energy or protein) on vegetation biomass, both for small and large grazers, is drawn in Fig. 17.5. This has important consequences for interactions between grazing species and, consequently, for understanding species-richness of grazers. At low biomass, the qualitative result is the same as for instantaneous food intake, namely a competitive advantage for species with smaller body mass because of higher intake per unit body mass. However, at a higher vegetation biomass, the situation is different. Small species might not be able to realize a positive net intake rate, unless this high biomass is removed by larger species, thereby increasing the quality of the vegetation to a level where it can be utilized by the smaller species. Summarizing, we have strong theoretical arguments that facilitation

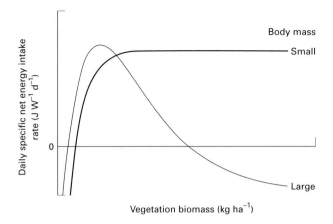

FIG. 17.5. Predicted pattern of functional response of daily specific net energy intake of grazers towards vegetation biomass for a small and a large herbivore, assuming a negative relationship between vegetation biomass and forage quality, and larger problems in the small species in coping with poorer quality forage.

plays an important role in understanding coexistence of grazers with different body mass.

Annual changes in population numbers as dependent on food availability: numerical response patterns

Besides the possible effects of predation, reproduction and mortality to a large extent depend on reserves (of energy and protein) stored in the body of the herbivore. Population dynamics of herbivores are thus determined by the quantity and quality of the food over a period long enough to enable the animal to deposit such reserves. Reproductive rates and survival therefore reflect an integrated measure of daily net intake over an extended period. On an annual time-scale, the energetic and nutrient costs of reproduction also have to be taken into account. From the literature (not reviewed here but see e.g. Pratchett, Capper, Light *et al.* 1977, Severinghaus 1979, Herd & Sprott 1986, Prins 1996) a coherent picture emerges that protein content in the tropics and energy contents in the temperate areas are closely related to population recruitment of grazers.

Given this information of food quality and reproduction, one would expect that seasonal food intake would be related to reproduction but we have not been able to find much information about the quantity of food intake and energy or protein demands in the wild in relation to reproduction. Grimsdell (1969) observed that lactating buffalo eat approximately 15% more than non-lactating females but how this has been measured is not clear to us. If reproduction is linked to food quantity and food quality, one would expect a relation between net population growth and food quality and/or quantity if mortality is affected in the same manner as reproduction is. In black-tailed deer the principal cause of mortality was starvation, not caused by a quantitative lack of food, but rather by a seasonal dip in the quality of the available forage (Taber & Dasman 1957). Viability and digestive function of rumen bacteria were not impaired by starvation since *in vitro* digestion of dry matter, cell wall and cell contents stayed the same (De Calesta, Nagy & Bailey 1974). Also in red deer, under-nutrition has been found to be a cause of mortality in the wild (Mitchell & Staines 1976, cf. Sinclair 1977). Hirst (1969) conducted an exhaustive study on causes of mortality. He found in the Transvaal that starvation might be an important cause of death but that this differs between different ungulate species (Table 17.1). From this information we concluded that seasonal intake, through condition, determines both the number of births and the number of deaths in the population of grazers.

Plant–herbivore interactions on the population level and on an annual time-scale can be studied theoretically by analysing systems of coupled differential equations (see e.g. DeAngelis 1992, Van de Koppel *et al.* 1996). When P denotes the plant density (kg ha^{-1}), and H denotes the herbivore biomass density (kg ha^{-1}) for a single species, respectively, the net rate of change of both populations is represented by the differential equations:

$$dP/dt = f(P) - c(P)H \tag{5a}$$

$$dH/dt = g(P)H \tag{5b}$$

TABLE 17.1. Causes of mortality of ungulates in Transvaal (South Africa). (Compiled from Hirst 1969; tables 3, 6, 9, 12, 17, 19, 22.)

Species	Percentage of known deaths			
	Starvation	Predation	Accidents	n
Warthog	70.8	29.2	0.0	25
Giraffe	58.7	33.9	7.4	317
Impala	41.9	54.5	3.5	747
Greater kudu	34.9	63.2	1.9	110
Waterbuck	18.2	81.8	0.0	45
Wildebeest	0.7	98.7	0.7	322
Zebra	0.0	90.2	9.8	52

where $f(P)$ describes plant growth as a function of plant density, $c(P)$ is the per capita (per unit biomass) consumption rate of the herbivore population (a function of plant density, also called the functional response), and $g(P)$ is the net per capita growth rate of the herbivore population (the numerical response). The herbivore biomass density H can also be written in terms of numbers per unit area (N) and body mass (W), as:

$$H = WN. \tag{6}$$

Using this approach, consumption and its effects on growth can be studied either in terms of nutrients or in terms of energy. Analysis of the dynamics and equilibria in this class of models proceeds by choosing specific forms for the functions $f(P), c(P)$ and $g(P)$.

Such a system of equations was recently studied by Van de Koppel et al. (1996) who investigated the consequences of either a decreasing per capita intake rate of herbivores at higher vegetation biomass (lower foraging efficiency), or a decreasing growth per unit consumed biomass (lower utilization efficiency) at higher vegetation biomass. When a Michaelis–Menten type function is chosen twice for $f(P)$ and $c(P)$, again an optimum is found in the functional response (Fig. 17.6). The population dynamic consequences of a functional response which intercepts the X-axis twice, is that such a population is very vulnerable to going extinct, especially when primary productivity is high (Van de Koppel et al. 1996). At low primary productivity the herbivore can keep the plant biomass at its favourable level. When productivity increases, two stable equilibria are found, depending on the starting conditions, one where the herbivore population 'keeps' the vegetation biomass at a low level (with e.g. high food quality) and one where the herbivore has gone extinct and a high vegetation biomass is found. At even higher productivity the herbivore population can never 'invade' in a high biomass vegetation.

Assembly rules for communities of grazers

Let us now return to a situation at intermediate productivity where a population of small herbivores maintains the biomass at a low level. This population is expected to be very sensitive to annual productivity fluctuations; indeed, a wet year might suddenly increase the vegetation biomass to a level where the forage quality is too low to allow for positive population growth rate. This would then result in a positive feedback loop where the reduced consumption by fewer grazers leads to more vegetation biomass (and standing dead) accumulation, decreasing forage quality, and, consequently, even fewer grazers. This process can be prevented when a population of another grazing species less sensitive to decreasing quality is present. Recalling that

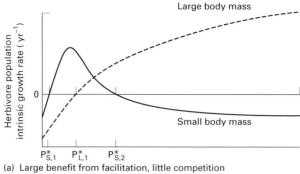
(a) Large benefit from facilitation, little competition

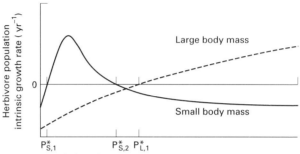
(b) Little benefit from facilitation, weak competition

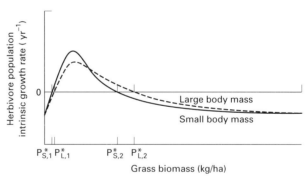
(c) Little benefit from facilitation, strong competition

FIG. 17.6. Predicted herbivore numerical response (annual net change in numbers) towards vegetation biomass, for a small and a large grazer, for a small (c), intermediate (a) and large (b) difference in body mass, leading to a differences in the importance of competition and facilitation (see text). P* values indicate plant biomass equilibria where herbivore births are balanced with mortality.

herbivores more sensitive to quality were most likely species with smaller body mass, the presence of a large species might prevent a small species from going extinct (i.e. facilitation takes place). This leads us to the formulation of 'assembly rules' for communities of grazing species. We hypothesize that when species are 'too similar' in body mass, they might not profit sufficiently from facilitation interactions; this would lead to competitive displacement of probably the larger one by the smaller one (Fig. 17.6c). When on the other hand species are 'too different' in body mass, the population of large grazers will keep the vegetation biomass at an equilibrium at which the vegetation has too low a quality for the smaller grazers. In this later scenario facilitation cannot occur (Fig. 17.6b). Some 'optimal' body mass difference can be depicted where the population of larger herbivores facilitates for the smaller ones (Fig. 17.6a). Under constant conditions, this would lead to a competitive replacement of the larger species by the smaller one. However, especially under fluctuating conditions (e.g. annual variation in rainfall) the competitive balance might vary between years, allowing coexistence. The 'optimal' body mass difference for a given habitat is likely to depend on primary productivity, vegetation biomass–quality relationships, and environmental fluctuations. Its importance lies in the relation to species packing; it might determine how many species can co-occur in a herbivore assemblage.

At this point, we conclude that functional response, competition and facilitation are important on an instantaneous, a daily and an annual time-scale, and we have tried so far to give various insights into important processes operating at each scale. Much work still needs to be done on integration across different time-scales to link instantaneous foraging processes and decisions to population dynamics. Nevertheless, from these theoretical considerations an important insight emerges about an expected regularity in the interspecific differences in body mass of grazers within an assemblage. Therefore, we explore some patterns of body-mass variation in the next part of this chapter but we first review evidence for the occurrence of facilitation and resource competition between grazers.

RESOURCE COMPETITION AND FACILITATION IN AFRICAN GRAZERS

From the literature we compiled a list of African vertebrate species heavier than or equal to 2 kg that feed to a larger or lesser extent on grass (Table 17.2). We have taken as average body mass of each species the mid-points of quoted weight ranges and averaged male and female body weights. We took data from Estes (1991, marked with *), and from Haltenorth and Diller (1979), except for giant eland from Kingdon (1982), and Sharpe's grysbok,

TABLE 17.2. List of 'grazers' (herbivores that include grasses in their diet) with a body mass larger than or equal to 2 kg, occurring in continental Africa, and the size of their geographical distribution range. See the text for sources; – = no information.

Species	Rank	Body mass	Geographical range (10^6 km)
Whyte's hare	1	2.0	13.95
Smith's hare	2	2.0	3.09
Cape hare	3	2.2	20.56
Rabbit	4	2.2	0.43
Egyptian goose	5	2.3	19.37
Swayne's dikdik	6	2.3	0.54
Uganda grass hare	7	2.5	0.84
Scrub hare	8	2.5	3.35
Ethiopian hare	9	2.5	2.17
Bush hyrax	10	2.6	8.68
Tree hyrax	11	2.9	6.21
Red-bellied dikdik	12	3.1	0.80
Spring hare	13	3.2	5.05
Salt's dikdik	14	3.3	0.36
Rock hyrax	15	3.6	15.57
Guenther's dikdik	16	4.6	0.75
Savanna cane rat	17	4.8	8.58
Kirk's dikdik*	18	5.3	1.53
Spur-winged goose	19	5.4	17.24
Cane rat	20	6.8	15.51
Barbary ape	21	7.0	0.14
Leopard tortoise	22	8.0	—
Steenbok*	23	11.1	4.31
Klipspringer*	24	11.9	5.16
Oribi*	25	14.1	7.75
Harvey's red duiker	26	14.5	0.56
Drill	27	15.0	0.16
Hamadryas baboon	28	15.0	1.02
Gelada	29	15.0	0.20
Black-fronted duiker	30	15.0	3.31
Ruwenzori red duiker	31	15.0	0.06
Dorcas gazelle	32	17.5	1.45
Mandrill	33	18.8	0.63
Speke's gazelle	34	20.0	0.22
Mountain gazelle	35	22.5	0.43
Thomson's gazelle*	36	24.9	0.34
Rhim	37	25.0	0.57
Vaal rhebok*	38	25.8	0.44
Red-fronted gazelle	39	26.3	2.23
Mountain reedbuck*	40	29.5	0.83
Springbok*	41	29.5	0.57
Baboon	42	29.5	16.13
Soemering's gazelle	43	40.0	1.12
Bohor reedbuck*	44	44.8	6.24
Bushbuck*	45	51.3	15.03
Impala*	46	52.5	4.30
Grant's gazelle*	47	55.0	1.17
Abbott's duiker	48	56.0	0.11

(Continued.)

TABLE 17.2. *Continued.*

Species	Rank	Body mass	Geographical range (10^6 km)
Dama	49	57.5	0.46
Common reedbuck*	50	58.0	5.20
Jentink's duiker	51	60.0	0.16
Yellow-backed duiker	52	62.5	4.91
Nubian ibex	53	65.0	0.14
Blesbok*	54	66.5	0.50
Bushpig	55	70.0	10.70
Puku*	56	71.5	1.02
Warthog*	57	73.5	13.98
Sitatunga*	58	76.8	5.92
Kob*	59	78.5	3.77
Barbary sheep	60	83.8	0.67
Nyala*	61	86.0	0.28
Nile lechwe	62	90.0	0.23
Lechwe*	63	91.0	0.39
Addax	64	93.8	0.24
Wild boar	65	95.0	0.58
Abyssinian ibex	66	101.3	0.04
Topi*	67	119.0	3.72
Ostrich	68	120.0	11.74
Black wildebeest*	69	132.3	0.07
Hartebeest*	70	134.0	2.98
Barbary stag	71	156.0	0.04
Blaauwbok	72	157.5	0.05
Oryx*	73	169.0	3.29
Scimitar-horned oryx	74	190.0	0.19
Mountain nyala	75	195.0	0.08
Giant forest hog*	76	205.0	1.69
Waterbuck*	77	211.0	9.33
Greater kudu*	78	213.0	5.98
Common wildebeest*	79	226.0	3.13
Sable*	80	227.5	3.73
Plains zebra*	81	235.0	5.01
Pigmy hippopotamus	82	237.0	0.29
Mountain zebra*	83	262.0	0.54
Bongo*	84	270.0	1.79
Roan*	85	270.0	7.46
African wild ass	86	275.0	0.53
Buffalo (nanus)	87	300.0	3.64
Grevy's zebra*	88	408.0	0.15
Lichtenstein's hartebeest	89	465.5	3.94
Common Eland*	90	471.3	3.87
Giant eland	91	559.0	1.03
Buffalo (caffer)*	92	631.0	6.82
White rhinoceros	93	1875.0	0.35
Hippopotamus*	94	1900.0	11.49
Forest elephant	95	2575.0	4.44
Savanna elephant	96	3550.0	11.24

*Estes (1991); see p. 465 for other sources.

Lichtenstein's hartebeest, and African wild ass from Alden, Estes, Schlitter & McBride (1995). The division of red duiker into Ruwenzori red duiker, Harvey's red duiker (both of which feed on grass) and Natal red duiker (which does not), with their weights, follows Kingdon (1982). We have not adopted Alden *et al.*'s (1995) separation of bushpig into two species (namely savanna bushpig, *Potamochoerus larvatus* and forest bush pig, *P. porcus*) but we have made a distinction between forest elephant *Loxodonta africana cyclotis* and savanna elephant *L. a. africana* and forest buffalo *Syncerus caffer nanus* and savanna buffalo (*S. c. caffer* and *S. c. brachyceros*) due to the wide difference in body masses of the forest races compared to those of the savanna. We have not been able to find body masses for the recently extinct quagga, *Equus quagga*, or for the very rare hirola (Hunter's hartebeest) *Damaliscus hunteri*. Bird weights are from Brown, Urban and Newman (1982).

Resource competition

What evidence is there for resource competition? The best test for proving the existence of competition is to find out whether there is a negative influence of one species on the fitness of individuals of another species. For large grazing herbivores evidence at the level of individual fitness is non-existent but the observations from Lake Manyara National Park (Tanzania) on the effects of removal of African elephant (through poaching) on changes in population growth rates of individual African buffalo herds (Prins 1996) are very revealing. The increased population growth rate of buffalo upon removal of elephant (Prins 1996, fig. 5.7) suggests a strong effect of resource competition on the fitness of buffalo.

Ultimately, resource competition may result in the extinction of the species which is negatively affected. A number of revealing 'experiments' have been conducted where species were, quite often simultaneously, reintroduced into their original habitat. These experiments were not intended for discovering whether resource competition took place since they were aimed at restoring National Parks to their former splendour. These reintroductions, however, can be interpreted as experiments because 'competition should become apparent *only* when an established system is challenged by some perturbation of species composition or relative density' (Putman 1996, p. 116; emphasis added), and we explore these to discover whether they can be interpreted in terms of competitive exclusion or not.

In many instances, species were lost due to overhunting and were later reintroduced. For example, in Kruger National Park area (South Africa), white rhinoceros became extinct due to overhunting at the end of the 19th century. Their reintroduction went very well (Pienaar 1963, Penzhorn 1971,

Novellie & Knight 1994). Although white rhinoceros and hippopotamus are in the same weight range (1875 and 1900 kg, respectively) the success of the reintroduction shows that there is no evidence for resource competition between these two species. A different case is the oribi, which went extinct in the Kruger N.P. without a clear cause in the 1940s. They were reintroduced in 1962 but this failed (Pienaar 1963; see also Plug 1989). Oribi (14.1 kg) are in the same weight class as the still-occurring klipspringer (11.9 kg) and steenbok (11.1 kg). Here resource competition may have played a role.

In the very small Golden Gate Highlands National Park (South Africa, 48 km^2), reintroductions (Rautenbach 1976, Avery, Rautenbach & Randall 1990, Novellie & Knight 1994) resulted in the following observations: of species triplets or pairs that have nearly the same body mass, two or one species could expand in population numbers while the intermediate or alternative species failed to establish itself. Cases in point are: (i) hartebeest (134 kg, failed) — black wildebeest (132.3 kg, succeeded); (2) warthog (73.5 kg, failed) — blesbok (66.5 kg, succeeded) — common reedbuck (58 kg, failed); and (iii) springbok (29.5 kg, succeeded) — mountain reedbuck (29.5 kg, failed).

From another small protected area, the Bontebok National Park (South Africa) (Stuart & Braack 1978, Novellie & Knight 1994), we know that common reedbuck failed to establish itself. Though Novellie and Knight (1994) consider common reedbuck a case of translocation, and not of reintroduction to this park, this is not very sure (*loc. cit.*). Its failure might have been the result of its small niche separation from blesbok (weights 58.0 and 66.5 kg, respectively). It would have been very interesting to have been able to observe whether the success of springbok, a species non-native to the area (Novellie & Knight 1994), would have resulted in its outcompeting the vaal rhebok (weight 29.5 kg and 25.5 kg, respectively) but the removal of springbok by the management precludes the outcome of this test.

The third case, from the Mountain Zebra National Park (South Africa) (Nel & Pretorius 1971, Novellie & Knight 1994), provides only information on the reintroduced species and not on the species still present. This makes it less easy to interpret. The reintroduction of eland (471 kg) was a success but there were no competitors in its weight class since there were no heavier species and the next species in line was mountain zebra (262 kg). The next size-group of introduced species comprised oryx (Gemsbok) (169 kg), which failed, hartebeest (134 kg), which succeeded, and black wildebeest (132.3 kg), which also succeeded. The interpretation might thus be that oryx suffered from resource competition. A smaller size-group comprised blesbok (66.5 kg) of which the reintroduction met with success, and the common reedbuck (58 kg), which failed. The smallest size-group consisted of mountain reedbuck (29.5 kg) and baboon (28.8 kg) for which no information was provided about their change in numbers, and the reintroduced springbok (29.5 kg) which increased in numbers.

A number of data sets from 'shrinking ecosystems' through insularization are available from Africa. In those cases where species loss has occurred, this loss and the patterns of loss can be related to resource competition. We discuss these cases in the section 'Effects of ecosystem fragmentation'.

Facilitation

Facilitation has been frequently deduced in African grazing studies. Often these studies only showed that intake rates could have been affected through the grazing pressures of another species (De Boer & Prins 1990). However, the number of studies that show that fitness was improved or that population numbers were affected through facilitation by another species is limited. Failed reintroductions in South African National Parks could, perhaps, be interpreted in the light of the effects of absence of facilitation. For example, buffalo were introduced into the Bontebok National Park to control the coarse *Pentaschistis* grass species so that burning would not be necessary any more. They did not really do this and jumped the fence too frequently. Although their condition was judged to be good, their recruitment was very low apparently due to disease (Van der Walt, Van Zyl & De Graaff 1976b). Eland also showed a low recruitment rate after reintroduction into this park, their condition was poor, they apparently did not get enough protein from their food, and dental wear was too high from feeding on coarse grasses (De Graaf, Van der Walt & Van Zyl 1976). Hartebeest also showed too low a recruitment; again their condition was poor and apparently they also did not obtain enough protein from their grass food (Van der Walt, De Graaff & Van Zyl 1976a). These three cases might indicate that the necessary facilitation of the 'proper' grass condition did not take place, perhaps due to the absence of an even larger grazer.

A good example of facilitation is provided in the Ngorongoro Crater where cattle, donkeys and small stock were removed in 1974. Since that time plains zebra, common wildebeest, common eland, hartebeest, and Grant's gazelle all declined in numbers (Runyoro, Hofer, Chausi & Moehlman 1995). The situation, however, is not simple, as buffalo sharply increased in numbers after livestock removal (Runyoro *et al.* 1995). The interpretation might be that cattle and buffalo showed competitive exclusion while the other herbivores were favoured by facilitation. It is, however, not clear why Grant's gazelle would not have increased in numbers after removal of the small stock (sheep and goats) nor plains zebra after the removal of donkeys.

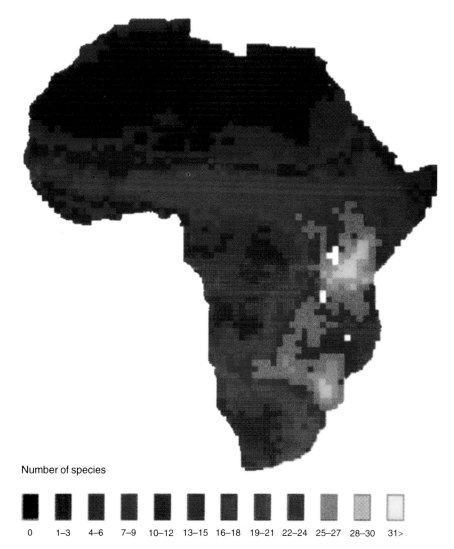

PLATE 17.1. Spatial variation in the regional species pool of large grazers larger than 2 kg for continental Africa (see Table 17.2), produced by making an overlay of all distribution maps. For each grid cell (1 × 1 degree) it was calculated for how many species that cell has included in the range of distribution. Distribution data around 1970 were used, as given by Haltenorth and Diller (1979) and Brown *et al.* (1982).

[facing page 470]

Species-richness

The best example to our knowledge for facilitation is from the extensive studies done in Uganda where, in the Queen Elizabeth National Park, it was decided to cull hippopotamus because their density was considered too high. Investigations carried out there (Lock 1967, Field 1968, Grimsdell 1969, Eltringham 1974) together show that with a *gradual* removal of hippopotamus buffalo increased, but with a near-instantaneous removal of hippopotamus grass production could not be matched by buffalo and the few remaining hippopotamus together any more, and the grass (especially *Hyparrhenia pyramidalis*) became too tall for buffalo: the result was a decline in buffalo. So before the removal of hippopotamus, these facilitated to some extent for buffalo although resource competition played a role too.

Although different interpretations of these anecdotal cases are perhaps possible, the observations underscore the deductions made on the basis of the interplay between the functional and numerical response curves of grazers with different body masses. We thus conclude that both facilitation and resource competition are important factors structuring assemblages of grazers.

RELATIONSHIP BETWEEN WEIGHT RATIO AND SPECIES-PACKING

To determine how similar competing species can be if they are to remain in an equilibrium community we assessed how closely African grazing species can be packed in a natural environment. It has been determined that species can be packed closer or wider if the environmental variations are smaller or larger (May 1973, p. 139) so we first investigated species-packing. Hutchinson (1959) suggested indirectly that character displacement among sympatric species leads to sequences in which each species is roughly twice as massive as the next. If the natural logarithm of body mass of species is plotted against the sequence rank-number of these species, with the lightest species in an assemblage assigned rank number zero, the one-but-lightest rank number one, etc., then the slope is expected to be ln 2 (= 0.693) if there is a sequence where each species is exactly twice as heavy as the next. In that case the weight ratio (May 1973, p. 166 *et seq.*) equals $e^{\ln 2}$ is 2. The ln body mass of the *i*-th species (W_i) in a ranked assemblage is thus expected to depend on its rank number (R_i) as:

$$\ln(W_i) = aR_i + b \qquad (7)$$

where the weight ratio $WR = e^a$, and e^b is the (interpolated) lowest body mass in the assemblage. Species-richness of an actual assemblage (SR) can then be described by the combination of density of species packing (higher WR)

and the body mass range over which this species-packing occurs (that is, maximum mass difference between the largest and the smallest species in the assemblage, MD).

It should be noted, however, that May and Hutchinson refer to a different kind of resource axis from that which might be appropriate here. Most studies on species-packing are concerned with discrete, particulate resources (such as seeds), where different species with different body mass have specialized on certain food size classes. The utilization of resources of a certain size class has only consequences for species which also utilize that particular size class (i.e. with an overlapping niche) but does not affect the abundance of other size classes. In fact, the resource axis in this case is composed of various resources. The graminaceous resource is a single resource dimension; when high biomass is consumed, the 'abundance' of low biomass patches in the area increases. This means we have to apply a different theoretical framework (based on competition and facilitation, as outlined before) when looking at 'rules' in weight ratios for grazers.

When ranked ln body mass is plotted against species rank number, a priori a relationship is expected, and it is important to recognize whether the observed body mass differences between subsequent species deviate from a randomness. Suppose that for a given body mass trajectory, delimited by a minimum body mass (W_{min}) and a maximum body mass (W_{max}), that every body mass has an equal probability of being found or to have evolved in the past; W follows then a uniform distribution. The expected difference between species of subsequent body mass in an assemblage will be (($W_{max} - W_{min}$)/SR), where SR is the number of species in the assemblage. The expected value of the body mass of the i-th species (W_i) with species rank number R_i can then be found as:

$$W_i = W_{min} + (R_i - 1)/((W_{max} - W_{min})/(SR - 1)) \qquad (8a)$$

or, by taking the natural logarithm of both sides,

$$\ln(W_i) = \ln(W_{min} + (R_i - 1))/((W_{max} - W_{min})/(SR - 1)). \qquad (8b)$$

According to eqn 8b, $\ln(W_i)$ is a non-linear function of R_i. Whenever a linear function (with slope WR) is found in a real assemblage (eqn 7) this implies that, on average, each species in the community is some constant proportion bigger than the next smaller species. Or, said otherwise, the body masses of species in a community are not a random sample from a uniform distribution: when $\ln(W_i)$ is plotted against species rank, a significant correlation is expected a priori, due to the ranking procedure. All regression lines shown in this chapter were significant ($P < 0.05$) but it is merely the size of the r^2 values that are of interest here. Using the body mass range observed in this study (from hare (2 kg) to savanna elephant (3550 kg)), 10 random draws of 100 different body masses resulted in linear regressions with an average r^2 of 0.75 ($\pm$ 0.05 SD). When much higher r^2 values are found, it can thus be concluded that the aforementioned 'proportional regularity' exists.

We have graphed ln body mass against species rank number for the whole of continental Africa, for whole ecosystems and for some small areas, and investigated whether the observed relationships showed strong linearity. Where this is so, the causes of variation in WR and MD, together describing species-richness, are discussed. When relating differences in species-richness between communities to WR we also have to take into account other causes of variation in species-richness. Other potentially important explanations for variation in species-richness of grazer assemblages are: area size, cultivation, habitat fragmentation (isolation) and spatial scale (whole vs. parts of ecosystems). Specific hypotheses on the way in which these factors are thought to affect species-richness are outlined in the next section.

Predictions of patterns of species-richness of grazers

1 More species can be packed into a large area than in a small one because the distribution of individual species across Africa is not equal. This is not due to competition but due to historical and geographical processes determining the distribution of individual species. Geographical isolation of populations in the past may have resulted in the evolution of different species due to habitat divergence or founder effects.

2 If ecosystems have approximately the same size, then at very high productivity species-richness of grazers is low because small species cannot cope with the relative poor quality during some periods of the year; at very low productivity, species-richness is also low because large species cannot procure sufficient food. Species-richness is thus highest at intermediate productivity where facilitation interactions are strongest.

3 An ecosystem that becomes smaller due to areas being put under cultivation (where wild grazers cannot occur any more) will show a decrease in species-packing (i.e. higher WR). This is the result of increased competition although random local extinctions may play a role too.

4 Only if an ecosystem stays unfragmented and larger than about 12 000 km^2, or if an area of that size is unaffected by human changes, stochastic processes or competition are not likely to be the cause of species loss (see Burkey 1995). In sufficiently large areas, population densities will be high enough to prevent population extinctions in bad years (or periods), and the number and area of different habitats is probably large enough to prevent

too much niche overlap. If, nevertheless, species loss occurs, it is likely that overhunting plays a role.

5 A part of an ecosystem has a lower species-packing than the whole ecosystem even potentially if all the constituting species of the ecosystem's assemblage can occur together. This decrease in species-packing is the result of competitive exclusion.

To test whether species-packing results from sympatric occurrence of species, we made a comparison of equal-sized areas in different parts of Africa.

Species-packing of grazers on a continental scale

The number of grazing species heavier than 2 kg in Africa is close to 100 (Fig. 17.7 and Table 17.2). From the relationship between the ln body mass vs. species rank number (Fig. 17.7) it is clear that the average increment in body mass is *c.* 7% between sequential species corresponding with a weight ratio of 1.066. The sequence in the figure generally fits a linear function very well. The pattern therefore deviates strongly from the pattern expected from a

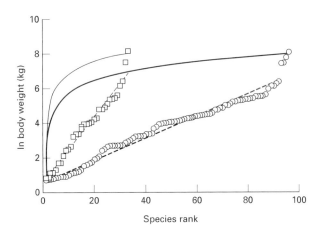

FIG. 17.7. Ranked ln body mass of grazers (W_i) plotted against rank number (R_i) for whole of Africa (circles, thick lines, 96 species, see Table 17.1 for species list) and the Serengeti ecosystem (squares, thin lines, 33 species). For each assemblage the observed patterns (circles, squares) with the linear regression lines (dashed lines) are given, as well as the patterns predicted assuming an equal probability of each body mass being found (solid lines). Observed pattern for whole of Africa: $\ln(W_i) = 0.0643 R_i + 0.6272$, $r^2 = 0.97$, predicted $\ln(W_i) = \ln(2 + (R_i - 1)/37.35)$. Observed pattern Serengeti: $\ln(W_i) = 0.169x + 0.4883$, $r^2 = 0.97$, predicted $\ln(W_i) = \ln(2 + (R_i - 1)/110.87)$ (see eqn 8b). Parameters for whole of Africa: $WR = 1.06$, $MD = 7.48$, $SR = 96$; Serengeti: $WR = 1.21$, $MD = 7.41$, $SR = 33$ (see text for abbreviations). (Data source for whole Africa see Table 17.1, for the Serengeti see legend to Fig. 17.13.)

uniform distribution (Fig. 17.7; and see section 'Relationship between weight ratio and species-packing'). Beyond 1000 kg, the range where herbivores are classified as 'mega-herbivores' (Owen-Smith 1988), there are apparently too few grazers in the general African assemblage. This apparent lack of mega-herbivores could possibly be explained if very large grazers had a wider distribution across Africa than smaller ones. This, however, is not true; there is no relationship between the size of the distribution area of grazing species and their body mass (Fig. 17.8; based on measured distribution areas as published by Haltenorth & Diller (1979)—see Table 17.2). This implies that there could be a lack of mega-grazers in African grazer assemblages—the number of 'missing species' heavier than the savanna buffalo can be estimated to be approximately six to ten judging from the general slope and using the savanna elephant as end-point in the weight range. Even though Africa experienced few extinctions of mega-herbivores at the *end* of the Pleistocene (Owen-Smith 1988), a number of species went extinct before: the proboscid *Deinotherium bozasi* during the Early Pleistocene with mainly a browse diet but living in a savanna-like habitat (Harris 1983) suggesting a diet more-or-less like that of the present-day African elephant, *Elephas recki, E. iolensis* and *E. zulu* (Beden 1983) of which *E. recki* was a typical grazer (Klein 1988) very similar to the other two *Elephas* species (Beden 1983), and *Loxodonta atlantica* with dentition resembling that of the present-day African elephant (Beden 1983). *E. recki* went extinct c. 1 million years ago, and *E. iolensis, E. zulu* and *L. atlantica* c. 400 000 years ago. Other

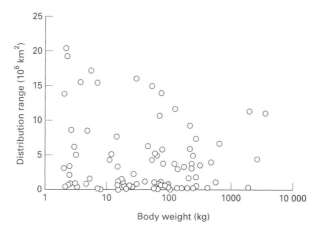

FIG. 17.8. Relationship between the distribution area size and body mass of all African grazers ≥ 2 kg. Linear regression: $y = 0.0011 \log_{10}(x) + 3.7067$, $r^2 = 0.01$, $n = 96$. (See Table 17.2 for the complete data set including species names.)

mega-grazers that went extinct during the Pleistocene were *Hippopotamus gorgops*, the giant hartebeest *Megalotragus priscus* (extinct 12 000 years ago) and the giant buffalo *Pelorovis antiquus*, which went extinct only 4000 years ago (Klein 1988, Owen-Smith 1988, p. 282). *Equus capensis* and *Hippotragus gigas*, which also went extinct during the end of the Pleistocene, were too small to fall in the category of very large grazers (Klein 1988, Klein & Cruz-Urbine 1991). Hence, the 'missing very large herbivores' in Fig. 17.7 could be the four mentioned elephant species, the extinct hippopotamus, the giant buffalo and perhaps the giant hartebeest. The predicted number of missing species (six to ten) tallies reasonably well with the number of recently extinct species (namely, seven). We have not been able to find published body mass estimations of these latter extinct species, but it would be very revealing to be able to include these estimations in the weight ratio calculations to find how these recently extinct species would have fitted in the body mass spectrum of extant African species. Research for extinct African grazers along the lines of Janis (1990) and Scott (1990) would fill the gap in our knowledge.

The pattern for the whole African assemblage is different from the results presented by Maiorana (1990). She found a polymodality in the size distribution of foliage-feeding animals in East Africa with 'a striking periodicity with successive peaks visually occurring at approximately four times the [body] mass of the previous peak' (Maiorana 1990, p. 76). However, we do not find in the present data set (Fig. 17.9). Maiorana tried to explain this periodicity by competitive displacement, postulating that species in adjacent peaks can (by assumption) competitively coexist on the basis of size alone, whereas those that fall into the same peak presumably differ in some other

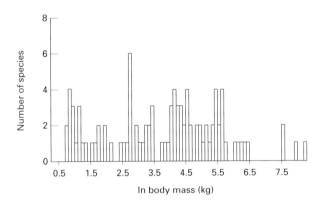

FIG. 17.9. Frequency distribution of ln-transformed body masses of African grazers ≥ 2 kg. (See Table 17.2 for the complete data set.)

dimension of niche space. The distance between peaks is the hypothetical average difference in size necessary for the coexistence of species that are otherwise ecologically indistinguishable. A precise mechanism for the four-fold mass differences affecting competition is unclear (Maiorana 1990, p. 79). A problem with this analysis, as indicated by Maiorana herself (Maiorana 1990, p. 80) is that the four-fold difference reflects competitive displacements of *sets* of species rather than single species. We have conducted a spectral analysis (Chatfield 1980) on the frequency distribution of African grazers (with an interval of 0.1 on the natural logarithm of body mass); this analysis does not show larger periodicity in body weights of these grazers (Fig. 17.10) than the observed relative increase of 6.6% of successive species (Fig. 17.7).

Regional variation in species-richness of grazers

We made a number of predictions regarding species-packing. The first one was that 'in a large area more species can be packed than in a small one because the distribution of individual species across Africa is not equal'. The distribution of species is a complex result of population interactions in the past, geographic patterns and other historical processes such as shrinking and expansion of (rain) forests due to climatic variations (see Moreau 1972, Klein 1980). These processes have resulted at present in a 'regional species-pool' for a certain area. This species-pool is the total array of species of which

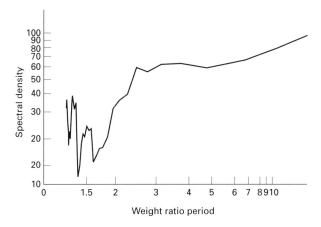

FIG. 17.10. The relative contribution (spectral density) of each weight ratio period to the total variation in body mass of all 96 African grazers ≥2 kg (see Table 17.2). A peak in spectral density indicates that some differences in ln-transformed body mass (weight ratio) occur more frequently than others.

a local assemblage can be a sub-set of. Plate 17.1 (p. 470) shows species-richness of African grazers across the whole continent. The plate represents the aggregated distribution maps of the different species approximately at 1970. For the distribution of species we used the maps published by Haltenorth and Diller (1979) and Brown *et al.* (1982). We ignored any part of the distribution of a species outside continental Africa (Madagascar and the Sinai or beyond). We assumed the maps to be of an equal area projection. The maps were digitized with a computer scanner, and overlays of the maps were made using the GIS-software IDRISI (Eastman 1992). The spatial resolution of the maps is rather coarse. Therefore, these data are only suitable for studying species distribution in relation to environmental factors on a very large scale (e.g. in relation to climatic gradients). It should be noted that the 1970 distributions are partly the result of local extinctions. This is especially true for the area north of the Sahara and in the Republic of South Africa; this implies that for these two regions the data shown in Plate 17.1 are too low. However, for the rest of Africa we believe that the distribution of the different grazing species reflects quite accurately their distribution from before the time man had a severe impact.

A number of conclusions can be drawn from the regional variation in species-richness (Plate 17.1). First, there are two 'hot spots', namely the

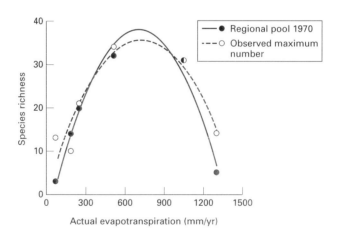

FIG. 17.11. Relationship between species-richness of whole ecosystems and average annual actual evapotranspiration. Both observed species-richness (open symbols, dashed line, $y = -9.10^{-5}x^2 + 0.1243x - 6.0446$, $r^2 = 0.97$) and species-richness predicted from the regional species pool (filled circles, solid line, $y = -6.10^{-5}x^2 + 0.0923x + 1.8305$) are given. Actual evapotranspiration was calculated from the IIASA Climate Database (Leemans & Cramer 1991) which forms an interpolation of weather station data. From left to right (data sources have been referred to in the text): Tassili n'Ajjer (Algeria), Southern Kalahari (from Mills 1990), Waza Lagone (Cameroon), Serengeti, Kruger, Rabi Field (Gabon).

Acacia savanna of northern Tanzania and Kenya in the north and the Lowveld area of southern Zimbabwe, southern Mozambique and northern South Africa. Second, the desert areas of the Sahara and the Namib show a low species-richness in grazers. Third, species-richness in the Guinea savanna of West Africa (an area that is characterized by high grass production and low quality; Fig. 17.2) is much lower than that in the hot spots. Fourth, the rainforest of West Africa and the Congo Basin has a very low species-richness.

These results appear to confirm the second prediction we made, namely that 'if ecosystems have approximately the same size, then at very high productivity, species-richness of grazers is low because small species cannot cope with the relative poor quality during some periods of the year; at very low productivity, species-richness is low too because large species cannot procure sufficient food. Species-richness is thus highest at intermediate productivity'. For the low species-richness of grazers in the rainforests we cannot yet provide an unequivocal explanation. Although primary productivity may be high, soils are generally poor, resulting in low food quality. Or, there might in general be scarcity of the graminaceous resource because of shading by trees. Finally, the continuously wet conditions might prevent fires that remove accumulated standing dead grass, again leading to decreased food quality. The picture can be simplified as shown in Fig. 17.11 in which the relationship between actual evapotranspiration (see Leemans & Cramer 1991) and species-richness on a whole ecosystem scale is shown. With increasing evapotranspiration, which is causally linked to increased primary production, there is an increase in the number of grazing species that potentially can *and* in reality do occur. However, beyond an actual evapotranspiration of *c.* 700 mm there is a decline in species-richness of grazers. Similar decreases but then in the biomass of large herbivores in high rainfall areas have been reported by Botkin, Mellilo and Wu (1981) and East (1984). This reflects, we think, the reduction in primary production of the grass resource due to increased production of woody species and, simultaneously, the high productivity and low food quality of the graminaceous resource in forest gaps.

From Plate 17.1 (p. 470) it can be postulated that the maximum number of grazing species in an ecosystem depends on the location of that ecosystem, since location determines what the regional species pool will be from which the ecosystem can draw its constituting species. Figure 17.11 shows the correspondence between the number of species in different African ecosystems and the maximum number that could have been drawn from the regional species pool. The deviations in Fig. 17.11 are from Tassili n'Ajjer in southern Algeria (Le Berre 1991, Le Houérou 1991) and from the Rabi area in Gabon (Prins & Reitsma 1989, De Bie *et al.* 1989); in both areas more species are found than expected from the distribution maps. In Tassili n'Ajjer this can be explained by the fact that the published distribution maps that we used do not reflect vagrant species that still occur occasionally in the northern Sahara but which we included in the actual species pool, while the Rabi area includes a narrow coastal savanna where species such as the cane rat or reedbuck occur that otherwise would not have been expected in this rainforest-dominated ecosystem. Yet, the striking correspondence between expected and observed species richness in Fig. 17.11 indicates that at the ecosystem level regional saturation occurs or has occurred in the recent past.

Effects of ecosystem fragmentation

Our third prediction that 'an ecosystem that becomes smaller due to areas being put under cultivation (where wild grazers cannot occur any more) will show a decrease in species-packing, that is, will have a higher weight ratio'. In other words, large areas within a native fauna have lost species in many instances when the area that became protected was only a fraction of the ecosystem. This has been compared to insularization of submerged pieces of land, and has led to the formulation of hypotheses about species loss in the light of island theory of MacArthur and Wilson (1967). Thus National Parks are considered 'landbridge islands' (Newmark 1995; see also Newmark 1996). We have studied cases were the fauna was known from before the isolation and from after this event. Even though most parks will not have lost all their species since there is no equilibrium yet (Western 1979), the patterns of species loss might be indicative of resource competition but random local extinctions may play a role too.

From a number of protected areas we have calculated what has happened when the ecosystem was put to a large extent under cultivation and only small protected areas remained. In northern Cameroon the Waza-Lagone area (Tchamba & Elkan 1995, M. Tchamba & H. Njiforti personal communication) is a case in point. About 50 years ago the weight ratio of the grazing species was 1.38 while at present it has increased to 1.58. The area is now surrounded by cultivation and high cattle numbers. Waza-Lagone lost five grazing species and the change in weight ratio shows that species-packing became less dense. Some species did *not* go extinct due to resource competition (buffalo due to rinderpest, sitatunga due to dam building) but for one reason or another: (i) roan (270 kg) survived but waterbuck (211 kg) did not; (ii) topi (119 kg) survived but hartebeest did not (134 kg); and (iii) bohor reedbuck (44.8 kg) survived but bushbuck (51.3 kg) did not. This could be indicative of extinctions due to resource competition. Likewise, in Tanzania Lake Manyara National Park about 50 years ago the weight ratio

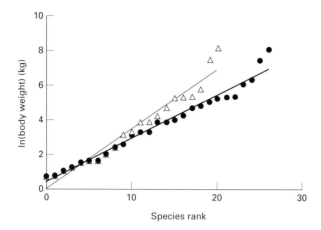

FIG. 17.12. Effect of increased isolation on species packing of grazers ≥ 2 kg in Lake Manyara National Park, Tanzania. Data of ln body mass against species body mass rank are given for the herbivore assemblage found in 1995 (solid circles, $y = 0.256x + 0.432$, $r^2 = 0.97$) and 1955 (open triangles, $y = 0.347x + 0.098$, $r^2 = 0.96$). (Data from Prins & Douglas-Hamilton 1990, personal observations and interviews.)

of grazing species was 1.30 while at present it is 1.41 with a loss of six grazing species of the original 27 (Fig. 17.12). The area is now surrounded by cultivation and the exchange of individuals between this park and other protected areas is minimal. Except for the extinction of black rhinoceros, species loss is not due to overhunting. Oryx (169 kg) and hartebeest (134 kg) extinctions may not be very good examples of extinction due to resource competition since they were in the 1930–50s only vagrants in the area that is presently a protected park (A. Seif personal communication). However, (i) the now-extinct eland (471.3 kg) may have competed with savanna buffalo (631 kg); (ii) common reedbuck (58 kg) went extinct while impala (52.5 kg) thrived and bushbuck (51.3 kg) maintained itself; (iii) the baboon (29.5 kg) population increased but mountain reedbuck (29.5 kg) and Thomson's gazelle (24.9 kg) went extinct even when there was still a very large Thomson's gazelle population just outside the park; and finally (iv) oribi (14.1 kg) went extinct while klipspringer (11.9 kg) survived.

It is interesting to note that, in contrast to what was found for recent North American mammalian extinctions (Newmark 1995), individual occurrences of species extinction in Africa are apparently not related to body size *per se*. We have not encountered local extinctions of vertebrates smaller than c. 10 kg while most species studied by Newmark (1995) were smaller than this value. It may be that in African savannas, population fluctuations of the 'small' species studied by us are much smaller than in North America; for example the proneness to extinction of lagomorphs in North America was ascribed to their wide amplitude in population fluctuations.

Whatever the exact cause of local extinctions, we think that the African examples collated here strongly point in the direction of resource competition between grazing species of approximately the same body mass as a potentially strong factor that shapes the structure of a local herbivore assemblage. The conclusion appears to be warranted since, due to the decrease in area available for the populations of wild grazers, the effects of resource competition cannot be mitigated through spatial heterogeneity any more. This conclusion is strengthened by observations on very large ecosystems that are not fragmented. Indeed, the next prediction was that 'only if the ecosystem area stays unfragmentedly larger than $c.$ 12000 km^2, or if an area of that size is unaffected by human changes, stochastic processes or competition are not likely to be the cause of species loss (see Burkey 1995) and conclusions may be drawn about overhunting as cause of species loss'. Kruger National Park ($c.$ 16000 km^2) lost two species, namely oribi and white rhinoceros. The latter was hunted to extinction at the end of the 19th century but the oribi's extinction is not understood. Before the two recent extinctions the species' weight ratio was 1.24 with 34 grazing species and afterwards it was 1.25 with 32 species. White rhinoceros was successfully reintroduced into Kruger National Park, but the reintroduction of oribi failed (Pienaar 1963).

A very large ecosystem in southern Algeria ($c.$ 80000 km^2) namely Tassili n'Ajjer (Le Berre 1991, Le Houérou 1991) showed a large species loss. The species that went extinct during this century were all from the top range of the body mass scale, namely scimitar-horned oryx (190.0 kg; rank number 1), hartebeest (134 kg; 2), ostrich (120 kg; 3), ibex (101.3 kg; 4), wild boar (95 kg; 5), dama (57.5 kg; 8) and red-fronted gazelle (26.3 kg; 9). Hence, *MD* (the maximum mass difference) decreased. Addax (93.8 kg; 6) and barbary sheep (83.8 kg; 7) still survive, but in very low numbers. This extinction pattern has, most likely, nothing to do with the 'land-bridge island phenomenon' (Burkey 1995) but reflects extinction due to overhunting. At the end of the Pleistocene many other grazing species went extinct in Tassili n'Ajjer as well: white rhinoceros, hippopotamus, the dromedary *Camelus dromedarius*, African buffalo, an extinct buffalo *Pelorovis antiquus*, the aurochs *Bos primigenius*, wild ass, plains zebra and *Equus mauritanicus*, an extinct hartebeest species *Alcelaphus antiquus*, common wildebeest, warthog and baboon (Le Berre 1991). Since we analysed modern species-richness, we did not take these species into account when constructing the niche-packing relationship for this area, since these Pleistocene extinctions are probably related to climatic change. The weight ratio of grazing species in Tassili n'Ajjer at the end of the 19th century was 1.39. This ratio is comparable to that for very species-rich areas in East Africa. At present, the weight ratio increased to 2.27. This is higher (i.e. less species) than what would be expected for such a large area. If feral donkeys (275 kg) and goats (27.5 kg) (Le Berre 1991) are taken into account, the weight ratio of the grazing species becomes 1.93 ($r = 0.978$).

The process of species loss cannot always be deduced from the weight ratio statistic. When all large herbivores go extinct and all small ones survive, the species-packing of the small herbivores may be the same as that of the original assemblage. A case in point here is Golden Gate Highlands National Park (South Africa, 50 km^2). All species heavier than baboon went extinct due to hunting and, again, MD decreased substantially as in Tassili n'Ajjer. The original weight ratio was 1.39 ($r = 0.983$ with 22 species) while in the 1940s, when the area became a park, the weight ratio was 1.44 ($r = 0.950$ with only nine small species). Although the number of species declines from 22 to nine only, the weight ratio of the species did not change and, apparently, niche separation of the *small* ones remains unaltered. An extreme case is West Coast National Park (187 km^2, South Africa) where eight of the 15 herbivores went extinct, all of them being heavier than baboon. The original weight ratio was 1.71 ($r = 0.990$ with 15 species) while in the 1940s the weight ratio had even decreased to 1.54 ($r = 0.977$ with seven species). If steenbok is also extinct (which is not sure) then the weight ratio was 1.67 during the 1940s ($r = 0.960$) and thus virtually unaltered as compared to the original.

The final prediction we made was that 'a part of an ecosystem has a lower species packing than the whole ecosystem even if all the constituting species of the ecosystem's assemblage potentially can occur together'. A good case is provided by an analysis of the Serengeti Ecosystem of Tanzania and Kenya. The whole ecosystem comprises *c.* 27 000 km^2 and the weight ratio of grazing species increases when smaller parts of the area are analysed (Fig. 17.13). The same applies to Kruger National Park: for two small areas within this park the weight ratio is 1.63 (Sabie River, $MD = 7.41$, species richness $SR = 14$) and 1.70 (Nwaswithsaka River, $MD = 7.41$, $SR = 13$) (species lists from Maclean 1985, Bailey 1993) but for the whole system of *c.* 16 000 km^2 $WR = 1.24$ with $MD = 7.48$ and $SR = 34$ (species lists, including white rhinoceros and oribi, from Pienaar 1963, 1964, Maclean 1985).

CONCLUSIONS

Species-richness of assemblages of African grazers in different ecosystems at present can be explained by a small number of parameters (Table 17.2). Observed patterns of interspecific body mass variation on all spatial scales deviates strongly from the pattern expected when each body mass has an

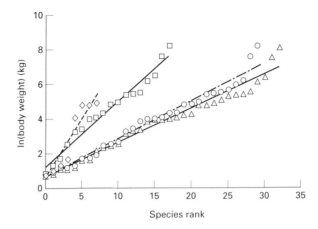

FIG. 17.13. Comparison of whole-ecosystem species packing of grazers vs. ecosystem parts of the Serengeti ecosystem, Tanzania and Kenya. Only grazers ≥ 2 kg were incorporated. Data of ln body mass against species body mass rank are given for the whole Serengeti–Mara ecosystem (triangles, $y = 0.196x + 0.684$, $r^2 = 0.97$, $WR = 1.21$, $MD = 7.41$, $SR = 33$), and for increasingly smaller parts of the Serengeti–Mara ecosystem, namely Masai Mara (circles, $y = 0.223x + 0.533$, $r^2 = 0.97$, $WR = 1.46$, $MD = 7.41$, $SR = 30$), Ngorongoro (squares, $y = 0.378x + 1.197$, $r^2 = 0.9695$, $WR = 1.46$, $MD = 7.35$, $SR = 18$), and Soit ol Modison kopjes (diamonds, $y = 0.725x + 0.362$, $r^2 = 0.89$, $WR = 2.06$, $MD = 4.13$, $SR = 8$). See text for variable abbreviations. (Data sources for whole Serengeti: Houston 1970, Williams 1967 personal observations; Masai Mara: Broten & Snip 1995, Williams 1967; Ngorongoro: Runyoro et al. 1995, Williams 1967, personal observations; and Soit ol Modison kopjes: Hendrichs & Hendrichs 1971.)

equal probability of occurrence. The weight ratio concept is applicable at all scales and in different areas; that is, the average body mass ratio between subsequent grazing species (if ordered from light to heavy) is always constant within each assemblage. This parameter reflects niche differentiation between grazers, and determines whether competitive displacement or facilitation occurs. At a large scale, WR can be low because spatial and temporal heterogeneity allows for a higher number of species coexisting in an area than at the smaller scale where facilitation and competition play a direct role. The weight ratio shows an optimum with increased primary production. At low primary productivity WR increases with primary productivity because of facilitation, which is both possible *and* necessary to maintain species with a low biomass in the assemblage. In areas with a very high primary productivity, vegetation quality decreases, thus lessening the potential for facilitation between grazing species, and leading to a decrease in WR. An indication of the importance of this parameter is the very good correlation between WR and species-richness (Fig. 17.14), which only to a small extent depends on the maximum difference between the smallest and the heaviest species.

Species-richness

TABLE 17.2. Conclusions on the effect of area size, isolation and primary productivity on species richness of grazer assemblages. These factors affect species richness through their effect on weight ratio and maximum mass differences of grazer assemblages (see text).

Changing factor	Effect on weight ratio (WR)	Effect on maximum mass differences (MD)	Consequence for species richness
Decreasing area size	Leads to more extinctions; lower weight ratio	No effect	Leads to lower species richness
Increasing isolation of an area	Leads of a lower immigration rate; lower weight ratio	Larger species first to go extinct; lower MD	Leads to lower species richness
Increasing grass productivity	Optimum WR at intermediate productivity; little facilitation at low or high productivity	Larger species require high intake rate; higher MD at higher productivity	Highest species richness at intermediate productivity

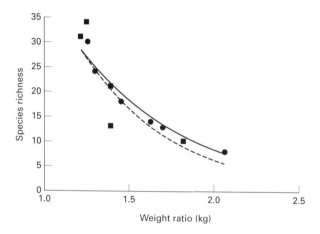

FIG. 17.14. The relationship between species-richness and the weight ratio of grazers ≥ 2 kg in parts (circles, $y = 182e^{-1.544x}$, $r^2 = 0.97$) and whole (squares, $y = 288e^{-1.190x}$, $r^2 = 0.76$) African ecosystems.

Fragmentation of protected areas also affects WR. A large area suffers less from random extinctions than a small one, and in small areas this loss of species from the assemblage results in an increased WR. Increased isolation has the same effect, since random extinctions are not compensated for by colonizations any more. An important danger of an increasing WR due to isolation may be that facilitation interactions are relaxed, leading to further loss of species.

The second most important parameter explaining species-richness is the maximum mass difference, MD, between the smallest and the largest grazer in the system. In areas with high primary production this parameter can have a higher value than in less productive areas since large grazers need more vegetation biomass to cover their intake requirements. MD is sensitive to isolation of protected areas and, especially, overhunting and cultivation: the large species are more prone to extinction. Fragmentation has no effect on MD per se.

These two parameters together determine species-richness at the present. Together with the effects of processes in the past, species-richness of grazers across Africa shows a striking pattern with highest values in the savannas of eastern and southern Africa. Resource competition and facilitation appear to form the basis for understanding this pattern, and thus provide a causal explanation.

ACKNOWLEDGEMENTS

We thank Rik Leemans for providing the data on actual evapotranspiration, and Sip van Wieren and Ignas Heitkönig for helpful discussions, Martin Tchamba and Hanson Njiforti for missing data from Cameroon, and Mzee Adam Seif for historical information on northern Tanzania. Three referees and Rory Putman helped us to improve the text. We are grateful for their input.

REFERENCES

Alden, P.C., Estes, R.D., Schlitter, D. & McBride, B. (1995). *National Audubon Society Field Guide to African Wildlife*. Knopf, New York.

Avery, D.M., Rautenbach, I.L. & Randall, R.M. (1990). An annotated check list of the land mammal fauna of the West Coast National Park. *Koedoe*, 33, 1–18.

Bailey, T.N. (1993). *The African Leopard: Ecology and Behavior of a Solitary Felid*. Colombia University Press, New York.

Beden, M. (1983). Family Elephantidae In *Koobi Fora Research Project*, Vol. 2. *The Fossil Ungulates: Proboscidea, Perissodactyla, and Suidae* (Ed. by J.M. Harris), pp. 40–129, Clarendon Press, Oxford.

Begon, M., Harper, J.L. & Townsend, C.R. (1990). *Ecology. Individuals, Populations and Communities*. Blackwell Scientific Publications, Oxford.

Botkin, D.B., Mellilo, J.M. & Wu, L.S.Y. (1981). How ecosystems processes are linked to large mammal population dynamics. In *Dynamics of Large Mammal Populations* (Ed. by C.W. Fowler & T.D. Smith), pp. 373–387, Wiley, New York.

Breman, H. & De Wit, C.T. (1983). Rangeland productivity and exploitation in the sahel. *Science*, 221, 1341–1347.

Broten, M.D. & Snip, M. (1995). Population trends of ungulates in and around Kenya's Masai Mara Reserve. *In Serengeti II: Dynamics, Management and Conservation of an*

Ecosystem. (Ed. by A.R.E. Sinctair & P. Arcese), pp. 169–193. Chicago University Press, Chicago.

Brown, L.H., Urban, E.K. & Newman, K. (1982). *The Birds of Africa,* Vol. I. Academic Press, London.

Burkey, T.V. (1995). Extinction rates in archipelagoes: implications for populations in fragmented habitats. *Conservation Biology,* **9**, 527–541.

Case, T.J. (1981). Niche separation and resource scaling. *American Naturalist,* **118**, 554–560.

Chatfield, C. (1980). *The Analysis of Time Series.* Chapman & Hall, London.

Crampton, E.W. & Harris, J.E. (1969). *Applied Animal Nutrition: The Use of Feedstuffs in the Formulation of Livestock Rations.* W.H. Freeman, San Francisco.

DeAngelis, D.L. (1992). *Dynamics of Nutrient Cycling and Food Webs.* Chapman & Hall, London.

De Bie, S. (1991). *Wildlife Resources of the West African Savanna.* Wageningen Agricultural University Papers, **91–2**. Wageningen.

De Bie, S., Geerling, C., Prins, H.H.T., Reitsma, J.M. & Zieren, M. (1989). *Ecological Baseline Study for Oilfield Development at Rabi, Gabon.* Report. Wageningen Agricultural University, Wageningen.

De Boer, W.F. & Prins, H.H.T. (1990). Large herbivores that strive mightily but eat and drink as friends. *Oecologia,* **82**, 264–274.

De Calesta, D.S., Nagy, J.G. & Bailey, J.A. (1974). Some effects of starvation on mule deer rumen bacteria. *Journal of Wildlife Management,* **38**, 815–822.

De Graaff, G., Van der Walt, P.T. & Van Zyl, L.J. (1976). Lewensloop van 'n elanbevolking *Taurotragus oryx* in die Bontebok Nasionale Park. *Koedoe,* **19**, 185–188.

De Kroon, H. & Olff, H. (1995). On the use of the guild concept in plant ecology. *Folia Geobotanica et Phytotaxonomica,* **30**, 519–528.

Delany, M.J. & Happold, D.C.D. (1979). *Ecology of African Mammals.* Longman, London.

Dorst, J. & Dandelot, P. (1972). *A Field Guide to the Larger Mammals of Africa.* Collins, London.

Du Toit, J.T. & Owen-Smith, N. (1989). Body size, population metabolism and habitat specialization among large African herbivores. *American Naturalist,* **133**, 736–740.

East, R. (1984). Rainfall, nutrient status and biomass of large African savannah mammals. *African Journal of Ecology,* **22**, 245–270.

Eastman, J.R. (1992). *IDRISI, Version 4.0.* Clark University, Worcester, MA.

Eltringham, S.K (1974). Changes in the large mammal community of Mweya Peninsula, Rwenzori National Park, Uganda, following removal of hippopotamus. *Journal of Applied Ecology,* **11**, 855–865.

Estes, R.D. (1991). *The Behavior Guide to African Mammals, Including Hoofed Mammals, Carnivores, Primates.* University of California Press, Berkeley.

Field, C.R. (1968). *The food habits of some wild ungulates in Uganda.* PhD thesis, Cambridge University, Cambridge.

Fryxell, J.M. (1991). Forage quality and aggregation by large herbivores. *American Naturalist,* **138**, 478–498.

Garland, T. (1983). Scaling the ecological costs of transport to body mass in terrestrial mammals. *American Naturalist,* **121**, 571–587.

Grimsdell, J.R.R. (1969). *The ecology of the buffalo* Syncerus caffer *in western Uganda.* PhD thesis, Cambridge University, Cambridge.

Gross, J.E., Shipley, L.A., Hobbs, N.T., Spalinger, D.E. & Wunder, B.A. (1993). Functional response of herbivores in food-concentrated patches: tests of a mechanistic model. *Ecology,* **74**, 778–791.

Haltenorth, T. & Diller, H. (1979). *Elsevier Gids van de Afrikaanse Zoogdieren (revised and enlarged by C. Smeenk).* Elsevier, Amsterdam.

Harris, J.M. (1983). Family Deinotheriidae. In *Koobi Fora Research Project*, Vol. 2. *The Fossil Ungulates: Proboscidea, Perissodactyla, and Suidae* (Ed. by J.M. Harris), pp. 22–39. Clarendon Press, Oxford.

Hendrichs, H. & Hendrichs, U. (1971). *Dikdik und Elefanten: Ökologie und Soziologie zweier Afrikanischer Huftiere.* R. Piper, München.

Herd, D.B. & Sprott, L.R. (1986). *Body Condition, Nutrition, and Reproduction of Beef Cows.* Texas Agricultural Extension Service B-1526. Texas A & M University Coll. Station, Texas.

Hirst, S.M. (1969). Populations in a Transvaal lowland reserve. *Zoologica Africana*, **4**, 199–230.

Hogeweg, P. & Richter, A.F. (1982). INSTAR, a discrete event model for simulating population dynamics. *Hydrobiologia*, **95**, 275–285.

Houston, D.C. (1979). The adaptations of scavengers. In *Serengeti: Dynamics of an Ecosystem* (Ed. by A.R.E. Sinctair & M. Norton Griffiths), pp. 263–286. Chicago University Press, Chicago.

Hutchinson, G.E. (1959). Homage to Santa Rosalia or why are there so may kinds of animals? *American Naturalist*, **93**, 275–285.

Illius, A.W. & Gordon, I.J. (1987). The allometry of food intake in grazing ruminants. *Journal of Animal Ecology*, **56**, 989–999.

Janis, C.M. (1990). Correlation of cranial and dental variables with body size in ungulates and macropoids. In *Body Size in Mammalian Paleobiology: Estimation and Biological Implications* (Ed. by J. Damuth & B.J. McFadden), pp. 255–299, Cambridge University Press, Cambridge.

Kingdon, J. (1982). *East African Mammals: An Atlas of Speciation*, Vol. III Part C (*Bovids*). Academic Press, London.

Klein, R.G. (1980). Environmental and ecological implications of large mammals from Upper Pleistocene and Holocene sites in Southern Africa. *Annals of the South African Museum*, **81**, 223–283.

Klein, R.G. (1988). The archaeological significance of animal bones from Acheulean sites in southern Africa. *African Archeological Review*, **6**, 3–25.

Klein, R.G. & Cruz-Urbine, K. (1991). The bovids from Elandsfontein, Southern Africa, and their implications for the age, palaeoenvironment, and origin of the site. *African Archives Review*, **9**, 21–97.

Le Berre, M. (1991). The role of the Tassili n'Ajjer (Algeria) in the conservation of the great mammalian fauna in the Central Sahara. In *Mammals in the Palaearctic Desert: Status and Trends in the Sahara–Gobian Region* (Ed. by J.A. McNeely & V.M. Neronov), pp. 181–192. The Russian Academy of Sciences, the Russian Committee for the UNESCO Programme on Man and the Biosphere (MAB), Moscow.

Leemans, R. & Cramer, W.P. (1991). *The IIASA Database for Mean Monthly Values of Temperature, Precipitation, and Cloudiness on a Global Terrestrial Grid. Report RR-91-18*. International Institute for Applied Systems Analysis, Luxemburg.

Le Houérou, H.N. (1991). Outline of a biological history of the Sahara. In *Mammals in the Palaearctic Desert: Status and Trends in the Sahara–Gobian Region* (Ed. by J.A. McNeely & V.M. Neronov), pp. 147–174. The Russian Academy of Sciences, the Russian Committee for the UNESCO Programme on Man and the Biosphere (MAB), Moscow.

Lock, J.M. (1967). *Vegetation in relation to grazing and soils in Queen Elizabeth National Park, Uganda*. PhD thesis, Cambridge University, Cambridge.

MacArthur, R.H. & Levins, R. (1967). The limiting similarity, convergence and divergence of coexisting species. *American Naturalist*, **101**, 377–385.

MacArthur, R.H. & Wilson, E.O. (1969). *The Theory of Island Biogeography*. Princeton University Press, Princeton.

Maclean, G.L. (1985). *Robert's Birds of Southern Africa*. John Voelcker Bird Book Fund, Cape Town.

Maiorana, V.C. (1990). Evolutionary strategies and body size in a guild of mammals. In *Body Size in Mammalian Paleobiology: Estimation and Biological Implications* (Ed. by J. Damuth & B.J. McFadden), pp. 69-102. Cambridge University Press, Cambridge.

Mattson, W.J. (1980). Herbivory in relation to plant nitrogen content. *Annual Review of Ecology and Systematics*, 11, 119-161.

May, R.M. (1973). *Stability and Complexity in Model Ecosystems*. Princeton University Press, Princeton.

Mills, M.G.L. (1990). *Kalahari Hyenas: The Comparative Behavioral Ecology of Two Species*. Chapman & Hall, London.

Mitchell, B. & Staines, B.W. (1976). An example of natural winter mortality in Scottish red deer. *Deer*, 3, 549-552.

Moen, A.N. (1973). *Wildlife Ecology: An Analytical Approach*. Freeman, San Francisco.

Moreau, R.E. (1972). *The Palaearctic-African Bird Migration System*. Academic Press, London.

Nel, J.A.J. & Pretorius, J.J.L. (1971). A note on the smaller mammals of the Mountain Zebra National Park. *Koedoe*, 14, 99-110.

Newmark, W.D. (1995). Extinction of mammal populations in western North American National Parks. *Conservation Biology*, 9, 512-526.

Newmark, W.D. (1996). Insularization of Tanzanian parks and the local extinction of large mammals. *Conservation Biology*, 10, 1549-1556.

Novellie, P.A. & Knight, M. (1994). Repatriation and translocation of ungulates into South African national parks: an assessment of past attempts. *Koedoe*, 37, 115-119.

Owen-Smith, N. (1979). *Factors Influencing the Transfer of Plant Products into Large Herbivore Populations*. Cyclostyled, University of Witwatersrand, Johannesburg.

Owen-Smith, R.N. (1988). *Megaherbivores: The Influence of Very Large Body Size on Ecology*. Cambridge University Press, Cambridge.

Penzhorn, B.L. (1971). A summary of the re-introduction of ungulates into South African National Parks (to 31 December 1970). *Koedoe*, 14, 145-159.

Pienaar, U. de V. (1963). The large mammals of the Kruger National Park — their distribution and present-day status. *Koedoe*, 6, 1-263.

Pienaar, U. de V. (1964). The small mammals of the Kruger National Park — a systematic list and zoogeography. *Koedoe*, 7, 1-25.

Platt, W.J. & Weiss, I.M. (1977). Resource partitioning and competition within a guild of fugitive prairie plants. *American Naturalist*, 111, 479-513.

Plug, I. (1989). Notes on distribution and relative abundances of some animal species, and on climate in the Kruger National Park during prehistoric times. *Koedoe*, 32, 101-119.

Pratchett, D., Capper, B.G., Light, D.E., Miller, M.D., Rutherford, A.S., Rennie, T.W. et al. (1977). Factors limiting live weight gain of beef cattle on rangeland in Botswana. *Journal of Rangeland Management*, 30, 442-445.

Prins, H.H.T. (1996). *Ecology and Behaviour of the African Buffalo: Social Inequality and Decision Making*. Chapman & Hall, London.

Prins, H.H.T. & Douglas-Hamilton, I. (1990). Stability in a multi-species assemblage of large herbivores in East Africa. *Oecologia*, 83, 392-400.

Prins, H.H.T. & Reitsma, J.M. (1989). Mammalian biomass in an African equatorial rain forest. *Journal of Animal Ecology*, 58, 851-861.

Putman, R.J. (1996). *Competition and Resource Partitioning in Temperate Ungulate Assemblies*. Chapman & Hall, London.

Rautenbach, I.L. (1976). A survey of the mammals occurring in the Golden Gate Highlands National Park. *Koedoe*, 19, 133-144.

Root, R. (1967). The niche exploitation pattern of the blue-grey gnatcatcher. *Ecological Monographs*, 37, 317-350.

Runyoro, V.A., Hofer, H., Chausi, E.B. & Moehlman, P.D. (1995). Long-term trends in the herbivore populations of the Ngorongoro Crater, Tanzania. In *Serengeti II: Dynamics, Management, and Conservation of an Ecosystem* (Ed. by A.R.E. Sinclair & P. Arcese), pp. 146–168. Chicago University Press, Chicago.

Schoener, T.W. (1974). Resource partitioning in ecological communities. *Science*, **185**, 27–39.

Scott, K.M. (1990). Postcranial dimensions of ungulates as predictors of body mass. In *Body Size in Mammalian Paleobiology: Estimation and Biological Implications* (Ed. by J. Damuth & B.J. McFadden), pp. 300–335. Cambridge University Press, Cambridge.

Severinghaus, C.W. (1979). Weights of white-tailed deer in relation to range condition in New York. *New York Fish and Game Journal*, **26**, 162–187.

Short, J. (1985). The functional response of kangaroos, sheep and rabbits in an arid grazing system. *Journal of Applied Ecology*, **22**, 435–447.

Sibbald, A.R., Maxwell, T.J. & Eadie, J. (1979). A conceptual approach to the modelling of herbage intake by hill sheep. *Agricultural Systems*, **4**, 119–134.

Simberloff, D. & Dayan, T. (1991). The guild concept and the structure of ecological communities. *Annual Review of Ecology and Systematics*, **22**, 115–143.

Sinclair, A.R.E. (1977). *The African Buffalo: A Study in Resource Limitations of Populations.* Chicago University Press, Chicago.

Spalinger, D.E. & Hobs, N.T. (1992). Mechanisms of foraging in mammalian herbivores: new models of functional response. *American Naturalist*, **140**, 325–348.

Stobbs, T.H. (1973a). The effects of plant structure on the intake of tropical pastures. I. Variation in the bite size of grazing cattle. *Australian Journal of Agricultural Research*, **24**, 821–824.

Stobbs, T.H. (1973b). The effects of plant structure on the intake of tropical pastures. II. Differences in sward structure, nutritive value and bite size of animals grazing *Setaria anceps* and *Chloris gayana* at various stages of growth. *Australian Journal of Agricultural Research*, **24**, 821–829.

Stuart, C.T. & Braack, H.H. (1978). Preliminary notes on the mammals of the Bontebok National Park. *Koedoe*, **21**, 111–117.

Taber, R.D. & Dasman, R.F. (1957). The dynamics of three natural populations of the deer *Odocoileus hemionus columbianus*. *Ecology*, **38**, 233–246.

Tchamba, M.N. & Elkan, P. (1995). Status and trends of some large mammals and ostriches in Waza National Park, Cameroon. *African Journal of Ecology*, **33**, 366–376.

Van de Koppel, H., Huisman, J., Van de Wal, R. & Olff, H. (1996). Primary productivity and the collapse of simple plant–herbivore systems. *Ecology*, **77**, 736–745.

Van der Walt, P.T., De Graaff, G. & Van Zyl, L.J. (1976a). Lewensloop van 'n rooihartbeesbevolking *Alcelaphus buselaphus caama* in die Bontebok Nasionale Park. *Koedoe*, **19**, 181–184.

Van der Walt, P.T., Van Zyl, L.J. & De Graaff, G.(1976b). Lewensloop van 'n Kaapse buffelbevolking *Syncerus caffer* in die Bontebok Nasionale Park. *Koedoe*, **19**, 189–191.

Van Soest, P.J. (1982). *Nutritional Ecology of the Ruminant: Ruminant Metabolism, Nutritional Strategies, the Cellulolytic Fermentation and the Chemistry of Forages and Plant Fibres.* O. & B. Books, Corvallis.

Van Wijngaarden, W. (1985). *Elephants–Trees–Grass–Grazers: Relationships Between Climate, Soils, Vegetation and Large Herbivores in a Semi-arid Savanna Ecosystem (Tsavo, Kenya).* ITC Publication, **4**. Enschede.

Western, D. (1979). Size, life history and ecology and mammals. *African Journal of Ecology*, **17**, 185–204.

Williams, J.G. (1967). *A Field Guide to the National Parks of East Africa.* Collins, London.

Wilson, D.S. (1975). The adequacy of body size as a niche difference. *American Naturalist*, **109**, 769–784.

18. NICHE SPECIFICITY AMONG TROPICAL TREES: A QUESTION OF SCALES

P. S. ASHTON

The Arnold Arboretum of Harvard University, 22 Divinity Avenue, Cambridge, Massachusetts 02138, USA

SUMMARY

1 It is obvious that stochastic processes and events play an important part in determining the structure, including floristic composition, of plant communities. But to what extent and on what scales, in space and time, are patterns observed in nature thereby determined; and to what extent are they the consequence of deterministic forces, specifically those which influence gene frequencies within species populations through selection rather than drift?

2 Of particular relevance is the renewed interest in the striking structural similarities that exist in forests on different continents that share a common habitat. With appropriate sampling design, such global comparisons of forest dynamics may yield a general theory which will be of importance to foresters and conservation biologists as much as to theoreticians. Do the relative roles of stochastic and deterministic processes also change with habitat?

INTRODUCTION

No-one denies that species are ecologically differentiated at certain broad scales, notably at trophic levels, and also in a more or less continuous fashion in relation to dynamic stages of communities, such as the three phases of the forest cycle, originally recognized by Watt (1924) in English beechwoods. The real issue is the extent to which there is ecological differentiation among species, sufficient to explain coexistence through equilibrating competition within the guilds so defined.

Species-rich communities such as rainforests share two relevant attributes which challenge explanation. First, they are rich in species occurring in exceptionally low population densities ('rare species') (Ashton 1984). Indeed, variation in species-richness between rainforests is mostly attributable to variation in the number of rare species. Does this variation correlate positively or negatively with the strength of selection? Does it reflect variation in the frequency of catastrophic extinction-causing events; or in selective fine tuning, the opportunities for which may be expected to increase the more equable the environment? Or are both factors influential, in which case does the relative importance of each vary? This last question is likely to be the most realistic.

A second shared phenomenon of species-rich communities is the existence of series of closely related taxa, coexisting within the community and generally sharing the same structural and functional guild (Fedorov 1966). These series are usually identified because taxonomists consider them congeneric, but they may often occur within finer taxonomic categories. The number and size of these series is correlated with overall species diversity (Table 18.1), broadly even irrespective of climate (Fig. 18.1) and perforce therefore with the abundance of rare species. Here again, such series could reflect lowered selection, in which case selection at the guild level would be likely to favour generalists and the species would exhibit high ecological complementarity. Alternatively series members, being genetically, structurally and chemically the most similar taxa in the community, may experience the most stringent diversifying selection (Gause 1932).

During the 1980s there was a renewed interest in the role of stochastic processes in the maintenance of mixed-species assemblages. Chesson and Warner (1981) showed that species mixtures can be sustained in the absence of equilibrating selection provided that the species share similar levels of fecundity, that fecundity fluctuates between seasons within populations yet

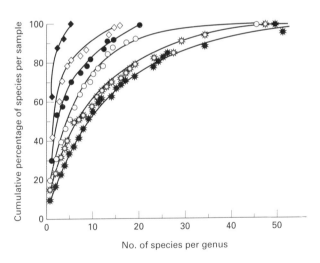

FIG. 18.1. Relative representation of congeneric species series in some tropical forests. ♦, Mudumalai; ◇, Barro Colorado Island; ✳, Ducke; ●, Sinharaja; ○, Pasoh; ✵, Lambir. Sample areas cited in Table 18.2; list for the whole reserve cited at Ducke, otherwise lists from the demographic plots.

Niche specificity among tropical trees 493

TABLE 18.1. Genera represented by at least four species in three groups of 4–5, 0.6-ha permanent plots in Sarawak.

Bako National Park (4 plots). Mean mineral soil P: 31 ppm; Mg: 100 ppm; K: 700 ppm		No. of species	Lambir National Park (4 plots) P: 81 ppm; Mg: 715 ppm; K: 2500 ppm		No. of species	Bukit Mersing (5 plots) P: 1325 ppm; Mg: 4692 ppm; K: 5292 ppm		No. of species
Eugenia*	Myrtaceae	21	Shorea*	Dipterocarpaceae	28	Eugenia*	Myrtaceae	13
Shorea*	Dipterocarpaceae	17	Diospyros*	Ebenaceae	15	Shorea*	Dipterocarpaceae	9
Garcinia*	Clusiaceae	13	Eugenia*	Myrtaceae	13	Aglaia	Meliaceae	8
Diospyros*	Ebenaceae	12	Calophyllum*	Clusiaceae	8	Diospyros*	Ebenaceae	6
Palaquium	Sapotaceae	9	Knema	Myristicaceae	8	Knema	Myristicaceae	6
Vatica	Dipterocarpaceae	8	Artocarpus	Moraceae	7	Litsea	Lauraceae	5
Knema	Myristicaceae	7	Baccaurea	Euphorbiaceae	7	Xanthophyllum	Polygalaceae	4
Litsea	Lauraceae	7	Canarium	Burseraceae	7			
Xanthophyllum	Polygalaceae	7	Santiria	Burseraceae	7			
Baccaurea	Euphorbiaceae	6	Dacryodes	Burseraceae	6			
Hopea*	Dipterocarpaceae	6	Memecylon*	Melastomataceae	6			
Horsfieldia	Myristicaceae	6	Xanthophyllum	Polygalaceae	6			
Memecylon*	Melastomataceae	6	Durio	Bombacaceae	5			
Elaeocarpus	Elaeocarpaceae	5	Garcinia*	Clusiaceae	5			
Santiria	Burseraceae	5	Gonystylus	Gonystylaceae	5			
Dacryodes	Burseraceae	4	Hopea*	Dipterocarpaceae	5			
Gonystylus	Gonystylaceae	4	Horsfieldia	Myristicaceae	5			
Lithocarpus	Fagaceae	4	Pentace	Tiliaceae	5			
Gluta	Anacardiaceae	4	Vatica	Dipterocarpaceae	5			
Mesua*	Clusiaceae	4	Aglaia	Meliaceae	4			
			Dipterocarpus	Dipterocarpaceae	4			
			Hydnocarpus	Flacourtiaceae	4			
			Litsea	Lauraceae	4			
			Mesua*	Clusiaceae	4			
			Myristica	Myristicaceae	4			
			Palaquium	Sapotaceae	4			
Total number of large genera		20	Total number of large genera		26	Total number of large genera		7
Total number of species in plots		326	Total number of species in plot		427	Total number of species in plots		230

* Genera in which adventive embryony is known or inferred to occur.

to some extent independently among the species, and that longevity of individuals exceeds the intervals between reproductive events. Hubbell and Foster (1983) and Hubbell (1997) argued that species-rich communities can be sustained by the stochastic processes of immigration, extinction and speciation from local to regional levels in continuous habitats and communities following predictions of the Theory of Island Biogeography (MacArthur & Wilson 1967). Dispersal distances will influence rates of immigration and add stochasticity to the species composition of local communities.

In Hubbell's words, the design of rigorous field tests of stochastic models is 'non-trivial'. Evidence for stochastic processes is, by its nature, negative and difficult to argue from field study. Likewise, it is unlikely that we will ever obtain direct evidence of species-wise competition among populations of tropical trees. Our demonstration of a non-random flowering sequence among six members of *Shorea* section *Mutica*, Dipterocarpaceae (Ashton, Givnish & Appanah 1988) seems to represent the only evidence to date for interspecific competition, albeit inferential and through the mediation of a mobile link, the shared pollinator. Competition generally simultaneously involves several non-linear interactions.

The intent of this chapter is to discuss several tests of deterministic forces which it has been suggested may mediate community structure, in light of the very preliminary data now available, and to suggest priorities for future research. Above all, I make a plea for research which integrates leaf-level physiology with whole plant life-history attributes, and both with their demography; and, finally, for comparisons among the members of those coexisting congeneric series that are so characteristic of species-rich tropical evergreen forests. I emphasize the need for continuity, and long-term monitoring of populations. A current attempt by the Center for Tropical Forest Science (CTFS) of the Smithsonian's Tropical Research Institute (STRI) to negotiate a worldwide partnership and network of research sites, at each of which is a plot large enough to permit demographic analysis of the majority of tree species, and in which common sampling methods are used, will provide future researchers with the opportunity to address questions hitherto out of reach. The degree to which species are ecological specialists and their competitive interactions can be predicted has relevance for silviculture, while the extent to which the theory of island biogeography can predict changes in extinction rates, consequent on reduction in area and increase in isolation of individual forests, has obvious implications for conservation planning and management.

AREA VS. HABITAT AS DETERMINANT OF RICHNESS

Theories that are non-equilibrium, in the context of species composition and relative abundances, predict that community species-richness will principally be determined by the balance achieved between rates of extinction and rates of immigration. Rates of extinction will mainly be influenced by the area of the community, rates of immigration by the proximity of a source. Speciation rates are unlikely to increase species-richness substantially, because stand density does not change with the number of species; the island biogeographic model is a 'zero-sum game' (Hubbell 1997). Addition of new species must therefore decrease the density of those already there, thereby increasing their extinction probabilities. It is well known from the quaternary studies of Davis (1976) and others that dispersal rates are slow in forests worldwide, and appear to be particularly so in climax tropical forest species. An equilibrium between extinction and immigration rates may take thousands of years to achieve, and history must therefore be taken into account in comparisons of species-richness.

Data of sufficient scale for useful comparisons are now available from a few large demographic plots in Asia and the Neotropics, and some similar data are also available (Table 18.2, Fig. 18.1). The greater part of tree species-richness can be captured in samples of 10-ha where trees ≥ 10 cm dbh have been censused (P. Hall unpublished data), or samples of >10000 trees (Condit, Loo de Lao, Leigh et al. 1996). The samples illustrated here exceed these numbers. Sinharaja forest, representative of an island of aseasonal lowland rain forest, which at its greatest Holocene extent would not have exceeded $10000 \mathrm{km}^2$, shows markedly depressed richness compared with Pasoh and Lambir which share the same aseasonal wet climate, but in these plots from formerly continuous continental forests there remains great variation in richness which is clearly habitat correlated. At a regional scale, species-richness declines with the length of the dry season in Asia. The very high and remarkably similar Lambir and Ducke, Manaus, values (Fig. 18.1) are associated with similar and high soil heterogeneity, in both cases a mosaic of humult and udult ultisols and entisols with humult soils predominating; but whereas Lambir has no dry season, Manaus experiences 3 months of dryness. (Admittedly, the Ducke list represents the whole 20 × 20 km Reserve in theory, but in practice collecting was concentrated in a few small areas.) Remarkably, Lambir, an ecological island at the margins of a continental forest, still appears to support more large congeneric series and perhaps a richer flora overall than the ultimate continental site in the Central Amazon. Nevertheless, the greater richness of Ducke over Barro

TABLE 18.2. Tree species- and genus-richness in some large plots in tropical forests in Asia and the Neotropics. Total species lists for reserves are included where used for Fig. 18.1.

Site	Forest type	Mean annual rainfall (mm)	Number of dry months (less than 100 mm rainfall)	Soil and topographic variability	Area of plot (ha)	Number of tree species (and genera) > 1 cm φ in plot	Total number of tree species (and genera) recorded from reserve
Mudumalai, Karnataka	Dry deciduous	900	6	Low	50	71 (58)	
Huai Kha Khaeng, NW Thailand	Closed dry evergreen and semi-deciduous	1600	5	Low	50	253 (161)	
Barro Colorado Island, Panama	Evergreen	2300	3	Low	50	309 (186)	
Ducke, Manaus, Brazil	Seasonal evergreen	1800	2	Low			1002 (289)
Pasoh, Peninsular Malaysia	Evergreen	2200	0	Low	50	814 (290)	
Lambir, Borneo	Evergreen	2900	0	High (2 types)	52	1175 (277)	
Sinharaja, Sri Lanka	(Insular) Evergreen	3900	0	Moderate	25	209 (112)	300 (149)

Colorado Island (BCI) must have a historic and island biogeographic explanation. The floristic variability of mixed tropical forests in relation to soils is now well known (cf. Ashton 1964, Austin Ashton & Greig-Smith 1972, Duivenvoorden & Lips 1995). This remarkable intercontinental similarity is, as yet, unverified with further data although a plot approaching completion at Huai Kha Khaeng Wildlife Sanctuary, Thailand, with a dry season almost 2 months longer than that at BCI has $c.$ 280 species, as might be anticipated. Comparison of stand structure and dynamics indicate that forests in seasonal climates have consistently fewer trees in the smaller size classes, and higher yet fluctuating mortality among them. This has been shown to be substantially due to mortality following an unusually severe dry season during the first period of measurement at BCI (Condit, Hubbell & Foster 1992) and is therefore consistent with the view that equability supports high species-richness.

The proportion of species in congeneric series broadly increases with species-richness (Fig. 18.1). Of particular interest though is the higher proportion in the island forest yet equable climate of Sinharaja compared to BCI which has higher species-richness. We infer that species attrition in less equable climates is not random, but differentially reduces those that are ecologically the most similar, even within guilds. This implies that co-occurrence of congenerics does in some measure depend on fine-tuning between the narrow niches which would be most dependent on climatic equability.

The well-known floristic boundary near the Malaysian–Thai border is correlated with the onset of a short, <3 month dry season. Evidence from floras implies that community species-richness is more than halved with the onset of seasonality there. Other changes include replacement of supra-annual mass canopy flowering with annual flowering, albeit fluctuating in intensity, during the dry season, and entrainment of fruitfall to the onset of the monsoon. Differences in seed predation across this boundary, and of recruitment and mortality in relation to variability in the onset of the rains and the length of the first dry season following establishment, would repay study.

Species distributions differ in relation to mineral soils nutrient concentrations, notably of phosphorus and magnesium, in the mixed dipterocarp forests of northwest Borneo (Baillie, Ashton, Court et al. 1987). Species-richness on a local scale in northwest Borneo varies in relation to soil nutrients. There is a humped distribution along gradients of total mineral soil magnesium (Fig. 18.2) and phosphorus. Here too, much of the variation is concentrated among the congeneric series (Table 18.1). The hump coincides with the generally sharp transition from soils with an acid fibrous root-matted surface organic horizon to those without (generally humult to udult

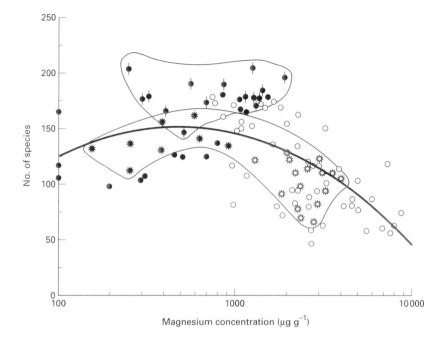

FIG. 18.2. Patterns of species richness in mixed dipterocarp forests of NW Borneo in relation to total mineral soil magnesium concentration (at 20–30 cm depth), as represented in 105 plots, each of 0.6 ha. Where Ns = number of species and Lmg = $\log_e$ of magnesium concentration, the fitted curve was given by $N_s = -220 + 145 \,(Lmg) - 11.8 \,(Lmg)^2$, $F = 28.0$, df = 2, 104, $P < 0.001$, $r^2 = 0.362$). Solid symbols, plots on humult ultisols; open symbols, plots on udult ultisols and inceptisols; starred plots, located in the Lambir Hills complex. Encircled plots: ●, the Bukit Iju rhyolite; ✳, Bukit Lambir sandstone, (ecological islands of c. 800 km² and 30 km², respectively); ✣, plots on the zonal udult soils of the Lower Tinjar and confluent southern Lambir Hills.

ultisols in the current United States Department of Agriculture classification). The peak could therefore have a deterministic explanation and is consistent with Tilman's (1982) resource allocation hypothesis, but it could also reflect stochastic immigration and survival of individuals outside their preferred habitat through subsidy across the ecotone in plots (Fig. 18.3) (Pulliam 1988), or simply a sampling artefact created by high within-plot soil heterogeneity at the ecotone. A 52-ha tree demography plot at Lambir National Park, Sarawak is placed across a sharp humult–udult ecotone, and has been placed to facilitate research towards resolving this problem.

Excepting the widespread coastal histosols with their distinctive flora, the organic lowland soils of northern Borneo form islands within a zonal matrix of udult ultisols and inceptisols. The relative floristic poverty of the heath forests, on spodosols, is therefore consistent with predictions from island biography theory as well as their low soil nutrient availability and proneness to water stress. But the mixed dipterocarp forests on humult ultisols, which share only c. 25% of their species with the adjacent mixed dipterocarp forests on the udult zonal soils (and less with the heath forests; see Ashton 1964) actually support a generally richer flora than the matrix notwithstanding that their soils are less nutrient rich and more drought prone (Fig. 18.2). It is difficult to come up with a non-equilibrium hypothesis that could explain this fact, and there is recent evidence in one species that population density on humult ultisols is correlated with depth of raw humus, opening the possibility that a small-scale habitat mosaic may differentially influence juvenile survival, thereby increasing overall local species diversity through increased β diversity (E. Kaplan unpublished data), again consistent with Tilman's (1982) model.

RANK ORDER OF ABUNDANCE

Gentry (1988) pointed out the remarkable consistency in rank order of abundance of representation of leading tree families in Neotropical and Old World forests, and the dominance of Dipterocarpaceae in Asian lowland rainforests has long been accepted (e.g. Whitmore 1984). I know of no good data which can test the predictability of rank order of abundance of species in a given habitat at more than local scale, particularly because such large sample sizes are required to estimate the densities of the less abundant majority. On a regional geographical scale, the consistent presence of certain, mainly dipterocarp species among the five most abundant in groups of five 0.6-ha plots from 14 distinct localities, on two different soils, in Sarawak and Brunei (Table 18.3), indicates that some predictability of rank order exists among these. Thirty of the 70 positions for the leading five in the 14 localities were occupied by one or other of 11 species. Highest relative dominance is achieved on soils with exceptionally high as well as exceptionally low nutrients.

Variance in growth rates, and maximum growth rates increase with mineral soil nutrient concentrations. This is accompanied by increasing dominance of a few fast growing canopy species (Ashton & Hall 1992). Unlike on low nutrient soils, there is some evidence on high nutrient soils that dominance is correlated, inversely, with overall species diversity as predicated by work in temperate vegetation and on theoretical grounds by Tilman (1982) (Fig. 18.3).

TABLE 18.3. Consistency of occurrence among the most abundant five species in groups of four 0.6-ha plots for different localities in Sarawak, E. Malaysia. Forests on humult (five groups) and udult (four groups) soils separately.

Humult soils	Udult soils
Dryobalanops beccarii (in all 5 groups)	*Dryobalanops lanceolata* (2)
Shorea macroptera ssp. *baillonii* (5)	*Millettia vasta* (2)
Allantospermum borneense (4)	*Diospyros curraniopsis* (2)
Dryobalanops aromatica (2)	*Koilodepas longifolia* (2)
Shorea falcifera (2)	
Swintonia schwenkii (2)	
Shorea balanocarpoides (2)	

A surprising number of species appear to occur consistently in low population densities. Table 18.4 shows mean density of 46 canopy species which were found both in Pasoh and on similar soils scattered throughout Sarawak and Brunei over 1000 km to the east of Peninsular Malaysia. Half (23 species) had similar densities in both samples and, of these, 15 occurred at densities of <1 ha^{-1}. Consistently rare tree species include habitat specialists, including successional species, particularly those which grow to large size

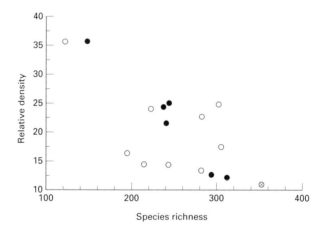

FIG. 18.3. Relationship between relative density (per cent) and species-richness in forest plots in NW Borneo. Relative density of the five most abundant species in groups of four 0.6-ha plots from 15 sites in mixed dipterocarp forest and one in heath forest. Richness represents number of species in 1000 individuals from these plots. Black circles, relatively high nutrient udult ultisols; open circles, relatively low nutrient humult ultisols. The upper left open circle represents heath forest; the lower right hand, crossed, a plot on the humult—udult ecotone. **, $P<0.01$; *, $P<0.05$ levels of significance: for five most abundant species $r^2 = 0.57$* for all plots, 0.89** for high nutrient plots, 0.42* for low nutrient plots; for most abundant species only, $r^2 =$ 0.45*, 0.93**, 0.13 respectively.

TABLE 18.4. *Canopy species common to Pasoh Peninsula Malaysia and some Sarawak plots (mean, individuals ha^{-1}). Species showing the same abundance classes asterisked (Classes: <0.5, 0.5–0.99, 1–1.99, > 2.0 trees ha^{-1}).*

Name		Pasoh	Sarawak
Apocynaceae: *Alstonia angustiloba*	*	0.38	0.10
Dyera costulata		2.10	0.16
Anacardiaceae: *Mangifera macrocarpa*	*	0.10	0.08
M. quadrifida	*	0.26	0.02
Bombacaceae: *Durio oxleyanus*	*	0.40	0.02
Burseraceae: *Canarium apertum*	*	0.22	0.16
Triomma malaccensis	*	1.76	1.33
Combretaceae: *Terminalia citrina*	*	0.26	0.13
T. subspathulata	*	0.08	0.02
Dilleniaceae: *Dillenia reticulata*	*	0.10	0.19
Dipterocarpaceae: *Anisoptera costata*	*	0.06	0.08
A. laevis		0.88	0.37
Dipterocarpus crinitus		0.58	1.38
D. costulatus		0.98	0.03
Hopea dryobalanoides		1.30	3.14
H. sangal	*	0.08	0.02
Shorea acuminata/ S. quadrinervis	*	3.82	2.24
S. bracteolata		2.34	0.75
S. dasyphylla	*	0.24	0.40
S. hopeifolia		0.14	1.84
S. leprosula		7.12	0.78
S. macroptera	*	4.04	11.00
S. maxwelliana	*	11.22	2.37
S. multiflora		0.06	2.00
S. parvifolia	*	5.40	3.68
S. pauciflora	*	5.40	2.67
Fagaceae: *Quercus argentata*		5.10	0.24
Irvingiaceae: *Irvingia malayana*		1.02	0.24
Lauraceae: *Litsea nidularis*		0.06	0.60
Leguminosae: *Dialium procerum*		1.06	0.05
Koompassia malaccensis		4.52	1.29
Millettia atropurpurea	*	5.70	6.65
Parkia speciosa		1.58	0.16
Sindora coriacea		1.26	0.29
Lecythidaceae: *Barringtonia pendula*	*	0.86	0.68
Magnoliaceae: *Aramodendron elegans*	*	0.28	0.11

(Continued on p. 502.)

TABLE 18.4. *Continued.*

Name		Pasoh	Sarawak
Moraceae: *Artocarpus rigidus*		0.94	0.02
A. nitidus		0.64	1.67
Parartocarpus bracteatus	*	0.34	0.06
Myristicaceae: *Gymnacranthera eugeniifolia*		0.76	0.30
G. forbesii		2.28	1.89
Myrtaceae: *Eugenia virens*		0.18	0.78
Sapotaceae: *Palaquium hexandrum*		2.76	0.54
Sterculiaceae: *Heritiera simplicifolia*		0.08	2.03
Pterocymbium tubulatum	*	0.04	0.26
Scaphium macropodum	*	2.08	3.81

such as *Octomeles*, *Alstonia* and *Pterocymbium* in Asia. Those of us who have collected widely are also familiar with climax species which, in their preferred habitats, seem to exist in consistently very low populations. Many others, of course, are relatively common in some parts of their ecological range, and less so in others. An interesting case is a group of species which are abundant in the seasonal tropics but occur, often widely, as scattered trees in the forests of the equatorial Far East and become rare in northern Borneo. Examples among dipterocarps include *Dipterocarpus gracilis*, *Hopea sangal*, *Anisoptera costata* and *Shorea guiso*, all of which occur on the zonal udult ultisols. It would be tempting to suggest that they reflect extraordinarily slow rates of extinction following expansion of aseasonal climates at the beginning of the Holocene, but no evidence of Pleistocene climatic change in northern Borneo exists, and inferential biogeographic evidence suggests climatic constancy (Ashton 1972).

Another prominent group of consistently rare species consists of canopy trees, often with large fruit, whose shared characteristic appears to be that their pollinator is long-lived, of limited fecundity, and depends on a steady sequence of different flowering trees as food sources through the year: a fragile system clearly dependent on climatic equability (cf. Start 1974). Pollinators in the Far East include the nectarivorous bat *Eonycteris spelaea*, and carpenter bees, *Xylocopa*. Tree genera include many canopy legumes, also *Neesia*, *Coelostegia* and *Durio*, which are all likely *Eonycteris* pollinated. *Durio* is of particular interest. There are two subgenera. Flowers of species in subgenus *Durio*, which are main canopy or emergent trees, are adapted to bat pollination. In my experience, their populations are always in low densities. But several species in subgenus *Boschia* such as *Durio griffithii* and *D. acutifolius*, small subcanopy to main canopy trees which are pollinated by meliponid bees, are abundant.

There also appears to be a relationship between seed size, abundance and habitat quality. In a dramatic example, Borneo floodplains are rich in large-fruited species; mixed peat swamp includes a few large-fruited arillate species such as *Durio carinatus*, *Gonstylus forbesii* and *G. bancanus* as well as many oil-fruited Sapotaceae; but the first three species only penetrate the peat swamps, going no further, and on the top of the swamp dome the six remaining tree species all have small fruits. Four are dry, one has sugary pulp (*Garcinia rostrata*) and one, a *Litsea*, is oily. Alluvium and valley sites support dense populations of some large-fruited species, notably *Eusideroxylon zwageri* which consistently occur as rare scattered individuals, with *Mangifera* and other large-fruited species, on upland soils. It is possible that soil resource availability limits fecundity and thus indirectly population density.

The existence of species with stable low population densities would be inconsistent with non-equilibrium theory. The very large plots of the CTFS network (Condit 1995, Ashton 1997) provide the opportunity for comparative demographic study, and for examination of their reproductive biology and breeding systems.

ECOLOGICAL PLASTICITY

At first sight, lottery and other non-equilibrium models would predict that selection should favour ecological generalists within guilds because, though chance will play a major part in securing a site, more habitat space would be available to generalists in which to establish. Chesson (1991) has pointed out that ecological differentiation is a requirement of lottery models, but ecological neutrality among guild members is assumed in Hubbell's (1997) model. Ecological specialization is a necessity in equilibrium models (e.g. Ashton 1969 for rainforest). Several recent community (e.g. Welden, Hewett, Hubbell & Foster 1991) and comparative ecophysiological studies (Brown & Whitmore 1992, Barker & Press 1995, Press, Brown, Barker & Zipperlen 1996, Whitmore & Brown 1996, Zipperlen & Press 1996) have concluded that, although broad groupings in relation to response to light may occur between and to some extent within pioneer and climax tropical tree guilds, there is great overlap in species' response and little evidence for partitioning of the light climate among juveniles of different species in gaps. 'Occupation is nine points of the law' (Poore 1968); that is, individuals established before or shortly after gap formation are those most likely to succeed. It is certainly impossible that the wealth of trees in many tropical forest communities could be ecologically partitioned along the single gradient of light, complex though that gradient is. Yet other workers are finding ecological differentiation, sometimes quite remarkable, among species sharing a common guild (Chazdon 1986, Dayanandan, Attygala, Abeygunasekera *et al*. 1990, Rogstad 1990, Turner 1990, 1991, Ashton & Berlyn 1992, Moad 1992, Mulkey, Wright & Smith 1993, Burslem, Grubb & Turner 1995, Ashton 1995, Ashton, Gunatilleke & Gunatilleke 1995, Gunatilleke *et al*. 1996, Davies, Ashton & Lee 1997).

The reason for the differing results, it seems clear, lies in differences of approach. Those that found the most clear-cut differentiation, implying niche differentiation, chose species from within congeneric series; their detailed comparisons were undertaken in the controlled environments of shade houses or growth chambers, and field transplant experiments and monitored wild populations were used to verify *ex situ* results. Most importantly, the primary interest of these workers and their collaborators has been, first, to compare species response to a variety of habitat variables including light, soil water, nutrients, and in one case temperature; and second, to measure response in relation to allocation to leaf, stem and branches, and root. Davies (1996, 1997) has related differentials in allocation, including fecundity, to the demographic characteristics of large wild sample populations, using the CTFS' 52-ha plot at Lambir National Park, Sarawak. The Gunatillekes and their colleagues (Ashton 1995, Ashton *et al*. 1995, Gunatilleke *et al*. 1996, Gunatilleke, Gunatilleke, Perara *et al*. 1997) working with eight members of a clade, section Doona of the climax genus *Shorea*, in the mixed dipterocarp forests at the CTFS Sinharaja site in southwestern Sri Lanka, and Davies (1996, 1997), who is comparing 12 species of the pioneer taxon *Macaranga* section Pachystemon, have found that each species is quite distinct when compared on the basis of resource allocation. These differences are consistent both with overall habitat differentials in the field, and individual species' demographic characteristics. But this is not to say that ecological distributions do not manifest high overlap, in relation to both soil and light. Competition experiments are currently underway for *Shorea* section Doona species. In spite of the manifest ecological differences between these congeneric species, it remains moot whether competition experiments will produce consistent winners and losers under specific environmental regimes.

Testing the existence of interspecific competition among trees in mixed tropical forests, even in large demographic plots, is daunting. Bearing in mind the large size and extraordinary numbers of co-occurring species in tropical forests, and their variable and often intermittent reproduction and limited dispersal, it is unrealistic to expect that any but the most abundant species have a reasonable chance of consistent species-specific direct competition over successive generations. Even mast fruiting dipterocarp species vary differentially in fruiting intensity in successive fruiting seasons (Curran 1994). Tropical trees therefore well meet the reproductive requirements of Chesson and Warner's (1981) lottery model. Why, then, does careful study reveal close relatives to be so distinctly differentiated?

First, many life-history parameters are functionally interdependent; their independent variation may incur selective costs. In a simple case, evolutionary change in longevity will influence life-time reproductive output, and thereby have impact on seed production per reproductive event and thus phenology and inflorescence size. Put another way, the attributes required of a guild member can be achieved by a large number of permutations of life-history attributes. Second, lack of rainfall seasonality, which appears to be so closely associated with the proliferation of long congeneric species series, increases opportunities for temporal differences in flowering times and local variation in flowering intensity between individuals and populations, both of which will tend to fragment populations and increase opportunities for genetic drift. Although several studies of common species have found evidence of ecological differentiation among congenerics which is consistent with selective competitive differentiation (Ashton *et al.* 1988. Rogstad 1990, Ashton 1995, Ashton *et al.* 1995), in other cases differentiation of an integrated suite of life-history traits could be accomplished through random drift in small populations as originally suggested by the late Andrei Fedorov (1966) (with whom I then so vehemently disagreed! (Ashton 1969)).

Both stochastic and deterministic, competitive forces are likely to moderate both interspecific differentiation and coexistence, and it is their relative roles in different environments that needs to be understood. Long-term demographic comparisons of congeneric, ecologically sympatric species are now needed. Again, particularly rewarding might be comparisons of phenology, reproductive output, establishment and seedling demography within congeneric series across the narrow climatic frontier, the Kangar–Pattani line of Whitmore (1984) which separates the aseasonal, Malaysian flora from the seasonal, Indoburmese. Comparisons of plasticity, and genetic variability, of species which occur together with series of congenerics in Malaysia, and alone in their genus in Indo-Burmese forests could provide inferential evidence of competitive niche differentiation in the former.

RARE SPECIES

The variation in the number of species in exceptionally low population densities, especially in members of congeneric species series, is the chief cause of variation in species richness with habitat. Chesson and Warner (1981) predict that rare species have a reproductive advantage over abundant species, leading to temporal and spatial fluctuations in abundance. Lack of constancy in relative abundance is predicted by Hubbell and Foster (1986) also. The stability of tree populations, relative to species diversity and habitat stability, and its causes therefore provides a critical test of the pre-eminence of equilibrium forces in the maintenance of species diversity. Large demographic plots now provide a means to monitor them.

There are several potential causes of rarity (Rabinowitz 1981). In tropical forests these certainly include specialists of rare habitats including gaps, gulleys and summits, also species on the edge of their ecological distribution including individuals which by chance establish beyond ecotones at the edge of their preferred habitat, known as the source–sink effect (Pulliam 1988). One example is provided by the map of populations of the 11 species of *Shorea* section Mutica co-occurring in the 10-ha research forest at Semengoh Forest Reserve, Sarawak, a heterogeneous infertile area; these were compared with the six species of the same section in Pasoh forest (Ashton 1988) which had been shown to compete for the same pollinator (Ashton *et al.* 1988). The four leading species at Semengoh were of comparable abundance to the leading four at Pasoh. Of the least abundant seven, three (two from udult soils, one a podsol species) were clearly at the edge or beyond their normal edaphic range, and three seem always to occur in low densities, and one was rare, both for unknown reasons. The exceptional number of species at Lambir and Manaus seem to be due to a combination of site heterogeneity and source–sink effects, and will be available for study once these plots are ready.

Then there may be species, as discussed, kept in low population densities by constraints on fecundity. Constraints on cross-pollination imposed by spatial isolation will act to increase extinction rates among outbreeders at low population densities. Hubbell and Foster (1986) showed a decline in the proportion of dioecious species at low population densities. Nevertheless, tropical forests show a greater number of rare species than expected from Preston's (1948) log normal distribution of species abundances, a pattern which Hubbell (1997) predicts on the basis of his stochastic model. He points out that density dependence should lead to a peak of species population densities close to the lower density limits at which these factors operate, for which there seems to be no evidence. But the Pasoh plot captures in 50 ha an estimated 50% of the available flora of its forest type (Kochummen LaFrankie & Manokaran 1992) and this appears to be true of other plots. This suggests that, for a given forest type, there should indeed be a decline in the number of species in classes of decreasing density beyond scales measurable in the 50-ha plot. It is possible that the unexpectedly high number of rare species at the scale of 50 ha represents parts of the predicted peak caused by density-dependent factors.

POPULATION GENETICS

Given that species-richness in tropical forests reaches its maximum in the most equable (i.e. predictable) climates, a feature common to mediterranean mixed shrubland communities (Cowling, Holmes & Rebelo 1992), and bearing in mind the spatial isolation of individuals and the likelihood, with limited seed dispersal, of neighbours being near-sibs, there would appear to be little advantage to maintaining genetic variation within populations if species-wise competition is rare or absent. Yet the overwhelming evidence to date indicates that outbreeding is favoured in tropical trees (Bawa & Opler 1975, Bawa, Perry & Beach 1985, Bawa & Hadley 1990, Bawa 1992, Murawski, Gunatilleke & Bawa 1994, Murawski 1995), that outcrossed progeny are selected for over selfed (Murawski *et al.* 1994), and that genetic variability within populations is at least as high as among temperate trees (Gan 1976, Gan, Robertson, Soepadmo *et al.* 1977, Hamrick & Loveless 1989, Hamrick & Murawski 1990, Hamrick 1994, Murawski 1995). Facultative self-compatibility is common, and there is some evidence that rates of selfing are inversely related to population density (Murawski & Hamrick 1991, 1992). Apomixis, through adventive embryony, sometimes apparently obligate, is now well known among rainforest trees (Kaur, Ha, Jong & Soepadmo 1978, Kaur, Jong, Sands & Manokaran 1986, Ha, Sands, Soepadmo & Jong 1988), but its prevalence and ecological and evolutionary role remains unknown. How can outbreeding, in view of its extraordinary cost and wastefulness among dispersed individuals in low-nutrient ecosystems like tropical forests, confer a selective advantage in a predictable habitat except in the presence of species-wise selection, which would favour niche specificity?

MOBILE LINKS

Mobile links may be the final resort for the selectionist explanation for the maintenance of tree species-richness in species-rich ecosystems. These interactions are density dependent, and can therefore explain why dominance is rarely achieved but not how exceptionally rare species are sustained, if they are, in stable densities. Mobile links may be species specific, but most are probably not; some may operate at higher taxonomic (phylogenetic) levels (see also Ashton 1988).

Tests of the existence of density dependence have inevitably been concentrated on its influence on the performance and mortality of individual tree species, through the agency of seed and seedling predators and pathogens (Augspurger 1983a,b, Hubbell & Foster 1990, Hubbell, Condit & Foster 1990). Results have been mixed, and Hubbell *et al.* (1990) concluded that its strength is insufficient to explain the population densities of all but the most abundant species. But Hubbell *et al.* (1990) confined this analysis to juveniles >1 cm dbh and it is likely that most predation and pathogenic mortality occurs among smaller individuals. Also, juvenile predation and pathogenicity is likely to be species-specific in its intensity and, more important, experience major temporal and spatial fluctuations, exacerbated by the clumped distributions of so many tree populations, due to Lotka–Volterra predator–prey cycles between the predators and pathogens and their own assailants. For example S. Thomas (personal communication) has observed high mortality, apparently pathogen related, in a single tight clump of ballistically dispersed *Mallotus* sp. at Pasoh forest, Malaysia, and I (unpublished) have documented such mortality in similarly dispersed *Pimeleodendron griffithianum* in plots at Bako National Park, Sarawak, East Malaysia. Such species are particularly good candidates for study. Here again, a true assessment will require much larger sample sizes, and over longer periods.

It is also likely that mobile links such as pollinators and seed dispersers may influence population densities by setting a ceiling on fecundity. Start (1974) showed that the foraging flights of the nectarivorous bat *Eonycteris spelaea* can exceed 50 km. Seasonally flowering tree species pollinated or dispersed by wide-ranging organisms, such as some bats and bees, could potentially survive in very low yet stable numbers if fecundity is maintained and the forces of extinction are weak. Large demographic plots are now available to start addressing such linkages.

CONCLUSIONS

This chapter is a logical sequel of an early article (Ashton 1969). On the one hand, progress in our understanding of the floristic structure of biodiverse tropical forests since then has been impressive; but on the other, it has become clear that important phenomena are being missed because our efforts have been too modest in scale, in time as well as space.

Our sample designs, for single or comparative species studies have often been too small. We are only beginning to obtain estimates of the change in abundance of species populations in time and in space, and as yet have no useful comparative data, particularly for the rare majority of species. Shade houses have proven critical to control physical habitat variables, but results from them must be tested against very large field samples. Species must be compared on the basis of suites of critical life history characteristics, rather than single characteristics. To understand the significance of results in the context of species and community dynamics it is clear that they must be compared with demographic population samples, and particularly with adequate samples of those few individuals, identifiable only after monitoring, which are performing at or near optimum. Phylogeny should be taken into account: The most informative evidence for ecological differentiation among species comes from comparison between those which are most similar overall.

Most gradients along which species may partition resources are non-linear and multivariate. Competition for the services of dispersers is a rare exception. It is obvious that single resource gradients can explain but a fraction of the ecological differences between species. By contrast, comparisons of species based on a variety of ecological and resource gradients, and more particularly a range of life-history parameters is proving very informative. It is becoming increasingly apparent that interactions between tree species and their mobile links, particularly pathogens and predators, may hold the key to understanding how population numbers may be maintained. This is a particularly demanding field, requiring long-term monitoring of population samples, particularly series of juvenile cohorts, replicated on large spatial scales.

The majority of species occur in low population densities. Understandably, most research has been concentrated on the common minority. But it is clearly the rare species that must now receive our attention. Is there a 'guild' of climax edaphic generalists which occur in consistently low densities and, if so, what limits their populations? What are the limits to density dependence? How is fecundity maintained in exceptionally low populations? What is their genetic structure?

Major applications of research into niche specificity and the maintenance of biodiversity are in silviculture and conservation planning and management. To what extent is it feasible to manipulate a mixed rainforest community in order to increase the abundance of single species of exceptional value? We must admit that we are no further on in solving this problem than were early Malayan foresters who tried (unsuccessfully) to favour regeneration of the durable heavy hardwood chengal, *Neobalanocarpus heimii*, at the turn of the century. We will not know until more demographic data are available, but evidence increasingly suggests that manipulation will at best be able to favour groups of ecologically similar species, such as the dipterocarp light hardwoods which benefited from stringent canopy opening combined with minimal disturbance of advance regeneration under the Malayan Uniform System of management (Wyatt-Smith 1963).

To what extent does the theory of island biogeography predict changes in extinction rates from reduction of forest area in a purely geographic sense? Here, the evidence from trees alone, also still preliminary, suggests that species-richness in forest communities beyond relatively small areas of c. 5000 km^2 are unaffected within time-spans which may exceed 10 000 years. But it is the mobile links and their predators whose critical minimum areas for survival will determine the survival of tropical forest communities; many may exist successfully in buffers and corridors of logged forest or even, in some cases, home gardens and other agroforestry systems.

ACKNOWLEDGEMENTS

This essay greatly benefited from comments by Stuart Davies, Elizabeth Losos and Peter Grubb. Pamela Hall prepared the analysis and graph for Fig. 18.2.

REFERENCES

Ashton, P.M.S. (1995). Seedling growth of co-occurring *Shorea* species in the simulated light environments of a rain forest. *Forest Ecology and Management*, **72**, 1–12.

Ashton, P.M.S. & Berlyn, G.P. (1992). Leaf adaptations of some *Shorea* species to sun and shade. *New Phytologist*, **121**, 401–412.

Ashton, P.M.S., Gunatilleke, C.V.S. & Gunatilleke, I.A.U.N. (1995). Seedling survival and growth of four *Shorea* species in a Sri Lankan rain forest. *Journal of Tropical Ecology*, **11**, 263–279.

Ashton, P.S. (1964). *Ecological Studies in the Mixed Dipterocarp Forests of Brunei State*. Oxford Forestry Memoirs, **25**.

Ashton, P.S. (1969). Speciation among tropical forest trees: Some deductions in the light of recent evidence. *Biological Journal of the Linnaean Society*, **1**, 155–196.

Ashton, P.S. (1972). The quaternary geomorphological history of western Malesia and lowland forest phytogeography. In *Transactions of the Second Aberdeen–Hull Symposium on Malesian Ecology: The Quaternary Era in Malesia* (Ed. by P. & M. Ashton), pp. 35–49. Hull Geography Department Miscellaneous Series **13**. University of Hull.

Ashton, P.S. (1984). Biosystematics of tropical forest plants: A problem of rare species. In *Plant Biosystematics* (Ed. by W.F. Grant), pp. 497–518. Academic Press, Toronto.

Ashton, P.S. (1988). Dipterocarp biology as a window to the understanding of tropical forest structure. *Annual Review of Systematics & Ecology*, **19**, 347–370.

Ashton, P.S. (1998). A global network of plots for understanding tree species diversity in tropical forests. In *Measuring and Monitoring Forest Biodiversity* (Ed. by F. Dallmeier). Smithsonian Institution and the UNESCO Man and the Biosphere Program, in press.

Ashton, P.S. & Hall, P. (1992). Comparisons of structure and dynamics among mixed dipterocarp forests of northwestern Borneo. *Journal of Ecology*, **80**, 459–481.

Ashton, P.S., Givnish, T. & Appanah, S. (1988). Staggered flowering in the Dipterocarpaceae:

New insights into floral induction and the evolution of mast flowering in the aseasonal tropics. *American Naturalist*, **132**, 44–66.

Augspurger, C.K. (1983a). Offspring recruitment around tropical trees: Changes in cohort distance with time. *Oikos*, **20**, 189–196.

Augspurger, C.K. (1983b). Seed dispersal of the tropical tree *Platypodium elegans* and the escape of its seedlings from fungal pathogens. *Journal of Ecology*, **71**, 759–771.

Austin, M.P., Ashton, P.S. & Greig-Smith, P. (1972). The application of quantitative methods to vegetation survey. III: A re-examination of rain forest data from Brunei. *Journal of Ecology*, **60**, 309–324.

Baillie, I.H., Ashton, P.S, Court, M.N., Anderson, J.A.R., Fitzpatrick, E.A. & Tinsley, J. (1987). Site characteristics and the distribution of tree species in mixed dipterocarp forest on tertiary sediments in central Sarawak, Malaysia. *Journal of Tropical Ecology*, **3**, 201–220.

Barker, M.G. & Press, M.C. (1995). Photosynthetic characteristics of dipterocarp seedlings in three light environments in tropical forest in Malaysia. *Bulletin of the Ecological Society of America*, **76**, 299.

Bawa, K.S. (1992). Mating systems, genetic differentiation and speciation in tropical rain forest plants. *Biotropica*, **24**, 250–255.

Bawa, K.S. & Hadley, M. (Eds) (1990). *Reproductive Ecology of Tropical Forest Plants*. UNESCO and Parthenon, Paris & Carnforth, England.

Bawa, K.S. & Opler, P.A. (1975). Dioecism in tropical forest trees. *Evolution*, **29**, 167–179.

Bawa, K.S., Perry, D.R. & Beach, J.H. (1985). Reproductive biology of tropical lowland rain forest trees. I: Sexual systems and self-incompatibility mechanisms. *American Journal of Botany*, **72**, 346–356.

Brown, N.D. & Whitmore, T.C. (1992). Do dipterocarp seedlings really partition tropical rain forest gaps? *Philosophical Transactions of the Royal Society, Series B*, **335**, 369–378.

Burslem, D.F. R.P., Grubb, P.J. & Turner, I.M. (1995). Responses to nutrient addition among shade-tolerant tree seedlings of lowland tropical rain forest in Singapore. *Journal of Ecology*, **83**, 113–122.

Chazdon, R.L. (1986). Physiological and morphological basis of shade tolerance in rain forest understory palms. *Principes*, **30**, 92–99.

Chesson, P. (1991). A need for niches? *Trends in Ecology and Evolution*, **6**, 26–28.

Chesson, P. & Warner, R.R. (1981). Environmental variability promotes coexistence in lottery competitive systems. *American Naturalist*, **117**, 923–943.

Condit, R. (1995). Research in large, long-term tropical forest plots. *Trends in Ecology and Evolution*, **10**, 18–21.

Condit, R., Hubbell, S.P. & Foster, R.B. (1992). Stability and change of a neotropical forest over a decade. *Bioscience*, **42**, 822–828.

Condit, R., Loo de Lao, S., Leigh, E.G., Sukumar, R., Manokaran, N., LaFrankie, J.V. *et al.* (1996). Species–area and species–individual relationships for tropical trees: A comparison of three 50 ha plots. *Journal of Ecology*, **84**, 549–562.

Cowling, R.M., Holmes, P.M. & Rebelo, A.G. (1992). Plant diversity and endemism. In *The Ecology of Fynbos: Nutrients, Fire, and Diversity* (Ed. by R.M. Cowling), pp. 62–112. Oxford University Press, Capetown.

Curran, L.M. (1994). *The ecology and evolution of mast-fruiting in Bornean Dipterocarpaceae: a general ectomycorrhizal theory*. PhD thesis, Princeton University, Princeton.

Davis, M.B. (1976). Pleistocene biogeography of temperate deciduous forests. *Geoscience and Man*, **13**, 13–26.

Davies, S.J. (1996). *The comparative ecology of* Macaranga *(Euphorbiaceae)*. PhD thesis, Harvard University, Cambridge Massachusetts.

Davies, S.J. (1997). Ecological and systematic studies of the diverse pioneer tree genus *Macaranga*. In *Proceedings of a Workshop on Long-term Ecological Research in Relation to Forest Ecosystem Management* (Ed. by H.S. Lee & A. Hamid). Sarawak Forest Department, Kuching, in press.

Davies, S.J., Ashton, P.S. & Lee, H.S. (1997). Life-history diversity in tropical pioneer trees: Comparative demography and allometry of 11 species of *Macaranga* from Borneo. In *Proceedings of a Workshop on Long-term Ecological Research in Relation to Forest Ecosystem Management* (Ed. by H.S. Lee & A. Hamid). Sarawak Forest Department, Kuching, in press.

Dayanandan, S., Attygala, D.N.C., Abeygunasekera, A.W.W.L., Gunatilleke, I.A.U.N. & Gunatilleke, C.V.S. (1990). Phenology and floral morphology in relation to pollination of some Sri Lankan dipterocarps. In *Reproductive Biology of Tropical Forest Plants* (Ed. by K.S. Bawa & M. Hadley), pp. 103–133. UNESCO/MAB/Parthenon, Paris.

Duivenvoorden, J.F. & Lips, J.M. (1995). A land-ecological study of soils, vegetation, and plant diversity in Colombian Amazonia. *Tropenbos Series*, 12.

Fedorov, A.A. (1966). The structure of the tropical rain forest and speciation in the humid tropics. *Journal of Ecology*, 54, 1–11.

Gan, Y.Y. (1976). Population and phylogenetic studies on species of Malaysian rain forest trees. PhD thesis, University of Aberdeen, Aberdeen.

Gan, Y.Y., Robertson, F.W., Soepadmo, E., Lee, D.W. & Ashton, P.S. (1977). Genetic variation in wild populations of rain forest trees. *Nature*, 269, 323–325.

Gause, G.F. (1932). The influence of ecological factors on the size of populations. *American Naturalist*, 65, 70–76.

Gentry, A.H. (1988). Changes in plant community diversity and floristic composition on environmental and geographical gradients. *Annals of the Missouri Botanical Garden*, 75, 1–34.

Gunatilleke, C.V.S., Perera, G.A.D., Ashton, P.M.S., Ashton, P.S. & Gunitalleke, I.A.U.N. (1996). Seedling growth of *Shorea* section Doona on soils from different forest sites of southwestern Sri Lanka. In *The Ecology of Tropical Forest Tree Seedlings* (Ed. by M.D. Swaine), pp. 245–265. UNESCO Man and the Biosphere Series, Paris.

Gunatilleke, C.V.S., Gunatilleke, I.A.U.N., Perera, G.A.D. Burslem, D.F.R.P., Ashton, P.M.S. & Ashton, P.S. (1997). Responses to nutrient addition among seedlings of eight closely related species of *Shorea* in Sri Lanka. *Journal of Ecology*, 85, 301–311.

Ha, C.O., Sands, V.E., Soepadmo, E. & Jong, K. (1988). Reproductive patterns of selected understorey trees in the Malaysian rain forest: the apomictic species. *Botanical Journal of the Linnean Society*, 97, 317–331.

Hamrick, J.L. (1994). Genetic diversity and conservation in tropical forests. In *Proceedings of the International Symposium on Genetic Conservation and Production of Tropical Tree Seed* (Ed. by R.M. Drysdale, S.E.T. John & A.C. Yapa), pp. 1–9. ASEAN, Canada Forest Tree Seed Centre, Mauk Lek, Saraburi, Thailand.

Hamrick, J.L. & Loveless, M.D. (1989). The genetic structure of tropical tree populations: Associations with reproductive biology. In *The Evolutionary Ecology of Plants* (Ed. by J.H. Bock & Y.B. Linhart), pp. 129–149. Westview, Boulder, CO.

Hamrick, J.L. & Murawski, D.A. (1990). The breeding structure of tropical trees. *Plant Species Biology*, 5, 157–165.

Hubbell, S.P. (1997). The maintenance of diversity in a neotropical tree community: conceptual issues, current evidence, and challenges ahead. In *Measuring and Monitoring Forest Biodiversity* (Ed. by F. Dallmeier). Smithsonian Institution and UNESCO Man and the Biosphere Program, in press.

Hubbell, S.P. & Foster, R.B. (1983). Diversity of canopy trees in a neotropical forest and implications for conservation. In *Tropical Rain Forest: Ecology and Management* (Ed. by S.L. Sutton, T.C. Whitmore & A.C. Chadwick), pp. 25–41. Blackwell Scientific Publications, Oxford.

Hubbell, S.P. & Foster, R.B. (1986). Commonness and rarity in a neotropical forest: Implications

for tropical tree conservation. In *Conservation Biology: The Science of Scarcity and Diversity* (Ed. by M.E. Soule), pp. 205–231. Sinauer, Sunderland, MA.
Hubbell, S.P. & Foster, R.B. (1990). The fate of juvenile trees in a neotropical forest: implications for the natural maintenance of tropical tree diversity. In *Reproductive Ecology of Tropical Forest Plants* (Ed. by K.S. Bawa & M. Hadley), pp. 317–341. UNESCO/Parthenon, Paris & Carnforth, England.
Hubbell, S.P., Condit, R. & Foster, R.B. (1990). Presence and absence of diversity dependence in a neotropical tree community. *Proceedings of the Royal Society of London, B*, **300**, 269–281.
Kaur, A., Ha, C.O., Jong, K., Sands, V., Chan, H.T., Soepadmo, E. *et al.* (1978). Apomixis may be widespread among trees of the climax rain forest. *Nature*, **271**, 440–441.
Kaur, A., Jong, K., Sands, V.E. & Soepadmo, E. (1986). Cytoembryology of some Malaysian dipterocarps, with some evidence of apomixis. *Botanical Journal of the Linnean Society*, **92**, 75–88.
Kochummen, K.M., LaFrankie, J.V. & Manokaran, N. (1992). Floristic composition of Pasoh Forest Reserve, a lowland forest in Peninsular Malaysia. In *In Harmony with Nature: Proceedings of the International Conference on Conservation of Tropical Biodiversity* (Ed. by S.K. Yap & S.W. Lee), pp. 545–554. Malayan Nature Society, Kuala Lumpur.
MacArthur, R.H. & Wilson, E.O. (1967). *The Theory of Island Biogeography*. Princeton University Press, Princeton.
Moad, A.S. (1992). *Dipterocarp sapling growth and understorey light availability in tropical lowland forest, Malaysia*. PhD thesis, Harvard University, Cambridge, MA.
Mulkey, S.S., Wright, S.J. & Smith, A.P. (1993). Comparative physiology and demography of three neotropical forest shrubs: Alternative shade-adaptive character syndromes. *Oecologia*, **96**, 526–536.
Murawski, D.A. (1995). Reproductive biology and genetics of tropical trees from a canopy perspective. In *Forest Canopies* (Ed. by M. Lowman & F. Halle), pp. 457–493. Academic Press, New York.
Murawski, D.A. & Hamrick, J.L. (1991). The effects of the density of flowering individuals on the mating systems of nine tropical tree species. *Heredity*, **67**, 167–174.
Murawski, D.A. & Hamrick, J.L. (1992). The mating system of *Cavanillesia platanifolia* under extremes of flowering tree density. *Biotropica*, **24**, 99–101.
Murawski, D.A., Gunatilleke, I.A.U.N. & Bawa, K.S. (1994). The effects of selective logging on inbreeding in *Shorea megistophylla* (Dipterocarpaceae) from Sri Lanka. *Conservation Biology*, **8**, 997–1002.
Poore, M.E.D. (1968). Studies in Malaysian rain forest. I: The forest on the Triassic sediments in Jengka Forest Reserve. *Journal of Ecology*, **56**, 143–196.
Press, M.C., Brown, N.D., Barker, M.G. & Zipperlen, S.W. (1996). Photosynthesic response to light in tropical rain forest seedlings. In *The Ecology of Tropical Tree Seedlings* (Ed. by M.D. Swaine), pp. 41–54. UNESCO–MAB Series 17, Parthenon, Carnforth, England.
Preston, F.W. (1948). The commonness, and rarity, of species. *Ecology*, **29**, 254–283.
Pulliam, H.R. (1988). Sources, sinks and population regulation. *American Naturalist*, **132**, 652–661.
Rabinowitz, D. (1981). Seven forms of rarity. *The Biological Aspects of Rare Plant Conservation* (Ed. by H. Synge), pp. 207–217. Wiley, Chichester.
Rogstad, S.H. (1990). The biosystematics and evolution of the *Polyalthia hypoleuca* species complex (Annonaceae) in Malesia. II: Comparative distributional ecology. *Journal of Tropical Ecology*, **6**, 387–408.
Start, A.K. (1974). *The feeding biology in relation to food sources of nectarivorous bats (Chiroptera: Macroglossinae) in Malaysia*. PhD thesis, University of Aberdeen.
Tilman, D. (1982). *Resource Competition and Community Structure*. Princeton University Press, Princeton.

Turner, I.M. (1990). The seedling survivorship and growth of three *Shorea* species in a Malaysian tropical rain forest. *Journal of Tropical Ecology*, **6**, 469–478.

Turner, I.M. (1991). Effects of shade and fertilizer addition on the seedlings of two tropical woody pioneer species. *Tropical Ecology*, **32**, 24–29.

Watt, A.S. (1924). On the ecology of British beechwoods with special reference to their regeneration. II. *Journal of Ecology*, **12**, 145–204.

Welden, C.W., Hewett, S.W., Hubbell, S.P. & Foster, R.B. (1991). Survival, growth, and recruitment of saplings in canopy gaps and forest understorey on Barro Colorado Island, Panama. *Ecology*, **72**, 35–50.

Whitmore, T.C. (1984). *Tropical Rain Forests of the Far East.* Clarendon Press, Oxford.

Whitmore, T.C. & Brown, N.D. (1996). Dipterocarp seedling growth in rain forest canopy gaps during six and a half years. *Philosophical Transactions of the Royal Society, Series B*, **351**, 1195–1203.

Wyatt-Smith, J. (1963). *Manual of Malayan Silviculture for Inland Forests* (2 Vols). *Malayan Forestry Records*, **23**.

Zipperlen, S.W. & Press, M.C. (1996). Photosynthesis in relation to growth and seedling ecology of two dipterocarp rain forest tree species. *Journal of Ecology*, **84**, 863–876.

19. DISTURBANCE AND SUCCESSION ON THE KRAKATAU ISLANDS, INDONESIA

S. F. SCHMITT and R. J. WHITTAKER

School of Geography, Oxford University, Mansfield Road, Oxford OX1 3TB, UK

SUMMARY

1 The development of Krakatau's forest ecosystems since apparent sterilization in 1883 has been the subject of numerous botanical studies. As reviewed here, these studies have emphasized: (i) the development of successional pathways; (ii) the role of dispersal systems and vectors; (iii) the influence of volcanism on diversification and community development; and (iv) the differing status of Rakata island (undisturbed by volcanism since 1883) compared with the other, volcanically disturbed islands. A number of different lines of evidence are presented which suggest a re-appraisal of parts of this pre-existing framework of ideas. An alternative successional model for Krakatau is presented in diagrammatic form.

2 All islands were found to have a more active disturbance regime during the most recent phase of volcanic activity than in the preceding inactive phase. Although Rakata does not receive ashfall and blast damage, it may experience more disturbances in the form of landslides, lightning strikes and windthrows, from increased numbers of earth tremors and local thunderstorms during volcanic activity.

3 Local and regional dispersal constraints appear a major factor influencing successional development, and species performance is in general negatively affected by ashfall, although this may vary according to interspecific differences in tolerance to this form of pollution. *Dysoxylum gaudichaudianum* saplings have been shown to be relatively tolerant both of ashfall and drought conditions.

4 Alongside the volcanic influences, particular attention has been drawn to the importance of extreme climatic events in influencing successional processes, particularly at the regeneration stage. Human impacts should also be included in the evaluation of successional processes on the islands. The confounding effects of volcanism and climate in particular are such as to complicate the identification of unique impacts of any single environmental forcing factor. Moreover, the varying role of disturbance, dispersal constraints and differential species performance make the prediction of path-

ways very difficult. A hierarchical model of succession, following that proposed by Pickett *et al.* (1987) is advocated as a means of clarifying the complex interrelationships between the different causes and mechanisms of succession.

INTRODUCTION

The recovery of ecosystems following disturbance has been a significant field within ecology since Clement's seminal (1916) paper set in train so many strands of debate concerning the patterns and processes of ecosystem change. Many of the classical ideas of succession theory stem from an essentially 'temperate zone' view of ecology and their application to tropical forest successions has proven particularly problematic. Tropical ecology has seen a pronounced turnover of ideas in recent decades as, for instance, notions of tropical stability and constancy have been rejected. The particular case study with which we are concerned is perhaps unique in that it spans all of the developments of theory alluded to, indeed the first botanical paper pre-dates Clement's article by nearly 30 years (Treub 1888). Several major themes have run as continuing threads between the contributions of different authors, notwithstanding the passing fashions of ecological theory. Our first aim in this chapter is to sketch out briefly some of the major themes and concerns of the earlier contributors, in order to identify how they may have shaped our understanding of this story.

The Krakatau islands (6°06′S, 105°25′E; Fig. 19.1) in the Sunda Straits between Java and Sumatra achieved their notoriety when a sequence of volcanic eruptions culminated in a catastrophic event on 27 August 1883, in which 36 000 Straits inhabitants perished (Thornton 1996). The three greatly altered and reshaped islands of the group were rendered barren, and so colonization of the islands began thereafter from nearby mainland and island sources. The first botanist to visit the islands in 1886, the 'distinguished and genial' Dr Melchior Treub, thus set out to study '... how a volcanic island, which had lost the whole of its flora... acquires a new vegetation... and further by what successive stages the new floral elements appear on the island and by what external agencies the colonisation is effected...' (Ernst 1908, p. 5). Much of the initial value of the site related to larger questions in plant geography, strongly connected to the efficacies of various means of plant dispersal in colonizing remote territories, and thus resting to a large degree on the status of the site as a prisere. Later, a fierce debate took place between Backer (1929), arguing against the sterilization hypothesis and (principally) Docters van Leeuwen (1936), who defended it in his lengthy monograph (see also Ernst 1934). The conflict between the two was both sci-

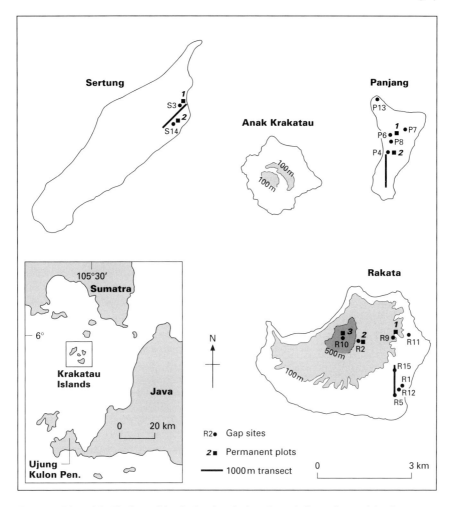

FIG. 19.1. Map of the Krakatau islands showing the locations of all gap sites and the three 1000-m 'cross-country' transects, and remeasured permanent plots.

entific in nature and personal and was not solely connected with Krakatau. Backer's (1929, p. 286) work closes with a dramatic allusion to the studies of Krakatau constituting 'wagon-loads of rubbish' largely because sterilization could not be proven. Docters van Leeuwen responded (1936, p. 4) that the value of the site did not depend on proof of absolute sterilization, but none the less mounted a spirited case for the status of the site as a prisere.

The initial findings on dispersal ecology were influential in the development of the subject and the topic has remained a key theme (Ernst 1908, 1934, Ridley 1930, Docters van Leeuwen 1936) as pattern in dispersal attributes of colonists underlies much of the pattern in the succession of communities and the turnover of species (Whittaker, Bush & Richards 1989, Whittaker & Jones 1994a,b). Another key theme of the early literature is the attention given to the development of plant associations. For example, Ernst (1908, p. 68) wrote that '. . . The comparatively advanced state of differentiation of the vegetation into plant-associations or formations is no less surprising than the remarkably large number and the variety of species in the new flora of Krakatau. As Penzig showed, the development of plant-associations had already begun in 1897 . . .'. Ernst's paper seems to presume the existence of plant formations, and while recognizing that they were loosely structured initially, he predicted that they would become closed through competition as more species colonized. The early pioneering and strand-line communities (the *Barringtonia* association and the *Pescaprae* association), when examined in relation to other locations in the region, do appear to be fairly predictable in their composition (Ernst 1908, Docters van Leeuwen 1936), but seres of the interior may be expected to become much less so as succession proceeds. This was appreciated by Docters van Leeuwen (1936, p. 262–263) who having considered the successions of coast and interior wrote: 'Of all these smaller and larger plant-associations, strictly speaking only two are more or less constant: 1. the *Pescaprae* association, and 2. the *Barringtonia* associations . . . All other associations in the Krakatau islands are of a temporary nature: they change or are crowded out'. Although he noticed differences between the three islands, he considered that in general terms the successions and associations which had developed in the lower parts of Rakata and on Panjang and Sertung were alike and the plants that were dominant in the various associations occurred in all islands in about equal numbers (p. 265).

The Krakatau story was brought to a wider audience through Richards' (1952) classic text *The Tropical Rain Forest*, in which the successions up to the 1930s were recounted and represented in diagrammatic form. In Richards' model the successions were shown as representing edaphic climaxes on the coast, and as leading towards climax lowland and climax submontane rainforest in the island interiors. Whilst the somewhat arbitrary distinction between primary and secondary successions was noted (Richards 1952, p. 269), this treatment was consistent with the earlier emphasis on the prisere status and with the classical paradigms of succession theory and of the diversity–stability view of the tropical forests. After 1934, apart from a single study *c*. 1951–52, there was a hiatus until 1979, when a new phase of research activity began (Flenley & Richards 1982).

In the last two years of Docters van Leeuwen's work on the islands (1931–32), renewed disturbance from the newly formed volcano Anak Krakatau resulted in extensive damage to the vegetation of Sertung and Panjang, opening up the canopy of the developing forests and thus setting back the successions on these islands. Subsequent damage has been documented (e.g. Borssum Waalkes 1960) and inferred from a variety of sources of evidence, in particular from complex stratigraphies of post-1930 volcanic ejecta on Sertung and Panjang (Whittaker et al. 1989, Whittaker, Walden & Hill 1992a, Bush, Whittaker & Partomihardjo 1992). It was found in the early 1980s that the vegetation of both islands had developed along rather different lines from Rakata (Bush, Richards & Jones 1986, Tagawa, Suzuki, Partomihardjo & Suriadarma 1985) and it has been argued that these two islands no longer represent uninterrupted priseres, whereas Rakata has continued an essentially uninterrupted succession from 1883 to the present day. Whittaker et al. (1989) presented two historical schema, following Richards' (1952) model but now split into two figures, representing in summary form these differences and the principal successional pathways on (i) Rakata, and (ii) Sertung and Panjang. Whilst these diagrams provide a reasonable first approximation to the documented (or inferred) historical patterns, they none the less must be seen as summarizing a much more complex reality. This chapter provides an opportunity to re-examine the story they tell and whether the 'prisere status' of Rakata might have become something of a straitjacket to the story-teller. Our recent emphasis on the differential impact of the volcano (e.g. Bush et al. 1992, Whittaker et al. 1992a, Whittaker, Bush, Partomihardjo et al. 1992b) is encapsulated in Fig. 19.2, which recognizes the role of forcing factors and of population ecological processes in a hierarchical model of succession, following the format of Pickett et al. (1987). The re-appraisal offered here is presented in Fig. 19.3, and the rest of the chapter presents lines of evidence which explain its change in emphasis.

Pickett and colleagues' (1987) model proposes site availability, species availability and differential species performance as the 'general causes of succession', forming the highest level of a hierarchy of successional causes (Table 19.1). The next level is termed 'Contributing processes or conditions' and includes the mechanisms of change of the highest level, for instance, coarse-scale disturbances; species dispersal and the propagule pool; and ecophysiology and stochastic environmental stress in relation to the three general causes in turn. 'Defining factors' constitute the particular factors that determine the outcome of the intermediate-level processes. The data presented in this chapter are of necessity a limited subset. They include landscape-scale variation in community and disturbance phenomena,

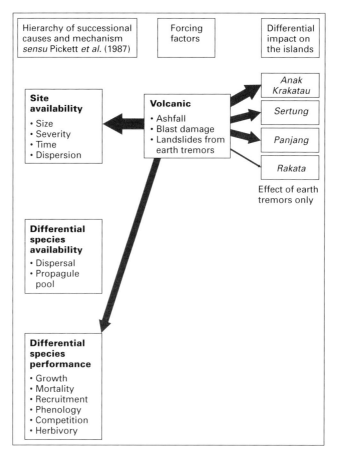

FIG. 19.2. A hierarchical representation of successional processes recognizing the special significance of volcanic disturbance in the Krakatau context. (After Pickett *et al.* 1987, with kind permission of Kluwer Academic Publishers.)

together with very local data on the dynamics of the regeneration layer in canopy gaps. They were gathered during a phase of active volcanism and of pronounced oscillations in climatic phenomena.

Successional history

The basic framework of vegetation succession and community types of Krakatau has been set out by Whittaker *et al.* (1989, 1992b) and Bush *et al.* (1992) and will only be outlined briefly here. The eruptions left three islands,

Disturbance and succession

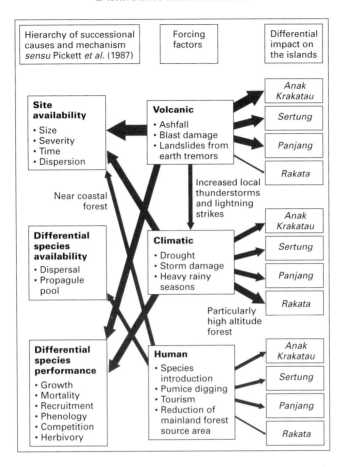

FIG. 19.3. A revised hierarchical representation of successional processes on Krakatau, emphasizing the significance of other external forcing factors in addition to volcanism.

Rakata, Sertung and Panjang, each greatly altered in area and shape, and mantled in pyroclastic deposits. Recolonization of the strand-lines was fairly swift as most of the strand-flora is of course sea-dispersed. The succession of these communities appears to have been fairly simple, with the first woodlands soon establishing behind the beaches and thereafter not changing greatly in general composition. The succession of the interior communities has been rather more interesting. Setting aside the strand-lines, the first phase of colonists were wind-dispersed pioneering composites, grasses and ferns, which developed a cover of increasingly dense grasslands. From around the turn of the century, bird- and bat-dispersed shrubs and trees

TABLE 19.1. *A hierarchy of successional causes* sensu *Pickett* et al. *(1987). The highest level of the hierarchy represents the broadest, minimal defining phenomena: general causes of succession. The second or intermediate level contains mechanisms of change: contributing processes or conditions. The third and lowest level of the hierarchy consist of the particular factors that determine the outcome of the intermediate-level processes. (Adapted from Pickett* et al. *1987, with kind permission of Kluwer Academic Publishers.)*

General causes of succession	Contributing processes or conditions	Defining factors
Site availability	Coarse-scale disturbance	Size, severity, time, dispersion
Differential species availability	Dispersal Propagule pool	Landscape configuration, Dispersal agents Time since disturbance, land use
Differential species performance	Resource availability	Microclimate, site history, soil conditions, topography
	Ecophysiology	Germination requirements, growth rates, population differentiation
	Life-history strategy	Allocation pattern, reproductive timing, reproductive mode
	Stochastic environmental stress	Climate cycles, site history, prior occupants
	Competition	Presence of competitors, identity of competitors, resource base
	Allelopathy	Soil characteristics, neighbouring plants
	Herbivory, disease and predation	Climate cycles, consumer cycles, plant vigour, plant defence, community composition, patchiness

established into these extensive pioneering communities, forming isolated clumps, which gradually coalesced so that during the 1920s the savanna gave way almost entirely to forest on each of the islands. Although the successions in the lowland interiors of each island were broadly similar, some differences were noted in the successions between and within islands due to variations in environment and to chance differences in landfall of bridgehead populations (Docters van Leeuwen 1936, Whittaker *et al.* 1989).

In 1930, renewed volcanism led to the formation of a new island in the centre of the group, Anak Krakatau, which has been active intermittently to the present day, depositing well in excess of 1 m of volcanic ejecta over most of Panjang and Sertung (Whittaker *et al.* 1992a). The heaviest falls appear to have been c. 1932 and 1952. In part — probably in large part — because of these events the forests of Panjang and Sertung developed a distinctive character. Rakata has been unaffected by ash falls and most of this island has, since the 1930s, been covered in forest dominated by an early successional wind-dispersed tree, *Neonauclea calycina*. On Panjang and Sertung, however, the successions have been set-back in a pattern of repeated patchy disturbance, and a mosaic of forest stands of differing ages and compositions has developed, in which two species stand out as particularly significant, *Timonius compressicaulis* and *Dysoxylum gaudichaudianum*. The patchwork reflects the timing and severity of disturbance and past availability of seed and dispersal agents. The particular contribution of disturbance events, underlying habitat differences, and differential species response to the evolution of the present mosaic structure cannot be fully reconstructed: each undoubtedly has had some role to play (Whittaker *et al.* 1989, Thornton 1996). Typically, the three species just mentioned dominate extensive tracts of the forests of these islands, and in the six permanent plots established in 1989 (Whittaker, Asquith & Bush 1990, Bush *et al.* 1992, Fig. 19.1) one or the other accounts for >40% canopy cover in each plot.

STUDY SITE AND METHODS

The Krakatau islands are mantled in pyroclastic deposits, which have been deeply dissected to form a sort of 'bad-lands' topography of often sinuous, generally steep-sided gullies, which commonly range from 10 to 30m in depth, frequently separated from one another by the narrowest of ridges (see Whittaker *et al.* 1989, for further background). This topography presents practical constraints on most types of field work within the islands.

Disturbance and gap formation

Disturbance frequency was quantified in 1994, using transects aligned in two ways: (i) six were 'cross-country' transects which ran from a chosen starting point along a compass bearing; and (ii) nine were 'gully transects' which followed gullies, mainly for ease of survey. They ranged in length from 450 to 1000m (total length surveyed: 5850m on Rakata; 3480m on Panjang; 3000m on Sertung). All canopy disturbances within approximately 10m either side of the transect line and of greater than 10m^2 were recorded. The disturbances, which included treefalls, lightning strikes, and landslides, were then classified into four age-classes to enable assessment of recent patterns of gap creation within and between islands (Schmitt & Partomihardjo 1998).

Six small permanent plots were established on the islands in 1989 (Whittaker *et al.* 1990, Bush *et al.* 1992; Fig. 19.1). They were resurveyed in

1992, when two others were installed, and all were examined for mortalities in 1994–95. This allows calculation of an albeit short baseline for mortality rate from the tree layer (all stems >5.0 cm dbh) in a period of volcanic inactivity (1989–92), which can be compared with that for the active period (1992–95).

Differences in inter- and intra-island diversity and compositional patchiness

In 1994, on each island, one 1000-m 'cross-country' transect from the disturbance survey was also used to enumerate species presence/absence (Fig. 19.1). Scoring of woody species (excluding climbers) in the understorey (*c.* 0.5–5 m), lower canopy (*c.* 5–15 m, depending on total height of canopy) and canopy (>15 m) took place for every 10-m section of the approximately 20-m wide transects. Species-area curves for all strata were calculated based on 0.1-ha sections derived by lumping five 10 m × 20 m sections together (Schmitt & Partomihardjo 1998).

As described in Schmitt and Partomihardjo (1998) this fast, if somewhat crude, method of survey has the advantage of being less liable to the subjective bias affecting site positioning with plot-based methods, and enables small-scale plot-based studies to be put into the context of medium-scale forest patterning. In the present chapter the larger scale patterns are further explored by using multivariate ordination and classification, in the form respectively of Detrended Correspondence Analysis (DCA, as provided by the CANOCO package of Ter Braak 1988), and the polythetic divisive classification technique Two Way Indicator Species Analysis (TWINSPAN; Hill 1979). For each 0.1-ha section each species was awarded a score between 0 and 5 as a function of presence in each 10-m section. These scores were used as input for the DCA and TWINSPAN analyses. For the latter, the pseudo-species cut-levels (Hill 1979, Kent & Coker 1992) corresponded to the scores.

Sapling performance in gaps

A study of sapling (>1.5 m height and between 1 and 10 cm dbh) performance in gaps was initiated in 1994. It was hoped to quantify species composition and performance of the gap regeneration layer, and how they would relate to island and gap size. In total, 15 sites of varying size were located in the lowlands of each island, and from the three main forest types previously identified; that is, canopies rich in one or other of *Neonauclea calycina*, *Timonius compressicaulis* and *Dysoxylum gaudichaudianum*. In each site, two transect lines were set out through the gap centre, one along the long-axis, and the other at right angles to it. The area of enumeration was determined by connecting the end-points of the transect lines. This yielded a rhomboid sample area, proportional to gap size. Gap size was approximated to that of an ellipse using the transect lengths as diameters, and it ranged from 361 to 3016 m^2. Seven sites were located on Rakata, six on Panjang and two on Sertung. All saplings were tagged, mapped, and marked with red paint at 1.3 m height. Sapling identities, dbh, height, mortality and recruitment were recorded in 1994 and 1995.

Mortality, recruitment and growth differences between gaps of different islands, and size categories, were tested using Student's t-test. For growth this compared mean performance in plots on Rakata vs. the ash-affected islands Panjang and Sertung. The data sets of the ash-affected islands were combined due to the small sample size of only two plots on Sertung. The analyses were undertaken for mean growth performance of all species occurring in the plots, and separately for the mean performance of *Dysoxylum gaudichaudianum*, which occurs in all gaps, and most frequently achieves 'common' species status (defined as density of ≥ 1.25 individuals 100 m^{-2}). To enable proper comparison with '*D. gaudichaudianum* only' results, t-tests were also performed on all saplings excluding this species. Percentage frequency of each species was calculated as a measure of relative abundance, allowing comparison between gaps of different size and thus of different sample area. From percentage frequency data of sapling species in gaps, rank abundance plots (Whittaker 1965, 1975, Hubbell 1979) were computed as a method of showing short-term dynamics in changes of species importance, and of evenness and diversity of the sapling layer as a whole, between years, and between gaps.

RESULTS

Disturbance and gap formation

Historically, direct volcanic impacts since 1930 have been somewhat greater on Sertung than on Panjang, and have not been evident on Rakata (Whittaker *et al.* 1992a). After several years of inactivity, Anak Krakatau commenced a continuous eruption phase from November 1992 to August 1995 (see also Thornton, Partomihardjo & Yukawa 1994, Thornton 1996). Between our visits of August 1992 and June 1993, a large number of new disturbances to the canopies of each island had taken place, coincident with this renewed volcanism. Ash fell on Sertung and to a lesser extent on Panjang, but not Rakata, throughout this period. All three islands were affected by other forms of disturbance, such as windthrows, lightning strikes and landslips. The islands were also affected by drought conditions in the extreme dry season of 1994, causing widespread wilting, leaf loss and mortality of both shrubs and trees.

It is difficult to judge what might be an appropriate baseline figure for the frequency of new disturbances (class 1) at times when Anak Krakatau is not active. Our expectation was that the volcano would have caused increased gap formation, and that this would be evident in greater proportions of class 1 disturbances in a trend from Rakata to Panjang to Sertung. The results of the disturbance survey using all survey data combined support this expectation (Schmitt & Partomihardjo 1998; Fig. 19.4a). However, for

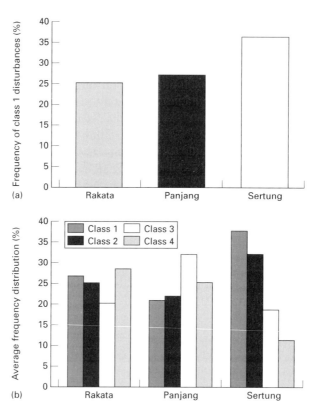

FIG. 19.4. Results of the 1994 disturbance survey using 'cross-country' and 'gully transects'. Age-classes: 1 = <2 years; 2 = 2–3 years, 3 = 3–4 years; 4 = >4 years. (a) Class-1 disturbances depicted as percentage of total number of disturbances recorded for all transects surveyed on Rakata, Panjang and Sertung, respectively. (b) Average percentage frequency distribution of disturbances of all age-classes for Rakata, Panjang and Sertung, using gully transects only. (Redrawn from Schmitt & Partomihardjo 1998, with kind permission of the editor, *Third International Flora Malesiana Symposium Proceedings*, Royal Botanic Gardens, Kew.)

TABLE 19.2. *Mortality (%) from the tree layer (dbh ≥ 5.0 cm) in permanent plots in a period without (1989–92) and with (1992–95) volcanic activity. The plots were established in 1989 (Bush et al. 1992), apart from Rakata 3 which was set up in 1992. Percentage mortality is calculated on the 1989 baseline except for Rakata 3, which uses the 1992 baseline.*

Plot	(m asl)	1989–92	1992–95*
Rakata	1 (110)	8	21
	2 (470)	9	46
	3 (680)	–	43
Panjang	1 (75)	6.5	20
	2 (130)	8	25
Sertung	1 (80)	14	37
	2 (140)	11	24

*For Rakata 2, the second period was 1992–94.

'gully-transects' only, Panjang's distribution of age-classes of disturbance, contrary to expectation, has a lower frequency of recent disturbances than both Sertung and Rakata (Fig. 19.4b). This indicates a role for events other than direct volcanic impacts in patterning of disturbance phenomena within these islands (see below).

Another means of approaching this is via the mortality data from the permanent plots (Table 19.2). These data indicate increased mortality and disturbance to forest canopies coincident with the period of Anak Krakatau's activity. Indeed, permanent plots on all islands showed raised percentage mortality of the trees enumerated in 1989, by a factor of 1.2–3.9, over the period 1992–95 compared with 1989–92. This applied notably within the larger size classes of tree, and the three plots with the greatest mortality in 1992–95 in declining order were Rakata 2, Sertung 1 and Panjang 1. On Rakata, the transect data discussed above were from regions below 100 m asl. In contrast to which, our observations from the higher regions (>300 m asl), suggest a phase of accelerated canopy disruption coincident with the eruptive period. These disturbances have frequently taken the form of lightning strikes, followed by subsequent extension of the gaps into large areas by further windthrows and by the erosion of the soil surface leading in turn to undercutting and landslips. Illustration of the local impacts in upland Rakata comes from both upland permanent plots, each of which serendipitously received a lightning strike. Plot Rakata 2 at *c.* 470 m asl experienced 9% mortality between 1989 and 1992, mostly of smaller stems, and at the time the forest structure suggested a fairly lengthy period without major disruption. By 1994, as a result of the form of storm damage described, mortality had risen to 46% judged by the 1989 baseline. These losses included several of the largest trees. The second site, at $c.$ 680 m asl, a small plot of mostly short-stature multiple-stemmed trees, experienced a massive 43% mortality between 1992 and 1995, associated with continued slipping and disturbance of the soil surface. This degree of disruption appeared fairly typical of the disturbance to much of the summit region.

Differences in inter- and intra-island diversity and compositional patchiness

The Rakata 1000-m transect has the highest cumulative species richness for the 2-ha area, followed by Panjang and then by Sertung. The curves for Panjang and Sertung also flatten earlier than that of Rakata, from which they differ significantly (Schmitt & Partomihardjo 1998; Fig. 19.5). These observations are consistent with the view that past disturbance to Sertung and Panjang has resulted in successional set-backs and has restricted the process of diversification (Whittaker *et al.* 1989). These data provide the first meso-scale evidence of this pattern. Whilst this has been suggested previously from whole-island species-richness and from plot-based analyses, in the latter case the varying form of the species–area curves, depending on which end of the transect is used to start the plot, demonstrates the danger of relying on very small plots for such analyses.

The results of the multivariate analyses of the transect data are pre-

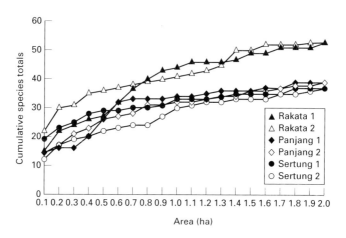

FIG. 19.5. Species–area curves for woody species (excluding climbers) of all strata. One 1000-m transect was enumerated on Rakata, Sertung and Panjang respectively. Approximate area covered was 20 m × 1000 m (2 ha). The alternative forms of the curves were derived by calculating the cumulative species totals in turn from each end of the three transects. (From Schmitt & Partomihardjo 1998, with kind permission of the editors, *Third International Flora Malesiana Symposium Proceedings*, Royal Botanic Gardens, Kew.)

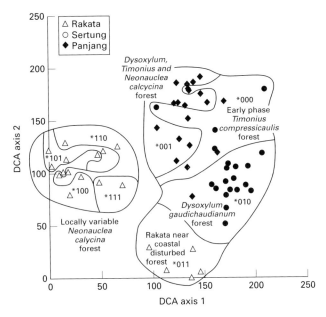

FIG. 19.6. DCA ordination and TWINSPAN classification of 0.1-ha sections of 1000-m 'cross-country' transects on Rakata, Sertung, and Panjang. The TWINSPAN groups are shown at the third level of the classification (eight groups).

sented in Fig. 19.6. TWINSPAN groupings at the third divisive level are superimposed on to the first two DCA axes. The principal forest types distinguished are: (i) *000 early phase *Timonius* forest, characterized by typical gap-fill species such as *Macaranga tanarius*, *Pipturus argenteus*, *Ficus septica* and in which *Bridelia monoica* is also frequent; (ii) *001 forest in which *Timonius compressicaulis* and *Neonauclea calycina* share the canopy in fairly even proportions, and in which other common species include *Dysoxylum gaudichaudianum*, *Buchanania arborescens*, *Oncosperma tigillarium*, and *Ardisia humilis*; (iii) *010 Sertung *Dysoxylum* forest, in which *D. gaudichaudianum* strongly dominates the canopy, and in which *Terminalia catappa*, *F. septica*, *Bridelia monoica*, and *Morinda citrifolia* are also consistent components; (iv) *011 disturbed near-coastal forest on Rakata in which *D. gaudichaudianum* forms much of the canopy, and consistent species include *Arthrophyllum javanicum* and *Ficus ampelas*; (v–viii) *100, 101, 110, 111 represent local variation within *Neonauclea* forest, in which several generally rare species, such as *Planchonella dulcitan* and *Horsfieldia glabra*, occur locally along a transect which has been subdivided to a greater extent in the TWINSPAN analyses than those of Panjang and Sertung.

Dysoxylum gaudichaudianum was the most frequently occurring species within each of the transects, occurring in more than 95 of the 100 sections of Sertung and Panjang and 82 sections of the Rakata transect. In the Rakata transect, seven species occur in >50 sections, in declining order of occurrence they are: *D. gaudichaudianum, Leea sambucina, Antidesma montanum, Neonauclea calycina, Ficus fistulosa, Buchanania arborescens* and *F. ampelas*. The equivalent set for Panjang is: *D. gaudichaudianum, Timonius compressicaulis, Ardisia humilis, Antidesma montanum, Buchanania arborescens* and *Neonauclea calycina*. For Sertung, there are only three species in the set, *D. gaudichaudianum, Bridelia monoica* and *F. septica*. Of the species listed, only *D. gaudichaudianum, Neonauclea calycina, Buchanania arborescens* and *Timonius compressicaulis* are canopy species. Within the Rakata transect, only in the sites of group *011 is *Dysoxylum* currently a major canopy species, but the data show that it is consistently present in the understorey and/or lower canopy. The transect is richer in species than the others, as noted earlier, and the patchy occurrence of canopy species such as *Litsea noronhae, Oroxylum indicum, Planchonella dulcitan, Syzygium polyanthum* and *Horsfieldia glabra* and of other components explains the greater degree of subdivision in the TWINSPAN analysis than of the other transects (see Table 19.3). Both Rakata and Panjang transects are locally patchy, as indicated by the higher number of transitions from one forest type to another along the transect line. The Sertung transect sections display a greater spread of points in the DCA ordination space than Panjang (Table 19.3, Fig. 19.6). Yet, this is mainly determined by two extreme points in group *000, and all the Sertung transect sections have been placed into one of only two TWINSPAN groups, with only one transition between groups occurring along the transect line. This indicates that while the extremes of the *Dysoxylum* forest, occupying three-quarters of the transect, and the

TABLE 19.3. *Summary statistics from the multivariate analyses of the 1000-m transects. DCA values are in standard deviations of species turnover × 100 (Hill 1979), providing a standardized measure of community variation within islands. TWINSPAN groupings and transitions between groups encountered in a traverse of each transect are given for the eight-group level of the analysis (as Fig. 19.6).*

Island	DCA				TWINSPAN	
	Axis 1	Axis 2	Axis 3	Axis 4	Groups	Transitions
Rakata	146	130	146	64	5	6
Sertung	106	128	96	111	2	1
Panjang	66	114	63	103	3	8
Eigenvalues	0.302	0.177	0.095	0.072	—	—

Disturbance and succession 531

Timonius forest block which makes up the remainder, must be fairly distinct, little patchiness exists within either stand type. Another notable feature of these analyses is the identification of a forest type on Panjang in which more or less equal proportions of *Neonauclea calycina* and *Timonius compressicaulis* characterize the canopy. This is a forest type which we now consider to be of sufficient areal extent to be recognized as an additional forest type on Panjang, but which has previously been neglected in plot-based sampling by Whittaker *et al.* (1989) and Bush *et al.* (1992).

Sapling performance in gaps

The data collected from the gap sites provide quantification over a 1-year period of: first, dbh and height changes; and second, mortality and recruitment of saplings. The comparison between mean height and dbh growth of 'all saplings' on Rakata with the values for the ash-affected islands of Panjang and Sertung did not return the expected result of significantly better growth on the former (Table 19.4a). However, on Rakata one gap site displayed anomalously low mean dbh and height growth: this was a size-3 site (R12) which was located in the flat, near-coastal area in the southeast of the island. Its low growth may reflect measurement problems particular to the site, but the area was also affected particularly strongly by drought. With the exclusion of this extreme site significantly lower dbh and height growth of all saplings on the ash-affected islands could be confirmed. The same result was obtained when the comparison was carried out for 'all saplings excluding *D. gaudichaudianum*' (Table 19.4b), the common gap sapling that was investigated separately (Table 19.4c).

A comparison of growth performance between different gap size-classes showed that results of the island comparisons are not confounded by gap size: no significant difference in mean dbh and height growth increment could be found between the two gap size-classes (Table 19.4a,b). Panjang and Rakata have more size-2 gaps than any other size, and in a further check on the effect of gap size on inter-island differences, a significant reduction of both mean dbh and height growth for sites on Panjang, as one of the ash-affected islands, was confirmed.

The between-island comparison for *D. gaudichaudianum* produced a mixed result when the extreme site R12 was excluded (Table 19.4c). A suppression in growth could only be confirmed for mean dbh increment whereas *D. gaudichaudianum* saplings showed no significant difference in height growth between Rakata and the ash-affected islands. This contrasts with the result for 'all saplings excluding *D. gaudichaudianum*', which showed significantly less height growth on the ash-affected islands.

TABLE 19.4. Tall sapling performance for the period 1994–95, from 15 gap sites (Rakata: $n = 7$, Panjang: $n = 6$, Sertung: $n = 2$). Size 1 (small) = <400 m²; size 2 (medium) = 400–1000 m²; size 3 (large) = 1000–2000 m²; size 4 (very large) = >2000 m²; sizes 1 and 2: $n = 11$, sizes 3 and 4: $n = 4$; (N.B. there were no size-2 gaps on Sertung). One-tailed Student's t-tests were used throughout to compare mean dbh and height increments. The samples of Panjang and Sertung are combined to form P + S the category 'ash-affected' islands. For (a)–(c) the island and gap-size analyses are additionally carried out with the extreme site R12 excluded. This was the only size-3 site on Rakata, and the number of size-3 and -4 sites was thus reduced to $n = 3$, with all of the large gaps located on Panjang and Sertung.

(a) Comparison of mean dbh and height increment of 'all sapling species' in gap sites on Rakata, Sertung and Panjang.

	Dbh growth	Height growth
Island comparison	R vs. P + S; NS	R vs. P + S; NS
Island comparison excluding R12	R > P + S*	R > P + S*
Gap size	Size 1 + 2 vs. size 3 + 4; NS	Size 1 + 2 vs. size 3 + 4; NS
Gap size excluding size-3 gap RG12	Size 1 + 2 vs. size 3 + 4; NS	Size 1 + 2 vs. size 3 + 4; NS
Size-2 gaps	R > P*	R > P**

*, $P \leq 0.05$; **, $P \leq 0.01$; NS, not significant ($P > 0.05$).

(b) Comparison of the mean dbh and height increment of all saplings excluding *Dysoxylum gaudichaudianum* in gap sites on Rakata, Panjang and Sertung.

	Dbh growth	Height growth
Island comparison	R > P + S*	R vs. P + S; NS
Island comparison excluding R12	R > P + S*	R > P + S*
Gap size	Size 1 + 2 vs. size 3 + 4; NS	Size 1 + 2 vs. size 3 + 4; NS
Gap size excluding size-3 gap RG12	Size 1 + 2 vs. size 3 + 4; NS	Size 1 + 2 vs. size 3 + 4; NS
Size-2 gaps	R > P*	R > P**

*, $P \leq 0.05$; **, $P \leq 0.01$; NS, not significant ($P > 0.05$).

(c) Comparison of the mean dbh and height increment of the common *Dysoxylum gaudichaudianum* in gap sites on Rakata, Panjang and Sertung. Two Rakata sites (size-2) were also excluded as they only contained one *D. gaudichaudianum* sapling each. Therefore the number of gap sites on Rakata was reduced to $n = 5$.

	Dbh growth	Height growth
Island comparison	R vs. P + S; NS	R vs. P + S; NS
Island comparison excluding R12	R > P + S*	R vs. P + S; NS
Gap size	Size 1 + 2 vs. size 3 + 4; NS	Size 1 + 2 vs. size 3 + 4; NS
Gap size excluding size-3 gap RG12	Size 1 + 2 vs. size 3 + 4; NS	Size 1 + 2 < size 3 + 4*
Size-2 gaps	R > P*	R vs. P; NS

*, $P < 0.05$; NS, not significant ($P > 0.05$).

Comparison of mean growth performance of *D. gaudichaudianum* between the two size-class categories of gaps also returned different results from that of 'all saplings excluding *D. gaudichaudianum*' (Table 19.4b,c). *D. gaudichaudianum* grew significantly more in mean height in the (larger) size-3 and -4 gaps, but dbh growth differences were non-significant as for all other sapling species. Again this result was obtained with the exclusion of the extreme size-3 site R12, and meant, in effect, that all size-3 and -4 gaps were then located on Panjang and Sertung. Thus, there is an indication that *D. gaudichaudianum* is not suffering from suppressed height growth on the ash-affected islands in general, and that in particular in larger gap size classes (for which all sites are found on Panjang and Sertung) it is showing significantly greater height growth. A comparison of height increment of size-2 gaps only (Table 19.4c), however, demonstrates that these height growth results for *D. gaudichaudianum* are not solely due to the greater number of large gaps on the ash-affected islands. Therefore, this species appears to perform better on the ash-affected islands, at least in height growth.

Another facet of these height growth differences is displayed in Fig. 19.7a, which shows that a higher proportion of *D. gaudichaudianum* saplings grew between 0 and 3 m on the ash-affected islands than on Rakata. Furthermore, a smaller proportion of the saplings experienced zero or negative 'growth' on Panjang and Sertung (ash-affected) than on Rakata. Moreover, when height growth on the ash-affected islands was compared between *D. gaudichaudianum* and all other sapling species, similar trends were displayed (Fig. 19.7b). Therefore, *D. gaudichaudianum's* height growth performance was not only better on the ash-affected islands, but in general its saplings performed better under the stress of ash than did the other sapling species (collectively).

We also posited that the increased stress from ash-fall should be reflected in the mortality and recruitment data, which in turn might provide some insight into the performance data discussed above. For this analysis, Sertung and Panjang sites were again treated as a single set, representing the ash-affected islands. Both percentage mortality and recruitment in Rakata gap sites ($n = 7$) were found to be not significantly different from Panjang and Sertung gap sites (one-tailed t-test; $n = 8$). However, in terms of the net change (i.e. gross recruitment minus mortality), Rakata sites were found to have significantly higher net recruitment (one-tailed t-test $P < 0.01$). Only one Sertung and one Panjang site returned a net increase in sapling numbers from 1994 to 1995. Further analyses revealed that percentage mortality per plot and net change in sapling number showed no significant differences between gaps of different size across islands.

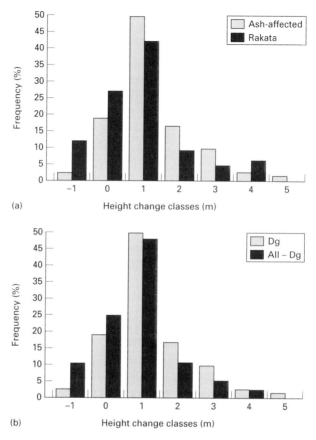

FIG. 19.7. (a) A comparison of proportional height change distribution of *Dysoxylum gaudichaudianum* saplings growing in gap sites on the ash-affected islands of Panjang and Sertung and the unaffected island of Rakata between 1994 and 1995. It demonstrates that a higher proportion of *D. gaudichaudianum* saplings displayed positive growth between 0 and 3 m, and at the same time a smaller proportion suffered negative 'growth' than saplings of the same species growing on Rakata. Negative 'growth' is defined as a reduction in height between measurements. This can be due to die-off of the top shoot and/or physical damage to the sapling. (b) A comparison of proportional height change distribution of *D. gaudichaudianum* (Dg), and all saplings excluding *D. gaudichaudianum* (all − Dg) growing in gaps on the ash-affected islands of Panjang and Sertung between 1994 and 1995. It demonstrates that a greater proportion of *D. gaudichaudianum* showed positive growth, as well as a smaller proportion of this species suffered negative 'growth', than 'all − Dg' growing under ash-stress.

The indication of *D. gaudichaudianum*'s better height growth performance under stress needs to be set in the context of the changing frequencies of each species from 1994 to 1995. The mortality-recruitment data provide evidence that *D. gaudichaudianum* has gained significantly in relative importance in sites on Panjang and Sertung in comparison to Rakata sites (one-tailed t-test; $P < 0.05$). It has lost in importance in all but one Rakata site. Its best gains in relative abundance were 11.6% and 7.1%, achieved in gaps on Sertung and Panjang respectively. In the former, *D. gaudichaudianum* became the numerical dominant of the sapling layer, whereas in the latter its dominance was merely reinforced. This trend seemed apparently unrelated to size of gap, since both were in the large size category (sizes 3 and 4).

In short, it appears that in terms of net change in frequency, as well as height growth increments of survivors, *D. gaudichaudianum* has done better on the ash-affected islands in comparison to Rakata. We postulate that *D. gaudichaudianum* possesses a better-than-average capacity to tolerate the polluting and physically damaging effects of ashfall. The deposition of these ejecta has led to enhanced mortality and/or reduced recruitment of some other species, so it is possible that part of the explanation for the apparent improved growth of surviving *D. gaudichaudianum* saplings lies in the reduction of interspecific competition. Percentage net recruitment of saplings of all species (collectively) was generally negative for Panjang and Sertung sites, and significantly lower for these islands than for Rakata (see above). Another contributory factor could be reduced herbivory on particular species. Some evidence for this has emerged from a pilot study in 1995 (L.T. White personal communication), which found that less herbivorous ant activity, and fewer stem entry holes occurred on *D. gaudichaudianum* in the ash-affected forests of Panjang, and Sertung, in comparison to Rakata.

Species compositional changes within the sapling layer during gap fill can be dramatic. Two sites are chosen to illustrate this. Both appeared severely stressed—Sertung site SG3 by very heavy ashfall, in combination with the drought, and Rakata site RG5 by a combination of the severe drought, and hydrological conditions which appeared to under-supply this area of the near-coastal plain of the southeast Rakata. Even large trees suffered mortality in this vicinity. A comparison of the rank-abundance curves for 1994 and 1995 for SG3 shows that *Pipturus argenteus* moved from first to the last rank, as all 19 individuals died, and only one recruited (Fig. 19.8). This allowed *Buchanania arborescens* to become the most abundant sapling species in this gap. Two species were lost from the site and one gained, with only minor alteration in the steepness and shape of the rank-abundance curve, thus evenness and diversity did not change very much between 1994 and 1995. This contrasts with RG5 in which the most abundant species, *Macaranga tanarius*, increased in importance from 25% in 1994 to 40% in

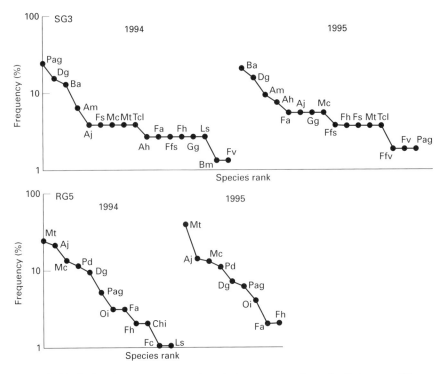

FIG. 19.8. Rank-abundance curve comparisons between 1994 and 1995 for Sertung gap SG3 and Rakata gap RG5. Ah, *Ardisia humilis*; Am, *Antidesma montanum*; Aj, *Arthrophyllum javanicum*; Ba, *Buchanania arborescens*; Bm, *Bridelia monoica*; Chi, *Canarium hirsutum*; Dg, *Dysoxylum gaudichaudianum*; Fa, *Ficus ampelas*; Fc, *Ficus callosa*; Fh, *Ficus hispida*; Ffs, *Ficus fistulosa*; Ffv, *Ficus fulva*; Fs, *Ficus septica*; Fv, *Ficus variegata*; Gg, *Gnetum gnemon*; Ls, *Leea sambucina*; Mc, *Morinda citrifolia*; Mt, *Macaranga tanarius*; Oi, *Oroxylum indicum*; Pag, *Pipturus argenteus*; Pd, *Planchonella dulcitan*; Tcl, *Timonius compressicaulis*.

1995 despite all but one of its 1994 representatives dying. In this case the rank order does not look very different between the two years, but it disguises a high amount of turnover and dynamism in the recruitment layer from one year to the next. Moreover, as indicated by the considerable steepening of the rank-abundance curve, it also disguises considerable change in evenness and diversity that is caused by the increased dominance of *M. tanarius*, and the loss of three (*Canarium hirsutum*, *Ficus callosa* and *Leea sambucina*) out of 12 species by 1995.

The dynamics of gap regeneration on Krakatau have thus been found to be responsive to a variety of controls. Marked differences can be found between and within islands. Although variations in factors such as initial species composition, gap geometry, soil conditions, further canopy disturbances, herbivory and hydrology can render each site in some way distinctive, volcanic pollution and regional climatic fluctuations appear to be overriding forcing factors. Thus, in these analyses they appear to be more readily detectable than differences related to gap size, type and so forth.

DISCUSSION

Over the past several decades, the successional development of the forest communities of the Krakatau islands has been analysed in terms of: (i) the development of successional pathways; (ii) the role of dispersal systems and vectors; and (iii) the influence of volcanism on diversification and community development (Ernst 1934, Docters van Leeuwen 1936, Richards 1952, Whittaker *et al.* 1989, 1992b, Whittaker & Jones 1994a,b). In the present chapter we have presented several different lines of evidence which require a re-appraisal of parts of this pre-existing framework. The vehicle for this discussion is provided by Fig. 19.3, following the hierarchical model of Pickett *et al.* (1987) which emphasizes process rather than pattern of succession.

Site availability

In Fig. 19.3, 'site availability' is shown as one of the 'general causes of succession' within Pickett *et al.*'s (1987) hierarchy of successional causes and mechanisms. In the period since its emergence in 1930, Anak Krakatau's activities are known to have impacted on Sertung and Panjang, principally through ash deposition, notably in 1932–33 and 1952–53, when large areas were defoliated. Lighter falls of ash, similar to those of 1992–95, have occurred intermittently throughout the post-1930 period (Whittaker *et al.* 1992a). The data on spatial frequency of gaps, and of mortalities within the permanent plots, provide evidence for an important role for the volcano in gap creation during the 1992–95 eruption period. However, the expectation of much greater increases in canopy opening on both Sertung and Panjang relative to Rakata was only partially supported by the data, pointing to the significance of another forcing factor, climate. The period of volcanism coincided with pronounced climatic oscillations associated with El Niño–Southern Oscillation (ENSO) events. For instance, in 1994 Indonesia experienced severe drought that led to large-scale forest fires in Kalimantan and Sumatra. Indonesia and the Pacific region are regularly affected by ENSO events (Ropelewski & Halpert 1987, 1989, Mueller-Dombois 1995), whereby generally in a biennial mode an El Niño event of below average rainfall is often preceded and/or followed by a pluvial period (La Niña; Nicholls 1993).

Observational evidence suggested an increase in mortality attributable to drought weakening trees and reducing canopy cover. In general, treefalls are more likely during the wet season (Oldeman 1974, Brokaw 1982, Brandani, Hartshorn & Orians 1988), particularly on more susceptible, loose, and steep substrates, such as are found on the Krakatau islands. During the study period, storms in the wet season may well have been more extreme than usual, leading to windthrow and to erosion of exposed soils (cf. Nicholls 1993). Evidence of such a phase of enhanced geomorphological activity was found in the form of gully incision (up to 2 m observed locally on all islands) often leading to the undermining of tree roots, and in the form of deposition and in-fill with eroded material in some lowland gullies, also in cases causing tree deaths. The significance of climate as a forcing factor is thus recognised in our revised diagrammatic model (compare Figs 19.2 and 19.3).

Despite not receiving any significant quantities of ashfall, it is argued that the dynamism of Rakata's forests might be linked indirectly to Anak Krakatau's activity (Fig. 19.3). One form of tectonic activity that has affected Rakata during eruptive periods has been earth tremors (Docters van Leeuwen 1936, S.G. Compton personal communication) leading to landslips. Another possibility is that there may be a link between volcanic and climatic phenomena. For instance, storm and lightning damage to the forest, especially the upper regions, was intense and widespread during Anak Krakatau's active phase of 1992–95. The volcano was ejecting large quantities of eruptive material to heights in excess of 1000 m, often associated with strong electrical discharges in the eruption clouds (as Ernst 1934, S.F. Schmitt personal observation; Fig. 19.9). This may, as implied in Fig. 19.3, precipitate increased numbers, and more severe thunderstorms with associated lightning events and windthrow of trees within the Krakatau locality (A. Robock personal communication, S. Self personal communication). Cloud forms around the peak of Rakata on more or less a daily basis, and even during times when the volcano is relatively inactive, lightning flashes in the cloud band have been observed by visiting scientists (Borssum Waalkes 1960). As the summit of Rakata (c. 735 m asl) exceeds the maximum elevation of the other islands by about 500 m, lightning may more frequently discharge to the higher altitudes of this island. This seems to have been a feature of the 1992–95 period, although strikes were noted on each island. Distinguishing between a regional climatic signal, a volcanic explanation and a synergistic interaction between the two is likely to be problematic. However, research elsewhere has demonstrated that climatic disturbance phenomena are of general significance in rainforest dynamics, and that the role of extreme weather phenomena is confined neither to islands, nor to the recognized cyclone belt (e.g. Whitmore 1974, 1989, Johns 1986, Nelson,

FIG. 19.9. Photograph of the formation of Anak Krakatau in 1930 showing severe electrical discharges within the ash clouds (source Umbgrove 1949, with kind permission of Cambridge University Press).

Kapos, Adams *et al.* 1994, Mueller-Dombois 1995, Newbery, Songwe & Chuyong, Chapter 11, this volume, Whitmore & Burslem, Chapter 20, this volume).

The occurrence of human disturbance on Krakatau is noted in Fig. 19.3, although it has not hitherto been detailed in this chapter. In recent years there has been an upsurge in the mining of pumice in the near-coastal gullies of Panjang and Sertung. Evidence of occasional tree-felling on the islands can also be found. These activities, especially the former, add another locally severe element to the disturbance regime, setting back the succession to a pioneering stage.

In previous papers it has been suggested that while volcanism has at times set back the successions of Panjang and Sertung, Rakata may continue to be considered as an essentially uninterrupted prisere (Whittaker et al. 1989, 1992a, Bush et al. 1992). This binary divide has been shown to be oversimplistic; the notion of an 'uninterrupted prisere' should be revised. The differences between the islands are of scale and type of disturbance, each experiences an active disturbance regime and variable weather conditions. Tree mortality rates appear to be comparatively high when compared with other studies (cf. Matelson, Nadkarni & Solano 1995) and at present, site availability would not appear in general to present a major constraint on the spread of later successional species. On the other hand, disturbance in some areas may well have been strong enough to selectively weed out some components, thereby reducing diversity, and resulting in the disparities seen in the multivariate analyses of the transect data.

Differential species availability

At the second level of the Pickett et al. (1987) hierarchy, dispersal and the propagule pool influence species availability; indeed, dispersal constraints and a reduction in the mainland propagule pool could potentially slow successional development even were site availability not a problem. At the archipelago scale, propagule supply is without doubt a general constraint on succession on these islands (Whittaker & Jones 1994a,b). Furthermore, while many species are shared between all three islands some, for no obvious reason of their ecology, are restricted to a subset, and others, while locally abundant on an island, have yet to spread widely (Whittaker et al. 1989, 1992b, Bush & Whittaker 1995). For instance, approximately half of the 47 species encountered in the gap sites could be potential canopy species. Yet, more than one-third of the species enumerated occurred in only one or two sites. These might come to fill the gap in question, but are not likely to become of major importance on the Krakatau islands in the short term. Some of the rare species, such as *Planchonella dulcitan*, *Horsfieldia glabra* and *Semecarpus heterophylla*, are recruiting successfully locally, thus the major constraint in these cases appears to be one of dispersal. These constraints might be caused by the low density of dispersers, or the lack of a more specialized disperser. This might be relevant to the case of *S. heterophylla* (Anacardiaceae), a species which has a distinctly under-dispersed distribution, concentrated around two large parent trees in northeastern Panjang. In extreme cases the lack of a specific pollinator, as required by *Ficus* species, may also act to constrain species spread.

The role of people as dispersers again deserves a mention. During the first 50 years a number of weed or food species were introduced by temporary inhabitants of the islands; few survived subsequent forest closure (Whittaker *et al.* 1989). In recent years, the number of species introduced to Anak Krakatau by tourists has increased considerably (Partomihardjo, Mirmanto & Whittaker 1992) while local fishermen, pumice collectors, the reserve wardens (regularly resident on Sertung) and most probably scientists, have been responsible for introducing a number of additional species to the group.

The present study has, through the 1000-m transects, shown there to be more patchiness on the beta-scale on Rakata, than on Panjang, with Sertung showing the least varied mosaic pattern (see also Schmitt & Partomihardjo 1997). This is a reflection of interactions between disturbance histories and local species availability. The transect on Rakata contained more species in general, and in particular, more canopy species. There has been debate as to the future prominence of *Dysoxylum gaudichaudianum* in the lowlands of Rakata, in which *Neonauclea calycina* remains the numerical dominant. The transect and gap data of the present study demonstrate unequivocally that *D. gaudichaudianum* is widespread in the understorey of this area of lowland Rakata (it remains predominantly lowland on Rakata), which concords with the predictions of a rise in its importance by Tagawa *et al.* (1985) and Tagawa (1992). None the less, the lowland forests of Rakata contain more species on the meso-scale, and it appears that many of them are managing to spread at least locally, and thus to contribute to a richer forest than has as yet developed on Sertung, with a different mix of associated species (cf. Bush *et al.* 1992). Panjang's forests are, in this sense, intermediate in character. The present study has also identified an additional stand type on Panjang, in which *Neonauclea calycina* and *Timonius compressicaulis* share numerical dominance. If this forest type were to be recognized as a significant element of Panjang's forests, it would require further revision of the successional summary diagrams of Whittaker *et al.* (1989), which distinguish between pathways followed on Rakata on the one hand, and Sertung and Panjang considered jointly on the other. While this form of representation, as offered by Richards (1952), Whittaker *et al.* (1989) and Tagawa *et al.* (1985) has its uses, it must be recognized that such diagrams over-simplify, placing artificial stress on the development of community types, or 'associations' (as in the earlier 20th century tradition; Ernst 1908, Richards 1952), whereas, in practice, successional pathways vary from patch to patch and over time as a function of the sorts of factors identified in Fig. 19.3. There is, as yet, no evidence to support Ernst's (1908) prediction of interior forest associations becoming 'closed' through competition as more species have colonized.

Differential species performance

Following on from considerations of site availability and species distributions across the landscape, differential species performance is the third 'general cause of succession' in Pickett's hierarchical model. This aspect has been addressed on the local scale in the present study, in terms of a subset of the differentiating factors recognized in Fig. 19.3, focusing mostly on turnover and growth of saplings in gaps.

The sapling data for 'all species' showed, as expected, suppressed growth on Panjang and Sertung (ash-affected). However, different species exhibit varying responses to pollution and drought, and this was exemplified by the data for *D. gaudichaudianum*. It was found in all of the gaps studied, and was often the first- or second-ranking species in abundance. This underscored the results of the landscape-scale evaluation, providing further evidence of the continuing significance of the species in canopy fill in the lowland forests of Panjang and Sertung, and of its rise on Rakata. It is striking, however, that particularly in terms of net change in frequency, but also with respect to height growth of survivors, this species has done disproportionately well on the two ash-affected islands in comparison to its performance in gaps on Rakata. This was also shown for its performance in comparison to that of the survivors of other species when under ash-stress. Thinning of the sapling and tree layer during the extreme dry season of 1994, when plants were generally under greatest stress, might have been followed by the release from competition of survivors during the following wet season (there may even be a fertilisation effect?). This provides a mechanism to explain the success of *D. gaudichaudianum* on Panjang and Sertung, consistent with the volcanic disturbance-affected pathways of Whittaker *et al.* (1989). Other species appear to be relatively intolerant of ashfall, as exemplified by the data from SG3 in which all 19 members of the cohort of *Pipturus argenteus* died, with just one recruit appearing the following year. Further examples can be drawn from our permanent plot data, with *Bridelia monoica* and *Arthrophyllum javanicum* seemingly badly affected by ashfall. In contrast to which, both *D. gaudichaudianum* and *Timonius compressicaulis*, if not over-mature, appeared able to withstand ash damage relatively well, as anticipated from the analyses of past successional changes on Panjang and Sertung (Whittaker *et al.* 1989, Bush *et al.* 1992).

The first discovery of the differences in forest dominance between Rakata on the one hand and Panjang and Sertung on the other, was made in 1982 by Tagawa and colleagues (Tagawa *et al.* 1985, Tagawa 1992). Their work was followed in 1983 (onwards) by further studies by Bush and colleagues (Bush, Richards & Jones 1986, Whittaker *et al.* 1989). Tagawa suggested that the differences might be explained largely by the order of arrival of the dominant species, and within Panjang and Sertung, by differing habitat preferences of *T. compressicaulis* and *D. gaudichaudianum*. Whittaker *et al.* (1989) pointed out the limitations of survey data which make it difficult to place great reliance on the order of arrival of particular species, and raised the additional hypothesis that disturbance by volcanism had an important part to play in the successional pathways which unfolded. In practice, these factors are difficult to isolate, and each must have had some part to play (Whittaker *et al.* 1989, 1992a, Bush *et al.* 1992, Thornton 1996). Thornton (1996) in his review of the forest divergence debate, notes two suggestions of Tagawa (1992) concerning the survival of *Neonauclea calycina* seeds. First, it was suggested that not only was *N. calycina* the first of the three dominants to be recorded, but that its seeds may have survived the 1883 eruption in deep gullies, germinating when rain erosion had cut through to the old soil surface; second, that *T. compressicaulis* and *D. gaudichaudianum* seeds, being larger, are more resistant to burial under volcanic ash than those of *N. calycina* (the order of seed size is *Neonauclea* < *Timonius* < *Dysoxylum*). Thornton (1996) further suggests that in consequence disturbance would put *N. calycina* at a disadvantage. These speculative suggestions are to some extent contradictory. To deal with them in turn, while survival through the 1883 events is difficult to entirely rule out (Whittaker *et al.* 1995), there is no particular basis to claim it for *N. calycina*, which was not recorded until the third floristic survey in 1905. From two germination trials conducted on Krakatau topsoils (Whittaker *et al.* 1995; S.F. Schmitt unpublished data) and from observations of seedling and sapling distributions in the present work, the suggestion that larger seed size confers greater resistance to burial must be subject to qualification. *N. calycina* appears to germinate and establish principally in sites of open canopy and a bare mineral soil, conditions which also appear to be favourable for *T. compressicaulis* (C. Ridsdale personal communication, S.F. Schmitt personal observation). From our observations it seems that a greater degree of canopy opening is needed for both, than is the case for *D. gaudichaudianum*. *D. gaudichaudianum* was the only one of the three which did not germinate from either experimental germination trials, suggesting that unlike the other two it is recalcitrant; seeds either germinate or they die. In separate experiments on Rakata (Whittaker & Turner 1994) it has been observed that seeds of *D. gaudichaudianum* can germinate at the soil surface, and it seems likely that they do so if buried just below it. This is consistent with our observations of carpets of ash-covered current-year seedlings in areas of Sertung in receipt of considerable ashfall *c.* 1994. The vast majority of seedlings succumb, but in the event of ashfall ceasing at the right moment, the species has the potential to establish in abundance.

Establishment of seedlings of *N. calycina* and *T. compressicaulis* appears at present to occur preferentially in areas of bare mineral soils such as landslide scars or the root-zones of treefall sites. Significant ashfalls might therefore be anticipated to create appropriate surface conditions for establishment of both *N. calycina* and *T. compressicaulis*, but to do so, as with *D. gaudichaudianum*, a supply of seed must be available. While both *N. calycina* and *T. compressicaulis* germinated in recent topsoil germination trials (S.F. Schmitt unpublished data), and while *N. calycina* seedlings also germinated from one sample collected from depth (the interior of a crab burrow), it is not clear if seeds of either species will actually germinate in a buried position, such as might be produced by an ashfall carpeting the soil surface. In the light of these observations, we do not find it easy to place the three species in a simple series of susceptibility to disturbance or to seed burial, as which of these (and indeed other canopy) species benefits from a disturbance depends on the particular nature, extent, location and timing of the event.

It seems reasonable to suggest that physical forcing factors, such as volcanism and climatic fluctuations, alter the competitive balance between species within these islands, and must have done so in the past. Whereas, at least during the period of the present study, factors such as gap size and type did not have an obvious or simple relation to sapling species performance. These properties of the gap itself, rather than external forcing factors, are likely to be of greater significance in forests of a more mature character, in which a greater range of the successional spectrum of species (*sensu* Swaine & Whitmore 1988) are present.

That is not to say that the regeneration requirements of different canopy species on Krakatau are indistinguishable. The three most significant canopy dominants on these islands illustrate this point. Each is clearly an early successional species, but in contrast to *D. gaudichaudianum*, both *N. calycina* and *T. compressicaulis* were rare in the sapling layer of the gaps in this study. Of the six sapling species which achieved 'common' status (see Methods): *Antidesma montanum*, *D. gaudichaudianum*, *Ficus fistulosa*, *Leea sambucina*, *Macaranga tanarius* and *Pipturus argenteus*, only *D. gaudichaudianum* is a large canopy tree at maturity. *N. calycina*, in contrast, was represented by one or two individuals in each of eight sites in 1994, and was reduced to five sites in 1995. *T. compressicaulis* was found in only four sites, in each as just one or two individuals. It is important to take life-history characteristics into account in evaluating such data, as some species may have a high probability of survival once in the sapling layer (we suspect *Terminalia catappa* to be in this category) whereas others (e.g. *M. tanarius*) are ephemeral. However, observations from the permanent plots do not suggest that either *N. calycina* or *T. compressicaulis* have unusually high rates of survivorship from sapling to canopy layer. The *N. calycina* stands on Rakata show signs of senescence, as do some of the older *T. compressicaulis* stands on Panjang. In interaction with extreme weather conditions, such as droughts and storms, the older specimens of both species appear particularly vulnerable to mortality and windthrow (cf. Mueller-Dombois 1995). Reversal of their decline would appear to require a far more severe disruption to canopy cover than has been provided by the relatively modest ash deposition of 1992–95 (*c.* 10–15 cm). Even then, the outcome of such large-scale canopy opening would be dependent on the timing of disturbance and on seed supply coincident with gap availability, which is unlikely to mimic precisely that of the early phase of forest formation earlier this century.

CONCLUDING REMARKS: PROCESSES OR PATHWAYS?

In preparing this chapter we were invited to consider the predictability of vegetation dynamics in a tropical context. As will be evident from the foregoing, certain trends are detectable in the Krakatau forests and it is possible to make projections concerning the rise or fall of particular species, just as authors such as Docters van Leeuwen (1936) and van Borssum Waalkes (1960) were able with some measure of success in the past. The broad characteristics of seral communities, particularly functional attributes such as growth form and dispersal spectra, may also be fairly predictable (cf. Whittaker *et al.* 1989, 1992b, Whittaker & Jones 1994a,b). However, as prefaced in the quote from Docters van Leeuwen in the introduction, successional pathways of the inland forests are highly variable, the resulting assemblages ('association' *sic*!) temporary and subject to change in space and time. The simplification of such successional change into representations of the form of the 'floristic development of a classic tropical prisere', while of some utility and convenience, may ultimately obscure a finer appreciation of the processes controlling and directing the succession. The approach adopted here has been to focus on the processes and their inter-relations. For instance, the recognition of the importance of environmental forcing factors (climatic, tectonic, geomorphological) is an important step in understanding the dynamics of the forest systems in this study (cf. Whitmore & Burslem, Chapter 20, this volume). These predominantly external abiotic factors have influence upon the nature of the disturbance regime, and the differential availability of species, while the local changes in the regeneration layer over time reflect the differential species responses to their biotic and abiotic environment (cf. Newbery *et al.*, Chapter 11, this volume). Many of the ideas discussed here under the headings of the role of disturbance, dispersal constraints and differential species performance, are threads which run through the Krakatau literature from its earliest stages, their significance recognised by several different authors. It is these sorts of factors which make the task of predicting floristic pathways so difficult in other than broad terms. Understanding the succession is thus a question of understanding the processes, in which documenting pathways is but a small part. The scheme developed in the present chapter, based closely on Pickett *et al.*'s (1987) model places these processes into a hierarchical framework which serves both to provide a research agenda and to clarify the inter-relations thus outlined: we suggest that its utility deserves a wider application in tropical forest research.

ACKNOWLEDGEMENTS

We wish to thank the Indonesian Institute of Sciences (LIPI) for granting research permission and the Indonesian Directorate General of Forestry and Nature Conservation (PHPA) of the Ministry of Forestry for permission to work on the Krakatau islands. We are grateful to Dr T. Partomihardjo for invaluable advice and assistance, Dr Soetikno Wiryoadmodjo and Dr Deddy Darnedi for help with permits, and Dr Johannis Mogea for permission to work in the Herbarium in Bogor. We are grateful to all of those who assisted in the gathering of the field data between 1993 and 1995, and to S.G. Compton, C.E. Ridsdale, A. Robock and S. Self for information cited in the text. We also wish to thank B.D. Turner for his input and advice at all stages of the development of this study, and P. Jepson for advice and support with logistics, and three anonymous reviewers for their comments on the manuscript. This research was financed by grant GT4/92/30/L from the Natural Environmental Research Council (S.F.S.), and further funding from the Royal Society's South-East Asia Rain Forest Research Programme (Publication A/126), and from Meyer International. Krakatau Research Project Publication No. 50.

REFERENCES

Backer, C.A. (1929). *The Problem of Krakatoa as Seen by a Botanist.* Published by the author, Surabaya.

Borssum Waalkes, J. van (1960). Botanical observations on the Krakatau Islands in 1951 and 1952. *Annales Bogorienses,* **4,** 1–64.

Brandani, A., Hartshorn, G.S. & Orians, G.H. (1988). Internal heterogeneity of gaps and species richness in Costa Rican tropical wet forest. *Journal of Tropical Ecology,* **4,** 99–119.

Brokaw, N.V.L. (1982). The definition of treefall gap and its effects on measures of forest dynamics. *Biotropica*, **14**, 158–160.
Bush, M.B. & Whittaker, R.J. (1995). Colonisation and succession on Krakatau: an analysis of the guild of vining plants. *Biotropica*, **27**, 355–372.
Bush, M., Richards, K., Jones, P. (Eds) (1986). *The Krakatoa Centenary Expedition 1983: Final Report. Miscellaneous Series*, **33**. Department of Geography, University of Hull, Hull.
Bush, M.B., Whittaker, R.J. & Partomihardjo, T. (1992). Forest development on Rakata, Panjang and Sertung: Contemporary dynamics (1979–1989). *GeoJournal*, **28**, 185–199.
Clements, F.E. (1916). *Plant Succession: An Analysis of the Development of Vegetation. Carnegie Institute Publication*, **242**. Washington DC.
Docters van Leeuwen, W.M. (1936). Krakatau 1883–1993. *Annales du Jardin Botanique de Buitenzorg*, **46–47**, 1–506.
Ernst, A. (1908). *The New Flora of the Volcanic Island of Krakatau*. Translated by A.C. Seward. Cambridge University Press, Cambridge.
Ernst, A. (1934). Das biologische Krakatauproblem. *Vierteljahresschrift der Naturforschenden Gesellschaft in Zürich*, **22**, 1–187.
Flenley, J.R. & Richards, K. (1982). *The Krakatoa Centenary Expedition: Final Report*. Department of Geography, University of Hull, Hull.
Hill, M.O. (1979). *TWINSPAN—A FORTRAN Program for Arranging Multivariate Data in an Ordered Two-way Table by Classification of Individuals and Attributes*. Cornell University, Ithaca, NY.
Hubbell, S.P. (1979). Tree dispersion, abundance, and diversity in a tropical dry forest. *Science*, **203**, 1299–1309.
Johns, R.J. (1986). The instability of the tropical ecosystem in New Guinea. *Blumea*, **31**, 341–371.
Kent, M. & Coker, P. (1992). *Vegetation Description and Analysis: A Practical Approach*. Belhaven Press, London.
Matelson, T.J., Nadkarni, N.M. & Solano, R. (1995). Tree damage and annual mortality in a montane forest in Monteverde, Costa Rica. *Biotropica*, **27**, 441–447.
Mueller-Dombois, D. (1995). Biological diversity and disturbance regimes in island ecosystems. In *Islands: Biological Diversity and Ecosystem Function* (Ed. by P.M. Vitousek, L.L. Loope & H. Andersen), pp. 163–173. Springer, Berlin.
Nelson, B.W., Kapos, V., Adams, J.B., Oliveira, W.J., Braun, O.P.G. & do Amaral, I.L. (1994). Forest disturbance by large blowdowns in the Brazilian Amazon. *Ecology*, **75**, 853–858.
Nicholls, N. (1993). ENSO, drought and flooding rain in South-East Asia. In *South-East Asia's Environmental Future* (Ed. by H. Brookfield & Y. Byron), pp. 154–175. United Nations University Press, Tokyo.
Oldeman, R.A.A. (1974). L'architecture de la forêt Guyanaise. *Memoires ORSTOM*, **73**, 1–204.
Partomihardjo, T., Mirmanto, E. & Whittaker, R.J. (1992). Anak Krakatau's vegetation and flora circa 1991, with observations on a decade of development and change. *GeoJournal*, **28**, 233–248.
Pickett, S.T.A., Collins, S.L. & Armesto, J.J. (1987). A hierarchical consideration of causes and mechanisms of succession. *Vegetatio*, **69**, 109–114.
Richards, P.W. (1952). *The Tropical Rain Forest: An Ecological Study*. Cambridge University Press, Cambridge.
Ridley, H.N. (1930). *The Dispersal of Plants Throughout the World*. Reeve, Ashford.
Ropelewski, C.F. & Halpert, M.S. (1987). Global and regional scale precipitation patterns associated with the El Niño-Southern Oscillation. *Monthly Weather Review*, **115**, 1606–1626.
Ropelewski, C.F. & Halpert, M.S. (1989). Precipitation patterns associated with the high index phase of the Southern Oscillation. *Journal of Climate*, **2**, 268–284.
Schmitt, S.F. & Partomihardjo, T. (1998). Disturbance and its significance for forest succession

and diversification on the Krakatau islands, Indonesia. In *Proceedings of the Third Flora Malesiana Symposium* (Ed. by J. Dransfield, M. Cooke & D.A. Simpson), in press. Kew Gardens, Richmond, Surrey.

Swaine, M. & Whitmore, T.C. (1988). On the definition of ecological species groups in tropical rain forests. *Vegetatio*, **75**, 81–86.

Tagawa, H. (1992). Primary succession and the effect of first arrivals on subsequent development of forest types. *GeoJournal*, **28**, 175–183.

Tagawa, H., Suzuki, E., Partomihardjo, T. & Suriadarma, A. (1985). Vegetation succession on the Krakatau Islands, Indonesia. *Vegetatio*, **60**, 131–145.

Ter Braak, C.J.F. (1988). *CANOCO – A FORTRAN Program for Canonical Community Ordination by Partial Detrended Canonical Correspondence Analysis, Principal Components Analysis and Redundancy Analysis.* ITI-TNO, Wageningen.

Thornton, I.W.B. (1996). *Krakatau: The Destruction and Reassembly of an Island Ecosystem.* Harvard Cambridge, Massachusetts.

Thornton, I.W.B., Partomihardjo, T. & Yukawa, J. (1994). Observations on the effects, up to July 1993, of the current eruptive episode of Anak Krakatau. *Global Ecology and Biogeography Letters*, **4**, 88–94.

Treub, M. (1888). Notice sur la nouvelle flore de Krakatau. *Annales du Jardin Botanique de Buitenzorg*, **7**, 213–223.

Umbgrove, J.H.F. (1949). *Structural History of the East Indies.* Cambridge University Press, Cambridge.

Walden, J., Whittaker, R.J. & Hill, J. (1991). The use of mineral magnetic analyses as an aid in investigating the recent volcanic disturbance history of the Krakatau Islands, Indonesia. *The Holocene*, **1**, 262–268.

Whitmore, T.C. (1974). Change with Time and the Role of Cyclones in Tropical Rain Forest on Kolombangara, Solomon Islands. *Commonwealth Forestry Institute Paper*, **46**, Oxford.

Whitmore, T.C. (1989). Changes over twenty-one years in the Kolombangara rain forests. *Journal of Ecology*, **77**, 469–483.

Whittaker, R.H. (1965). Dominance and diversity in land plant communities. *Science*, **147**, 250–260.

Whittaker, R.H. (1975). *Communities and Ecosystems.* Macmillan, New York.

Whittaker, R.J. & Jones, S.H. (1994a). The role of frugivorous bats and birds in the rebuilding of a tropical forest ecosystem, Krakatau, Indonesia. *Journal of Biogeography*, **21**, 245–258.

Whittaker, R.J. & Jones, S.H. (1994b). Structure in re-building insular ecosystems: an empirically derived model. *Oikos*, **69**, 524–529.

Whittaker, R.J. & Turner, B.D. (1994). Dispersal, fruit utilization and seed predation of *Dysoxylum gaudichaudianum* in early successional rain forest, Krakatau, Indonesia. *Journal of Tropical Ecology*, **10**, 167–181.

Whittaker, R.J., Bush, M.B. & Richards, K. (1989). Plant recolonisation and vegetation succession on the Krakatau Islands, Indonesia. *Ecological Monographs*, **59**, 59–123.

Whittaker, R.J., Asquith, N.M. & Bush, M.B. (1990). Krakatau Research Project 1989 Expedition Report. Unpublished report. School of Geography, University of Oxford, Oxford.

Whittaker, R.J., Walden, J. & Hill J. (1992a). Post-1883 ash fall on Panjang and Sertung and its ecological impact. *GeoJournal*, **28**, 153–171.

Whittaker, R.J., Bush, M.B., Partomihardjo, T., Asquith, N.M. & Richards, K. (1992b). Ecological aspects of plant colonisation of the Krakatau islands. *GeoJournal*, **28**, 201–211.

Whittaker, R.J., Partomihardjo, T. & Riswan, S. (1995). Surface and buried seed banks from Krakatau, Indonesia: implications for the sterilization hypothesis. *Biotropica*, **27**, 346–354.

20. MAJOR DISTURBANCES IN TROPICAL RAINFORESTS

T. C. WHITMORE* and D. F. R. P. BURSLEM[†]

Geography Department, University of Cambridge, Cambridge CB2 3EN, UK; [†] *Department of Plant and Soil Science, University of Aberdeen, Aberdeen AB24 3UU, UK*

SUMMARY

1 Disturbances of various kinds influence the species composition of tropical rainforest in numerous ways. Community-wide, large-scale disturbance to the forest canopy at a frequency of once every one or few tree generations (that can range from decades to centuries depending on species) has now been observed or deduced at many places throughout the tropics. Causes are landslides, volcanoes, mobile rivers, wind, lightning, drought, fire and various human activities.

2 Factors differ in their impact and a broad distinction can be made between disturbances leading to forest recovery by primary succession, by secondary succession, or by resprouting and release of established seedlings. Detection is either direct, through species composition or population structure, or indirect, through, for example, charcoal, potsherds or a concentration of useful plants. There are implications for equilibrium theorists and biodiversity pundits, for perhaps most of the world's tropical rainforests experience large-scale, community-wide disturbances.

INTRODUCTION

Disturbance to a forest canopy, any change in the *status quo*, influences species composition in numerous ways. The focus here is on dryland lowland tropical rainforests, about which most is known, and on trees. Disturbance thus defined is a gigantic subject and only one kind is explored in detail, namely community-wide disturbance (*sensu* van der Maarel 1993) at a frequency of at least once every few tree generations, and that can range from decades to several centuries depending on species. This mode of disturbance needs first to be seen in its context.

The species composition of a tropical rainforest is determined by a range of factors that can be arranged roughly in a hierarchy of descending importance (Whitmore 1975). First, overriding all else, phytogeography determines the pool of available species. Phytogeography itself is determined by plate-tectonic history and palaeoclimates. Second in the hierarchy comes community-scale disturbance, the subject of the present analysis. Third in importance is geology, which influences topography and soil. Where there is a sharp lithological boundary there is usually a sudden change in flora. Different intermingling suites of species are commonly found on topographic catenas. Finally, at the bottom of the hierarchy, composition is determined by a whole complex of biological causes as well as by chance. The occurrence of a tree at any given site is a result of seed production, dispersal, survival and germination, followed by seedling establishment, persistence below closed canopy, and often then by release and upwards growth after development of a canopy gap. Seed and seedling biology are affected by many factors, including quantity and quality of solar radiation, herbivory, drought and physical damage. Species differ from each other in their safe sites for regeneration, for example in microtopographic and soil preferences. Both site and chance at many levels play a role. This bottom level of the hierarchy has received great attention over the past two decades, much of it summed up as 'gap-phase dynamics', because release of seedlings in a small canopy gap (patch-size disturbance *sensu* van der Maarel 1993) is central to the determination of species composition of that patch for the rest of the forest growth cycle. Moreover, at this scale it is possible to conduct experiments.

In fact, disturbance defined as change of the *status quo*, operates at all levels of this hierarchy. Plate tectonics cause change on the secular, evolutionary time-scale (Whitmore 1981, 1987). Climatic change operates on the time-scale of millennia to centuries, with big increases in the extent of rainforests since the last Glacial maximum 18000 years ago (Flenley 1979, Whitmore 1981, Whitmore & Prance 1987). There is contemporary interest in how global warming may cause future changes (Markham, 1998). Canopy disturbance is important on time-scales comparable to the life-span of a big rainforest tree: one to a few centuries. This is the scale at which population and community dynamics are affected, the subject of the present symposium. At this scale disturbance may initiate a new generation of trees, hence the species composition for the next turn of the forest growth cycle.

It will be noted from this analysis that two spatial scales of canopy disturbance can be distinguished (van der Maarel 1993). First, community-wide disturbance creates large canopy gaps and allows establishment of a forest of pioneer species (Swaine & Whitmore 1988). This will usually be a forest of different species composition from the preceding one, and will change at the following growth cycle back to a forest of climax or non-pioneer species

(Swaine & Whitmore 1988) unless similar disturbance is repeated. Community-wide disturbance ranks high in the hierarchy of forest-controlling factors because it can trigger a complete change in forest floristic composition. Second, and by contrast, patch disturbance, on the scale of just one to a few trees, comes lowest in the hierarchy and operates on the ecological class (guild) usually called climax, shade-tolerant, late successional, old-growth, or primary species whose seedlings can persist (as a 'seedling bank') below closed canopy. These species occur in the canopy in kaleidoscopic mixture due, as described above, to the interaction of numerous facets of their biology, plus the multiple roles of chance, from seed production onwards. The fluctuation of canopy composition in space as well as time was described by Aubréville long ago (e.g. Aubréville 1938), and Richards (1952, p. 49) publicized it as the mosaic or cyclical theory of regeneration. It has taken a human generation to unravel some of the intricacies. The framework is now well understood, but examples and embellishments continue to be added, many at this symposium, namely phenology and recruitment (Newbery, Songwe & Chuyong, Chapter 11) seed size (Grubb, Chapter 1), and dispersal (Forget, Milleron & Feer, Chapter 2), seedling architecture (Bongers & Sterck, Chapter 6), demography (Zagt & Werger, Chapter 8), release (Brown & Jennings, Chapter 4) and insect attack (Hammond & Brown, Chapter 3). By contrast, the pervasiveness and diversity of community-scale canopy disturbance is only now becoming fully realized.

LARGE-SCALE DISTURBANCE

Large-scale, or community-wide, disturbance can lead to recruitment of species that have no seedling bank below closed canopy and which arrive after gap creation either from the soil seed bank or in a 'seed rain'. In lowland tropical rainforests these species succeed in gaps of $c.$ 0.1 ha or more. This is the ecological species group now usually called pioneers. It too was recognized long ago and defined as 'biological nomads' (van Steenis 1958). Pioneer tree species have a syndrome of characters whose manifestation is variable (e.g. most, but not all, have small seeds, low density timber and large leaves; Swaine & Whitmore 1988, Whitmore 1990), but all do possess two characters which define the group, namely a requirement for full sunlight for germination and then for establishment and growth (Swaine & Whitmore 1988). Because of these two characters pioneers depend on repeated community-wide disturbance to regenerate, without which they cannot persist at a site. Longevity of pioneers varies from less than 10 years (Saulei 1984) to perhaps a century (Taylor 1957), and tree height varies in parallel from $c.$ 2 m to over 30 m. The long-lived big pioneers are mostly useful timber trees and South American rainforests in particular have many such species (Finegan 1992), of which the most famous is balsa, *Ochroma lagopus*. Although the persistence of pioneers at a locus depends on repeated extensive disturbances, for the big long-lived species disturbance need only be about once a century, long on the human time-scale. Thus, rare disturbances can be very important (Whitmore 1974, 1989). Research and development-aid projects span only a miniscule part of this time-scale and easily miss them. As scientific observations of tropical rainforests have increased over the past half century, more and more evidence has been found of big disturbances on the decadal to centennial scale.

KINDS OF LARGE-SCALE DISTURBANCE

Large-scale disturbances differ in their impacts on forest structure and species composition. A broad distinction can be made between, first, disturbances that are followed by primary succession and, second, disturbances followed by secondary succession. When the second sort of disturbance occurs at low intensity it is followed directly by regeneration of climax species by resprouts or from pre-established seedlings, instead of pioneers.

The first group of disturbances includes factors which destroy all vegetation and disrupt the soil profile, resulting in loss or deep burial of the soil seed bank. It includes landslides (e.g. Guariguata 1990), volcanoes (Whittaker, Bush & Richards 1989) and mobile rivers (Salo, Kalliola, Häkkinen *et al*. 1986). Some anthropogenic activities have similar effects, such as clearance for continuous cultivation or construction of logging roads and log landings.

The second group includes most other large-scale disturbance factors. Among these we recognize a difference between disturbance events that disrupt the soil profile as well as open the canopy (e.g. severe windstorms, shifting cultivation and logging) and those that open the canopy without disrupting the soil profile (e.g. lightning, drought, fires and epidemic herbivory).

Although there are very few long-term observations of tropical rainforest regeneration after large-scale disturbance, general observations plus the results of experiments on the response of seeds and seedlings to canopy removal and to soil disturbance suggests that the different kinds of major disturbance have different effects on the rate and pattern of forest regeneration.

Let us now therefore consider in turn the various disturbance factors that have been noted to create conditions for each of the two sorts of regeneration.

Disturbances followed by primary succession

These are the first sorts of disturbances.

Landslides

Extensive landslides follow earthquakes or periods of very high rainfall. New Guinea, which has a young, steeply mountainous landscape and lies on a continental plate margin, is very susceptible. Garwood, Janos and Brokaw (1979) drew attention to this, estimated that 8–16% of the land surface is affected per century, and made a contrast with Panama where, although earthquakes occur, they are less frequent and 2% is affected per century. Aerial photographs of 1965 show clear scars of 1918 and 1935 landslides following earthquakes (Johns 1986). At Madang in November 1972 an earthquake destroyed a quarter of the forest across an area of 240 km² (Johns 1986). Steep slopes are prone to landslides after heavy rain and this affects a further c. 3% of New Guinea per century. Five weeks of unprecedentedly heavy rain at Lambir, Sarawak, in early 1963 created huge landslides, still clearly visible on aerial photographs 18 years later, when further very heavy rain created more (Ohkubo, Maeda & Kato 1995). Plant colonization of landslides on Caribbean islands is limited by low nutrient supply and may include species that only rarely occur in other forest habitats (Guariguata 1990, Dalling & Tanner 1994). Landslides only occur in mountainous terrain so do not affect rainforests on flattish country such as the Amazonian *terra firme*.

Volcanoes

Where active volcanoes occur their eruption affects forests both directly and indirectly (Schmitt & Whittaker, Chapter 19, this volume). Species-poor forests dominated by the big pioneer tree species *Octomeles sumatrana* and *Paraserianthes falcataria* were recorded from the environs of Mt Victory in Papua 84 years after an eruption (Taylor 1957). The forests of the Krakatau group of islands between Sumatra and Java were destroyed in a famous huge eruption in 1883. A century later only a species-poor forest had redeveloped (Whittaker *et al*. 1989). On these islands long distance from a seed source and the poor dispersal capacity of many primary forest species has arrested forest recovery. The Krakatau eruption also caused a tsunami and ashfalls to which forest structure and composition on Ujung Kulon west Java, 70 km distant, still bear witness (Hommel 1987).

Mobile rivers

A sensation was created a decade ago with the demonstration that the rivers of the Amazonian floodplain in Peru move laterally across the landscape in the annual floods by as much as 180m (Salo et al. 1986); 12% of the part of Peru in Amazonia is affected (Nelson 1994) and these forests are in a continuous state of primary succession. A series of individually species-poor successional forest types develops. Species composition in the floodplain forests is thus under the dominance of river movement, and contributes to the total (gamma) diversity of the region (which also includes the *terra firme* forest of the interfluves). The same processes are at work on a smaller scale in New Guinea (Vogelkop Peninsula, Sepik, Ramu and Papua) where big rivers debouch from the precipitous mountainous interior (Johns 1986).

Disturbances followed by secondary succession or recovery

These are the second sort of disturbances and are followed by regrowth of pioneers or of climax species, dependent on intensity.

Wind

Cyclones occur in two belts, between 10° and 20° north and south of the equator and become less frequent at the edges of these zones. Islands of the Caribbean are hit on average every 15–20 years (Sugden 1992) so the forest is permanently influenced. However, the amount of damage, and hence species composition of the regrowth forest, varies. Hurricane Hugo in 1989 snapped or uprooted 20, 14 and 80% of stems in affected forests in Puerto Rico, Jamaica and Nicaragua respectively (Sugden 1992). In the Blue Mountains of Jamaica Hugo caused 7.1% mortality over a 5-year period, only slightly more than normal, with many trees, though stripped of foliage and with badly damaged crowns, subsequently recovering. Thus, species composition of the forest was little changed (Bellingham, Kapos & Varty *et al.* 1992). By contrast, a typhoon at Bislig Bay, east Mindanao, Philippines, flattened swathes of forests and regrowth was of pioneers (Ashton 1993). In Papua, extensive pure forests of the dipterocarp *Anisoptera thurifera* ssp. *polyandra* have grown up since a 1940s cyclone (Johns 1986). In Nicaragua after Hurricane Joan of 1988, recovery was by a mixture of resprouting primary forest trees and release from the seedling bank, supplemented by the pioneers *Cecropia obtusifolia* and *C. insignis* (some of which, however, soon died due to lack of their mutualist ant *Azteca*) (Ferguson, Boucher, Pizzi & Rivera 1995).

Canopy structure may be controlled by cyclones even where species composition is not affected, especially where they are frequent. Examples are the dense, low uniform canopy and gnarled crowns of the forests of the eastern Sierra Madre mountains of Luzon, Philippines, which are frequently struck (Ashton 1993); and the open canopy with scattered big trees clothed by climbers found along the ocean-facing slopes of the north Queensland rainforests (Webb 1958).

There is evidence of big winds influencing species composition outside the cyclone belts. A variety of lines of evidence has shown that rare wind storms influence the dipterocarp rainforests of Peninsular Malaysia, 2–6°N. In the 1950s the storm forest of Kelantan consisted of disturbance-favoured, light-demanding, climax, red meranti *Shorea* species and resulted from a windstorm in 1880 (Wyatt-Smith 1954, see fig. 7.20 in Whitmore 1990). The Sungai Menyala forest, under close study since 1948, has experienced two major windstorms in 1948 and 1958, that caused respectively 76 and 52 deaths of trees 10 cm diameter or more, by uprooting and snapping, on a 2-ha observation plot with *c.* 950 trees of this size, and continues to be dominated by disturbance-favoured *Shorea* and other dipterocarps (Whitmore 1975, Manokaran & Swaine 1994).

Anecdotal reports of windstorms destroying a few hectares, and of line squalls causing corridors of destruction, from around the peninsula (Whitmore 1984, Manokaran & Swaine 1994) provide evidence of the kind of disturbance that permits perpetuation of this group of dipterocarps as opposed to the species with more strongly shade-tolerant seedlings (Whitmore & Brown 1996).

Storm tracks in the form of corridors of treefalls are relatively common in the Ituri forest, northeastern Zaire (Hart, Hart, Dechamps *et al.* 1996).

Recently, in the Brazilian Amazon, inspection of satellite images followed by ground-truthing of a few have shown community-scale, fan-shaped areas of secondary forest that have filled canopy gaps created by downbursts of wind from moving convectional storms. These blowdowns total 900 km^2 in area across 3.8 million km^2 of the Amazonian rainforest, most densely (at up to 0.2–0.3% of a Landsat scene) in a north–south belt running through Manaus where rainfall and thunderstorms are most frequent. The largest is 3370 ha, but most are 30–100 ha in extent (Nelson 1994, Nelson, Kapos, Adams *et al.* 1994).

Lightning

Aerial photographs of mangrove forests show the smooth canopy is commonly pock-marked with small holes where patches of a few trees have been killed by lightning, which strikes a single tree and then travels through root connections to others adjacent which are either killed entirely or just on the facing side (Johns 1986). A similar pattern of death can sometimes be found in dryland evergreen rainforest (T.C. Whitmore personal observation) and has been seen on Krakatau island (R.J. Whittaker personal communication), but is not easily detected on aerial photographs because of the very uneven canopy top, so its frequency cannot be assessed. Community-wide lightning gaps have been recorded from New Guinea, namely a 50-m diameter hole in mangrove forest, and as circular holes in *Nothofagus* forest which also has root connections (Johns 1986), and on Krakatau island (R.J. Whittaker personal communication).

Drought and fire

Until about 1980 it was a truism that primary tropical rainforest does not catch fire (Richards 1952, Whitmore 1975). Then charcoal fragments from the upper Rio Negro in Colombia and Venezuela produced evidence of widespread fires from 6250 BP onwards, some of which could be associated with drier periods, while some were associated with pottery shards dated 3750–460 BP (Sanford, Saldarriaga, Clark *et al.* 1985, Saldarriaga & West 1986). More recently, charcoal from 90 km north of Manaus in the central Amazon has been dated to 1800–550 BP, and phytoliths from the same site show it has been forested since 4600 BP (Piperno & Becker 1997). Charcoal dated 2430 and 1110–1180 BP has been found below lowland rainforest at La Selva, Costa Rica, believed to result from fire set by human activity in an unusually dry spell (Horn & Sanford 1992). There are records of rare but extensive fires in Guyana shown by both charcoal and present-day species composition (Fanshawe 1954, Whitton 1962, T.C. Whitmore personal observation). Radiocarbon-dated charcoal from the soil below the Ituri forest, northeastern Zaire, shows a series of episodes of fire since 2290 BP when the climate is known to have become drier. The pattern suggests escaped human fires that occurred in unusually dry periods and spread a short distance into forest that was only marginally inflammable (Hart *et al.* 1996).

In Asia the 1983 Great Fire of Borneo, followed 18 months with only one-third of average rainfall. Three million hectares of rainforest were destroyed east of the 3500/3000 mm isohyet (Goldammer & Seibert 1990). Most of it had been logged so had a high fuel load of dry trash but 1.35 million ha were primary forest and of this drought itself, not subsequent fire, killed the big trees on 100 000 ha (Whitmore 1990). The 1982–83 drought was associated with a strong El Niño–Southern Oscillation (ENSO) event and later ENSO droughts led to further fires in 1991 and 1997. Subsequent examination of the historical record has produced evidence of droughts back to the 19th century (Whitmore 1990) and charcoal provides evidence of earlier fires from 17 510 to 350 BP, some of them persisting for years in superficial coal seams, spreading out at time of drought (Goldammer & Seibert 1990). Fires have been recorded in Papua New Guinea since 1885, many of them known to have followed drought (Johns 1986).

South of New Guinea in Queensland the boundary of tropical rainforest against wet sclerophyll forest dominated by *Eucalyptus grandis* is fire-controlled. Rainforest has expanded substantially since the late Pleistocene (Hopkins, Ash, Graham *et al.* 1993), most recently this century since burning by Aboriginal peoples has ceased (Adam 1992). Fire can also control the rainforest boundary at its limits in continental SE Asia (Ashton 1993, Stott, Goldammer & Werner 1990). On Indonesian mountains *Casuarina junghuhniana* and *Pinus merkusii* both have ranges that extend on to dry, fire-prone ridges and would have much smaller populations in the absence of periodic natural or man-set fire (van Steenis 1972).

The 1982–83 ENSO drought also increased mortality on Barro Colorado Island, Panama. Amongst stems ≥1 cm diameter there was 3% mortality during 1982–85 compared to 2% during 1985–90; about 70% of species were affected, but not equally. Canopy trees were significantly affected (cf. Borneo), as were species of moister sites, while treelets and shrubs showed little difference. Herbs were more sensitive than woody plants (Wright 1992, Condit, Hubbell & Foster 1995).

In the high mountains of New Guinea frosts are associated with drought periods, and can kill forest. Anthropogenic fire often follows and tree-free frost hollows are created (Johns 1986). Similar damage has been reported to upper montane rainforests in Sri Lanka by drought and frost followed by fire (Werner 1988).

In southwestern Amazonia, at 7–11°S, 66–74°W, is an area of 121 000 km² of rainforest dominated by bamboos, occupying much of central and east Acre State in Brazil and extending across the frontier into southeast Peru and northeast Bolivia (Nelson 1994). These bamboos are monocarpic, flowering and dying on a 26–29 year cycle as a slow wave moving across the area, and fire may follow their death, especially in ENSO dry years. The ecology of these vast and remote forests is unknown. In slightly seasonal tropical Asia dominant bamboo is a sign of disturbed semi-evergreen rainforest, for example parts of the forests of Vietnam recovering from the aerial spraying by Americans of the herbicide Agent Orange during the Vietnam War (Ashton 1986, Richards 1984). The edges of the bamboo forest are rounded or amoeboid in shape, suggestive of recent spread into adjacent forest after fire (Nelson 1994). It is possible that these bamboo forests are the result of fire and, perhaps also, of past human interference.

Epidemic herbivory

Several outbreaks of mass defoliation by lepidopteran larvae have been observed. There are two dramatic instances. Patches of several to many hectares of the almost pure stands of *Shorea albida* that form parts of the peat swamp forests of Sarawak were reported killed, and other species defoliated but not killed, by Anderson (1961). An area of 5–10 km^2 of *Excoecaria agollocha*, at the inland edge of a mangrove forest in north Sumatra, was likewise defoliated, but not killed (Whitten & Damanik 1986).

Man

Community-wide disturbances caused by human activity are very varied and are variously followed by either of the sorts of regeneration distinguished above. In addition to the human disturbances occurring today more and more signs are being discovered of extensive former human activity (Boerboom & Wiersum 1983), many in places now very sparsely inhabited or empty. This applies especially to the Amazon, which today has about a quarter million inhabitants compared to perhaps 10 million at the time of European contact, when Europe itself had a population of 12 million (Layrisse 1992), a reduction to 2.5%. Balée (1989), who reviews the evidence, believes 12.8% of the *terra firme* forests of Amazonian Brazil show signs of human influence. He reports 8000 km^2 of forests across the basin in which the Brazil nut tree (*Bertholletia excelsa*) is dominant. This is an economically valuable, long-lived, light-demanding species, likely to have been favoured, and perhaps planted, by indigenous communities. There are 100 000 km^2 of liane-dominated forests, mainly in the region of the big southern tributaries of the Xingu and Tocantins rivers. They occur as patches of a few up to hundreds of hectares. On sample plots many of the important tree species were known disturbance-indicators, including the palm babassu *Attalea speciosa* (*Orbignya phalerata*) and *Inga* spp. The soils are diverse and include black *terra preta do indio*. *Terra preta* are known to have been created by prolonged Amerindian cultivation. They usually contain potsherds. These occur scattered across the Amazon (and are sought out today by farmers). The conclusion drawn is that the liane forests result from swidden cultivation, and are either in a state of succession or are a plagioclimax (Balée 1989, Balée & Campbell 1990, Nelson 1994).

In the Serra Parima on the Brazil/Venezuela frontier (c. 2° 40′N, 64°W) there occur 600 km² of fern savannas dominated by *Pteridium aquilinum* var. *arachnoideum*. These have been created by burning by Yamoama Indians over many decades or centuries either to flush game or as 'recreational pyromania' (Nelson 1994).

Another sign of past human influence is forests with high concentrations of useful plants, fruit trees and medicinal species (Balée 1987). Such concentrations have also been found in Middle America, where present-day rainforest high in useful plants occurs on what was open farmland at the time of the Spanish Conquest (Budowski 1965, Gomez-Pompa & Kaus 1992). At Quintana Roo in Mexico it contains scattered giant mahogany (*Swietenia macrophylla*) trees (L. Snook personal communication). Another example is the palmunculus *Areca guppyi* in the Solomons which occurs outside its native range, on the floor of high forest, attesting to the former existence of pagan shrines.

The first detection of past extensive human influence in tall, apparently undisturbed, lowland rainforest was in Nigeria, where Jones (1955–56) was able to show that the abundant and commercially valuable Meliaceae in the upper part of the canopy were not replacing themselves and that they have resulted from colonization by forest of farmland abandoned during a civil war. Here too the soil contained charcoal and potsherds as testimony to the former human habitation.

The forests of north Kolombangara in the western Solomon Islands are similar to these of Nigeria in having an upper canopy of big tree species known to regenerate after disturbance (Whitmore 1974). Population structure, with the big trees not replacing themselves, suggests this forest has grown up since catastrophic disturbance about a century ago. Tree demography has been studied over 30 years, 1964–94. Cyclones, which occurred in 1967–70, 3.5–6 years after the study started, caused massive canopy damage (Whitmore 1974, 1989) but have not changed the relative abundance of the 12 common big tree species. The conclusion is that the catastrophe that preceded forest establishment was probably cultivation, not an earlier cyclone (D.F.R.P. Burslem & T.C. Whitmore unpublished data), and evidence has recently been found that these now empty landscapes were indeed once populated (Chaplin 1993).

In Asia the very extensive *Adinandra*- or *Ploiarium*-dominated forest of Singapore, nearby islands and south Peninsular Malaysia has regrown on areas exhaustively farmed for gambir (*Uncaria gambir*, a source of tannin) and pineapples, with soil degradation and loss of organic matter, up to 80 years ago. These vast secondary forests and the even more extensive ones spread through the Malay archipelago today, dominated by small pioneer trees, mainly *Macaranga, Mallotus, Trema, Commersonia* (and *Alphitonia* and *Trichospermum* in the east), have no natural analogue. They have arisen largely as a result of shifting agriculture and are becoming ever more extensive as logging opens fresh forests and shifting agriculture follows. Their constituent species would naturally have occurred only as small patches of a hectare or less on landslides and river banks. Similar anthropogenic secondary forests occur in Africa (*Musanga, Trema*) and South America (*Cecropia, Ochroma, Trema*). Succession back to primary forest is dependent partly on there being seed sources within dispersal range and on the continued presence of dispersers. In Singapore areas protected as the central water catchment since the 1950s have developed from *Macaranga*, etc. to a forest with only a few primary species, all of which have small fruits that the surviving bird, bat and squirrel species can disperse (Corlett & Turner, 1997). On Hong Kong Island, the subtropical forest regrown since fire was excluded in the late 1940s is likewise a small-seeded subset of the total because the animals that once dispersed large diaspores are now locally extinct (Dudgeon & Corlett 1994).

DISCUSSION

Thus, signs have now been found throughout the humid tropics of past widescale disturbance. Most rainforest trees probably only live one or a few centuries so the direct signs of past disturbance shown by species composition and population structure will disappear on that time-scale if seed and seed dispersers of climax species are available. Indirect signs, some but not all of which testify to human disturbance, last longer, most as charcoal or potsherds in the soil. The presence of species introduced or increased in abundance by people, or the failure to recolonize by poorly dispersing species, also provide long-persisting clues to past disturbance. Another long persistent sign of past human impact is species-poor forests where repeated cultivation and fire have led to soil degradation which is either irreversible or only ameliorates very slowly.

There are, by contrast, some tropical rainforests which show the classic signs of stability. In the lower Amazon the preponderance in a survey area of $100 000 km^2$ of big trees of many species most of which have extremely dense timber ($>700 kg m^{-3}$ at 15% moisture content) (Whitmore & Silva 1990) is a sign of a forest with a very fine canopy structural mosaic, patch-scale disturbance, the gaps tiny and resulting from single tree death, thus favouring regeneration of this species group. Scattered across the rainforests of Africa are areas dominated by big trees of one or a few species of caesalpinioid Leguminosae (White 1983) that have very dense timber and very shade-tolerant seedlings. For example, *Gilbertiodendron dewevrei* occurs on plateaux around the Zaire basin set within a more species-mixed forest with more canopy gaps (Hart, Hart & Murphy 1989). These two forest types have respectively 18 and 65 species of trees ≥10 cm diameter per hectare. No difference in soil can be found. *Gilbertiodendron* is dominant in the seedling bank and all larger size classes, and has big, poorly dispersed, seeds. Its continuing dominance of the forest is believed to be favoured by very small-scale canopy disturbance. Recent studies in the Täi forest of Côte d'Ivoire show that this too has tiny canopy gaps, of mean size 50 m^2. These form at 0.4% of the area per year so the turnover time is 240 years (Jans, Poorter, van Rompaey *et al.* 1993). The forests of western Malesia dominated by dense-timbered, shade-tolerant Dipterocarpaceae (e.g. *Neobalanocarpus heimii*, *Shorea* section *Shorea*) are also likely to be the result of minimal disturbance, *viz.* a fine-scale canopy mosaic, with the gaps created by single tree death.

There is some evidence, however, that on a long, historical time-scale, even these forests of shade-tolerant species, with strongly positive stand tables, that are currently perpetuating themselves *in situ*, may be unstable. The species identified from soil charcoal show that northeastern Zaire has had tropical rainforest in the central Ituri region for 4000 years, yet several species abundant in the charcoal flora are absent today, and *Gilbertiodendron dewevrei*, which currently forms consociations to 10 km across, was not identified in the charcoal flora (Hart *et al.* 1996). Newbery *et al.* (Chapter 11, this volume) cite similar changes amongst other caesalpinioid Leguminosae in the Cameroon rainforest. Unstable population structure was detected by Poore (1968) at Jengka, Malaysia, where *Shorea leprosula* showed an 'invasion front', and *Dipterocarpus crinitus* a negative stand table. The late Pleistocene record of the Queensland rainforest shows its vagility at regional level, in this case correlated with climatic fluctuations (Hopkins, Ash, Graham *et al.* 1993); but what disturbances have caused the African and Malaysian species fluxes just mentioned remain unknown.

CONCLUSIONS

Disturbance in its widest sense is all pervasive in the determination of the species composition of rainforests, as indeed of all vegetation (Sousa 1984). The two longest time-scale disturbances are due to continental drift and climatic change and all others occur within the framework they have created. At the other extreme lies small-scale or patch-size disturbance, related to mortality of one or a few trees. This has recently been subject to a great deal of study and analysis because it is amenable to experiment within the temporal and spatial scales at which scientists and foresters usually work.

Large-scale, or community-wide, disturbance is less easily studied except by observation. As observations have intensified its pervasiveness has been increasingly realized, adding colour to the long-running debate on whether rainforests are or are not equilibrium communities. It has numerous causes, including the actions of humans, and some of its signs can persist many centuries. An attempt has been made here to synthesize and codify the present state of knowledge. It is found that a distinction can usefully be made between disturbances followed by primary succession and by secondary succession, and that where the latter kind occur at low intensity there may be direct regrowth of primary species without succession. Human impact on tropical rainforests is very diverse and can result in either sort of succession, dependent on its intensity.

There is now so much evidence for community-wide disturbance that tropical rainforests whose current species composition is controlled instead by small patch-scale canopy disturbances may be the minority.

Today, human impact on tropical rainforests is intense and much discussed. Sometimes alarmist conclusions are drawn (e.g. Myers 1995). Rainforests are the contemporary 'forest frontier' and are suffering the attrition and fragmentation by human action earlier meted out sequentially in China, the Mediterranean, western Europe, temperate and boreal America, and Australasia (Mather 1990). But because of the power of modern technology this current onslaught is more rapid than the earlier ones. There is intense speculation on the possible consequences of loss of area and fragmentation on tropical rainforest biodiversity, which translates mainly into impact on species composition (Whitmore & Sayer 1992). It is perhaps helpful for the clarification of these issues to realize and analyse the dynamism in species composition that has occurred in the past. Contemporary human activity is the latest chapter in a long saga of disturbances, some of which this chapter has discussed.

ACKNOWLEDGEMENTS

T.C.W. is very grateful to everyone who continually sends him reprints and books, without which it would be impossible even to attempt to keep abreast of new developments in rainforest science.

REFERENCES

Adam, P. (1992). *Australian Rainforests.* Clarendon Press, Oxford.

Anderson, J.A.R. (1961). The destruction of *Shorea albida* forest by an unidentified insect. *Empire Forestry Review*, **40**, 19-29.
Ashton, P.S. (1986). Regeneration in inland lowland forests in south Vietnam one decade after aerial spraying by agent Orange as a defoliant. *Bois et Forêts des Tropiques*, **211**, 19-34.
Ashton, P.S. (1993). The community ecology of Asian rain forests, in relation to catastrophic events. *Journal of Biosciences*, **18**, 510-514.
Aubréville, A. (1938). La forêt coloniale; les forêts de l'Afrique occidentale française. *Annales Academie Sciences Coloniale*, **9**, 1-245. Also in translation (1971) Regeneration patterns in the closed forest of Ivory Coast. In *World Vegetation Types* (Ed. by S.A. Eyre), Chapter 2. Macmillan, London.
Balée, W. (1987). Cultural forests of the Amazon. *Garden*, **11**, 12-14, 32.
Balée, W. (1989). The culture of Amazonian forests. In Posey, D.A. & Balée, W. (Eds.) *Resource Management in Amazonia: Indigenous and Folk Strategies* (Ed. by D.A. Posey & W. Balée). *Advances in Economic Botany*, **7**, 1-21.
Balée, W. & Campbell, D.G. (1990). Evidence for successional status of liana forest (Xingu river basin, Amazonian Brazil). *Biotropica*, **22**, 36-47.
Bellingham, P.J., Kapos, V., Varty, N., Healey, J.R., Tanner, E.V.J., Kelley, D.L. *et al.* **(1992).** Catastrophic disturbances need not cause high mortality. The effects of a major hurricane on forests in Jamaica. *Journal of Tropical Ecology*, **8**, 217-223.
Boerboom, J.H.A. & Wiersum, K.F. (1983). Human impact on tropical moist forest. In *Man's Impact on Vegetation* (Ed. by W. Holzner, M.J.A. Werger & I. Ikusima), pp. 83-106. Junk, The Hague.
Budowski, G. (1965). Distribution of tropical American rain forest species in the light of successional processes. *Turrialba*, **15**, 40-42.
Chaplin, G. (1993). Silvicultural Manual for the Solomon Islands. *Solomon Islands Forest Record*, **8**.
Condit, R. Hubbell, S.P. & Foster, R.B. (1995). Mortality rates of 205 neotropical tree and shrub species and the impact of a severe drought. *Ecological Monographs*, **65**, 419-439.
Corlett, R.T. & Turner, I.M. (1997). Long-term survival in tropical forest remnants in Singapore and Hong Kong. In *Tropical Forest Remnants* (Ed. by W.F. Laurance & R. Bierregaard), pp. 333-345. University of Chicago Press, Chicago.
Dalling, J.W. & Tanner, E.V.J. (1994). An experimental study of regeneration on landslides in montane rain forest in Jamaica. *Journal of Ecology*, **83**, 55-64.
Dudgeon, D. & Corlett, R. (1994). *Hills and Streams: An Ecology of Hong Kong.* Hong Kong University Press, Hong Kong.
Fanshawe, D.B. (1954). The vegetation of British Guiana, a preliminary review. *Imperial Forestry Institute Oxford, Institute Paper*, **29**.
Ferguson, B.G., Boucher, D.H., Pizzi, M. & Rivera, C. (1995). Recruitment and decay of a pulse of *Cecropia* in Nicaraguan rain forest damaged by Hurricane Joan: relation to mutualism with *Azteca* ants. *Biotropica*, **27**, 455-460.
Finegan, B. (1992). The management potential of neotropical secondary lowland rain forest. *Forest Ecology & Management*, **47**, 295-321.
Flenley, J.R. (1979). *The Equatorial Rain Forest: A Geological History.* Butterworth, London.
Garwood, N.C., Janos, D.P. & Brokaw, N. (1979). Earthquake-caused landslides: a major disturbance to tropical forests. *Science*, **205**, 997-999.
Goldammer, J.G. & Seibert, B. (1990). The impact of drought and forest fires on tropical lowland rain forest of East Kalimantan. In *Fire in the Tropical Biota* (Ed. by J.G. Goldammer), pp. 11-31. Springer, New York.
Gomez-Pompa, A. & Kaus, A. (1992). Traditional management of tropical forests in Mexico. In *Alternatives to Deforestation* (Ed. by A.B. Anderson), pp. 47-61. Colombia University Press, New York.

Guariguata, M.R. (1990). Landslide disturbance and forest regeneration in the Upper Luquillo mountains of Puerto Rico. *Journal of Ecology*, **78**, 814–832.

Hart, T.B., Hart, J.A. & Murphy, P.G. (1989). Monodominant and species-rich forests of the humid tropics: causes for their co-occurrence. *American Naturalist*, **133**, 613–633.

Hart, T.B., Hart, J.A., Dechamps, R., Fournier, M. & Ataholo, M. (1996). Changes in forest composition over the last 4000 years in the Ituri Basin, Zaire. In *The Biodiversity of African Plants* (Ed. by L.J.G. van der Maesen, X.M. van der Burgt & J.M. van Medenbach de Rooy), pp. 545–563. Proceedings 14th AETFAT Congress. Kluwer, Dordrecht.

Hommel, P.W.F.M. (1987). *Landscape Ecology of Ujung Kulon (West Java, Indonesia)*. Soil Survey Institute, Wageningen.

Hopkins, M.S., Ash, J., Graham, A.W., Head, J. & Hewett, R.K. (1993). Charcoal evidence of the spatial extent of the *Eucalyptus* woodland expansion and rain forest contractions in north Queensland during the late Pleistocene. *Journal of Biogeography*, **20**, 357–372.

Horn, S.P. & Sanford R.L. Jr (1992). Holocene fires in Costa Rica. *Biotropica*, **24**, 354–361.

Jans, L., Poorter, L., van Rompaey, S.A.R. & Bongers, F. (1993). Gaps and forest zones in tropical moist forest in Ivory Coast. *Biotropica*, **25**, 258–289.

Johns, R.J. (1986). The instability of the tropical ecosystem in New Guinea. *Blumea*, **31**, 341–371.

Jones, E.W. (1955–56). Ecological studies on the rain forest of southern Nigeria. IV. The plateau forest of the Okomu Forest Reserve (contd.). *Journal of Ecology*, **43**, 564–594; **44**, 83–117.

Layrisse, M. (1992). The 'holocaust' of the Amerindians. *Interciencia*, **17**, 274.

Manokaran, N. & Swaine, M.D. (1994). Population Dynamics of Trees in Dipterocarp Forests of Peninsular Malaysia. *Malayan Forest Records*, **40**.

Maarel, van der, E. (1993). Some remarks on disturbance and its relations to diversity and stability. *Journal of Vegetation Science*, **4**, 733–736.

Markham, A. (Ed.) (1998). Tropical forests special issue. *Climatic Change*, in press.

Mather, A.S. (1990). *Global Forest Resources*. Bellhaven Press, London.

Myers, N. (1995). Environmental unknowns. *Science*, **269**, 358–360.

Nelson, B.W. (1994). Natural forest disturbance and change in the Brazilian Amazon. *Remote Sensing Reviews*, **10**, 105–125.

Nelson, B.W., Kapos, V., Adams, J.B., Oliviera, W.J., Braun, O.P.G. & do Amaral, I.L. (1994). Forest disturbance by large blow downs in the Brazilian Amazon. *Ecology*, **75**, 853–858.

Ohkubo, T, Maeda, T., Kato, T., Tani, M., Yamakura, T., Lee, H.S. *et al.* (1995). Landscape scars in canopy mosaic structure as a large scale disturbance to a mixed dipterocarp forest at Lambir Hills National Park, Sarawak. In *Long Term Ecological Research of Tropical Rain Forest in Sarawak* (Ed. by H.S. Lee, P.S. Ashton & K. Ogino), pp. 172–184. Ehime University, Matsuyama.

Piperno, D.R. & Becker, P. (1997). A vegetational history of a site in the central Amazon basin derived from phytolith and charcoal records from natural soils. *Quaternary Research*, **45**, 202–209.

Poore, M.E.D. (1968). Studies in Malaysian rain forest. I. The forest on Triassic sediments in Jengka forest reserve. *Journal of Ecology*, **56**, 143–196.

Richards, P.W. (1952). *The Tropical Rain Forest*. Cambridge University Press, Cambridge.

Richards, P.W. (1984). The forests of Vietnam in 1971–72: a personal account. *Environmental Conservation*, **11**, 147–153.

Saldarriaga, J.G. & West, D.C. (1986). Holocene fires in the northern Amazon basin. *Quaternary Research*, **26**, 358–366.

Salo, J., Kalliola, R., Häkkinen, I., Mäkinen, Y., Niemalä, P., Puhakka, M. *et al.* (1986). River dynamics and the diversity of Amazon lowland forest. *Nature*, **322**, 254–258.

Sanford, R.L., Saldarriaga, J., Clark, K.E., Uhl, C. & Herrera, R. (1985). Amazon rain forest fires. *Science*, **277**, 53–55.

Saulei, S.M. (1984). Natural regeneration following clear-fell logging operations in the Gogol Valley, Papua New Guinea. *Ambio*, **13**, 351–354.
Sousa, W.P. (1984). The role of disturbance in natural communities. *Annual Review of Ecology and Systematics*, **15**, 353–391.
Steenis, C.G.G.J. van (1958). Rejuvenation as a factor for judging the status of vegetation types. The biological nomad theory. In *Proceedings of the Symposium on Humid Tropics Vegetation, Kandy*, pp. 212–218. UNESCO, Paris.
Steenis, C.G.G.J. van (1972). *The Mountain Flora of Java*. Brill, Leiden.
Stott, P.A., Goldammer, J.G. & Werner, W.L. (1990). The role of fire in the tropical lowland deciduous forests of Asia. In *Fire in the Tropical Biota* (Ed. by J.G. Goldammer), pp. 32–44. Springer, Berlin.
Sugden, A.M. (1992). Hurricanes in tropical forests. *Trends in Ecology and Evolution*, **7**, 146–147.
Swaine, M.D. & Whitmore, T.C. (1988). On the definition of ecological species groups in tropical rain forests. *Vegetatio*, **75**, 81–86.
Taylor, B.W. (1957). Plant succession on recent volcanoes in Papua. *Journal of Ecology*, **45**, 233–243.
Webb, L.J. (1958). Cyclones as an ecological factor in tropical lowland rain forest, north Queensland. *Australian Journal of Botany*, **6**, 220–228.
Werner, W.L. (1988). Canopy die back in the upper montane rain forests of Sri Lanka. *Geojournal*, **17**, 245–248.
White, F. (1983). *The Vegetation of Africa: A Descriptive Memoir to Accompany the UNESCO/AETFAT/UNSO Vegetation Map of Africa*. UNESCO, Paris.
Whitmore, T.C. (1974). Change with time and the role of cyclones in tropical rain forest on Kolombangara, Solomon Islands. *Commonwealth Forestry Institute Paper*, **46**.
Whitmore, T.C. (1975). *Tropical Rain Forests of the Far East*. Clarendon Press, Oxford.
Whitmore, T.C. (Ed.) (1981). *Wallace's Line and Plate Tectonics*. Clarendon Press, Oxford.
Whitmore, T.C. (1984). *Tropical Rain Forests of the Far East*, 2nd edn. Clarendon Press, Oxford.
Whitmore, T.C. (Ed.) (1987). *Biogeographical Evolution of the Malay Archipelago*. Clarendon Press, Oxford.
Whitmore, T.C. (1989). Changes over 21 years in the Kolombangara rain forests. *Journal of Ecology*, **77**, 469–483.
Whitmore, T.C. (1990). *An Introduction to Tropical Rain Forests*. Clarendon Press, Oxford.
Whitmore, T.C. & Brown, N.D. (1996). Dipterocarp seedling growth in rain forest canopy gaps during six and a half years. *Philosophical Transactions Royal Society, Series B*, **351**, 1195–1203.
Whitmore, T.C. & Prance, G.T. (Eds) (1987). *Biogeography and Quaternary History in Tropical America*. Clarendon Press, Oxford.
Whitmore, T.C. & Sayer, J.A. (Eds) (1992). *Tropical Deforestation and Species Extinction*. Chapman & Hall, London.
Whitmore, T.C. & Silva, J.N.M. (1990). Brazilian rain forest timbers are mostly very dense. *Commonwealth Forestry Review*, **69**, 87–90.
Whittaker, R.J., Bush, M.B. & Richards, K. (1989). Plant recolonisation and vegetation succession on the Krakatau islands, Indonesia. *Ecological Monographs*, **59**, 59–123.
Whitten, A.J. & Damanik, S.J. (1986). Mass defoliation of mangroves in Sumatra, Indonesia. *Biotropica*, **18**, 176.
Whitton, B.A. (1962). Forests and dominant legumes of the Amatuk region, British Guinea. *Caribbean Forester*, **23**, 1–24.
Wright, S.J. (1992). Seasonal drought, soil fertility and the species density of tropical forest plant communities. *Trends in Ecology and Evolution*, **7**, 260–263.
Wyatt-Smith, J. (1954). Storm forest in Kelantan. *Malayan Forester*, **17**, 5–11.

21. THE IMPACT OF TRADITIONAL AND MODERN CULTIVATION PRACTICES, INCLUDING FORESTRY, ON LEPIDOPTERA DIVERSITY IN MALAYSIA AND INDONESIA

J. D. HOLLOWAY

Department of Entomology, The Natural History Museum, Cromwell Road, London SW7 5BD, UK

SUMMARY

1 Quantitative samples of moths made using light traps are providing interesting data on the response of natural associations of species in rainforest to various types of disturbance and conversion to managed systems.

2 Results already published or in press on the effects of logging and conversion to softwood plantation on moth diversity in Malaysia are reviewed.

3 New analyses are presented, in particular of transects across the interface between natural forest systems and two types of cultivation: traditional swidden in Seram; and modern field crops and plantation systems in Sulawesi. These show that, whilst conversion to modern cropping systems results in a major reduction in the level of moth diversity and its quality in terms of faunistic composition, endemism, etc., the cyclic nature of sustainable swidden can operate within the natural regeneration cycle of the forest type where it occurs, permitting diversity to recover to near its original level from an initial drop after clearance for cultivation.

INTRODUCTION

The forests of the Indo-Australian tropics are increasingly being modified by man in a variety of ways (Whitmore 1984). Some, such as forest farming (swidden) and shifting cultivation, have been going on for centuries. Others, such as logging, clearance for timber monoculture or cash crop plantation, and conversion to field agriculture, have only in the past century become sufficiently intense to threaten the very survival of extensive lowland tracts of tropical forest.

The impact on the forests themselves has been extensively documented (e.g. Aiken & Leigh 1992, Bryant, Rigg & Stott 1993), together with the social and political issues behind these pressures. Floristic and structural changes are often obvious to the casual observer, as are the effects of these on vertebrate groups (e.g. Johns 1992, Lambert 1992). Very much less is known of the less apparent but possibly more serious costs to the major part of the biodiversity of such systems, the micro-organisms and invertebrates, particularly insects, and to the role they play in the continuing health of the forests and the survival of their components. The very diversity of such groups and the finer grain of their biogeographic and ecological patterns render them much more suitable than vertebrates as indicators of the state of forest systems and for monitoring the impact of changes. The importance of insects and their value as indicators is discussed by Holloway (1983), Brown (1991), Sutton and Collins (1991) and Holloway and Stork (1991).

A pointer to their potential value as environmental indicators, currently anecdotal and in need of further investigation, is the observation over the past few years of unusually low levels of butterfly and moth populations in Peninsular Malaysia to as far north as Penang (Saari & Storey 1997, H.S. Barlow personal communication). One hypothesis is that this is in some way due to the intense atmospheric pollution caused by smoke from recent forest fires in Borneo.

A number of insect groups are currently championed for use as indicators in tropical ecosystems, but no single group appears paramount when assessed in relation to criteria for a good indicator group, such as: (i) ease and objectivity in sampling; (ii) taxonomic tractability; (iii) ecological generality combined with fine-grained habitat fidelity (including low blurring of pattern through mobility); and (iv) rapid response to disturbance (see references cited immediately above). Night-flying moths satisfy most of these criteria. They can be collected in large numbers by light-trapping, a method where inherent bias (Holloway 1977) is probably no worse than in many other mass-sampling methods. With experience, rapid field sorting to species is feasible, and good taxonomic characters in wing markings and genital structure permit accurate cross-referencing between samples, with a moderate expenditure of laboratory time. The taxonomic foundation, in the form of modern reference works and well-curated collection resources is already substantial and expanding rapidly (see references in Robinson, Tuck & Shaffer 1994), enabling biogeographic and taxonomic diversity components to be incorporated in field survey data. Moths are found in numbers in most vegetation types. (Kozlov (1996) illustrated the use of moths as indicators within the Arctic Circle) and, as herbivores, showed a varying degree of specificity along floristic lines, though not necessarily with a precise correlation between moth diversity and that of the associated vegetation type (Holloway 1989, 1993a). The mobility of the sampled, adult stage tends to blur pattern, but this is probably no worse than in most vertebrate groups, and it is possible to select groups, such as the Geometroidea, where mobility is relatively low (Holloway 1984, 1985, Intachat 1995). Response time (as perceived in adult moth populations) to a disturbance event or climatic change is relatively fast, noted at between 1 and 3 months in the SE Asian tropics (Intachat 1995, Kato, Inoue, Hamid *et al.* 1995) and after 3 months in the subtropics (Holloway 1977).

A prerequisite for assessment of human impact on tropical forest systems is an adequate baseline set of samples from undisturbed forest types. These are available from a wide range of forest types in Malaysia (Peninsular, Intachat 1995; Bornean, Holloway 1987, Holloway, Kirk-Spriggs & Chey 1992) and Indonesia (Sulawesi, Holloway 1987, Holloway, Robinson & Tuck 1990); Seram, Holloway 1987, 1993b). Most of these surveys involved altitude transects and, as will be seen, covered a variety of disturbed habitats contemporaneously.

Four main types of human impact on forest ecosystems and their associated moth fauna are assessed: selective logging followed by regeneration within the natural forest cycle (Malaysia); softwood plantation following complete clearance (Malaysia: Sabah); a traditional swidden cycle (Indonesia: Seram); and conversion to modern cash or field crop systems (Indonesia: Sulawesi).

Assessment is made in respect of three aspects of diversity: (i) species richness; (ii) higher taxonomic diversity; and (iii) biogeographic diversity (degree of localization of species: endemic vs. widespread). It also focuses on the extent to which the changes are reversible, for example through natural successional processes, and finally considers them within the context of landscape mosaics.

METHODS

Several different types of light-trap or method of collecting at light have been used in these studies so care is necessary when comparing results. A standard Robinson-pattern 125-W mercury vapour light-trap was used for the author's studies. Intachat (1995) demonstrated that results obtained by this type of trap and a Rothamsted trap with 250-W bulb with a standard filament yielded samples from closed lowland forest in Peninsular Malaysia that were not significantly different from each other. The diversity values were also similar to those obtained by a different experiment in the same forest, samples being collected off a sheet with a tungsten lamp suspended in front of it, the method employed by Chey (1994) in sampling from softwood plantations (see below). Consistency of sampling effort across sites was attempted where possible, but the data were not standardized for sample size in analyses (Holloway 1979, pp. 162–163), nor in generating the contour diagrams in Figs 21.7–21.10 and Figs 21.14–21.16.

Field counts of some readily identified species were made in Sulawesi and Seram, but the bulk of the samples was retained, sorted and identified with reference to published guides such as the author's monographs on the Bornean fauna (Holloway (1996a) is the most recent of eight published), to the Heterocera Sumatrana series (for example, Sommerer 1995) or to the collections of The Natural History Museum, London. Conspecificity within series of specimens was checked by genitalic dissection in cases where full identity could not be established.

The study areas in Malaysia do not experience marked seasonality of climate. Those in Sulawesi and Seram have definite dry seasons, though these have greater impact near coasts than inland in both places (e.g. as discussed for Sulawesi by Knight & Holloway (1990)). Sampling at different times of the year in the same localities yielded very similar diversity values. The lunar cycle can also cause significant fluctuations in sample size, but a pilot study in southern Sulawesi (Bulansyarih unpublished data) has indicated that these are not accompanied by significant fluctuations in diversity values.

The diversity measure for species-richness to be used throughout this discussion is the α-statistic of Fisher, Corbet and Williams (1943). Justification for this on grounds of the frequent approximation of light-trap moth samples to a log-series distribution of abundance among the species is given by Taylor, Kempton and Woiwod (1976) and, within a SE Asian context, by Barlow and Woiwod (1989). Wolda (1983) demonstrated that this statistic was the most sample-size independent of a number of frequently used diversity measures. However, this measure only indicates the diversity of samples and, as will be seen, these usually represent mixtures of species drawn from a number of different ecological associations, including successional stages. Indeed, as sampling at a site progresses through an annual cycle, the value of α for the cumulative sample can increase progressively (Barlow & Woiwod 1989, Holloway & Barlow 1992, Chey 1994), hence care must be taken when comparing diversity values for samples made over a few days with those made cumulatively over a longer period.

The type of two-way table of quantitative data that accrues from a transect survey can be analysed in two complementary ways: (i) site samples can be classified in terms of faunal composition (Q-mode); and (ii) species can be classified according to their representation across the samples (R-mode). A range of clustering, ordination and other methods is available for performing such classifications, and a number have been applied comparatively to moth data (Holloway 1977, 1979).

Cluster analysis offers a sensitive, if sometimes cumbersome, means of mapping and assessing structure in such data. In the Q-mode it and other sorting methods can give a broad assessment of segregation of samples, for example zonation with altitude (Holloway 1984, 1985, 1993b, Holloway *et al.* 1990), but, because samples often represent mixtures of overlapping associations, Q-mode analyses usually provide only part of the picture.

R-mode analysis enables the extent to which species form at any one time relatively discrete associations, rather than the ecological continua to be gauged. Hengeveld (1990, p. 130) described the spatial distribution (including population density) of a species as an optimum-response surface in relation to variation in environmental parameters. The abundance levels of a species in different samples from a transect provide an approximate estimate of such a surface, and R-mode analysis enables assessment of the extent to which such species surfaces overlap.

In any set of samples, it will be progressively more difficult to assign the rarer species reliably to any associations recognized. This is as much a statistical problem, which can be overcome by increasing the intensity of sampling, as an ecological one, given the possibility that, whilst the more abundant species may show aggregation in their distributions, the rarer ones may 'fill gaps' between these distributions and will therefore possibly be more evenly dispersed across the transect. Few data sets approach a sufficient sample size to enable this to be tested. In a much simpler ecological system on Norfolk Island, very large samples have indicated a tendency for at least some of the rarer species to occupy peripheral positions in relation to the main clusters in a species association analysis (Holloway 1996b). On the other hand, whilst the author was examining some of the data from the four contrasting forest sites sampled through a whole year by Intachat (Holloway & Speight 1997), it was apparent that species of what might be termed an intermediate level of abundance mostly nested well within strong clusters recognized in an analysis of the most abundant ones.

Once species are assigned to ecological associations, one can examine each for other aspects of diversity in addition to richness. Higher taxonomic diversity and biogeographic diversity are to some extent complementary, and together permit estimates to be made of quality of diversity (taxic diversity of an area in relation to the degree of endemism it supports; e.g. Vane-Wright, Humphries & Williams (1991)) and relative vulnerabilities of whole higher taxonomic groups, for example loss of biodiversity due to forest clearance will be particularly serious in groups with a high proportion of endemics restricted to lowland forest (Holloway & Barlow 1992, Holloway 1994). A negative relationship between geographic range of butterfly species and the ecological maturity of the forest habitats in which they are found was noted in Vietnam by Spitzer, Novotný, Tonner and Lepš (1993) and in Sumba, Indonesia, by Hamer Hill, Lace and Langon (1997). Hence, disturbance may result in an increase in the proportion of widespread species relative to endemics, and thus a decrease in biogeographic diversity.

RESULTS

Logging

Much of the natural forest estate of the SE Asian lowlands is intended to be managed for timber harvest in sustainable cycles through natural or enriched regeneration (Whitmore 1984). Such management systems ideally operate within the natural disturbance regime of such forests with its mosaic of gap, building and mature phases modified towards the predominance of the early successional stages. Modern logging methods with heavy machinery can cause considerable damage to the soils and remnant stands, and may lead to the development of extensive areas of a floristically distinct secondary forest.

Observations of the impact of such silvicultural practices on moth diversity are currently very limited. Holloway et al. (1992) made pilot comparisons in the Danum Valley area of Sabah between unlogged forest and an extensive area of forest logged selectively 7 years previously by modern mechanized methods. They noted a significant decline in moth diversity to about two-thirds of that of the primary forest, together with changes in the dominant moth species recorded. The diversity level recorded in the logged forest was comparable with that noted in lowland softwood plantations and in alluvial forest regenerating after 10 years on abandoned farmland in northern Sarawak. However, the fauna was generally more similar to that of the undisturbed forest than to that recorded in the plantations.

In the Indonesian island of Buru, butterfly diversity (richness, abundance and evenness) and taxonomic distinctivness (a biogeographic measure) were significantly reduced in selectively logged vs. unlogged forest (Hill, Hamer, Lace & Banham 1995), an observation similar to that of Spitzer et al. (1993) mentioned earlier.

The effects on moth diversity of more traditional selective logging methods, such as the Malayan Uniform System described by Whitmore (1984), have been studied by Intachat (1995). This work has still to be published and so will not be described in detail here. In comparisons of paired sites of different forest types, undisturbed vs. logged two to three decades previously, Intachat has found no significant differences in overall α-diversity, but there were differences in abundance levels and faunistic composition. In another sampling experiment she identified a number of potential indicator species for dipterocarps (Interchat *et al.* 1997). These were mostly present in the unlogged forest sample for each pair, but much rarer in, or absent from, the logged forest sample. As dipterocarps were the primary timber species logged initially, this may indicate that their regeneration subsequently has been limited, and that the overall character of the forest has been altered.

Softwood plantations

Considerable areas of lowland forest in SE Asia are being converted to softwood plantations. In Sabah the principal species are *Acacia mangium*, *Paraserianthes falcataria*, *Gmelina arborea*, *Pinus caribaea* and *Eucalyptus deglupta*. The role of these plantations in conservation of lowland forest biodiversity has probably been underestimated. Chey (1994) made a monthly series of moth samples through one annual cycle in each plantation type and in a representative example of secondary forest in the Brumas area of southeastern Sabah. This work is described in more detail by Chey, Holloway and Speight (1997).

The majority of these plantations supported a 'short-term' (monthly) sample diversity of less than half that recorded in a range of Bornean primary forest types (Holloway *et al.* 1992), and the *Eucalyptus deglupta* plantation surveyed had a diversity as high as that recorded for the secondary forest sample (Fig. 21.1). Cumulative samples for the same sites over a full, annual cycle had very much higher diversity values (Fig. 21.1) but, for reasons given in Methods, these are not strictly comparable with the primary forest samples that were made over periods of 1–4 days. The problems of augmentation of diversity values by 'tourist' individuals of species not resident in the ecosystem being sampled are addressed by Chey *et al.* (1997). This work included sampling along transects through boundaries between plantations and secondary forest. The R-mode analyses of species associations using clustering methods can facilitate recognition of tourists and residents across such transects. It is also possible to select groups with higher-than-average habitat fidelity (Holloway 1984, 1985).

Few of the species recorded could be associated definitely with the tree crop. The two legumes (*Acacia* and *Paraserianthes*), not surprisingly, supported the most, as the Leguminosae are the preferred host family for many groups of Lepidoptera (Holloway 1989). A much smaller number of species was recorded specifically in association with *Gmelina* and *Pinus*, and there were no apparent *Eucalyptus* associations. However, the analysis of species associations indicated that a major component of this diversity was of

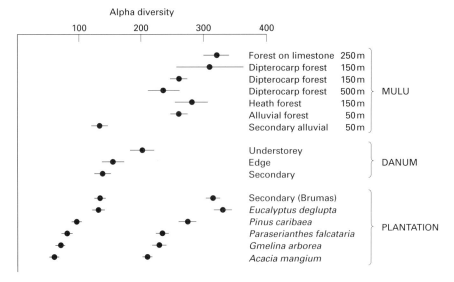

FIG. 21.1. Diversity values for moth samples (all macrolepidoptera) from primary, secondary and plantation forests in Borneo as discussed in the text. The two sets of values for the plantation and secondary forests represent monthly pooled samples (lower values) and samples pooled over a whole annual cycle. The values for Mulu and Danum samples were obtained from pooling two to four nights of samples using a more powerful mercury vapour (ultra-violet) light-trap. The horizontal lines indicate 95% confidence limits.

species general to all plantation types and also found in the secondary forest. This was augmented in the *Eucalyptus* plantation and secondary forest by a further suite of species exclusive to them.

In all these plantations there develops a mixed understorey of shrubs, forbs and pioneer tree species. This was particularly diverse in the *Eucalyptus* plantation. The host requirements of the majority of species in the general association indicate that it is this understorey that is the basis of the moth diversity. A similar phenomenon was noted for butterfly diversity in *Araucaria* plantations in Papua New Guinea (Parsons 1991, p. 16). However, the proportions of various families in these plantations were suggested by Holloway *et al.* (1992) to be somewhat different from those in undisturbed forest, particularly the representation of some of the more minor families. So there may be a reduction of higher taxonomic diversity in plantations in addition to that of overall species-richness.

It may be possible, therefore, to manage softwood plantations to optimize their biodiversity, without affecting yield, by allowing development of such an understorey. This may also prove beneficial in acting as a reservoir for natural enemies effective in pest control (Chey 1994), and in enhancing soil quality. The extent to which this is feasible will depend on factors such as the nature of the seed bank remaining from the original forest, the effects of clearance at each harvest cycle, and the distribution of patches of natural forest as reservoirs of diversity in the landscape mosaic in which such plantations occur.

Shifting cultivation (swidden)

The observations in this section were made in 1987 with Operation Raleigh in the environs of Kanikeh village in the central enclave of the Manusela National Park in Seram, where several such villages occur in an undulating, broad valley between the ridges of Gunung Kobipoto and Gunung Binaiya. The samples were made as part of a much larger survey of moths in natural habitats that included samples from a range of altitudes from sea level to 1470 m on the summit ridge of Kobipoto and to 2800 m on Binaiya (Fig. 21.2). This survey is described in detail by Holloway (1993b), where the profile of diversity with altitude was illustrated, proportions of different moth higher taxa were assessed, and Q-mode classifications of the sample sites were made, based on the full data and taxonomic subsets. A subjective attempt was made to identify associations of species relating to the different site groupings recognized in the Q-mode approach. Subsequently, a number of analyses of species associations have been made (J.D. Holloway unpublished data), and these are reported here.

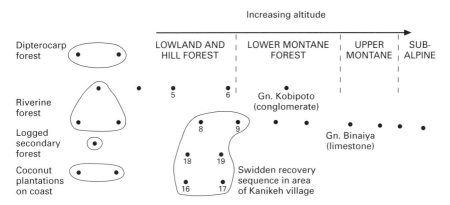

FIG. 21.2. Diagram of sites and vegetation types sampled during the moth survey of Seram. This provides a proforma for the diagrams for species associations in Figs 21.7–21.10. The dots representing the sites are arranged with this in mind, though there is an altitude sequence from left to right and, at the left, increasing disturbance from top to bottom. Full details of the sites are given by Holloway (1993b).

Traditional swidden agriculture in Seram, when practised at low population densities, is sustainable and operates cyclically within the earlier secondary stages of forest regrowth often within enclaves in, or at the margins of, primary forest (Ellen 1985, 1993). It is often characterized by dense stands of giant bamboo in the fallow phase of the cycle.

During the survey, two sites (16, 17) under current cultivation and one fallow site (19) with bamboo or secondary forest were sampled. Site 18 was a mosaic of different stages of the swidden cycle, but with some recent burning. In addition, samples were made in a bamboo and secondary site of longer abandonment (8) and one in an area of advanced secondary forest that had been farmed in the past (9). Equivalent undisturbed forest sites within the altitude range of the disturbed sites were 5 (570 m) and 6 (900 m), the latter just above the boundary between lowland hill forest and lower montane forest. The highest diversity was recorded at site 5 and the second highest at site 6 (Holloway 1993b). Diversity levels in the cultivated and early fallow samples were approximately half these natural forest levels, that at site 19, the fallow of longest standing, higher than that in the other three. In sites 8 and 9 diversity levels were only slightly lower than those at site 6 at a similar altitude. A diagram of the sites sampled and the diversity profile obtained are given in Figs 21.2 and 21.3.

Independent diversity profiles for the major families Noctuidae and Geometridae showed a similar pattern, and the sites with swidden distur-

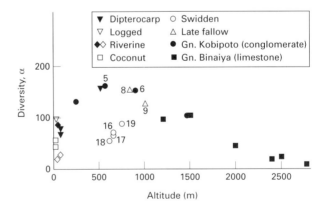

FIG. 21.3. Diversity values for moth samples (all macrolepidoptera) from Seram plotted against altitude. Where samples were made over two nights, these have been pooled: they were kept separate in the plots figured by Holloway and Stork (1991) and Holloway (1993b). Different vegetation types are indicated by symbols: for the riverine category, the open symbol represents river bank sites, the closed one an alluvial forest interior site. Sites referred to in the text and in Figs 21.2 and 21.11 are numbered.

bance had similar proportions of higher taxonomic groups to the undisturbed sites at the same altitudes (Holloway 1993b).

R-mode cluster analyses were performed on four major taxonomic subsets of the data: Geometridae; Noctuidae; Arctiidae; and miscellaneous (other Noctuoidea and Geometroidea, Bombycoidea, Cossoidea, Zygaenoidea). All species represented by 10 or more individuals were included, approximately 30% of the total of 1125 recorded during the survey. In addition, an analysis was performed for all species represented by 40 or more individuals.

Single-link cluster analyses are performed on an array of coefficients between every possible pair of species, the coefficients being a measure of overlap or association of the two species across the samples (see Holloway (1977, 1979) for a more detailed description). Single-link analyses give only a first impression of clustering structure: the building-up of linkage diagrams enables the strength of each cluster to be assessed in terms of completeness of linkage amongst its members relative to external links. Some single-link clusters prove to have a tenuous chaining structure or to consist of several more tightly structured subclusters. Only strong clusters, those where each member is linked with other members by a greater number of links than the majority have with taxa outside the cluster, are taken to indicate species associations: the non-hierarchic clustering approach of Jardine and Sibson (1968), described in more detail as applied to moth sample data by Holloway (1977, 1979).

An example is given in Figs 21.4 and 21.5 for the Noctuidae analysis from the Seram data. The single-link dendrogram is shown in Fig. 21.4. Most of the major clusters proved to have good, tight clustering structure in the linkage diagram of Fig. 21.5. The large lower montane cluster (open circles) has several subclusters on the dendrogram, but these appear weakly only in the linkage diagram. Members of the major clusters in the linkage diagram are indicated by different symbols on the dendrogram.

The associations indicated by these clusters form an altitudinal sequence from two lowland hill forest groupings (open squares, solid circles) through a major lower montane forest cluster (open circles) to a small upper montane forest associations (solid triangles) and a larger subalpine association (solid squares) that includes a number of species from more Palaearctic genera such as *Agrotis* and *Diarsia*. Species strongly associated with the swidden sites are few (5, 85).

The linkage diagram for the Geometridae is shown in Fig. 21.6, the species being numbered as in Table 21.1. The clustering structure is somewhat more complex than for the Noctuidae, but a comparable altitudinal sequence of associations is evident, and also several clusters with species

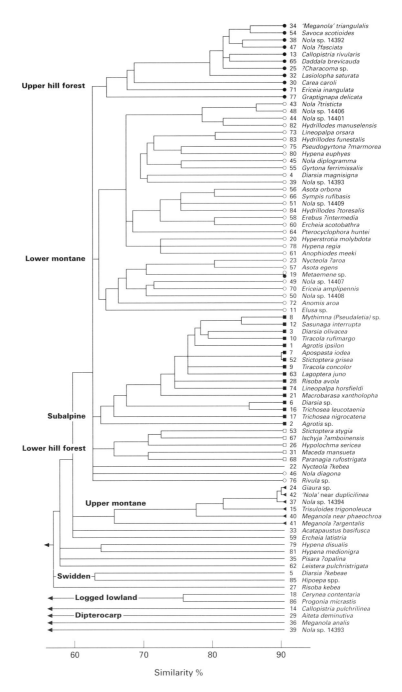

FIG. 21.4. This is an R-mode single-link cluster analysis dendrogram for Noctuidae species represented by 10 or more individuals in the Seram moth survey. Species associations recognized from clusters in the linkage diagram of Fig. 21.5 are indicated on the dendrogram by the same symbols in both figures. Outlying species are without symbols. Taxon 19 has two symbols as it falls within two clusters.

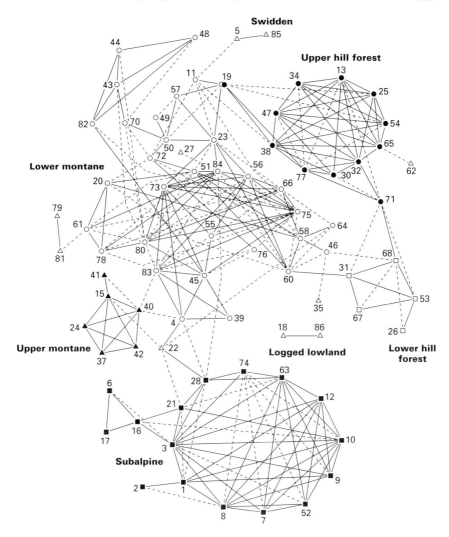

FIG. 21.5. Linkage diagram for the R-mode analysis of Seram Noctuidae (as in Fig. 21.4, with symbols marking the clusters recognized as the same in each, though open triangles indicate outlying species without symbols). Links of 60% similarity and above are indicated by solid lines, and those of 55–59% similarity are indicated by broken lines: choice of these levels is determined when hand-clustering from a high to low level in a manner so as to indicate most clearly the clustering structure. Line length has no significance, and the points are distributed to ensure the diagram is clear, rather than being placed through an ordination method.

580   J. D. HOLLOWAY

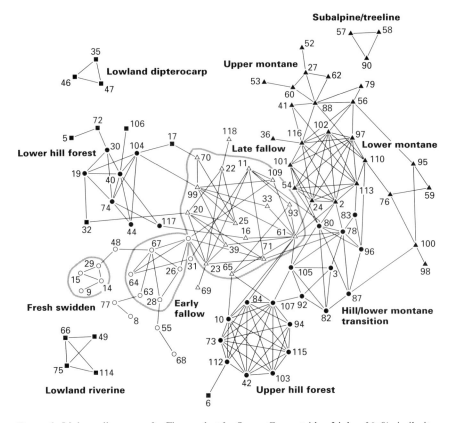

FIG. 21.6. Linkage diagram as for Fig. 21.5 but for Seram Geometridae. Links of 60% similarity and above are indicated. The three swidden-succession association clusters are enclosed by stippled lines. Swidden and early fallow associations are indicated by open circles, the late fallow one by open triangles and the various lowland hill forest associations within the zone in which the swidden occurs are indicated by solid circles. This convention is followed in the rank/abundance plots in Fig. 21.11. Montane species and associations are indicated by solid triangles and other lowland species and categories by solid squares. The species are numbered as in Table 21.1.

particularly strongly represented in the swidden sites. Before discussing this and the other analyses, a method of illustrating the character of each association should help appreciation of the rather complex situation.

The strength of representation of each association in the various samples can be depicted by summing up the individuals for each site from all the component species and plotting them with contours on the site diagram of Fig. 21.2. This is done for all the taxonomic subset analyses in Figs 21.7–21.10. All follow a similar pattern except the highest association for the

TABLE 21.1. *Geometridae species included in the cluster analysis for Seram, numbers 1–118 as in Fig. 21.6. Larger numbers refer to genitalia slide numbers in The Natural History Museum.*

1	*Sarcinodes holzi*	41	*Maxates pervicax*	81	*Hypomecis cladara*
2	*Alex continuaria*	42	*Thalassodes curiosa*	82	*Gasterocome* sp.
3	*Ozola ramifascia*	43	*Thalassodes* sp. 12526	83	*Myrioblephara flexilinea*
4	*Aeolochroma viridimedia*	44	*Thalassodes* sp. 12523	84	*Myrioblephara simplaria*
5	*Hypodoxa* sp.	45	*Hemithea* sp. 12556	85	*Cleora repetita*
6	*Pingasa lariaria*	46	*Hemithea dorsiflavata*	86	*Cleora ?pupillata*
7	*Pingasa blanda*	47	*Hemithea* sp. 12559	87	*Ectropis pallidistriga*
8	*Pingasa porphyrocrostes*	48	*Symmacra solidaria*	88	*Alcis nigriscripta*
9	*Pingasa chlora*	49	*Zythos molybdina*	89	*Cleora subbarbaria*
10	*Tanaorhinus unipuncta*	50	*Problepsis metallopictata*	90	*Cleora ?meceoscia*
11	*Agathia ampla*	51	*Antitrygodes parvimacula*	91	*Hypomecis infaustaria*
12	*Agathia maculimargo*	52	*Anisodes flavissima*	92	*Hypomecis notaticosta*
13	*Agathia conjunctiva*	53	*Anisodes* sp. 12572	93	*Hypomecis ?transcissa*
14	*Agathiopsis maculata*	54	*Gonanticlea sublustris*	94	*Craspedosis ?ernestina*
15	*Agathiopsis basipuncta*	55	*Horisme boarmiata*	95	*Pseudeusemia posticiguta*
16	*Ornithospila odontogramma*	56	*Chaetolopha ornatipennis*	96	*Craspedosis tenebrica*
17	*Ornithospila psittacina*	57	*Tympanota arfakensis*	97	*Monocerotesa seriepunctata*
18	*Spaniocentra stictoschema*	58	*Tympanota* sp.	98	*Aplochlora* sp.
19	*Spaniocentra gibbosa*	59	*Chloroclystis craspedozona*	99	*Peratophyga oblectata*
20	*Protuliocnemis biplagiata*	60	*Chloroclystis eugerys*	100	*Scardamia klossi*
21	*Comibaena attenuata*	61	*Micrulia* sp.	101	*Lomographa luciferata*
22	*Eucyclodes albiceps*	62	*Calluga* sp.	102	*Orthocabera cymodegma*
23	*Eucyclodes ?aphrias*	63	*Hyposidra talaca*	103	*Clepsimelea phryganeoides*
24	*Prasinocyma vagilinea*	64	*Hyposidra incomptaria*	104	*Plutodes signifera*
25	*Eucyclodes absona*	65	*Hyposidra ?nivitacta*	105	*Plutodes ?connexa*
26	*Prasinocyma nivisparsa*	66	*Achrosis semifulva*	106	*Synegia prospera*
27	*Prasinocyma punctulata*	67	*Ctimene ochreata*	107	*Synegia nigrallata*
28	*Maxates ?orthodesma*	68	*Ctimene tricinctaria*	108	*Cypra* sp.
29	*Thalassodes viridifascia*	69	*Gonodontis* sp.	109	*Racotis cogens*
30	*Thalassodes nivestrota*	70	*Hypomecis tetragonata*	110	*Ectropis* sp.
31	*Maxates spumata*	71	*Nadagarodes* sp.	111	*Ectropis sabulosa*
32	*Metallochlora lineata*	72	*Eutoea heteroneurata*	112	*Ectropidia* sp. 12491
33	*Olerospila oleraria*	73	*Luxiaria subrasata*	113	*Cleora sevocata*
34	*Albinospila* sp.	74	*Luxiaria undulataria*	114	*Cleora tenebrata*
35	*Argyrocosma inductaria*	75	*Godonela avitusaria*	115	*Synegia* sp. 12505
36	*Comostola leucomerata*	76	*Abraxas punctifera*	116	*Synegia correspondens*
37	*Comostola laesaria*	77	*Iulotrichia semialbida*	117	*Luxiaria* sp. A
38	*Berta annulifera*	78	*Chorodna strixaria*	118	*Luxiaria* sp. B
39	*Comostola ?conchylias*	79	*Ruttellerona obsequens*		
40	*Comostola pyrrhogona*	80	*Ruttellerona lithina*		

Arctiidae is lower montane, and only the Noctuidae have a strong subalpine association, though there is a trio of geometrid species from treeline forest that could be equivalent. The analysis for all species with 40 or more individuals also yielded similar groupings. The strength of each association for each taxonomic group is variable and is discussed in more detail elsewhere. Associations with strong representation in the swidden samples are seen in all except the Noctuidae (Fig. 21.8), as mentioned above. The 'miscellaneous'

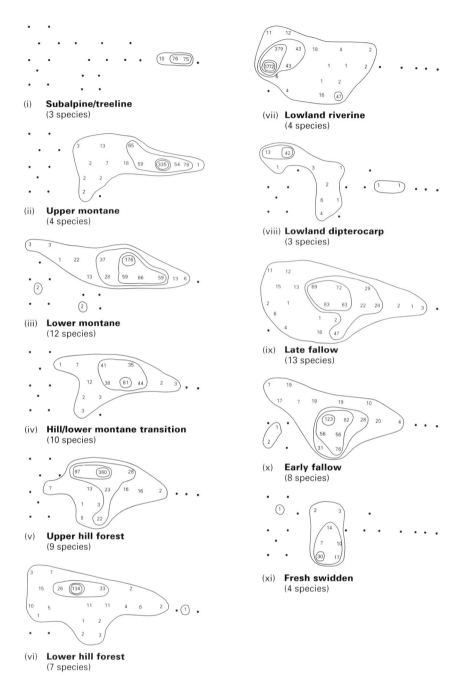

FIG. 21.7. Contour diagrams of species associations recognized for Seram Geometridae, indicating number of individuals sampled at each site and number of species in each association mapped on to the site diagram of Fig. 21.2. Associations characteristic of undisturbed forest are (i)–(viii) and those characteristic of swidden areas or other disturbed sites are (ix)–(xi), arranged in altitude sequence. Contours are set at levels where there is a major hiatus in site values.

Lepidoptera diversity and cultivation 583

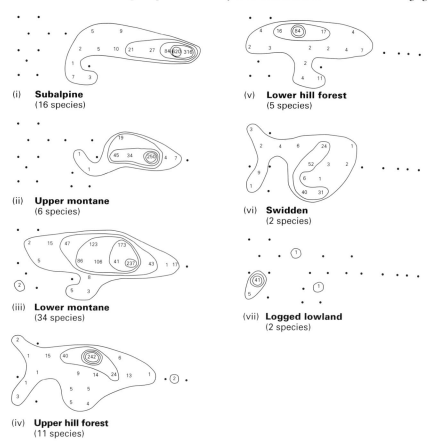

FIG. 21.8. Contour diagrams of species associations recognized for Seram Noctuidae. Conventions as for Fig. 21.7. Undisturbed: i–v; disturbed: vi, viii.

analysis (Fig. 21.9) yielded an association of five species, and that for the Arctiidae (Fig. 21.10) one of 12 that was, however, only weakly segregated from an association more strongly centred on undisturbed hill forest at site 6.

The Geometridae analysis (Figs 21.6. & 21.7) yielded a sequence of clusters whose species show, at one extreme, almost exclusive representation in the swidden sites, particularly those under current cultivation, through an intermediate cluster more evenly represented, though slightly more strongly in the fallow ones, to one where the species are more strongly represented in the long term fallow sites and the undisturbed hill forest sites 5 and 6. There is a parallel association to this last in the miscellaneous analysis.

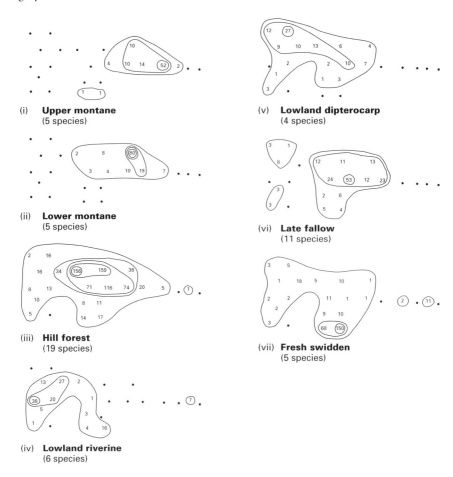

FIG. 21.9 Contour diagrams of species associations recognized for miscellaneous family groups from Seram. Conventions as for Fig. 21.7. Undisturbed: i–v; disturbed: vi, viii.

The swidden-associated clusters in all analyses tend to be much less compact, with a number of weakly linked outlying species, compared to the main primary forest associations. This is evident in the Geometridae linkage diagram (Fig. 21.6).

The sequence of these associations appears to be successional, correlating with the recovery of the forest system following partial clearance and cultivations. Rank/abundance curves plotted for Geometridae in three exemplary samples from this sequence (Fig. 21.11) support this. Samples 16 and 17 from currently cultivated sites were pooled, sample 9 was taken to

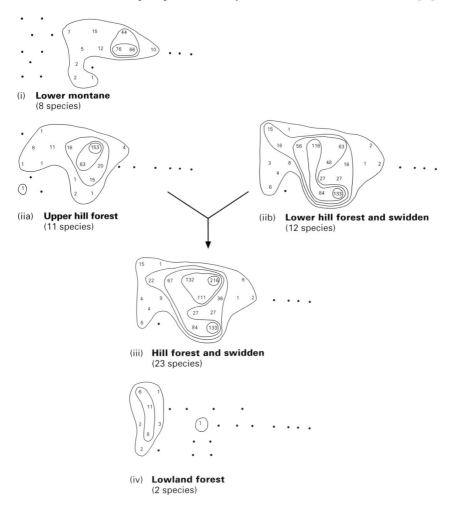

FIG. 21.10. Contour diagrams of species associations recognized for Seram Arctiidae. Conventions as for Fig. 21.7. The two hill forest associations were only weakly segregated and could be treated as a combined hill forest and swidden association.

exemplify the advanced bamboo and secondary forest fallow phase, and sample 6 the most appropriate undisturbed forest sample. Species from Swidden, fallow and undisturbed forest associations at the same altitude are indicated by symbols.

In the plot for site 6 (Fig. 21.11) the major part of the curve is relatively shallow, with numerous species represented by very few individuals, but

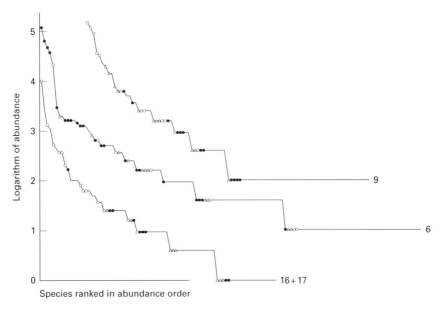

FIG. 21.11. Rank/abundance curves for samples of Geometridae from Seram representative of cultivated swidden (sites 16 and 17 pooled), late fallow recovery from swidden (site 9) and an equivalent undisturbed forest site (site 6). Swidden and early fallow association species are indicated by open circles, late fallow association species by open triangles, and species from undisturbed forest associations at similar altitude by closed circles. For clarity, the Y origin for the plots for sites 6 and 9 is displaced upwards by 1.0 and 2.0 respectively, and the X origin for that for site 9 is displaced to the right. The options of superimposing the curves or keeping them as separate figures were attempted and found to be less satisfactory.

there is a dominant group of five species, four of which belong to the primary forest associations and one from the swidden associations. The majority of the next 20 or so most abundant species are from the primary associations or intermediate one. The presence of several swidden species may be because the site was near a small landslip where the vegetation was more open, with some bamboo present.

The plot for the combined cultivated swidden sites (Fig. 21.11) is much steeper, with stronger dominance, lower equitability, with the more abundant species mostly drawn from the swidden association: only one primary forest species is represented in the most abundant 20, and no intermediate group species are. Four of the dominant species are drawn from a small lower-altitude association most characteristic of disturbed riverine floodplain vegetation.

The plot for the fallow site 9 is of steepness closer to that for site 6, with a greater proportion of rare species than for the previous one. However, it is still dominated by five of the swidden group species four of which are the same in each case (two *Ctimene* and two *Hyposidra* species). At middle levels of abundance the intermediate group is well represented and there is a slight increase in primary forest species.

These analyses broadly confirm as swidden specialists the group of species identified subjectively as such by Holloway (1993b). Knowledge of the host plants of even the larger moths in the Indo-Australian tropics is very incomplete, but the moths of the swidden association include species specialist on ferns, Araceae, Compositae, Boraginaceae, Leguminosae, Oleaceae, Rhamnaceae, Rubiaceae, Urticaceae and Vitaceae, as well as a number of polyphagous arboreal feeders. These plant families are perhaps more characteristic of disturbed secondary forest, and few feature strongly in the lists of tree family representation in the account of the forest types of Seram by Edwards, Proctor and Riswan (1993). Many of the swidden moth genera and species are known as adults to visit the eyes of large mammals to imbibe lachrymal secretions, though the significance of this requires investigation (Holloway 1993b); a possibility is that the adults need to supplement some general dietary deficiency in the foliage of pioneer plant species.

Shifting cultivation as practised in this part of Seram thus seems to operate largely within the natural successional cycle of the surrounding primary forest, albeit perhaps maintained at the early successional stage in areas that are more intensely cultivated in proximity to the villages. Further away, long-term abandonment (probably associated with a general fall in human population levels in the area (Edwards 1993, pp. 7–8)) has led to development of a secondary forest that is capable of supporting high moth diversity, albeit with early successional species still dominant. The potential of such traditionally cultivated zones within or adjacent to national park and other reserved areas for maintaining natural diversity is therefore moderate to high, provided the intensity of cultivation is monitored carefully within the context of the broader landscape mosaic.

Modern cash and field crop systems

A similar survey to that just described for Seram was undertaken by the author in northern Sulawesi as part of Project Wallace in 1985. A transect in natural forest types to 1870m was combined with a series of samples from disturbed forest edge sites, field crop systems (rice, soy, maize) and areas with mixtures of coconut, clove and a field crop or herbaceous ground cover.

These sites are described diagrammatically in Fig. 21.12, and listed by Holloway *et al.* (1990).

A general report on this survey, with Q-mode classifications of the samples, was published by Holloway *et al.* (1990) and the diversity profiles and higher taxonomic diversity of some of the samples have also been discussed by Holloway (1987) and Holloway and Stork (1991). The diversity profile is reproduced in Fig. 21.13, and it can be seen that the cultivated and secondary vegetation sites have considerably lower diversity than that of the natural forest at an equivalent altitude. The samples from cultivated sites also showed a marked decline in higher taxonomic diversity, with several families, particularly where the larvae are predominantly arboreal feeders, lost or present at very low levels. This effect was least marked in the most complex of the coconut and clove systems sampled.

R-mode analysis of species associations for three major taxonomic groupings are presented here to assess the situation further. The groupings are those examined in detail in the publications cited above: Arctiidae; trifine subfamilies of the Noctuidae; Geometridae. However, all Geometridae subfamilies are covered, rather than just the largest subfamily Ennominae.

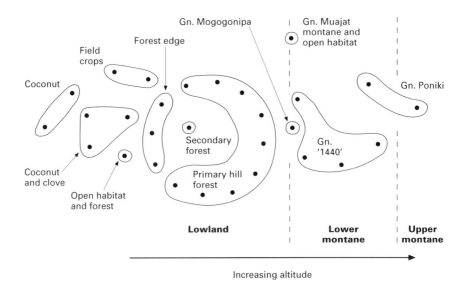

FIG. 21.12. Diagram of sites and vegetation types sampled during the moth survey in Sulawesi, designed to provide a proforma for the diagrams for species associations in Figs 21.14–21.16. Disturbed sites are on the left, the forested sites being arranged in an approximate altitudinal sequence to the right. Full details of the sites are given in Holloway *et al.* (1990).

Lepidoptera diversity and cultivation 589

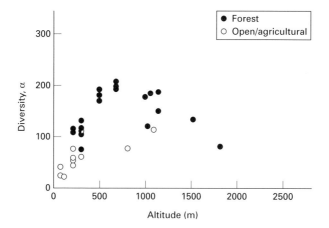

FIG. 21.13. Diversity values for moth samples (all macrolepidoptera) from Sulawesi plotted against altitude. Samples from undisturbed forest sites are indicated by solid circles and those from disturbed or cultivated sites by open circles. Modified from Holloway (1987).

The associations recognized are illustrated on the site diagram of Fig. 21.12, as described for the Seram analysis, in Figs. 21.14–21.16. The clustering structure for the Noctuidae indicated four distinct groupings (Fig. 12.14): three in lowland hill, lower montane and upper montane zones; and one, the largest, general across the disturbed and cultivated habitats. The species in the upper montane groups are of a similar taxonomic character to those of the higher noctuid associations recognized in Seram. The lower altitude groups show fewer parallels, but the bulk of the Seram fauna at these levels is from the non-trifine subfamilies, not included in the Sulawesi analysis.

The open habitat grouping consists of 16 species that include a large number of crop pests or otherwise open habitat specialists, for example grass feeders in *Mythimna* and *Sesamia*, fern feeders in *Callopistria*, one Compositae feeder in *Condica* and more polyphagous dicotyledonous forb feeders in *Athetis* and *Spodoptera*. Almost all are geographically widespread from India to Australia and often extending into the Pacific (the widespread geographic element illustrated in Holloway & Stork (1991), from Holloway (1979)). Such species were virtually absent from the swidden samples from Seram.

For the Arctiidae (Fig. 21.15), apart from a pair of species of upper montane character, montane representation consisted of a major grouping of 17 species that had their greatest abundance in a sample from the summit of Gunung Mogogonipa, a volcanic cone of 1000 m that was unusually mossy on its small summit plateau.

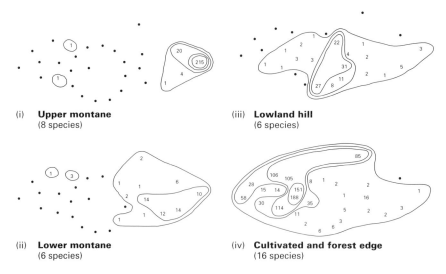

FIG. 21.14. Contour diagrams of species associations recognized for Sulawesi trifine Noctuidae mapped on to the site diagram of Fig. 21.12. Conventions as for Fig. 21.7.

There was a large lowland hill association of equivalent size and three small groupings associated with the disturbed habitats. One of five consisted of species broadly distributed over lowland forest, disturbed (edge) open and cultivated habitats; one of four was found more or less exclusively in open and cultivated habitats; and there was a rather loosely defined one of three very abundant in three forest edge sites. The edge and widespread groups combined a mix of species from genera common to the swidden arctiid group in Seram, but the exclusively open habitat quartet included two species of *Creatonotos* and *Utetheisa lotrix* Cramer, all three extremely widespread in the Oriental tropics and generally associated with open habitats; that is, equivalent to the major noctuid group.

The Geometridae analysis (Fig. 21.16) yielded two major composite clusters, one montane and one lowland. The former consisted of a major lower montane grouping of 28 species with a smaller satellite grouping of 10 species equivalent to the arctiid Mogogonipa lower montane association, and a smaller, more distinct, upper montane group of six species.

The major lowland cluster consisted of two partially distinct subclusters, though several species occupied an intermediate position between them. One (Fig. 21.15iv), of 15 species, was of equivalent character to the lowland hill associations of the other two analyses. The other (Fig. 21.15v) consisted of more widespread species with a much greater representation in disturbed

Lepidoptera diversity and cultivation   591

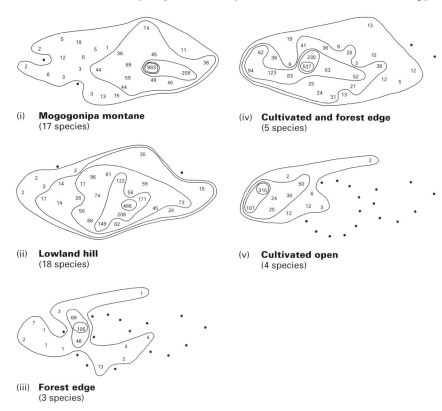

FIG. 21.15. Contour diagrams of species associations recognized for Sulawesi Arctiidae mapped on to the site diagram of Fig. 21.12. Conventions as for Fig. 21.7.

and edge-of-forest sites, though with little representation in cultivated areas. Only a single pair of species was specifically associated with open habitats: the widespread geometrine *Pamphlebia rubrolimbraria* Guenée, and a recently introduced Neotropical ennomine, *Macaria abydata* Guenée, that is associated with *Leucaena* and introduced, weedy *Mimosa* species.

The division of the lowland cluster into two appears, as in Seram, to be into a mature forest group and an early stage successional group. Indeed, there is significant overlap at a generic, and sometimes specific, level between the early successional associations in each island: *Zythos*, *Problepsis*, *Pingasa* (including *P. chlora* Stoll), *Thalassodes*, *Hypomecis*, *Racotis*, *Hyposidra talaca* Walker and *Hyposidra incomptaria* Walker. In Seram, the first two genera contained riverine species that contributed to the dominants in the swidden samples. *Pingasa*, *Thalassodes*, *Hypomecis* species

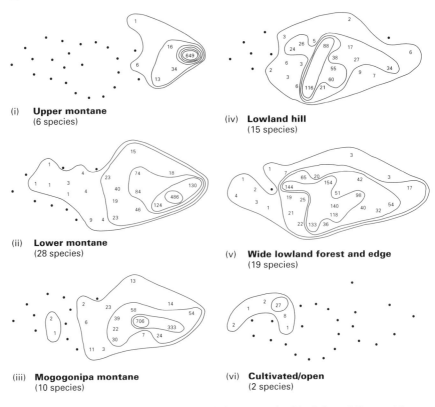

FIG. 21.16. Contour diagrams of species associations recognized for Sulawesi Geometridae mapped on to the site diagram of Fig. 21.12. Conventions as for Fig. 21.7.

and *Hyposidra talaca* also feature in the general plantation association recognised by Chey (1994).

Thus, the open and cultivated habitats in Sulawesi support an association of species distinct from those of disturbed and secondary forest, including fallow swidden. These are in the main geographically widespread, mobile, opportunistic species, usually with forb-feeding larvae, typical r-strategists and often agricultural pests. They tend to be known from rather specialist taxonomic sectors: the trifine noctuid subfamilies discussed here, acontiine, plusiine and some catocaline noctuids; a few arctiine arctiids; and, in the geometrids, a few Geometrinae and a number of Scopulini and Eupitheciini. The habitats in which they occur are thus characterized by low species-richness, considerably reduced higher taxonomic diversity and low biogeographic diversity with the predominance of widespread taxa.

DISCUSSION

The modern agricultural systems and the low diversity of moth herbivores they support are near one extreme of a spectrum of disturbance and modification of tropical ecosystems, with climax primary forest at the other extreme. Intensity of disturbance or modification of those primary systems appears to determine the position on the spectrum, with Lepidoptera herbivores more or less reflecting in their composition that of the flora, with climax forest species, a sequence of early- to late-stage successional forest species, and a distinct suite of specialist species of open vegetation associations, ephemeral in the past but, mainly through human activity, increasingly predominant.

Recovery or restoration of high diversity will be more difficult the further towards the low diversity extreme of the spectrum the system has been pushed, but even then it is not impossible if the degraded system is part of a landscape mosaic that retains at least some areas of climax vegetation. The area of Malaysia sampled by Barlow and Woiwod (1989) yielded over a cumulative annual cycle a moth diversity roughly equivalent to that encountered in short-term samples from undisturbed forest at similar altitudes. Today it consists of a mosaic of patches of primary and secondary forest and others under cultivation. Yet four decades earlier the valley immediately around the trapping site was cleared for vegetable growing and a tea plantation (Holloway & Barlow 1992). The persistence of primary forest further up the slopes has undoubtedly been a major factor in the recovery of forest cover and associated moth diversity once the cultivation was abandoned.

A further example is provided by a pair of samples from fringes of the Gunung Mulu National Park in Sarawak (Holloway 1984, Holloway et al. 1992) and reproduced in Fig. 21.1. Both were from an alluvial floodplain that would support climax alluvial forest. One was made in secondary forest that was developing on land that had been farmed up to a decade before. The moth sample from it had a diversity lower than that recorded by Chey (1994) in the softwood plantations discussed earlier. The other was made in what appeared to be primary alluvial forest but the area had probably been mostly under cultivation a century earlier. Moth diversity in this sample was at a level equivalent to that measured in hill dipterocarp forest nearby. Again, the proximity of that forest and also other, undisturbed alluvial forest sites in the vicinity was probably a major factor in the recovery of diversity to that level. This area has now again been completely cleared for cultivation (A.C. Jermy personal communication), and areas of surrounding forest not in the National Park have been extensively logged.

In all these situations where diversity has at least partially been restored, there has been a landscape dimension, with fragments of primary forest remaining to act as reservoirs of diversity. Only if such fragments remain, is it likely that successional processes, if given time, will proceed to completion. Effective conservation of diversity can only be managed at a landscape level over a mosaic of different vegetation systems, both natural and managed in various ways. This is often stressed in modern conservation literature, most specifically for insects by Samways (1994) and for a wider range of groups by, for example, Bowman, Woinarski, Sands *et al.* (1990). Within such integrated landscape mosaics, both forestry plantations and swidden cultivation can play a role in conservation of biodiversity. Less of a case can be made for, in descending order, agroforestry, cash crop plantations and field crop systems, though an understanding of the way these interact with more diverse components of the mosaic will be essential to reassure those whose livelihoods depend on such cultivation that there is no unacceptable increase in pest damage due to proximity of other, more diverse ecosystems.

Thus, much work is needed to assess the dynamics and behaviour of the various associations of insects at a landscape level in the tropics, particularly the dynamics of potential pest species and their natural enemies across forest/crop boundaries. There is also scope for sampling across temporal boundaries to monitor short-term responses of insect associations to disturbance events, such as to monitor moth populations in tropical forest for a period prior to a logging coupe and then for a period of years subsequently.

But we must stop using the so-called 'taxonomic impediment', the relative lack of expertise, guide books and collections facilities, as an excuse for not investigating the diversity of highly speciose invertebrate groups in the tropics. The advantages of grasping this nettle, I hope, are evident.

ACKNOWLEDGEMENTS

I thank Chey Vun Khen and Jurie Intachat for allowing me to present brief previews of their recent work in Malaysia. My field work in Sarawak, Sulawesi and Seram was funded variously by grants from the Royal Society, the British Ecological Society and the Royal Entomological Society, and work in Indonesia was with the support of the Indonesian Institute of Sciences (LIPI). The swidden samples in Seram were made by David Jones. This field work and subsequent collation and analysis of the data were undertaken partially as core research whilst with the International Institute of Entomology, but also to some extent self-funded and undertaken in leisure time. This paper is numbered 149 in the Results of Project Wallace. A preliminary account of the analyses of the Seram and Sulawesi data was presented at the Pacific Science Congress in Beijing in June, 1995.

The computer program for the association analyses was written by Gaden Robinson. I am grateful to my wife, Phillipa, for word-processing the manuscript and to Núria López Mercader for help with the illustrations. I am grateful for a number of constructive comments from three anonymous referees.

REFERENCES

Aiken, S.R. & Leigh, C.H. (1992). *Vanishing Rain Forests, the Ecological Transition in Malaysia.* Clarendon Press, Oxford.

Barlow, H.S. & Woiwod, I.P. (1989). Moth diversity of a tropical forest in Peninsular Malaysia. *Journal of Tropical Ecology*, **5**, 37–50.

Bowman, D.M.J.S., Woinarski, J.C.Z., Sands, D.P.A., Wells, A. & McShane, V.J. (1990). Slash-and-burn agriculture in the wet coastal lowlands of Papua New Guinea: response of birds, butterflies and reptiles. *Journal of Biogeography*, **17**, 227–289.

Brown, K.S. Jr (1991). Conservation of Neotropical environments: insects as indicators. In *The Conservation of Insects and their Habitats* (Ed. by N.M. Collins & J.A. Thomas), pp. 349–404. Academic Press, London.

Bryant, R.L., Rigg, J. & Stott, P. (Eds) (1993). The political ecology of southeast Asian forests: transdisciplinary discourses. *Global Ecology and Biogeography Letters*, **3**, 101–296.

Chey, V.K. (1994). *Comparison of biodiversity between plantation and natural forests in Sabah using moths as indicators.* Unpublished DPhil thesis, Oxford University.

Chey, V.K., Holloway, J.D. & Speight, M.R. (1997). Diversity of moths in forest plantations and natural forests in Sabah. *Bulletin of Entomological Research*, **87**, 371–385.

Edwards, I.D. (1993). Introduction. In *Natural History of Seram* (Ed. by I.D. Edwards, A.A. Macdonald & J. Proctor), pp. 1–12. Intercept, Andover.

Edwards, I.D., Proctor, J. & Riswan, S. (1993). Rain forest types in the Manusela National Park. In *Natural History of Seram* (Ed. by I.D. Edwards, A.A. Macdonald & J. Proctor), pp. 63–74. Intercept, Andover.

Ellen, R.F. (1985). Patterns of indigenous timber extraction from Moluccan rain forest fringes. *Journal of Biogeography*, **12**, 559–587.

Ellen, R.F. (1993). Human impact on the environment of Seram. In *Natural History of Seram* (Ed. by I.D. Edwards, A.A. Macdonald & J. Proctor), pp. 191–205. Intercept, Andover.

Fisher, R.A., Corbet, A.S. & Williams, C.B. (1943). The relation between the number of species and the number of individuals in a random sample of an animal population. *Journal of Animal Ecology*, **12**, 42–58.

Hamer, K.C., Hill, J.K., Lace, L.A. & Langan, A.M. (1997). Ecological and biogeographical effects of forest disturbance on tropical butterflies of Sumba, Indonesia. *Journal of Biogeography*, **24**, 67–74.

Hengeveld, R. (1990). *Dynamic Biogeography.* Cambridge University Press, Cambridge.

Hill, J.K., Hamer, K.C., Lace, L.A. & Banham, W.M.T. (1995). Effects of selective logging on tropical forest butterflies on Buru, Indonesia. *Journal of Applied Ecology*, **32**, 754–760.

Holloway, J.D. (1977). *The Lepidoptera of Norfolk Island, their Biogeography and Ecology.* Series Entomologica, **13**. Junk, The Hague.

Holloway, J.D. (1979). *A Survey of the Lepidoptera, Biogeography and Ecology of New Caledonia.* Series Entomologica, **15**. Junk, The Hague.

Holloway, J.D. (1983). Insect surveys – an approach to environmental monitoring. *Atti XII Congresso Nazionale Italiano di Entomologia, Roma,* **1980**, 239–261.

Holloway, J.D. (1984). The larger moths of Gunung Mulu National Park; a preliminary assessment of their distribution, ecology and potential as environmental indicators. In *Gunung Mulu National Park, Sarawak, Part II* (Ed. by A.C. Jermy & K.P. Kavanagh), pp. 149–190. *Sarawak Museum Journal,* **30**. Special Issue 2.

Holloway, J.D. (1985). Moths as indicator organisms for categorising rain forest and monitoring changes and regenerating processes. In *Tropical Rain-Forest. The Leeds Symposium* (Ed. by A.C. Chadwick & S.L. Sutton), pp. 235–242. Special Publication, Leeds Philosophical and Literary Society.

Holloway, J.D. (1987). Macrolepidoptera diversity in the Indo-Australian tropics: geographic, biotopic and taxonomic variations. *Biological Journal of the Linnean Society,* **30**, 325–341.

Holloway, J.D. (1989). Moths. In *Tropical Rain Forest Ecosystems of the World,* 14B (Ed. by H. Lieth & M.J.A. Werger), pp. 437–453. Elsevier, Amsterdam.

Holloway, J.D. (1993a). Lepidoptera in New Caledonia: diversity and endemism in a plant-feeding insect group. *Biodiversity Letters,* **1**, 92–101.

Holloway, J.D. (1993b). Aspects of the biogeography and ecology of the Seram moth fauna. In *The Natural History of Seram* (Ed. by I.D. Edwards, A.A. Macdonald & J. Proctor), pp. 91–114. Intercept, Andover.

Holloway, J.D. (1994). The relative vulnerabilities of moth higher taxa to habitat change in Borneo. In *Systematics and Conservation Evaluation* (Ed. by P.L. Forey, C.J. Humphries & R.I. Vane-Wright), pp. 197–205. *Systematics Association Special Volume,* **50**. Clarendon Press, Oxford.

Holloway, J.D. (1996a). The moths of Borneo: family Geometridae, subfamilies Oenochrominae, Desmobathrinae and Geometrinae. *Malayan Nature Journal,* **49:1**, 147–326.

Holloway, J.D. (1996b). The Lepidoptera of Norfolk Island, actual and potential, their origins and dynamics. In *The Origin and Evolution of Pacific Island Biotas, New Guinea to Eastern Polynesia: Patterns and Processes* (Ed. by J.A. Keast & S.E. Miller), pp. 123–151. SPB Academic Publishing, Amsterdam.

Holloway, J.D. & Barlow, H.S. (1992). Potential for loss of biodiversity in Malaysia, illustrated by the moth fauna. In *Pest Management and the Environment in 2000* (Ed. by H.S. Barlow & A. Aziz Kadir), pp. 293–311. CAB International and Agricultural Institute of Malaysia, Kuala Lumpur.

Holloway, J.D. & Stork, N.E. (1991). The dimensions of biodiversity: the use of invertebrates as indicators of man's impact. In *The Biodiversity of Microorganisms and Invertebrates: Its Role in Sustainable Agriculture* (Ed. by D.L. Hawksworth), pp 37–62. CAB International, Wallingford.

Holloway, J.D., Robinson, G.S. & Tuck, K.R. (1990). Zonation in the Lepidoptera of northern Sulawesi. In *Insects and the Rain Forests of South-East Asia (Wallacea)* (Ed. by W.J. Knight and J.D. Holloway), pp. 153–166. Royal Entomological Society, London.

Holloway, J.D., Kirk-Spriggs, A.H. & Chey, V.K. (1992). The response of some rain forest insect groups to logging and conversion to plantation. *Philosophical Transactions of the Royal Society, Series B,* **335**, 425–436.

Intachat, J. (1995). *Assessment of moth diversity in natural and managed forests in Peninsular Malaysia.* Unpublished DPhil thesis, Oxford University.

Intachat, J., Holloway, J.D. & Speight, M.R. (1997). The effects of different forest management practices on geometrical moth populations and their diversity in Peninsular Malaysia. *Journal of Tropical Forest Science,* **9**, 411–430.

Jardine, N. & Sibson, R. (1968). The construction of hierarchic and non-hierarchic classifications. *Computer Journal,* **11**, 177–184.

Johns, A.D. (1992). Vertebrate responses to selective logging: implications for the design of logging systems. *Philosophical Transactions of the Royal Society, Series B*, **335**, 437–442.

Kato, M., Inoue, T., Hamid, A.A., Nagamitsu, T., Merdek, M.B., Nona, A.R. et al. (1995). Seasonality and vertical structure of light-attracted insect communities in a dipterocarp forest in Sarawak. *Researches on Population Ecology*, **37**, 59–79.

Knight, W.J. & Holloway, J.D. (1990). Introduction. In *Insects and the Rain Forests of South-East Asia (Wallacea)* (Ed. by W.J. Knight and J.D. Holloway), pp. 1–5. Royal Entomological Society, London.

Kozlov, M.V. (1996). Subalpine and alpine assemblages of Lepidoptera in the surroundings of a powerful smelter on the Kola Peninsula, NW Russia. *Nota Lepidopterologica*, **18**, 17–37.

Lambert, F.R. (1992). The consequences of selective logging for Bornean lowland forest birds. *Philosophical Transactions of the Royal Society, Series B*, **335**, 443–457.

Parsons, M. (1991). *Butterflies of the Bulolo-Wau Valley. Wau Ecology Institute Handbook*, **12**. Bishop Museum Press, Honolulu.

Robinson, G.S., Tuck, K.R. & Shaffer, M. (1994). *A Field Guide to the Smaller Moths of Southeast Asia*. Malayan Nature Society, Kuala Lumpur.

Saari, G. & Storey, H.R.M. (1997). To get out of this sinful city. Malayan Naturalist, **50(4)**, 12–18.

Samways, M.J. (1994). *Insect Conservation Biology*. Chapman & Hall, London.

Sommerer, M. (1995). The Oenochrominae (auct.) of Sumatra (Lep., Geometridae). *Heterocera Sumatrana*, **9**, 1–77.

Spitzer, K., Novotný, V., Tonner, M. & Lepš, J. (1993). Habitat preferences, distribution and seasonality of the butterflies (Lepidoptera, Papilionoidea) in a montane tropical rain forest, Vietnam. *Journal of Biogeography*, **20**, 109–121.

Sutton, S.L. & Collins, N.M. (1991). Insects and tropical forest conservation. In *The Conservation of Insects and their Habitats* (Ed. by N.M. Collins & J.A. Thomas), pp. 405–424. Academic Press, London.

Taylor, L.R., Kempton, R.A. & Woiwod, I.P. (1976). Diversity statistics and the log-series model. *Journal of Animal Ecology*, **45**, 255–271.

Vane-Wright, R.I., Humphries, C.J. & Williams, P.H. (1991). What to protect? Systematics and the agony of choice. *Biological Conservation*, **55**, 235–254.

Whitmore, T.C. (1984). *Tropical Rain Forests of the Far East*, 2nd edn. Clarendon Press, Oxford.

Wolda, H. (1983). Diversity, diversity indices and tropical cockroaches. *Oecologia (Berlin)*, **58**, 290–298.

22. TROPICAL FORESTS – SPATIAL PATTERN AND CHANGE WITH TIME, AS ASSESSED BY REMOTE SENSING

E.V.J. TANNER*, V. KAPOS† and J. ADAMS‡

*Department of Plant Sciences, University of Cambridge, Downing Street, Cambridge CB2 3EA, UK; † World Conservation Monitoring Centre, 219 Huntingdon Road, Cambridge, CB3 0DL, UK; ‡ Department of Geological Sciences, University of Washington, Seattle WA 98195, USA

SUMMARY

1 Remote sensing (RS) offers a way to investigate many ecological questions at the scale of hundreds to thousands of square kilometres. What RS has contributed to answering questions in tropical ecology, what it can and cannot do now, and what it might do in the future are explored.

2 Vegetation disturbance, both anthropogenic and natural, has been measured by RS. Estimates of deforestation are now routinely done for Brazil by RS and could be done for other tropical countries. Natural disturbance such as blowdowns in the Amazon were first detected by RS; hurricane damage has also been measured by RS. Spectral mixture modelling has proved to be useful for studying land use and land cover change on a 200-km^2 ranch in central Brazil.

3 Some very distinct vegetation types can been distinguished by remote sensing, for example bamboo forest and deciduous forest, but in general it has not proved possible to separate forests differing in more subtle ways, such as *terra firma* forests on different types of soils, which have very different species compositions.

4 Phenological changes are measurable by RS, and these may soon help to detect interannual vegetation responses to climate fluctuations such as El Niño effects; in future such information could help to measure the effects of climate change.

5 In the near future it should be possible to detect effects such as selective logging due to changes in canopy structure and crown size. In the more distant future it may be possible to map the distributions of distinct emergent trees.

6 Ecologists will probably be reading a lot more about RS and many will be using RS because there are plans for new satellites producing much more data with both greater spatial and greater spectral resolution.

INTRODUCTION

The diversity of organisms in tropical forests, their distributions, associations into communities and changes in distribution have been intensively studied, as have the spatial and temporal variation in edaphic factors, disturbance processes and climate. However, logistics have with some exceptions limited studies of such phenomena to scales of a few hectares (famously to 50 ha, e.g. Condit, Hubbell & Foster 1996) and a few years (occasionally to several decades e.g. Manokaran & Swaine 1994). Variation over larger areas (tens to thousands of square kilometres) and longer times is of increasing interest to ecologists seeking to explain high biodiversity in the tropics and to identify requirements and constraints for its conservation. Remote sensing provides a tool for exploring variation at larger scales than hitherto possible, but it has yet to be much exploited in the ecological context. Few papers reporting the

TABLE 22.1. *The number of articles published in 16 leading ecological journals from 1993 to 1996, together with the number which include the words 'remote sensing' in the title, key words or abstract; search done on Bath Information and Data Services (BIDS) on 26 April 1996. Impact factors are from SCI Journal Citation Reports for 1994, section 'Ecology'. The Journal of Vegetation Science, Vol. 5(5) and Ecological Applications, Vol. 4(2) are both devoted to papers which use remote sensing.*

	Rank impact factor	Journal	Number of articles 1993–96	Number of articles with remote sensing in the title, key words, or abstract	% of articles with remote sensing
1	1	*Ecological Monographs*	61	1	1.6
2	8	*Ecology*	795	1	0.1
3	14	*Oikos*	601	0	0.0
4	15	*Journal of Ecology*	292	0	0.0
5	16	*Conservation Biology*	602	2	0.3
6	17	*Ecological Applications*	320	15	4.7
7	18	*Functional Ecology*	343	0	0.0
8	20	*Oecologia*	923	0	0.0
9	24	*Australian Journal of Ecology*	159	0	0.0
10	25	*Vegetatio*	279	3	1.1
11	28	*Journal of Applied Ecology*	216	4	1.9
12	31	*Biotropica*	218	0	0.0
13	35	*Journal of Biogeography*	186	3	1.6
14	39	*Biological Conservation*	439	2	0.5
15	40	*Journal of Vegetation Science*	315	8	2.5
16	42	*Journal Tropical Ecology*	171	0	0.0
Total			5920	39	0.7

results of research using remote sensing have appeared in the mainstream ecological journals despite many examples in the remote sensing literature reporting variation in vegetation; a notable exception is Foody and Curran (1994). Of 16 leading ecological journals in 1993–96, 39 out of 5920 articles (0.7%) used the word 'remote sensing' in the title, key word or abstract (Table 22.1). Two journals (*Journal of Vegetation Science* and *Ecological Applications*) had special issues on the topic. These notwithstanding, we think that tropical ecologists could and will make much more use of remote sensing; this chapter discusses what remote sensing has contributed to answering questions in tropical ecology, what it can and cannot do now, and what it might do in the future.

The chapter covers: (i) a brief description of the various sensors which have been used for remote sensing, and the approaches used to analyse the data; (ii) remote sensing-based studies of vegetation disturbance, both natural and anthropogenic; (iii) questions addressed using distinctions among communities and vegetation types caused by phenology and strong spectral contrast; (iv) suggestions of phenomena which seem not to be studied at present, but which could be using current technology — the near future; and (v) phenomena which might be studied in the future using new methods and/or new instruments which are currently under development — the more distant future.

INSTRUMENTS AND ANALYSES

The principal satellite sensor data of use to ecologists are those from the Landsat-MSS and Landsat-TM, the satellites for which were first launched in 1972, those from the NOAA-AVHRR launched in 1981, and data from SPOT-HRV which came into service in 1986 (Table 22.2). Radar satellite data have great potential for ecological use (Holmes 1992), especially in the tropics, because they are unaffected by cloud cover, but a number of methodological problems must be resolved before their potential is fully realized. The criteria that determine the usefulness for ecological work of data from a particular satellite include: spatial resolution (or pixel size), temporal resolution (or return time), and spectral resolution (or numbers and types of spectral bands in which data are recorded). Spatial resolution (ranging from $10m \times 10m$ to $1100m \times 1100m$) determines both the detail that can be detected and the extent or scale of investigation which is practicable (local/landscape/regional). Temporal resolution (return times of 1–44 days) determines not only the absolute frequency of observations possible, but more importantly in the tropics, the probability of obtaining cloud-free imagery for any given location. In general, the smaller the pixel size the

Table 22.2. Characteristics of remote sensors useful for vegetation analysis. ATSR-2, Along Track Scanning Radiometer-2; AVHRR, Advanced Very High Resolution Radiometer; ERS, European Remote Sensing Satellite; HRV, High Resolution Visible; HRVIR, High Resolution Visible and Infra-Red; IRS, Indian Remote Sensing Satellite; JERS, Japanese Earth Resources Satellite; LISS III, Linear Imaging Self Scanning System; MSS, Multispectral Scanning System; NOAA, National Oceanic and Atmospheric Administration; RADARSAT, Radar Satellite (Canadian); SAR, Synthetic Aperture Radar; SPOT, Systeme Probatoire d'Observation de la Terre. TM: Thematic Mapper. Costs: moderate, hundreds of $US per scene; high, thousands of $US per scene.

Satellite	Sensor	Start date	Swathe width (km)	Pixel size (m)	Number of Spectral bands	Return time (days)	Cost
Landsat 1–5	MSS	1972	185	80 × 80	4	18	Moderate
Landsat 4–5	TM	1982	185	30 × 30	6 + 1 thermal	16	High
NOAA 6–12	AVHRR	1981	3000	1100 × 1100, nadir*	2 useful for landcover	1	Moderate
SPOT 1–3	HRV	1986	60 + 60 with 3 km overlap	10 × 10 panchromatic 25 × 25 multispectral	1 panchromatic 3 multispectral	Variable	High
ERS 2	ATSR-2	1995	500	1000 × 1000, nadir Up to 2000, off nadir	4 useful for landcover	1	Moderate
SPOT 4	VEGETATION	1997	2200	1000 × 1000	4	26	
SPOT 4	HRVIR	1997	60 + 60 with 3 km overlap	20 × 20	4	26	
IRS 1C	LISS III	1995	142–148	23.5 × 23.5	4	24	
ERS 1–2	AMI-SAR	1991	100	30 × 30	1 C (wavelength 5.6 cm)	35	
JERS 1	SAR	1991	75	18 × 18	1 L (wavelength 24 cm)	44	
RADARSAT	SAR	1995	100 (standard)	25 × 25	1 C (wavelength 5.6 cm)	24	

* LAC mode of viewing

longer the return time (Table 22.1). Increasing spectral resolution usually increases the subtlety of differences among vegetation types that can be detected in the images.

The data from sensors are digital; sensors do not produce photographs, so all images are produced from some form of processing the numbers; the colours one sees on processed images are the subjective choice of the researcher (though often they are chosen to be similar to the colours that would be seen by a human observer). Thus, one should beware of 'pretty pictures' as they could be as much a result of image processing as of actual patterns in the surface cover of vegetation and soils. Very importantly, it must be remembered that the *structure* of both vegetation and soil gives rise to a great deal of the pattern seen in images and that differences in colour as seen by a human observer may contribute rather little to the signal.

Remote sensing data may be processed in a number of ways. The simplest is just to plot the radiance values in each band; band ratios are commonly used; the most important of these is probably the normalized difference vegetation index (NDVI) which is infrared minus red divided by infrared plus red. Band or band ratio data may be classified in unsupervised (i.e. with no constraints) or supervised (i.e. to limit possible classifications, or to search for pixels which are similar to pixels which represent certain types of land cover) ways, and finally data may be analysed by spectral mixture modelling. Many of the patterns discussed in this chapter are the result of analysing Landsat-TM data by spectral mixture modelling.

Spectral mixture analysis (SMA) can be applied to various sorts of remotely sensed data. As applied to Landsat-TM data the technique used data from six Landsat spectral bands to provide detailed information about pixel composition that is robust for comparison of changes through time (Adams, Sabol, Kapos *et al.* 1995). This is accomplished by comparing the spectral response of a pixel with spectra from its probable components, such as green vegetation, non-photosynthetic vegetation (e.g. wood, dry grass), soil and shade (note that shade is not a black hole and it emits radiation), which are known as 'endmembers' (Fig. 22.1). A modelling procedure provides best fit estimates of the composition of each pixel in terms of the chosen endmembers. The SMA approach is exceedingly useful for ecologists because they can make informed predictions of the composition of pixels based on their endmember composition and an understanding of the physical characteristics of different vegetation types even before any ground data are collected. This both helps to maximize the usefulness of field efforts and improves the potential for resolving misclassifications.

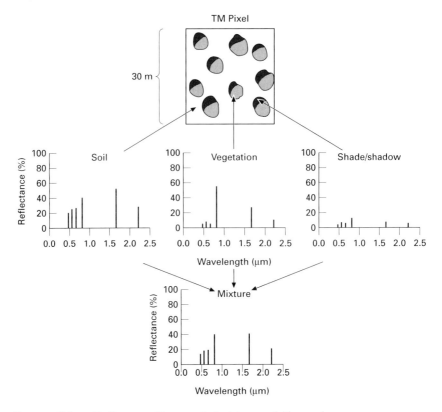

FIG. 22.1. Schematic diagrams of how spectral mixture modelling works.

VEGETATION DISTURBANCE

There are many issues of interest to ecologists which relate to vegetation disturbance. We have grouped them under two headings: *anthropogenic* disturbance, which frequently changes vegetation type (especially the destruction of rainforest); and *natural* disturbance, for example hurricanes, which cause damage and death of patches of trees. Vegetation disturbance, both anthropogenic and natural, is often easily mapped by remote sensing when the disturbance is dramatic and well defined (e.g. when forest is converted to pasture and edges between the two are distinct). This is due to the strong structural contrast (uneven-canopied trees vs. more even-canopied grassland) between the resulting different vegetation types. If the structural contrast is small and edges are not well defined then mapping disturbances by remote sensing is much more difficult and much less accurate. Despite the problems there are many examples of the use of remote sensing to measure vegetation disturbance at landscape and continental scales.

Anthropogenic disturbance

The best-known and developed application of remote sensing in tropical forests is the monitoring of deforestation. Without the use of remote sensing, estimates of deforestation can vary widely for the same countries or areas; there is no good agreement between the two major global data sets (Myers 1989, FAO 1993) published so far (Fig. 22.2). Part of the difference between the Myers and FAO data is due to their use of different definitions of forest, but part is in how deforested areas were measured. Neither Myers nor the FAO report much data that was collected by satellite sensors.

The Brazilians have successfully developed a programme of monitoring deforestation in the 'Legal Amazon' on an approximately annual basis using Landsat-TM data. Their estimates underwent a significant revision in the early stages of the programme, but they are now widely accepted and indicate a slow decline in deforestation rate after 1988 (INPE 1992).

At least two groups are working towards monitoring tropical deforestation on a global scale using satellite sensor data: the NASA Pathfinder project is producing pantropical coverage of Landsat sensor data; and the

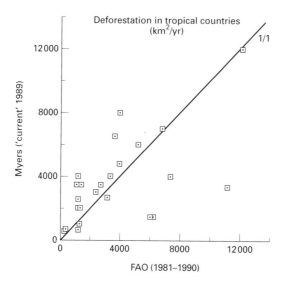

FIG. 22.2. Comparison of the rates of deforestation for the 25 tropical countries with the largest amount of extant tropical forest (excluding Brazil) from table 1 in Myers (1989) and table 8 in FAO (1993).

TREES project of the JRC at Ispra is using using AVHRR 1.1 km data. As processed images accumulate, the accuracy of deforestation estimates will improve because of the ability to construct time series of images. A pixel which was unquestionably deforested in one year will remain something other than primary forest far into future—though this raises the interesting question of whether and how long it takes regrowth to revert to 'primary forest' (Lucas, Honzak, Foody et al. 1993).

In addition to tracking simple rates of deforestation, remotely sensed data can be used to evaluate landscape changes that are important in predicting changes in the remaining forest areas. Skole and Tucker (1993) used Landsat-TM data to assess not only areas of the Legal Amazon subjected to complete deforestation, but also the area of forest which was within 1 km of deforested pixels and therefore potentially edge-affected, and the amount of isolated forest. They report that in 1988, of a total of 588 000 km² of forest affected by deforestation, 39% was actually deforested, 59% was within 1 km of deforestation and 2% was left as isolated patches (>100 km², and completely surrounded by a deforested corridor at least 1 km wide). The total area potentially affected was thus 2.5 times the area deforested. This approach can be combined with field studies of the effects of edges and isolation to produce refined estimates of deforestation-affected forest cover.

Investigating fragmentation processes at different scales and in greater detail can provide insight into how the changes in landscape configuration may affect ecological processes. Chatelain, Gautier and Spichiger (1996) used Landsat sensor data to track deforestation in a 10 000-km² area of the Ivory Coast and examined remaining forest fragments in a 970 km² subarea. They found that 79% of the larger area had been deforested in the past 20 years and that the longer an area had been deforested the less likely it was to retain any forest fragments. Most forest fragments in the area studied in detail were <1 ha and severely degraded. They tended to be smaller with increasing distance from remaining main forest blocks. Though too small to function as primary forest, such fragments could be important nuclei for forest regeneration, and could act as habitat for surviving populations and individuals of forest species, which could be a source of colonizers.

By exploiting more of the spectral information in Landsat sensor data, it is possible to detect more subtle changes in landcover within cleared areas. Spectral mixture analysis was used to provide a detailed assessment of changes in landcover and land use on Fazenda Dimona, a ranch c. 80 km north of Manaus in Brazil (Adams et al. 1995). Changes within the deforested area of the ranch (c. 20 000 ha) were followed using co-registered and intercalibrated Landsat sensor images taken in the dry seasons (August) of 1988, 1989, 1990 and 1991. Adding a temporal dimension to the interpretation of remotely sensed imagery in this way permits improved pixel classification because basic ecological knowledge of allowable transitions (e.g. a pixel cannot change from pasture to primary forest in only 3 years) can provide additional context.

The SMA-based classification of landcover at Dimona provided seven distinct robust landcover classes (as well as other less clear groups), and by following the temporal transitions among these classes in the light of basic ecological understanding it was possible to build up a very accurate picture of how land use in the ranch changed. Over the 4-year period, the images showed decreased maintenance of the pastures on the ranch and increasing regrowth in the unmaintained areas (Fig. 22.3), as well as some of the reclearing and burning used to restrain the regrowth in some areas. There was a net increase in regrowth and a net loss of pasture over the study period, reflecting the inherent unsustainability of cattle ranches in this area (Fearnside 1993).

Multitemporal analysis of satellite imagery thus permits investigation of the rates at which vegetation recovers from disturbance and different management regimes (see also Moran, Brondizio, Mausel & Wu 1994). It has also been used to track biomass accumulation in the context of regional carbon cycles (Lucas *et al.* 1993) and it could be used to generate predictions about recovery following different intensities of management.

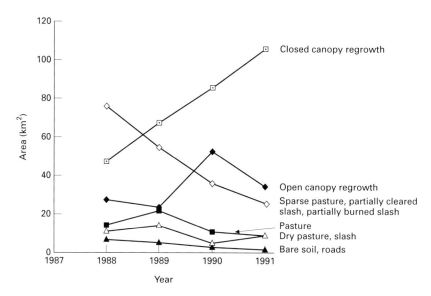

FIG. 22.3. Four-year time series for land use at Fazenda Dimona *c.* 70 km north of Manaus Brazil, based on fig. 8 in Adams *et al.* 1995.

Natural disturbance

Remote sensing can be used to evaluate some aspects of natural disturbance regimes, such as the location, extent, frequency, severity and probable cause of disturbance events, and the time-scale for recovery; but only for disturbances that affect patches greater than c. 1 ha. It can thus be used to address, and to guide field studies addressing, such questions about disturbance as: (i) what is its role in maintaining diversity; (ii) do certain species depend on it; and (iii) can its patterns explain species distributions?

Some forms of natural disturbance, such as hurricane damage, are well known but difficult to investigate on the ground, especially over large areas, for logistical reasons. Satellite sensor imagery potentially provides a tool for looking at the broader scale landscape patterns of hurricane damage to determine factors affecting forest vulnerability and, potentially, to generate and test hypotheses about likely patterns of variation in species composition and vegetation structure as a consequence of repeated hurricane impact.

We have begun to investigate the possibilities for such analysis using Landsat sensor data for the Blue Mountains of Jamaica acquired soon after the passage of Hurricane Gilbert in 1988. We have detailed ground data from relatively localized plot-based surveys and broader-scale information from photographs acquired from vantage points. Gilbert had a major impact on the Blue Mountains forests, defoliating large areas and felling trees in smaller areas (Bellingham, Kapos, Varty et al. 1992). In Landsat-TM images analysed by SMA areas of defoliation and downed trees can be detected by an increase in wood fraction in mixture modelled Landsat-TM data. In a 42-km^2 area of the Blue Mountains 82% of Landsat-TM pixels had more 'non-photosynthetic vegetation' (NPV) after the hurricane than before; NPV increased by ≥30% in 25% of these pixels (i.e. by 82%) and by ≥20% in 43% of these pixels.

The problem is exactly how to verify and quantify the magnitude and distribution of this signal. Complications arise from the steep topography, which means that shade is the dominant fraction in some parts of the image, reducing the contribution of other fractions to tiny amounts. Other problems arise from the extremely fine grain of the variation in damage. Where the human eye integrates an area of several hundred square metres on a hillside as being 'severely damaged' (e.g. see fig. 2 in Bellingham et al. 1992), each Landsat-TM pixel in that area has different spectral response, and the mean pixel spectral response has a large variance. One solution to this is to collect ground data over large areas in rather little detail, for example by surveying for blown-down trunks. Once this has confirmed the validity of the analyses, the image can be used to select sites for sampling to test predictions about tree composition based on damage class. In the longer term it may be possible to plot past hurricane tracks, assess where damage would probably have occurred, and then decades later compare the floristic composition in damaged and undamaged patches.

In a much more wide-ranging study of hurricanes in New England and Puerto Rico Boose, Foster and Fluet (1994) used aerial photographs to study the extent, distribution and intensity of hurricane effects on forests which were then compared to damage predicated from measured and modelled wind speed and directions. In Puerto Rico they found an uneven distribution of hurricane damage with 6% of the area undamaged, 75% slightly or moderately damaged, and 20% very damaged or destroyed. The aerial photography used in this assessment had finer spatial resolution than any of the current generation of satellite sensors, but it will be possible to measure many of the same things using satellite remote sensing technology now in development.

Other forms of natural disturbance can be detected using satellite remote sensing. In some cases, the very occurrence of such disturbances was poorly known prior to regional-scale investigation using remotely sensed data.

Landsat sensor images revealed fan-shaped patches of disturbance in Amazonian forest before they were discovered on the ground (Nelson, Kapos, Adams et al. 1994). These patches proved to correspond to large-scale blowdowns in the midst of undisturbed forest. Following their identification in a single Landsat scene (180 km × 180 km), a survey of 137 Landsat scenes covering most of Brazilian Amazonia found 330 such patches >31 ha in size (the largest was >3000 ha), covering a total of 90 000 ha (Nelson et al. 1994). The patches were concentrated in the western Amazon in an area roughly corresponding to the highest occurrence of heavy rainfall. They are thought to be caused by downbursts from moving convectional storms.

Spectral properties of the patches and of regenerating vegetation permitted some dating of the patches; patches <2 years old occupied 0.04% of the area where they were most concentrated. In this area, complete turnover of the forest by this mechanism would require 5000 years. Thus, large blowdowns are not a major factor in forest turnover, but they create locally unique conditions of large open areas, large inputs of nutrient-rich organic matter from downed trees and disruption of the habitat.

It is also possible to look for similar but smaller disturbances, though not at the regional scale. A detailed search of two Landsat scenes found roughly 10 times the number of blowdowns when patches 5–30 ha in size were included in the survey.

Fire, flood and other mortality events are readily detectable in satellite images because they kill trees (Nelson 1994). Fire killed $c.$ 500 km^2 of tree crowns in $c.$ 3.5×10^6 km^2 of the forested legal Amazon in 1983. Flooding killed trees on 2% of a 100-ha forest reserve north of Manaus in 1989. Understanding the distributions of these patches of mortality, in space and time, can shed light on the constraints on species that depend on them, on dispersal distances and patterns of gene flow. Long-term landscape-scale studies of disturbance and vegetation change using remote sensing (cf. Hall, Botkin, Strebel *et al.* 1991) can provide greater insight into the factors governing the distribution of natural vegetation types and their component species.

COMMUNITIES AND VEGETATION TYPES

There is a good deal of interest in describing the distribution and species composition of communities. Such results are useful, for example, to address questions of whether geographically separate communities, which are growing in otherwise similar conditions, are of similar composition (as they are for example in old floodplain forests in Manu; Terborgh, Foster & Nuñez 1996); or to assess beta diversity in order to select areas that might be most valuable as biodiversity reserves. Remote sensing can contribute to addressing such questions.

Remote sensing can distinguish between strongly contrasting vegetation types such as forest and savanna (Furley, Dargie & Place 1994). Analyses of the boundaries between such vegetation types (and their changes with time) may shed light on the factors determining their distribution.

Some distinctions between different types of forests can also be made using satellite sensor imagery. Coniferous forests can be differentiated from broadleaved ones (e.g. Hall, Botkin, Strebel *et al.* 1991). Phenological differences permit distinction between deciduous and evergreen forests; for example, patches of the deciduous *Peltogyne*-dominated forest can be distinguished among the mainly evergreen forest on Maracá Island in Brazil (Furley *et al.* 1994). In a much larger scale study Tucker, Townshend and Goff (1985) analysed AVHRR data showing the seasonal changes in vegetation across the whole of Africa over one year.

Other structural differences can facilitate detection of distinct forest types. A striking example of structurally distinct forest that can be detected in satellite images is the 121 000 km^2 of 'bamboo' forest in the southwestern Amazon mapped by Nelson (1994). Earlier vegetation maps of this part of the Amazon do not distinguish this forest from the surrounding vegetation despite the strong contrast in structure and composition. The bamboo forest is likely to be functionally as well as floristically very distinct; the episodic death of the bamboos following synchronous flowering will produce nutrient flushes and increase the risk of fire. Burning, of course, is more likely as settlement of the area increases; such a conflagration could be much larger than the great fire of Borneo in which 40000 km^2 of forest burned (Malingreau, Stephens & Fellows 1985).

In other parts of Amazonia it has proved possible to distinguish patches of campina in the midst of *terra firme* forest. These forests, which grow on white sand soils, have short trees which differ in their leaf colour and canopy structure from the *terra firme* forests that surround them. Investigation of the distribution of 'campinas' (or other distinctive forest types) using satellite sensor imagery should make it possible to quantify the distances involved in pollination and dispersal of campina species and therefore reproductive isolation (cf. Renner, Chapter 13, this volume). On this basis, areas can be selected for field study to investigate questions of floristic and genetic distinctiveness in relation to isolation from similar patches, which could throw light on speciation processes. The applied value of such distributional data in designing conservation networks and assessing threats to them is obvious.

Differentiating between less distinct forest types in the humid tropics has so far proved to be difficult or impossible. Most researchers have been unable to distinguish structurally similar communities in remotely sensed images, even when they know they exist from ground surveys. For example, Westman, Strong and Wilcox (1989) could not distinguish four closed canopy broadleaf forest types in the Mabira forest reserve in SE Uganda; though it was possible to distinguish coniferous plantations from broadleaved forest and from crops. In Landsat-TM images of Tambopata, southern Peru, Hill and Foody (1994) distinguished four broad groups of forests, which differed markedly in structure and species composition: (i) bamboo forest; (ii) flooded forest; (iii) lower floodplain forest grouped together with premontane forest; and (iv) all the rest, including *terra firme* forests on sandy and clayey soils (which are almost certain to have had striking floristic differences).

In complete contrast, Tuomisto, Ruokolainen, Kalliola *et al.* (1995) identified hundreds of 'biotypes' in Landsat sensor images of lowland Amazonian Peru. However, because the number of 'biotypes' was so large, it was impossible to collect ground data to check whether each 'biotype' represented a different vegetation type on the ground, or whether minor variations in, say, topography caused the same vegetation types to produce distinct spectral responses; we think that both of these possibilities are likely.

Thus, in general, distinctions can only be made between strongly contrasting vegetation types and it does not seem very likely that more subtle distinctions will be possible in the near future.

THE FUTURE OF REMOTE SENSING AND ECOLOGY

The near future

Phenological change investigated by comparing images between seasons allows the investigation of year-to-year variation in phenological responses as it relates to climatic variation and anomalies such as El Niño events; such analyses may provide a basis for predicting and detecting responses of tropical forests to climate change.

Deforestation, when it results in complete conversion from forest to pasture and when it has distinct edges, is already easily measured by remote sensing; it should soon become possible to monitor changes in the condition or quality of forest which is disturbed but not actually clear felled. Already it should be possible to use SMA to detect changes in forests due to logging, even if the forest is not completely cleared. Because selective logging usually removes the largest trees (precisely those that contribute most to the spectral response of the canopy), it changes the canopy structure of the forest, probably creating more shade, and exposing more soil, due to canopy removal and the presence of roads and skid trails.

Landscape analysis using remotely sensed data can be used to improve the design of networks of conservation areas through assessing isolation and connectivity. Monitoring programmes to improve the management of such areas and the control of threats to them can also be developed using remote sensing.

The more distant future

Further into the future when images with more spectral and spatial resolution are available it should be possible to measure crown sizes and frequencies of the emergent and canopy species. This information could contribute to assessing the impact of selective logging. (It has already been shown that tree crowns in lowland rainforest can be identified and measured by aerial photography (Vooren & Offermans 1985), but this has apparently not been followed up by quantitative comparisons of crowns in logged and unlogged forest.)

Similarly, it will also be possible to map certain very distinctive emergent or canopy species. If, by remote sensing, we could identify emergent trees over large areas (hundreds of square kilometres) we could ask whether trees were clumped and at what scale (cf. Hubbell 1979); if they were clumped we could then ask why; are they picking out patches with different soils and/or climate or are they picking out 'refugia', or some other pattern due to founder effects?

Mapping emergent or canopy trees species could be useful for selecting areas of conservation priority at the scale of hundreds or thousands of square kilometres, where species-richness in different groups is congruent; at smaller scales the congruence diminishes (Pomeroy 1993, Balmford & Long 1995).

It will also be possible to measure and map gaps or disturbance regimes in some detail, thus providing further insight into forest dynamics and a means of assessing aspects of habitat quality. Ultimately, the development of radar techniques may make it possible to detect disruption of the understorey as well as of the canopy. The effects of such disruption on forest quality can be severe; in the case of intense cultivation of understorey crops such as cocoa in West Africa and cardamon in Sri Lanka, the removal of all regeneration may mean that a standing forest is effectively doomed.

These developments are going to happen partly because there will be more satellites and much more data. Three companies in the US have commercial satellites under development; the first was to be launched in April 1996 (Fritz 1996). These satellite sensors will produce data with resolutions of 1–3 m, coverage every 1–3 days, and data available within 15 min of collection by the satellite. Tropical ecologists are going to be reading much more about, and many more of them will actually be doing work involving, remote sensing.

ACKNOWLEDGEMENTS

We are grateful to the following people who over the years have contributed through discussions to our understanding of remote sensing: R. Almeida Filho and other colleagues at INPE; S.A. Bohlman, F. Gerard (who wrote Table 22.2), R.M. Lucas, B.W. Nelson, D.A. Roberts, D.E. Sabol, M.O. Smith and B. Wyatt.

REFERENCES

Adams, J.B., Sabol, D.E., Kapos, V., Filho, R.A., Roberts, D.A., Smith, M.O. *et al.* (1995).
Classification of multispectral images based on fractions of endmembers: application to land-cover change in the Brazilian Amazon. *Remote Sensing of Environment*, **52**, 137–154.

Balmford, A. & Long, A. (1995). Across-country analyses of biodiversity congruence and current conservation effort in the tropics. *Conservation Biology*, **9**, 1539–1547.

Bellingham, P.J., Kapos, V., Varty, N., Healey, J.R., Tanner, E.V.J., Kelly, D.L. et al. (1992). Hurricanes need not cause high mortality: the effects of Hurricane Gilbert on forests in Jamaica. *Journal of Tropical Ecology*, **8**, 217–223.

Boose, E.R., Foster, D.R. & Fluet, M. (1994). Hurricane impacts to tropical and temperate forest landscapes. *Ecological Monographs*, **64**, 369–400.

Chatelain, C., Gautier, L. & Spichiger, R. (1996). A recent history of forest fragmentation in southwestern Ivory Coast. *Biodiversity and Conservation*, **5**, 37–53.

Condit, R., Hubbell, S.P. & Foster, R.B. (1996). Changes in tree species abundance in a neotropical forest: impact of climate change. *Journal of Tropical Ecology*, **12**, 231–256.

FAO (1993). Forest resources assessment 1990 tropical countries. *FAO Forestry Paper*, **112**.

Fearnside, P.M. (1993). Deforestation in Brazilian Amazonia: the effect of population and land tenure. *Ambio*, **22**, 537–545.

Foody, G.M. & Curran, P.J. (1994). Estimation of tropical forest extent and regenerative stage using remotely sensed data. *Journal of Biogeography*, **21**, 223–244.

Fritz, L.W. (1996), The era of commercial earth observation satellites. *Photogrammetric Engineering and Remote Sensing*, **62**, 39–45.

Furley, P.A., Dargie, T.C.D. & Place C.J. (1994). Remote sensing and the establishment of a geographic information system for resource management on and around Maraca Island. In *The Rainforest Edge: Plant and Soil Ecology of Maraca Island, Brazil* (Ed. by J. Hemming), pp. 115–133. Manchester University Press, Manchester.

Hall, F.G., Botkin, D.N., Strebel, D.E., Woods, K.D. & Goetz, S.J. (1991). Large scale patterns of forest succession as determined by remote sensing. *Ecology*, **72**, 628–640.

Hill, R.A. & Foody, M. (1994). Separability of tropical rain-forest types in the Tambopata–Candamo Reserved Zone, Peru. *International Journal of Remote Sensing*, **15**, 2687–2693.

Holmes, M.G. (1992). Monitoring vegetation in the future: radar. *Botanical Journal of the Linnean Society*, **108**, 93–109.

Hubbell, S.P. (1979). Tree dispersion abundance and diversity in tropical dry forest. *Science*, **203**, 1299–1309.

INPE (Instituto Nacional de Pesquisas Espaciais) (1992). *Deforestation in Brazilian Amazonia.* INPE, Sao Jose dos Campos, Brazil.

Lucas, R.M., Honzak, M., Foody, G.M., Curran, P.J. & Corves, C. (1993). Characterizing tropical secondary forests using multi-temporal Landsat sensor imagery. *International Journal of Remote Sensing*, **14**, 3061–3067.

Malingreau, J.P., Stephens, G. & Fellows, L. (1985). Remote sensing of forest fires in Kalimantan and North Borneo in 1982–83. *Ambio*, **14**, 314–321.

Manokaran, N. & Swaine, M.D. (1994). *Population dynamics of trees in dipterocarp forests of Peninsular Malaysia.* Forest Research Institute Malaysia.

Moran, E.F., Brondizio, E., Mausel, P. & Wu, Y. (1994). Integrating Amazonian vegetation, land-use, and satellite data. Attention to differential patterns and rates of secondary succession can inform future policies. *BioScience*, **44**, 329–338.

Myers, N. (1989). *Deforestation Rates in Tropical Forests and their Climate Implications.* Friends of the Earth, London.

Nelson, B.W. (1994). Natural forest disturbance and change in the Brazilian Amazon. *Remote Sensing Reviews*, **10**, 105–125.

Nelson, B.W., Kapos, V., Adams, J.B., Oliveira, W.J., Braun, O.P.G. & Amaral, I.L.D. (1994). Forest disturbance by large blowdowns in the Brazilian Amazon. *Ecology*, **75**, 853–858.

Pomeroy, D. (1993). Centers of high biodiversity in Africa. *Conservation Biology*, **7**, 901–907.

Skole, D. & Tucker, C. (1993). Tropical deforestation and habitat fragmentation in the Amazon: satellite data from 1978 to 1988. *Science*, **260**, 1905–1910.

Terborgh, J., Foster, R.B. & Nuñez, P. (1996). Tropical tree communities: a test of the nonequilibrium hypothesis. *Ecology*, **77**, 561–567.

Tucker, C.J., Townshend, J.R.G. & Goff, T.E. (1985). African land cover classification using satellite data. *Science*, **227**, 369–375.
Tuomisto, H., Ruokolainen, K., Kalliola, R., Linna, A., Danjoy, W. & Rodriguez, Z. (1995). Dissecting Amazonian biodiversity. *Science*, **269**, 63–66.
Vooren, A.P. & Offermans, D.M.J. (1985). An ultralight aircraft for low-cost, large-scale stereoscopic aerial photographs. *Biotropica*, **17**, 84–88.
Westman, W.E., Strong, L.L. & Wilcox, B.A. (1989). Tropical deforestation and species endangerment: the role of remote sensing. *Landscape Ecology*, **3**, 97–109.

LIST OF REVIEWERS

Adler, G.H.
Ashton, P.S.
Becker, P.
Bond, W.J.
Bongers, F.
Boot, R.
Brakefield, P.
Brown, N.D.
Chivers, D.
Clark, D.B.
Clutton-Brock, T.H.
Compton, S.G.
Corlett, R.T.
Cummins, C.P.
Deutsch, J.C.
Dunbar, R.
Duncan, P.
Eltringham, S.K.
Fenner, M.
Fishelson, L.
Foody, G.M.

Hall, J.
Hamilton, A.C.
Happold, D.C.D.
Harthorn, G.S.
Hladik, A.
Hopkins, M.S.
Inger, R.F.
Kapos, V.
Kellman, M.
Kemp, A.
King, D.
Koyama, T.
Lack, P.
Laird, L.M.
Leirs, H.
Longman, K.A.
Lowe-McConnell, R.H.
McCormick, M.
Milner-Gulland, E.J.

Newbery, D.M.
Newton, I.
Pascal, J.P.
Phillips, O.L.
Prance, G.
Prentice, I.C.
Prins, H.H.T.
Proctor, J.
Read, J.
Riera, B.
Shugart, H.H.
Tanner, E.V.J.
TerSteege, H.
Thompson, K.
Turner, G.M.
Vitt, L.J.
Werger, M.J.A.
Whittaker, R.J.
Woodward, F.
Young, M.

INDEX

Please note: page numbers in *italic* refer to figures and those in **bold** to tables

Abies spp., crown development 148
acidity, tropical rainforest soils **180**
Aciotis aequatorialis, Amazon floodplain 411
Africa
  birds
    migration 428–35
    species 427–8
  herbivores, species richness 449–90
  rainfall 423–5
    seasonal *425*
  savannas 426–7
    arboreal insectivorous birds 421–47
    southern, insectivorous birds **430**
  tropical rainforests 426
    geological background 310
    vegetation 426–7
  wildlife, exploitation 244–5
  see also South Africa; West Africa
agoutis, seed removal studies 29–30, 37–8, 40–1
Amazon floodplain
  beetles 406–7
  characteristics 385–6
  communities, evolution and diversity 385–419
  disturbance studies 609
  diversity 396–407
  fishes 400–6
    distribution *403*
    species numbers **402**
  geological background 408–9
  grasses 397–8
  habitats 386, 388–92
  macroevolution 407–16
  macrophytes 398–9
  oxygen, vertical profiles *394*
  palms 414–15
  plants 396–400
  processes 389–92
  remote sensing studies 611
  temperature, vertical profiles *394*
  water depth, seasonal variation *391*
  water quality studies 392–6
Anak Krakatau, activity 538, *539*
anurans
  opportunism 233–4
  reproduction tactics 222–3
  risk tactics, West Africa 221–42
  studies
    issues 233–9
    methods 223–4
    results 224–33
Arapaima spp., Amazon floodplain 401
Araucaria spp., plantations 574
Arctiidae
  diversity studies 577–83, 589–90
  species associations *585, 591*
Argentine, pollination studies *348*
Asia, trees, species richness and genus richness **496**
Aucoumea klaineana 314
Australia
  flesh/seed dry mass quotients *17*
  seed mass studies 1–24

Bactris tefensis, Amazon floodplain 414–15
bamboo
  cultivation 576
  remote sensing studies 610–11
Barcella odora, Amazon floodplain 414
Barro Colorado Island (Panama)
  geography 28–9
  pollination studies 353–4
  rodent populations 28–9
  rodent sightings *30*
  seed characteristics **31**
  seed removal studies 25–49
  seedling growth studies 197
  species studies 196
  tree mortality 167
bats, pollination **345**, 346, 352
BCI *see* Barro Colorado Island (Panama)
bees, pollination **345**, 348–9
beetles
  Amazon floodplain 406–7
  pollination 350
Bertholletia excelsa
  exploitation 558

leaf density studies *204*
biomass
  fishing studies **113**
  vs. fishing effort *112*
  vs. gross daily intake rate *457*
birds
  Afrotropical *428–32*
    competition *438–41*
    densities *437–8*
    movements *435–7*
    seasonal distribution *435*
  arboreal insectivorous, community dynamics *421–47*
  community dynamics, future research *443–4*
  densities, factors affecting *441–3*
  insectivorous
    Nigeria **429**
    seasonal changes *431*
    southern Africa **430**
  migration *421*
    Africa *428–35*
    systems *422–3*
  Palearctic *432–5*
    competition *438–41*
    densities *437–8, 442*
    movements *435–7*
    seasonal distribution *435*
    wintering strategies *432–5*
  pollination *342–5, 352–3*
  seed dispersal *8*
  species, Africa *427–8*
body mass
  frequency distribution, herbivores *476*
  and gross daily intake rate *458–60*
  and gross instantaneous consumption rate *455*
  vs. distribution range, herbivores *475*
  vs. species richness, herbivores *474*
Bolivia, leaf density studies *204*
Bontebok National Park (South Africa), herbivore studies *469*
Borneo
  forests, species richness *498*
  moths, diversity studies *574*
Brazil
  deforestation *605*
  disturbance studies *609*
  floodplain community studies *385–419*
  landcover studies *606–7*
  pollination studies *346–7, 349*
Brighamia spp., pollination studies *347*
Brosimum spp.
  seed fate *36*
  seed removal studies *35–8*
Buchanania arborescens, abundance *535*
Bucida spp., Amazon floodplain *399–400*

Bufo maculatus
  breeding *229–30*
  characteristics *224*
  female population levels *237*
  risk tactics *221–2, 235–6*
  tadpoles
    antipredation tactics *236*
    risks *228–30*
Bufo regularis, tadpoles, risks *228–30*
Bulu (Ndian)
  radiation *280–1*, **282**
  rainfall *280–1*, **282**
Buru, butterflies *572*
bush-shrikes, competition *439*
butterflies
  Buru *572*
  pollination **345**, *346–7*

caatingas
  trees
    average height *11*
    seed mass *10–14*
Calamus spp., Amazon floodplain *415*
Cameroon, rainforests, phenology and dynamics *267–308*
Cano do Mamirauá, vertical profiles *395*
canopies
  gaps
    importance of *80*
    light studies *87*
  openness
    effects on crown traits *134*
    parameters **132**
    and photosynthetically active radiation *84*
  structure, and plant distribution *88*
Carabidae, Amazon floodplain *406–7*
Caranx spp., predation rates *99*
Carapa spp., seed predation studies *63–4*
Carapa guianensis
  effects of light environment on *71*
  seed predation studies *71–2*
cash crops, Malaysia and Indonesia *587–93*
Catostemma spp., seed mass *13*
Cecropia spp.
  Amazon floodplain *414*
  crown development, response times *150*
  crown position index *129*
  gap-size niche differentiation studies *81*
  tree architecture *154*
    studies *127–35*
Cecropia membranacea, Amazon floodplain *413*
Ceiba pentandra, phenology *290*
Celtis mildbraedii, distribution, Ghana *172*
Central America
  Brosimum, seed dispersal *36–7*

Index

seed removal 44
Ceratophyllum spp., Amazon
  floodplain 409–10
Cercocebus spp., diet 316
Cercopithecus spp.
  diet 316
  habitats 331
Chlorocardium spp.
  seed mass 13
  seed predation studies 63, 67–9
  seed survival 70
Chlorocardium rodiei
  leaf density studies 204
  population size distribution studies 212
  survival 199
cichlids, predation of *Bufo* spp. 229
climate
  effects on tropical rainforest trees 166–7, 280–3
  and fruit availability 319–25, 327–9
  and mast fruiting 283–4
  and tree physiology 294
  see also radiation; rainfall; temperature
climate change
  effects on bird densities 441–3
  geological background 329–31
  and primates, future trends 331–2
  tropical rainforests 310
climax species
  densities 502
  in gaps, factors affecting 208–10
  regeneration 204–5
  seedlings, and gap formation 205–8
  tropical rainforest trees 81–2
Colobus spp., diet 316
Colossoma macropomum, Amazon
  floodplain 409
communities, remote sensing studies 610–12
community dynamics
  arboreal insectivorous birds
    future research 443–4
  savannas 421–47
community structure
  and competition 79–80
  and species interactions 80–1
  tropical rainforests, primary
    species 193–219
Comoé National Park (Côte d'Ivoire)
  anuran studies 221–42
  rainfall 223
compensatory growth
  and defoliation 67, 68
  model, seedling survival 67–9
competition
  algorithm **370**
  and community structure 79–80
  fish studies 97

hypotheses, coral reef fishes 97
interspecific 80
for light 89
plants, modelling 369
trees 504–5
coral reefs
  communities, predation 98–100
  fish populations
    factors affecting 98
    hypotheses 96–8
    non-fishing effects 115–16
  fishes
    predation rates 99–100
    pre-settlement life 96
  habitats, dendrogram *114*
  small-scale fishing effects 95–124
Costa Rica
  flesh/seed fresh mass quotients *17*
  pollination studies 342, 346
  species studies 196
Côte d'Ivoire
  anuran studies 221–42
  deforestation studies 606
  MUSE studies 361
CPI *see* crown position index
crown allometry 143–4
  definitions 143
  and light levels 152
  tropical rainforest trees *131*
crown development 147–51
  definitions 147
  path analysis 149–50
  response delays 150–1
  and shade tolerance 148–9
crown position index
  definitions **128**
  tropical rainforest trees *129*
crowns
  components 147–9
  light environment, measurements 200–2
crude protein concentration studies,
    grasses *456*
cuckoos, competition 439
cultivation
  effects on Lepidoptera, Indonesia and
    Malaysia 567–97
  and species distributions 164
Cupania spp., seed removal studies *33, 34*
Curimatidae, Amazon floodplain 406
Cynometra ananta
  distribution, Ghana *170*, 183–4
  relative growth rate *177*

Dacrydium cupressinum, phenology 289
damselfishes, competition studies 97
DCA *see* Detrended Correspondence
    Analysis

deer mice, seed removal studies 37
defoliation, and compensatory growth 67
deforestation
  rates 605
  remote sensing 605–6
demography
  tropical rainforests
    primary species 193–219
    seedlings 285–7
density/distance-dependent attack
  Janzen–Connell model 51–2
    evaluation 72–4
  on seeds and seedlings 51–78
Desmoncus spp., Amazon floodplain 415
Detrended Correspondence Analysis 524
  Krakatau Islands 529, **530**
  studies 528–30
Dicorynia spp.
  crown allometry *131*, 143–4
  crown development *133*, 148
  crown position index *129*
  height/diameter ratio *130*
  parameters **132**
  safety factors *133*
  traits *134*
  tree architecture studies 127–35
Dicymbe altsonii
  growth rate studies 207
  leaf density studies 204
  population size distribution studies 212
  seed distribution studies 206
  survival 199
Didelotia idae, phenology 288
Diospyros dendo
  fruit scores *321*
  fruiting patterns 320
Dipterocarpaceae, abundance 500
distribution range, vs. body mass, herbivores 475
disturbance
  community-wide 550–1
  definitions 550
  large-scale 551–2
    types of 552–60
  and succession
    primary 553–4
    secondary 554–60
  tropical rainforests 549–65
  see also human disturbance; vegetation disturbance; volcanic disturbance
Doliocarpus spp., seed removal studies 33, *34*
droughts
  adaptations, lowland rainforest **173**
  Ghana *171*
  seasonal, effects on tree species 173–9
  and tree mortality 166
  tree responses to *176*, 298–9

tree survival experiments 174–9
tropical rainforests 556–8
Duboscia macrocarpa
  flowering 328
  fruit scores *322*
  fruiting patterns 321
Dysoxylum gaudichaudianum
  abundance 542
  habitats 543
  height change distribution *534*
  sapling performance 524–5, 531–5

Echinochloa spp., Amazon floodplain 397
ecological niche, differentiation, tropical rainforest trees 79–94
ecological plasticity, trees 503–5
ecology, and remote sensing 612–13
ecosystems
  disturbance and succession, Indonesia 515–48
  diversity, Krakatau Islands 524, 528–31
  fragmentation, effects on herbivores 480–4
  modelling
    generic 380–1
    savannas 369–72
  recovery 516
El Niño Southern Oscillation
  and droughts 556–7
  effects on tree mortality 166
  remote sensing 612
  and volcanism 537–8
Elaeis oleifera, Amazon floodplain 414
electric fishes, Amazon floodplain 400
Electrophorus spp., Amazon floodplain 401
Eleocharis radicans, Amazon floodplain 411–12
Elisabetha princeps, compensatory growth 68
ENSO *see* El Niño Southern Oscillation
environment, modelling 368
environmental factors, large trees vs. seedlings **198**
Eperua spp., seed mass 13
Eperua grandifolia, phenology 288
equilibrium theories, tropical rainforests 193, 194–5
Eschweilera spp., phenology 288
estivation, *Hyperolius viridiflavus nitidulus* 232
Eucalyptus deglupta, plantations 573–4
Eugenia spp., seed removal studies 33, *34*
Eusideroxylon spp., seed mass 14
evapotranspiration
  Ghana *171*
  and species richness, herbivores 478
extinctions, herbivores 475–6, 481–2

Index

facilitation, herbivore studies 470–1
Fagus spp., phenology 289
Faramea spp.
  seed removal studies 34–5
  seed survival 59–60
Faramea occidentalis, seed survival 58
Fazenda Dimona
  land use 607
  landcover studies 606–7
Ficedula spp., competition 438
Ficus spp.
  as source of furanocoumarins 67
  wasp pollination 352
field crops, Malaysia and Indonesia 587–93
Fiji
  reefs
    dendrogram *114*
    fishing-grounds **105**
    maps *102*
    small-scale fishing effects 95–124
fires
  remote sensing studies 610
  and tree mortality 166–7
  tree resistance to, modelling 372–6
fishes
  Amazon floodplain 400–6
    distribution *403*
    species numbers **402**
  forest-dwelling 391
  mortality rates, small-scale fishing effects 100–1
  population fluctuations 96
  predatory, small-scale fishing effects 95–124
  prey, small-scale fishing effects 95–124
  species listing **106–11**
fishing
  effects on fish populations
    Fijian reefs 95–124
    variations 104–14
  effects on prey 118–19
  small-scale, differential effects 95–124
  studies
    biomass **113**
    data analysis 103–4
    habitat variables 113–14
    issues 115–20
    limitations 115–16
    methods 101–4
    species listing **106–11**
    statistics **105**
fishing effort
  studies 104
  vs. biomass *112*
flesh/seed dry mass quotients 2–3
  Australia *17*
  fruits 16–19

flesh/seed fresh mass quotients, Costa Rica *17*
flesh/seed-per-fruit mass quotients **18**
floodplain communities, evolution and diversity, Amazon 385–419
floodplains, habitats, terminology **388**
floods, remote sensing studies 610
forest dynamics
  Cameroon 267–308
  issues 300–2
  models 268
  overview 267–71
  studies 271–5
forestry, effects on Lepidoptera, Indonesia and Malaysia 567–97
forests
  human disturbance 567–9
  phenology, Cameroon 267–308
  species richness, Borneo *498*
  types
    Ghana *169*
    remote sensing studies 610–11
    tree distributions *172*
FORET 363
FORSKA 363
French Guiana, tree architecture studies 127–35
FRENCH model
  in MUSE 372–6
  results *374*
  spatial patterns *376*
frugivores, diet change 311
frugivory, and primate populations, Gabon 309–37
fruit availability
  annual patterns 317–18
  Gabon 315–16
  phenology and climatic variables 319–25, 327–9
  species numbers 320
  variations 317–19
fruit production, and radiation 290
fruit scores
  calculation 316
  monthly *319*, *321*, *322*, *323*
fruit tissues, for seed protection 14–16
fruits
  flesh/seed dry mass quotient 16–19
  fleshy, tropical rainforests 1–24
  lowland rainforest 16
furanocoumarins, toxicity 67

Gabon, primates, factors affecting 309–37
Ganophyllum giganteum
  fruit scores *321*
  fruiting patterns 320
gap formation

climax species seedling studies 205–10
and species richness 208–10
and volcanic disturbance 523–4
studies 525–8
gap models 363–4
gaps
remote sensing studies, future research 613
sapling performance 524–5, 531–7
gap-size niche differentiation
detection 82–8
evidence for 81–2
hypotheses 87
improved model 89–91
measurement issues 82–5
and species richness 195–6
tropical rainforest trees 79–94
genus richness
and species richness 492
trees, Sarawak **493**
Geometridae
abundance 586
diversity studies 576–87, 590–3
linkage diagrams 580
species, Seram **581**
species associations 582, 592
Ghana
climatic data 171
forest type 169
rainfall 171, 172–3
topsoil distribution 169
tree species distributions 168–73
maps 170
tree species profiles, droughting experiments **175**
Gilbertiodendron spp.
disturbance 561
phenology 288
seed distribution 206–7
Gnetum spp., Amazon floodplain 409–10
Golden Gate Highlands National Park (South Africa), herbivore studies 469
Gorilla spp., diet 316
grasses
Amazon floodplain 397–8
biomass, vs. herbivore population 464
crude protein concentration studies 456
distribution, and herbivore population dynamics 451–65
modelling 369–72
primary production
annual 379
model outputs 378
modelling 376–9
grazers *see* herbivores
gross daily intake rate
and body mass 458–60

response patterns 454–8
vs. biomass 457
gross instantaneous consumption rate
and body mass 455
calculation 453–4
growth performance
and light environment 207
and light level 85
seedlings, experiments 180–3
Guadua spp., Amazon floodplain 397
guild delimitation, tropical rainforests 194–8
Guinea, savannas, anuran studies 221–42
Gunung Mulu National Park (Sarawak), moth diversity studies 593–4
Gustavia spp.
seed fate 39
seed removal studies 31, 38–41
Guyana
growth rate studies 207
leaf density studies 204
population size distribution studies 212
seed distribution studies 206
seed predation studies 62–4
Gymnotidae, Amazon floodplain 400–1, 405–6
Gymnotus spp., Amazon floodplain 401
Gynerium spp., Amazon floodplain 397

habitats
Amazon floodplain 386, 388–92
change
effects on arboreal insectivorous birds 421–47
and pollination systems 340
coral reefs, dendrogram 114
degradation, effects on bird densities 441–3
floodplains, terminology **388**
fragmentation
effects on plant–pollinator interactions 339–60
and pollination systems 342–50
plasticity 409
and seed fate 43–4
survival, vs. human population increase 244
vs. area, effects on species richness 495–9
harvesting
mammals 243–65
timber, Malaysia and Indonesia 572–5
wildebeest
estimates 250–2
future research 261–2
illegal 257–8
legal 258–60
likelihood profiles 256
sustainable levels 256–7

Index

Hawaii, pollination studies 343, 347
height, and seedling survival *199*
Heliocarpus spp., crown development, response times 150
hemispherical canopy photography *84*, 85
  limitations 136
herbivores
  body mass
    frequency distribution *476*
    vs. distribution range *475*
  communities, assembly rules 463–5
  and ecosystem fragmentation 480–4
  extinctions 475–6, 481–2
  facilitation, studies 470–1
  geological background 475–6
  gross daily intake rate 454–8
    and body mass 458–60
  gross instantaneous consumption rate 453–4
  isolation effects *481*
  net daily intake rate 460–1
  population
    annual changes 461–3
    vs. grass biomass *464*
  population dynamics, and grass distribution 451–65
  resource competition 465–71
  resource utilisation 451–3
  species, distribution range **466–7**
  species richness
    Africa 449–90
    continental 474–7
    and evapotranspiration *478*
    factors affecting **485**
    predictions 473–4
    regional variations 477–80
    vs. body mass *474*
    and weight ratio 471–84, *485*
  weight ratio period, spectral densities *477*
herbivory, epidemic, tropical rainforests 558
Heritiera utilis, leaf water relations **178**
Hippolais spp.
  competition 438
  species segregation 440
Hippolais icterina
  density 437
  migration 433
Hirundo rustica, migration 433
honeycreepers, pollination 343
Hopea nervosa, light acquisition 89
Hoplias spp., Amazon floodplain 401
Hoplobatrachus occipitalis
  breeding 226–7
    opportunities 225
  characteristics 223–4
  egg production studies 227
  female population levels 237
  migration 227–8
  opportunism 233–4
  pool preference 226
  predation of *Bufo* spp. 229, 230
  risk tactics 221
  risk-sensitive behaviour 225–8
  tadpoles
    carnivorous nature of 224, 226
    survival *228*, 234–5
howler monkeys, and seed dispersal 36–7
human disturbance
  forests 567–9
  tropical rainforests 558–60
hummingbirds, pollination 342–3
Hurricane Gilbert, Jamaica 608
hurricanes, disturbance, remote sensing studies 608–9
Hyperolius viridiflavus nitidulus
  antipredation tactics 238
  breeding 230–1, 237
  characteristics 224
  egg production 231, 238
  estivation 232
  froglets, growth patterns 233
  population ecology 230–1
  risk tactics 222, 237–9
Hypsipyla, seed predation studies 71–2

igapó, use of term 388
India, pollination studies 350
Indonesia
  cash crops 587–93
  ecosystems, disturbance and succession 515–48
  field crops 587–93
  Lepidoptera, effects of cultivation on 567–97
  shifting swidden cultivation 575–87
  timber harvesting 572–5
insects
  as environmental indicators 568
  Janzen–Connell model studies **55–6**
  roles, in forest survival 568
  seed predation 61–2
Inter-Tropical Convergence Zone (ITCZ) 421
  positions *425*
  and rainfall patterns 423–5
invertebrates
  seed predation 52, 60–2
  seed selection/search parameters **61**
irradiance, and relative growth rate *182*
Ivory Coast
    see Côte d'Ivoire

J–C model *see* Janzen–Connell model
JABOWA 363

Jamaica, Hurricane Gilbert 608
Janzen–Connell model 51–2, 54–60
  discrepancies 60–1
  empirical findings 55–7
  graphical 53
  issues 53, 64–5
    theoretical 54–5
  juvenile recruitment studies 56–7
  mortality studies 58
  predictions
    evaluation 72–4
    factors affecting 57–60
  studies **55–6**
Rio Japurá
  evolution and diversity studies 385–419
  floodplain 387
  lake size distribution 390
  vertical profiles 394
journals, ecological, remote sensing content analysis **600**
Julbernardia spp., seed distribution 206–7

kangaroos, gross daily intake rate 457
Kenya
  birds, species segregation 440
  grasslands, crude protein concentration studies 456
Korup National Park (Cameroon)
  characteristics 269–70
  geological background 294–5
  historical development 299
  radiation 293
  rainforests, phenology and dynamics 267–308
  seedling survival 286, 287
Krakatau Islands (Indonesia)
  Detrended Correspondence Analysis 529
  ecosystems
    disturbance and succession 515–48
    diversity 524, 528–31
  human disturbance 539–40
  map 517
  sapling performance **532**
  site availability 537–40
  species
    differential availability 540–2
    differential performance 542–5
  species–area curves 528
  successional history 520–3
  trees, mortality **527**
  volcanic disturbance 516, 553
    frequency distributions 526
    and gap formation 523–4
    recovery 516–20
    successional processes 520, 521
Kruger National Park (South Africa), herbivore studies 468–9

LAI *see* leaf area index
Lake Manyara National Park (Tanzania), herbivore studies 468, *481*
lakes, size distribution studies 390
Lamto Research Station (Côte d'Ivoire), MUSE studies 361
land use, Fazenda Dimona 607
landcover, remote sensing studies 606–7
Landsat, remote sensing studies 601–3, 605, 606, 608–9
landslides, tropical rainforests 553
LAR *see* leaf area ratio
Laziacis spp., Amazon floodplain 397
leaf area index 369
leaf area ratio
  and leaf density 204
  measurement problems 202–3
leaf density, and leaf area ratio 204
leaf water relations, characteristics **178**
least-squares statistics, in seed predation studies 63–4
Lecythis spp., phenology 288
Leguminosae, as host for Lepidoptera 573
Lepidoptera
  effects of cultivation on, Indonesia and Malaysia 567–97
  human disturbance studies
    analysis techniques 570–1
    future research 594
    issues 593–4
    methods 569–72
    overview 567–9
    results 572–93
  see also butterflies; moths
Lepidosiren spp., Amazon floodplain 401, 408
Licania spp.
  seed fate 39
  seed removal studies 31, 38–41
light, competition for 89
light environment
  crowns, measurements 200–2
  effects on *Carapa guianensis* 71
  and growth performance 207
  and large trees 213
  and safety factors 152–3
  and seed predation 65
  and seed survival 70–2
  and seedling growth 85
  and tree architecture, studies 127–35
  and tree architecture models 151
  tropical rainforest trees 126–7
light levels
  and growth performance 85
  and seedling growth 86–7

and tree allometry 151–2
and tree architecture models 137–9
and tree height 135
tropical rainforest trees, architecture and development 125–62
variability 135–6
lightning, tropical rainforests 555–6
light-trapping
methods 569–70
moths 568
LMRF *see* lower montane rainforest
logging
Malayan Uniform System 572
Malaysia and Indonesia 572–3
Lopé Reserve (Gabon)
characteristics 312
climate change, geological background 329–31
fruit availability 315–16
future research 332–3
maps *313*
primates
diets *326*
factors affecting 309–37
species 311
rainfall *314*
vegetation 312–15
Lophira alata 314
dominance 166
relative growth rate *177*, *182*
lottery hypotheses, coral reef fishes 96–7
lower montane rainforest, seed mass 12–13
lowland rainforest
drought adaptations **173**
fruits 16
seed mass 12–13
tree species distributions, limits 163–91
LRF *see* lowland rainforest

Macaranga tanarius, abundance 536
macrophytes, Amazon floodplain 398–9
magnesium, as nutrient, in seedlings *183*
Malaysia
cash crops 587–93
field crops 587–93
Lepidoptera, effects of cultivation on 567–97
shifting swidden cultivation 575–87
timber harvesting 572–5
Mallotus spp., seed mass 9
mammals, harvesting 243–65
Mandrillus spp., diet 316
manganese, occurrence 182
Mansonia altissima
distribution, Ghana *170*, *172*
field experiments 184
relative growth rate *177*, *182*

Mara macho, leaf density studies *204*
mast fruiting 287–8
and climate 283–4
predator–satiation hypothesis 292
Mauritia flexuosa, Amazon floodplain 415
Mauritiella armata, Amazon floodplain 415
maximum likelihood statistics, in population dynamics model 252–3
maximum sustainable yield, wildebeest 256, 257
Melipona spp., Amazon floodplain 400
Metroxylon spp. 414
Microberlinia bisulcata
abundance 267
clustering 302
density studies **273**
distribution studies 272–5
frequency distribution *274*
fruiting patterns 278
growth rates 286
in Korup National Park 270–1
phenology 276–9, 290–2
regeneration 295–300
relative proportions 275
seedling bank studies **279**
size distributions 297–300
survival 287
microclimate, gap-size as surrogate for 83–5
migration
birds 421
Africa 428–35
systems 422–3
minimum distance rule, and seed fate 52–3
mistletoes, pollination 343
Montrichardia arborescens, Amazon floodplain 413
Mora spp., seed mass 13
Mora excelsa, leaf density studies *204*
moths
diversity studies
Borneo *574*
Seram *576*
Sulawesi *589*
effects of cultivation on, Indonesia and Malaysia 567–97
light-trapping 568
pollination **345**, 346–7
species associations 582–5
studies
Seram 575–87
Sulawesi 587–93
Mountain Zebra National Park (South Africa), herbivore studies 469
MSY *see* maximum sustainable yield
multifactor analysis, tree architecture 153–4
Multistrata Spatially Explicit model *see* MUSE

Mundemba *see* Ndian
MUSE 361–83
  adaptability 372
  development 363
  for ecosystem modelling 380–1
  future developments 379–81
  main loop 367
  models, library of **371**
  modules 372–9
  overview 364–72
  resource competition algorithm **370**
  structure 365–8
  variables 369

NAR *see* net assimilation rate
NASA Pathfinder project 605–6
Ndian
  oil palms, yields 284–5, 293
  radiation 280–1, **282**
  rainfall 280–1, **282**
NDVI *see* normalized difference vegetation index
Nearctic–Neotropical migration system 422–3
Nectarinia spp., migration 431–2
Neonauclea calycina, habitats 543
neotropics
  seed mass 6–7
  seed removal studies 25–49
  trees, species richness and genus richness **496**
net assimilation rate, measurement problems 202–3
net daily intake rate, functional responses 460–1
New England, hurricane disturbance studies 609
New Zealand, pollination studies 343
niche differentiation hypothesis *see* gap-size niche differentiation
niche specificity
  tropical trees 491–514
    future research 509–10
Nigeria
  birds
    densities 442
    insectivorous **429**, *431*
nitrogen, in seeds 14–16
Noctuidae
  dendrograms 578
  diversity studies 576–83, 589
  linkage diagrams 579
  species associations 583, 590
non-equilibrium theories, tropical rainforests 193, 195
normalized difference vegetation index 603
les Nouragues (French Guiana), tree architecture studies 127–35
nutrients
  mineral, seeds 2
  seedlings *183*
  tropical rainforest soils **180**
Nypa spp. 414
oil palms
  yields 284–5
    Ndian 293

ontogeny, and trunk allometry 151
Oriolus oriolus, migration 433
Oryza spp., Amazon floodplain 397
Osteoglossum spp., Amazon floodplain 401
oxygen, vertical profiles, Amazon 394–5

PACER *see* phenological and climatic ectomycorrhizal response
Palearctic–Afrotropical migration system 422–3, 428–35
paleotropics, seed mass 4–6
palm-rich forests
  trees
    average height 11
    seed mass 10–14
palms, Amazon floodplain 414–15
Palo Santo Colorado, leaf density studies 204
Pan spp., diet 316
Panama, seed removal studies 25–49
PAR *see* photosynthetically active radiation
Parana Aranapu, lake size distribution 390
paranas, use of term 389
Pasoh, trees, abundance **501–2**
passerines, migration 422–3
path analysis, crown development 149–50
Pentadesma spp., flowering 328
Pentadesma butyracea
  fruit scores 322
  fruiting patterns 321
perennial grass primary production model 377
Peru
  pollination studies 346
  remote sensing studies 611
phenological and climatic ectomycorrhizal response 293
phenology
  and fruit availability 319–25, 327–9
  and Janzen–Connell model predictions 57–60
  and primate populations, Gabon 309–37
  tropical rainforests 311
    Cameroon 267–308
    issues 287–95
    studies 276–9
phosphorus
  as nutrient

Index

in seedlings *183*
  in tropical rainforests 179–80
  in seeds 15–16
photosynthetically active radiation 82, 91
  and canopy openness *84*
  effects on seedling growth 86
  and gap-size measurement 83
  sensors 136
Phyllanthus spp., Amazon floodplain 410–11
Phylloscopus trochilus, density 437
Phyllyrea latifolia, phenology 289
Piper spp., seed mass 9
Pipturus argenteus, abundance 535
piscivores, depletion 116–18
plant distribution
  and canopy structure 88
  and plant competitive ability 90
plant–pollinator interactions **344–5**
  effects of habitat fragmentation on 339–60
  generalizations 350–6
  studies
    data 341–2
    issues 350–6
  tropical vs. temperate latitudes 354–6
plant shape, modelling 368
plant size, and light acquisition 89
plants
  Amazon floodplain 396–400
  interactions, modelling 368–9
  tropical rainforests 1–24
Plectropomus laevis, fishing effects 118
poaching, economic benefits,
    countermeasures 260
pollination systems
  bats **345**, 346, 352
  bees **345**, 348–9
  beetles 350
  birds 342–5, 352–3
  butterflies and moths **345**, 346–7
  and habitat change 340
  and habitat fragmentation 342–50
  studies, Argentine *348*
  wasps **345**, 350, 352
pollinator–flower mutualisms 355
pollinator–plant interactions *see*
    plant–pollinator interactions
poor soils, and seed mass 10–14
population dynamics model
  confidence bounds 255–6
  data fitting
    method 252–3
    results 253–5
  goodness-of-fit 255
  wildebeest 246–57
population recruitment curve 53
PORTOBRAS, water depth studies 391, **392**
Pourouma spp., crown position index *129*

PRC *see* population recruitment curve
predation
  coral reef communities 98–100
  rates
    coral reef fishes 99–100
    models 99–100
  stages 62
  see also seed predation
predator–satiation hypothesis, mast
    fruiting 292
PRF *see* palm-rich forests
primates
  and climate change, future trends 331–2
  diets 309, 316–17, 325–7
    analysis *326*
  factors affecting, Gabon 309–37
  food availability 310–11
  howler monkeys, and seed dispersal 36–7
  mobility 332
  studies
    issues 327–33
    methods 312–17
    results 317–27
Pterygota macrocarpa
  distribution, Ghana *172*
  droughting experiments 177–8
  leaf water relations **178**
Puerto Rico, hurricane disturbance
    studies 609
Pycnanthus angolensis
  distribution, Ghana 170, *172*
  drought sensitivity *184*
  relative growth rate *177*

Quercus spp., phenology 289

rabbits, gross daily intake rate 457
radiation
  and fruit production 290
  Korup National Park 293
  Ndian 280–1, **282**
rainfall
  Africa 423–5
  effects
    on arboreal insectivorous birds 421–47
    on tropical rainforest trees 166, *172*,
      280–3
  Gabon *314*
  Ghana *171*, 172–3
  Ndian 280–1, **282**
  seasonal, Africa 425
  and soil fertility, tree species 173, 179
rare species, causes of 505–7
redundancy model, seedling survival 69–72
reed frog *see Hyperolius viridiflavus nitidulus*
regeneration
  climax species 204–5

strategies 210–13
tropical rainforest trees 295–300
relative growth rate
  and irradiance *182*
  measurement problems 202–3
  saplings *177*
  and soil fertility 182, 184–5
remote sensing
  analyses 601–3
  applications 599
  in ecological journals, content analysis **600**
  future trends 612–13
  sensor characteristics **602**
  studies
    communities 610–12
    vegetation types 610–12
  techniques 601–3
  tropical rainforests 599–615
  vegetation disturbance 604–10
reproductive adult density, vs. seed crop size 58, 59
resource competition, herbivores 465–71
RGR *see* relative growth rate
Rio Solimões
  evolution and diversity studies 385–419
  floodplain *387*
  lake size distribution *390*
  vertical profiles *394*
rivers, mobile, tropical rainforests 554
rodents
  in Barro Colorado Island 29
  neotropical, seed removal 25–49
  seed dispersal 8
  sightings, Barro Colorado Island *30*
RS *see* remote sensing

Sabah
  softwood plantations 573
  timber harvesting 572
Sacoglottis gabonensis 314
safety factors
  and light environment 152–3
  trees *133*, 145–7
Saimiri spp., Amazon floodplain 400
saplings
  performance
    in gaps 524–5, 531–7
    Krakatau Islands **532**
  rank-abundance curves *536*
  survival and growth rate *177*
Sarawak
  moth diversity studies 593–4
  trees
    abundance **500**, **501–2**
    genus richness **493**
satellites
  remote sensing studies 601–2

future trends 613
savannas
  Africa 426–7
  anurans, risk tactics 221–42
  arboreal insectivorous birds, community dynamics 421–47
  characterization 362–3
  ecosystem models, generality 369–72
  functions and dynamics 362–3
    spatial model 361–83
  modelling 362–4
    future developments 379–80
    generic model *365*
    principles 363–4
  stability issues 373–4
Scleria secans, Amazon floodplain 415–16
seed attack *see* seed predation
seed caching
  rates 42
  vs. seed mass *41*
seed crop size
  and Janzen–Connell model predictions 57–60
  vs. reproductive adult density 58, 59
seed distribution, understorey 205–6
seed fate
  Barro Colorado Island *36*, *39*
  change in
    large seeds 38–41
    medium seeds 35–8
    small seeds 33–5
  and distance from parent 53
  escape 52–3, 64
    and juvenile recruitment 57–8
    and life-history characteristics 66
  light-dependent 51–2
  and habitat 43–4
  pathway diagrams 27
  post-dispersal 42
  and seed mass 25–49
  see also seed survival
seed mass
  caatinga vs. palm-rich forest *11*
  future research 20
  by genera numbers *5*
  interpretation, tropical rainforests 1–24
  light-demanding vs. shade-tolerant *10*
  and nitrogen concentration *15*
  and poor soils 10–14
  range 3–10
  and seed fate 25–49
  and seedling size 11–12
  and shade 7–8
  theoretical models 8–9
  vs. distance carried 42
  vs. seed caching *41*
  vs. seed survival *69*

Index

seed predation
  factors affecting 51–78
  field experiments, Guyana 62–4
  invertebrates 52
    alternative outcomes 64–72
    vs. vertebrates 60
  Janzen–Connell model studies **55–6**
  large seeds 14
  and light environment 65
  studies 62–4
    analysis 63–4
seed removal
  dispersal vs. predation 27
  future research 42–4
  by neotropical rodents 25–49
  studies
    analysis 32–3
    experimental design 31–2
    recommendations 41–4
    results 33–41
    species 30–1
seed size *see* seed mass
seed survival 58
  Chlorocardium spp. 70
  and light environment 70–2
  vs. seed mass 69
seed weight *see* seed mass
seedling banks, tropical rainforest trees, studies 279–80
seedling distribution, understorey 205–6
seedling growth
  and light environment 85
  and light levels 86–7
  understorey 207
seedling size, and seed mass 11–12
seedling survival
  compensatory growth model 67–9
  in gaps 209–10
  and height 199
  Korup National Park 286, 287
  redundancy model 69–72
  stochastic effects 207
  susceptibility model 66–7
  understorey 207
seedlings
  climax species, and gap formation 205–8
  environmental factors **198**
  establishment, rodent impact 27
  growth performance, experiments 180–3
  growth studies 197
  light acquisition 89
  mortality 199
    barriers to 66–7
  nutrients 183
  predation
    factors affecting 51–78
    Janzen–Connell model studies **55–6**
    responses, quantification 202–3
    species-specific differentiation 202–4
  studies
    future research 214
    scarcity 202
    stochastic effects 203–4
  tropical rainforests, demography 285–7
seeds
  characteristics, Barro Colorado Island **31**
  definitions 3
  dispersal 8
  mortality risks 8–9
  nitrogen concentrations, theories 14–16
  production patterns 57
  protection, by fruit tissues 14–16
  shade-tolerators 4, 6
    vs. light-demanders 7–8, 9–10
  trees, small 10–14
  tropical rainforests 1–24
    mineral nutrients 2
Seram
  moths
    Arctiidae studies 585
    diversity studies 576
    Geometridae studies 580, **581**, 582, 586
    Noctuidae studies 578–9, 583
    species associations 584
  shifting swidden cultivation, moth studies 575–87
  vegetation types 575
Serengeti, wildebeest, exploitation limits 243–65
Serrasalmus nattereri, Amazon floodplain 409
shade, and seed mass 7–8
sheep, gross daily intake rate 457
shifting swidden cultivation, Malaysia and Indonesia 575–87
Shorea albida, mass defoliation 558
site availability, and successional processes 537–40
slash and burn horticulture *see* shifting swidden cultivation
SMA *see* spectral mixture analysis
Smithsonian Tropical Research Institute, Center for Tropical Forest Science 494
Socratea spp., safety factors 146
softwood, plantations, Malaysia and Indonesia 573–5
soil chemistry
  and species distributions 167
  tropical rainforests **180**
soil fertility
  and rainfall, tree species 173, 179
  and relative growth rate 182, 184–5
  and species distributions 179–83

soil type, preference, tropical rainforest
    trees **181**
Sorghum intrans, dynamics, modelling 379
South Africa
    herbivore studies 468–9
    ungulates, mortality causes **462**
South Pacific, pollination studies 346
species
    differential availability, Krakatau
        Islands 540–2
    differential performance, Krakatau
        Islands 542–5
species–area curves, Krakatau Islands 528
species distributions
    and cultivation 164
    factors affecting 164–8
    Ghana 168–73
    limits
        environmental 167
        factors affecting 167–8
        issues 183–6
    lowland rainforest trees 163–91
    and soil fertility 179–83
species interactions, and community
    structure 80–1
species richness
    area vs. habitat 495–9
    diversity measures 570
    and evapotranspiration, herbivores 478
    forests, Borneo 498
    and gap formation 208–10
    and gap-size niche differentiation 195–6
    and genus richness 492
    geological background 450
    herbivores
        Africa 449–90
        factors affecting **485**
    and maximum mass difference 486
    stochastic effects 210–11
    tropical rainforests 194–8, 491–2, 549–50
    use of term 197
    vs. body mass, herbivores 474
    vs. relative density 499
    and weight ratio, herbivores 471–84, *485*
    see also genus richness
spectral mixture analysis
    in remote sensing 603
    schematic overview *604*
squirrelfishes, competition studies 97
Stenogyne kanehoana, pollination 343, 351–2
Sternobothrus spp., seed predation
    studies 67–9
STRI *see* Smithsonian Tropical Research
    Institute
subsoil, chemistry **180**
successional history, Krakatau Islands 520–3
successional processes

hierarchy **522**
Krakatau Islands 520, 521
and site availability 537–40
Sulawesi
    crops, moth studies 587–93
    moths
        Arctiidae studies *591*
        diversity studies *589*
        Geometridae studies *592*
        Noctuidae studies *590*
    vegetation types *588*
susceptibility model, seedling survival 66–7
Swartzia schomburgkii, compensatory
    growth 68
Sylvia spp.
    competition 439
    densities 442
    species segregation 440
Synbranchus spp., Amazon floodplain 401
Systema Mamirauá
    vertical profiles *395*
    water depth, seasonal variation *391*

Tanzania, herbivore studies 468, *481*
temperate regions, species distributions 165
temperature
    effects on tropical rainforest trees 165–6
    vertical profiles, Amazon *394–5*
Terminalia superba, droughting
    experiments 178
terra firme
    remote sensing studies 611
    use of term 388
Tetraberlinia bifoliolata
    abundance 267
    density studies **273**
    distribution studies 272–5
    frequency distribution *274*
    fruiting patterns 278
    growth rates 286
    phenology 276–9, 291–2
    regeneration 295–300
    relative proportions 275
    seedling bank studies **279**
    size distributions 297–300
Tetraberlinia moreliana
    abundance 267
    density studies **273**
    distribution studies 272–5
    frequency distribution *274*
    fruiting patterns 278
    growth rates 286–7
    phenology 276–9
    regeneration 295–300
    relative proportions 275
    seedling bank studies **279**
    size distributions 297–300

Index

timber, harvesting, Malaysia and Indonesia 572–5
Timonius compressicaulis, abundance 542
topsoil
  chemistry **180**
  distribution, Ghana *169*
Transvall (South Africa), ungulates, mortality causes **462**
tree allometry 139–44
  definitions 139
  large trees 199–202
  and light levels 151–2
  see also crown allometry; trunk allometry
tree architecture
  and light environment, studies 127–35
  and light levels 125–62
  models 136–9
    and light environment 151
  multifactor analysis 153–4
  use of term 127
  variability 154–5
  and virtual trees 139
tree crowns, components **130**
tree development, and light levels 125–62
tree height
  effects on crown traits *134*
  and light levels 135
  parameters **132**
  and stem diameter 125
  studies *130*
tree mortality, factors affecting 166–7
trees
  abundance
    Pasoh **501**–2
    rank order 499–503
    Sarawak **500**, **501**–2
  competition 504–5
  diameter measurements 199–200
  drought responses *176*
  ecological plasticity 503–5
  ecological profiles, droughting experiments **175**
  fire resistance 372–6
  growth models *140*
  growth strategies 139–40
  large
    allometry 199–202
    environmental factors **198**
    generalist nature of 198–202
    and light environment 213
    negative performance 200
    responses 198–9
  mechanical design **142**, 144–7
    studies 152–3
  physiology, and climate 294
  population dynamics 372–6
  population genetics 507

rare species 505–7
root systems 140
safety factors 145–7
seeds, small 10–14
species distributions
  effects of seasonal droughts on 173–9
  factors affecting **165**
  and forest types *172*
  West Africa 163–91
species richness
  area vs. habitat 495–9
  and genus richness **496**
  mobile links 507–8
stability models 153
tropical, niche specificity 491–514
tropical rainforests
  climate change effects 166–7
  gap-size niche differentiation 79–94
  light variations 125–62
  rainfall effects 166
  regeneration 295–300
  soil type preference **181**
  temperature sensitivity 165–6
  tree distribution limits 163–91
  see also virtual trees
Trema spp., tree architecture 154
Trilepidea adamsii, pollination 343
Triplochiton scleroxylon
  distribution, Ghana *172*
  phenology 288
  relative growth rate *182*
tropical rainforests
  Africa 426
  canopy gaps, importance of 80
  change 269
  climate change 310
  deforestation rates 605
  disturbance 549–65
  droughts 556–8
  epidemic herbivory 558
  equilibrium theories 193, 194–5
  genus richness 492
  geological background 310
  guild delimitation 194–8
  human disturbance 558–60
  landslides 553
  lightning 555–6
  lowland, tree distribution limitations 163–91
  mobile river disturbance 554
  non-equilibrium theories 193, 195
  phenology 311
    issues 287–95
  phenology and dynamics, Cameroon 267–308
  primary species, community structure and demography 193–219

seed mass 1–24
seedlings, demography 285–7
soil chemistry **180**
spatial patterns, remote sensing
 techniques 599–615
species richness 194–8, 491–2, 549–50
successions 518
trees
 abundance 499–503
 climax species 81–2
 crown allometry *131*
 crown position index *129*
 gap-size niche differentiation 79–94
 light environment 126–7
 light level responses 125–62
 mechanical design **142**
 niche specificity 491–514
 regeneration 295–300
volcanic disturbance 553
wind disturbance 554–5
see also caatingas; palm-rich forests
tropical reefs *see* coral reefs
trunk allometry 141–3
 definitions 141
Tsavo National Park (Kenya), grasslands,
 crude protein concentration
 studies *456*
TWINSPAN *see* Two Way Indicator Species
 Analysis
Two Way Indicator Species Analysis 524
 Krakatau Islands **530**
 studies 528–30

Uapaca guineensis
 flowering 327–8
 dry–sunny trigger hypothesis 324–5
 fruit scores *323*
 phenology 321–5
Uganda, remote sensing studies 611
UMRF *see* upper montane rainforest
understorey
 resource availability, variations 207–8
 seed and seedling distribution 205–6
 seedling growth and survival 206–7
underwater visual census, fish studies 95,
 101–3
ungulates, mortality, causes **462**
upper montane rainforest, seed mass 12–13
Urera baccifera, Amazon floodplain 413
UVC *see* underwater visual census

várzea, use of term 388
vegetation
 Africa 426–7
 Gabon 312–15
 maps *313*
 remote sensing **602**

types
 remote sensing studies 610–12
 Seram *575*
 Sulawesi *588*
vegetation disturbance
 anthropogenic 605–7
 natural 607–10
 remote sensing 604–10
vegetative tissues, for seed protection 14–15
Venezuela, seed mass studies 1–24
vertebrates
 Janzen–Connell model studies **55–6**
 seed predation 60–2
 seed removal, in neotropical rainforests 26
 seed selection/search parameters **61**
 small, and seed removal 28
Victoria amazonica, Amazon floodplain 412
Virola spp.
 seed fate *36*
 seed removal *31*
 seed removal studies 35–8
virtual trees, and tree architecture *139*
volcanic disturbance
 frequency distributions, Krakatau
 Islands *526*
 and gap formation 523–4
 studies 525–8
 Krakatau Islands 516
 tropical rainforests 553
volcanism
 effects on ecosystems 515
 and El Niño ecosystemsns, Krakata 537–8
Vouacapoua spp.
 crown allometry *131*, 143–4
 crown development 148
 crown position index *129*
 height/diameter ratio *130*
 parameters **132**
 safety factors *133*, 146, 152–3
 traits *134*
 tree architecture studies 127–35

Wappu Reserve (Guyana), seed predation
 studies 62–4
warblers
 competition 438–9
 densities *442*
wasps, pollination **345**, 350, 352
water budget, modelling 377
water depth, studies 391–2
water quality, spatial and temporal
 variations 392–6
weather *see* climate
weight ratio, and species richness,
 herbivores 471–84, *485*
West Africa
 anurans, risk tactics 221–42

droughts, and tree mortality 166–7
tree species distributions, limits 163–91
wildebeest
  calf survival rates 249
  exploitation limits, Serengeti 243–65
  exploitation studies
    data **247–8**
    data collection 246
    issues 257–62
    methods 246–53
    results 253–7
  harvesting
    estimates 250–2
  future research 261–2
  illegal 257–8
  legal 258–60
  likelihood profiles 256
  sex ratio 258
  maximum sustainable yield 256, 257
  population dynamics model 246–57
    issues 257
  population levels *254*
  survival *254*
wildlife, exploitation, Africa 244–5
wind, disturbance, tropical rainforests 554–5
WINPHOT, canopy openness calculation *84*